AF327360

Imaging Spectrometry

# Remote Sensing and Digital Image Processing

VOLUME 4


*Series Editor:*

Freek D. van der Meer, *Department of Earth Systems Analysis, International Institute for Geo-Information Science and Earth Observation (ITC), Enschede, The Netherlands & Department of Physical Geography, Faculty of Geosciences, Utrecht University, The Netherlands*

*Editorial Advisory Board:*

Michael Abrams, *NASA Jet Propulsion Laboratory, Pasadena, CA, U.S.A.*
Paul Curran, *University of Bournemouth, U.K.*
Arnold Dekker, *CSIRO, Land and Water Division, Canberra, Australia*
Steven M. de Jong, *Department of Physical Geography, Faculty of Geosciences, Utrecht University, The Netherlands*
Michael Schaepman, *Centre for Geo-Information, Wageningen UR, The Netherlands*

# IMAGING SPECTROMETRY

## Basic Principles and Prospective Applications

*edited by*

## FREEK D. VAN DER MEER

*Utrecht University, The Netherlands and the*
*International Institute for Geo-Information Science and Earth Observation (ITC),*
*Enschede, The Netherlands*

and

## STEVEN M. DE JONG

*Faculty of Geosciences, Utrecht University, The Netherlands*

*Including a CD Rom with color images and datasets*

 Springer

A C.I.P. Catalogue record for this book is available from the Library of Congress.

ISBN-10 1-4020-0194-0 (HB)
ISBN-13 978-1-4020-0194-9 (HB)
ISBN-10 0-306-47578-2 (e-book)
ISBN-13 978-0-306-47578-8 (e-book)

Published by Springer,
P.O. Box 17, 3300 AA Dordrecht, The Netherlands.

*www.springer.com*


*Printed on acid-free paper*

02-0703-300 ts
03-0506-250 ts

All Rights Reserved
© 2001, 2003, 2006 Springer
No part of this work may be reproduced, stored in a retrieval system, or transmitted
in any form or by any means, electronic, mechanical, photocopying, microfilming, recording
or otherwise, without written permission from the Publisher, with the exception
of any material supplied specifically for the purpose of being entered
and executed on a computer system, for exclusive use by the purchaser of the work.

Printed in the Netherlands.

# Table of Contents


Acknowledgments ............................................................ XIII
About the Editors ............................................................ XV
Contributors ................................................................ XVII
Introduction ................................................................ XXI

Part I: Basic principles of imaging spectrometry ............................ 1

Chapter 1: Basic physics of spectrometry by F.D. van der Meer ............... 3
  Introduction ............................................................. 3
  Radiation principles ..................................................... 4
  Surface scattering properties ............................................ 4
  Reflectance spectroscopy ................................................. 5
  Reflectance properties of materials ...................................... 6
    Minerals and Rocks ..................................................... 7
    Vegetation ............................................................ 10
    Soils ................................................................. 11
    Water ................................................................. 13
    Man-made and other materials .......................................... 14
    The effect of the atmosphere .......................................... 14
  Mixing problematics ..................................................... 15

Chapter 2: Imaging spectrometry: Basic analytical techniques by F.D. van der Meer, S.M. de Jong and W. Bakker ....................................... 17
  Introduction ............................................................ 17
  Imaging spectrometry: airborne systems .................................. 18
    Airborne Simulators ................................................... 20
  Imaging spectrometry: spaceborne instruments ............................ 21
  Spaceborne versus airborne data ......................................... 24
    Costs of a spaceborne system .......................................... 25
    Costs of an airborne system ........................................... 26
    Coverage .............................................................. 27
    Resolution ............................................................ 28
    Geometry .............................................................. 28
    Processing ............................................................ 29
    Flexibility ........................................................... 30
    Complementary data .................................................... 30
  Pre-processing .......................................................... 31
  Laboratory set-up of a pre-processing calibration facility .............. 31
  The spectral pre-processing chain ....................................... 32
  Spatial pre-processing .................................................. 35
  Quality control: signal to noise characterization ...................... 35
    The "homogeneous area method" ......................................... 36
    The "local means and local variances method" .......................... 36

VI

The "geostatistical method" ................................................................... 37
Noise Adjustment ................................................................................ 38
Atmospheric correction ............................................................................ 39
Relative Reflectance ........................................................................... 39
Absolute Reflectance ........................................................................... 40
Re-sampling and image simulation .......................................................... 41
Analytical processing techniques ............................................................ 44
Introduction ....................................................................................... 44
Binary encoding .................................................................................. 44
Waveform characterization .................................................................. 45
Spectral Feature Fitting (SFF) .............................................................. 47
Spectral Angle Mapping (SAM) ........................................................... 47
Spectral unmixing ............................................................................... 47
Iterative spectral unmixing .................................................................. 51
Constrained energy minimization (CEM) .............................................. 55
Foreground-background analysis ........................................................... 55
Classification ...................................................................................... 56
Cross Correlogram Spectral Matching (CCSM) ..................................... 56
Geophysical inversion .......................................................................... 57
Endmember selection for spectral unmixing and other feature finding
algorithms .............................................................................................. 60

Part II prospective applications of imaging spectrometry ............................. 63

Chapter 3: Imaging spectrometry for surveying and modelling land degradation
by S.M. de Jong and G.F. Epema ...................................................... 65
Introduction ............................................................................................ 65
Processes of land degradation ................................................................. 66
Soils and Degradation ........................................................................ 66
Vegetation and Degradation processes ................................................. 68
Spectrometry for land degradation .......................................................... 70
Soils ....................................................................................................... 71
Classical Soil Description Methods ....................................................... 71
Soil Spectral Properties ....................................................................... 72
Mapping soil degraded state by imaging spectrometry and spectral
unmixing ................................................................................................ 76
Mapping Soil Units using imaging spectrometry and Spectral Matching ..... 77
Vegetation ............................................................................................. 80
Spectral Reflectance of Vegetation ...................................................... 80
Conventional Remote Sensing and Vegetation ....................................... 81
Assessing Vegetation Properties for Erosion Models from spectroscopical
images ............................................................................................... 82
Contextual approaches to land cover mapping ......................................... 83
Conclusions ........................................................................................... 86

Chapter 4: Field and imaging spectrometry for identification and mapping of expansive soils by S. Chabrillat, A.F.H. Goetz, H.W. Olsen and L. Krosley ............................................................................... 87
  Introduction ................................................................................ 87
    Nature of expansive soils ......................................................... 87
    Spectroscopic indicators of clay minerals ................................. 91
  Field and laboratory analyses ....................................................... 95
    Expansive soils in the Front Range Urban Corridor (Colorado) ... 95
    Field sampling and laboratory analyses ................................... 97
    Relationships between reflectance, mineralogy and swelling potential ........ 98
  Hyperspectral image analysis ..................................................... 100
    Expansive clays in Colorado: remote sensing considerations ..... 100
    Images acquisition and analysis ............................................. 102
    Mapping results ..................................................................... 105
  Conclusions ............................................................................. 108

Chapter 5: Imaging spectrometry and vegetation science by L. Kumar, K. Schmidt, S. Dury and A. Skidmore ..................................................... 111
  Introduction .............................................................................. 111
  Spectroscopy versus spectrometry ............................................. 113
  Fundamental factors affecting vegetation reflectance .................. 115
    Leaf optical properties ............................................................ 115
    reflectance (400-700 nm.) ....................................................... 115
    The reflectance red-edge (690 - 720nm) .................................. 117
    The near-infrared region (700-1300nm) ................................... 117
    The mid-infrared region (1300 - 2500nm) ................................ 119
    BRDF ..................................................................................... 119
  Vegetation reflectance curve ...................................................... 121
    Pigments ................................................................................ 121
    Green Leaf Structure ............................................................. 124
  Leaf optical models ................................................................... 127
    Introduction ........................................................................... 127
    Ray tracing models ................................................................ 127
    Models based on the Kubelka-Munk (K-M) theory ................... 128
    The Plate models .................................................................... 129
    Stochastic models .................................................................. 130
  Vegetation biochemistry ............................................................ 130
    Chemical Compounds in Plants and methods of estimation ...... 130
    Empirical approach ................................................................ 133
  Reflectance models for foliar biochemical estimations ................. 137
    Radiative transfer modelling ................................................... 137
    Leaf radiative models ............................................................. 137
    Canopy reflectance models ..................................................... 138
    Spectral analogies between leaves and canopies ...................... 139
  Applications: Field spectroscopy for vegetation studies ............... 140
    Phenological studies ............................................................... 140
    Pigment correlations .............................................................. 141

VIII

Water status ......................................................................................... 142
Plant stress ......................................................................................... 144
Applications: Airborne imaging spectroscopy for vegetation studies. 146
   Extracting biophysical variables (e.g. LAI, $F_{APAR}$, Cover) ...................... 146
   Physically-based methods ................................................................. 149
Hyperspectral-BRDF inverse modelling ............................................. 150
   Introduction ....................................................................................... 150
   Biomass/Yield ................................................................................... 152
Other applications ............................................................................. 153
   Extracting Biochemical variables ...................................................... 153
   Carbon fluxes ................................................................................... 154
   Cover ................................................................................................. 154

Chapter 6: Imaging spectrometry for agricultural applications by J.P.G.W.
   Clevers and R. Jongschaap ............................................................... 157
Introduction ....................................................................................... 157
Role of imaging spectroscopy in agriculture ...................................... 158
   Introduction ....................................................................................... 158
   Spectral analysis (PCA) .................................................................... 159
   Case study ......................................................................................... 160
   MAC Europe campaign ..................................................................... 160
   Test site ............................................................................................. 160
   AVIRIS .............................................................................................. 160
   Results ............................................................................................... 161
   Conclusions ....................................................................................... 162
Red-edge index .................................................................................. 163
   Introduction ....................................................................................... 163
   Definition red-edge index ................................................................. 163
   Simulations using radiative transfer models ...................................... 163
   SAIL model ....................................................................................... 165
   PROSPECT model ............................................................................. 165
   Sensitivity analysis ........................................................................... 166
   Simulation results at the leaf level .................................................... 167
   Simulation results at the canopy level ............................................... 167
   Atmospheric influence ...................................................................... 173
   Discussion and conclusions ............................................................... 174
Case study I – Red-edge index and crop nitrogen status ..................... 175
   Introduction ....................................................................................... 175
   Estimation of nitrogen status ............................................................. 175
   Estimation of nitrogen deficiency ..................................................... 176
   Managing nitrogen stress ................................................................... 176
   Experimental data ............................................................................. 176
   Set-up of the potato trials ................................................................. 176
   Crop measurements ........................................................................... 177
   Results and discussion ...................................................................... 177
   Conclusions ....................................................................................... 178
Case study II – A framework for crop growth monitoring ................... 179

Introduction .......................................................................................... 179
Framework for yield prediction ......................................................... 179
Crop growth models ............................................................................ 179
Estimating LAI ................................................................................... 181
Estimating leaf angle distribution (LAD) ......................................... 182
Estimating leaf optical properties in the PAR region ....................... 182
Linking optical remote sensing with crop growth models ............... 183
Experimental data .............................................................................. 184
Introduction ........................................................................................ 184
Ground truth ....................................................................................... 185
Meteorological data ........................................................................... 185
Spectra of single leaves ..................................................................... 185
CropScan<sup>TM</sup> ground-based reflectances ............................................. 185
CAESAR ............................................................................................. 185
Results ................................................................................................. 186
Measurements of leaf optical properties ........................................... 186
Estimating LAD .................................................................................. 186
Estimating LAI ................................................................................... 187
Estimating leaf optical properties ..................................................... 187
Results calibration SUCROS ............................................................. 188
Conclusions ......................................................................................... 189
Case study III – Using MERIS for deriving the red-edge index ......... 190
Introduction ........................................................................................ 190
Data sets ............................................................................................. 191
Results and discussion ....................................................................... 192
AVIRIS spectra ................................................................................... 192
Red-edge index simulation with MERIS ........................................... 192
Upscaling to the MERIS resolution ................................................... 195
Conclusions ......................................................................................... 195
Conclusions ............................................................................................ 197

Chapter 7: Imaging spectrometry and geological applications by F. van der
   Meer, H. Yang and H. Lang ............................................................ 201
Introduction ............................................................................................ 201
Mineral mapping; surface mineralogy ................................................. 201
Mineral mapping; exploration ............................................................... 202
Mineral mapping; lithology .................................................................... 203
Vegetation stress and geobotany ........................................................... 204
Environmental geology .......................................................................... 204
Petroleum related studies ...................................................................... 205
Atmospheric effects resulting from geologic processes ...................... 205
Thermal infrared studies ....................................................................... 206
Case-study I: Petroleum case study: seepage detection at Bluff using
   mineral alteration ............................................................................. 206
Mineral alteration .............................................................................. 206
The Bluff area and petroleum geology ............................................... 207
Probe-1 imaging spectrometer data ................................................... 210

X

Mineral alteration mapping for microseepage detection at Bluff ............... 211
Quaternary eolian loess deposits (Qe) ..................................... 211
Dakota Sandstone (Kd) ..................................................... 211
Brushy Basin Member (Jmb) ................................................. 211
Recapture Member (Jmr) .................................................... 211
Bluff Sandstone (Jb) ...................................................... 213
Mapping gray-green colored rocks that may associate with hydrocarbon microseepage ............................................................... 213
Case-study II: mining ....................................................... 213
Background on the mining problem .......................................... 213
Data and analysis ......................................................... 215
Validation and interpretation ............................................. 217
Discussion .................................................................. 218

Chapter 8: Imaging spectrometry and petroleum geology by F. van der Meer, H. Yang, S. Kroonenberg, H. Lang, P. van Dijk, K. Scholte and H. van der Werff ................................................................... 219
Introduction ................................................................ 219
Spectra of organics ......................................................... 219
Hydrocarbon microseepage .................................................... 222
Introduction .............................................................. 222
Microbial effects and Hydrocarbon-induced surface manifestations .......... 224
Detecting hydrocarbon-induced surface manifestations by remote sensing ...................................................................... 225
Bleached red beds ......................................................... 226
Clay mineral alteration ................................................... 226
Carbonates ................................................................ 227
Geobotanical anomalies .................................................... 228
In Summary .................................................................. 231
Future trends ............................................................... 232
Trends in exploration ..................................................... 232
Trends in monitoring emissions ............................................ 233
A case study from Santa Barbara, southern California ........................ 233
Petroleum geology of the Southern Californian Basins ...................... 233
Hyperspectral data analysis ............................................... 237

Chapter 9: Imaging spectrometry for urban applications by E. Ben Dor ....... 243
Introduction ................................................................ 243
Remote Sensing for Urban Applications ....................................... 244
Aspects of Remote Sensing of the Urban Environment .......................... 245
HSR and Urban Applications .................................................. 246
Spectral Properties of Urban Material ....................................... 248
Building a Spectral Library of Urban Objects from the existing database .. 251
The PUSL Spectra .......................................................... 251
The CASL Spectra .......................................................... 252
Summary for Urban Library from Existing Database .......................... 259

Building a Spectral Library of Urban Objects from in Situ Measurements 260
Summary for Urban Library from in-Situ Measurements ..........................262
Spectral Pattern Recognition ..............................................................262
Examining the Spectral-Based Information for Urban Mapping in the VIS-NIR Region ...........................................................................................262
Examining the Spectral-Based Information for Urban Mapping in the VIS-NIR-SWIR Region ..............................................................................263
Examining the Spectral-Based Information for Urban Mapping, Using the TIR Region ......................................................................................264
Special Benefit of the HSR over Urban Areas: Asphalt and Shade ...........270
Summary and Conclusion for the Spectral Analysis ...............................271
Remote Sensing of the Urban Atmosphere using HSR ......................272
Recent HSR Urban Applications: A Discussion ...............................273
Recommendation for HSR Utilization over Urban Areas ...................276
Data Evaluation and Processing ...............................................277
Spectral Preprocessing steps: ....................................................277
Spatial Pre-processing ...............................................................279
Atmospheric Correction ............................................................279
Mixed Pixel Problem ................................................................280
General Summary .............................................................................280

Chapter 10: Imaging spectrometry in the Thermal Infrared by M.J. Abrams, S. Hook and M.C. Abrams .................................................................283
Introduction ....................................................................................283
Theory ...........................................................................................284
Thermal Emission .....................................................................284
Spectral Emissivity ....................................................................284
Emissivity of rocks and minerals ................................................284
Current TIR airborne systems ...........................................................287
Early scanners ..........................................................................287
TIMS .....................................................................................287
ATLAS, AMSS, AAS, MIVIS ....................................................294
Master .....................................................................................296
Sebass .....................................................................................298
Satellite instruments .......................................................................301
ASTER ....................................................................................301
MTI ........................................................................................302
Future of Hyperspectral TIR Imaging ...............................................303
Major Trends in Hyperspectral Remote Sensing .............................303
The Future – 1-5 Year Forecast: ..................................................305
The Future: 5-15 Year Forecast ..................................................306

Chapter 11: Imaging spectrometry of water by A.G. Dekker, V.E. Brando, J.M. Anstee, N. Pinnel, T. Kutser, E.J. Hoogeboom, S. Peters, R. Pasterkamp, R. Vos, C. Olbert and T.J.M. Malthus ...........................307
Introduction ....................................................................................307
Light in water .................................................................................308

Introduction to the theory................................................................308
Optically deep waters....................................................................310
Optical properties of the water column for optically deep waters ..............311
The inherent optical properties ........................................................311
Radiometric variables and apparent optical properties ....................313
The diffuse apparent optical properties............................................316
The two-flow model for irradiance ..................................................316
An analytical model for the irradiance reflectance ....................321
Optical shallow waters....................................................................323
Light above water............................................................................325
Water surface effects ......................................................................325
Atmospheric effects and atmospheric correction....................326
Optically deep and shallow waters: applications and case studies......327
Introduction....................................................................................327
Optically deep inland and estuarine waters....................................327
Imaging spectrometry of optically deep inland waters ....................327
Imaging spectrometry of optically deep estuaries............................334
Conclusions for imaging spectrometry of optically deep inland and estuarine
waters...............................................................................................340
Optically shallow waters................................................................341
Bathymetry and bright substrate mapping ......................................341
Macrophyte/seagrass and macro-algae mapping............................345
Conclusions imaging spectrometry of optically shallow waters................357
Conclusions................................................................................358

Acronyms ................................................................................361

Index  ....................................................................................365

References ...............................................................................371

*Including a CD-Rom with color images and datasets*

# 1    Acknowledgments

Freek van der Meer is grateful to both the Delft University of Technology and the International Institute for Aerospace Surveys and Earth Sciences (ITC) for providing the time to work on this book. I acknowledge the input of Wim Bakker on the differences and similarities of airborne and spaceborne sensor systems. The professional advice and continuous assistance by Mrs. Petra van Steenbergen from Kluwer Academic Publishers was highly appreciated by us.

Work by Michael Abrams and co-authors was performed at the Jet Propulsion Laboratory/California Institute of Technology under contract with the National Aeronautics and Space Administration.

Part of the Chapter by J. Clevers and R. Jongschaap was previously published in the ISPRS Journal of Photogrammetry & Remote Sensing, the book "Imaging Spectrometry – a Tool for Environmental Observations" (published by Kluwer Academic Publishers), and Remote Sensing of Environment.

S. Chabrillat and co-authors acknowledge that their work was performed under NASA/JPL contract No. 960983. They thank the AVIRIS team for the acquisition and support for the AVIRIS data. S. Chabrillat. expresses her thanks to Paul Boni for his help in the XRD analyses, and Dennis Eberl for his help in the interpretation of the XRD scans. The help of Eric Johnson, Bruce Kindel and Sara Martinez-Alonso was appreciated for their immediate availability for field work when needed. We are indebted to Dave Noe of the Colorado Geological Survey for his help throughout the whole project, and sharing his knowledge of Colorado swelling soils with us.

## 2   About the Editors

*Prof. Dr. F.D. van der Meer*; Prof. Dr. F. (Freek) D. van der Meer (1966) has a M.Sc.

in structural geology and tectonics of the Free University of Amsterdam (1989) and a Ph.D. in remote sensing from Wageningen Agricultural University (1995) both in the Netherlands. He started his career at Delft Geotechnics (now Geodelft) working on geophysical processing of ground penetrating radar data. In 1989 he was appointed lecturer in geology at the International Institute for Aerospace Surveys and Earth Sciences (ITC in Enschede, the Netherlands) where he worked too date in various positions (presently associate professor). His research is directed toward the use of hyperspectral remote sensing for geological applications with the specific aim of use geostatistical approaches to integrate airborne and field data into geologic models. He teaches at post graduate, master and Ph.D. level on these topics. On 1 November 1999, Dr. van der Meer was appointed part-time (two days a week) full professor of imaging spectrometry at the Delft University of Technology (Faculty of Civil Engineering and Geosciences). He combines this position with a 60% appointment at ITC. In 1999 he delivered his first Ph.D. graduate working on remote sensing for hydrocarbon microseepage detection, at present six related Ph.D. projects are starting/running under his guidance at ITC and DUT. Prof. Van der Meer published over 70 papers in international journals and authored more than 100 conference papers and reports. He is chairman of the Netherlands Society for Earth Observation and Geoinformatics, chairman of the special interest group geological remote sensing of EARSeL and associate editor for Terra Nova. Freek van der Meer can be reached at <u>vdmeer@itc.nl</u> or f.d.vandermeer@citg.tudelft.nl.

XVI

*Prof. Dr. S.M. de Jong*; From 1988 to 1998 Prof. Dr. Steven de Jong worked at the

Department of Physical Geography at Utrecht University in the Netherlands where he was responsible for the remote sensing curriculum and for the national and international contacts in this field.

Steven M. de Jong is head of the Laboratory for Geo-Information and Remote Sensing of the Department of Environmental Sciences of the Wageningen University in the Netherlands since 1998. He completed his PhD study on applications of 'GIS and Remote Sensing for Mediterranean Land Degradation Monitoring and Modelling' in 1994 at Utrecht University. In 1987 he received his MSc on 'Rural Land Evaluation Procedures using GIS' from Utrecht University in the Netherlands. Prof. Dr. de Jong worked as a visiting scientist at NASA's Jet Propulsion Laboratory (JPL) in Pasadena, California in 1995 and 1996. During his stay at JPL he contributed to the applied research initiatives of the AVIRIS group. AVIRIS is JPL's Airborne Visible Infra Red Imaging Spectrometer. He is chairman of the national remote sensing committee of the Dutch Remote Sensing Board for Agricultural and Environmental Applications. Furthermore, he is the National Representative of the IAHS-ICCE: International Commision on Continental Erosion. De Jong was or is member of the organising committees of a number of international conferences. He is having major scientific experiences in the field of image processing, GIS, geostatistics, soil science, soil erosion modelling and field surveys. Steven de Jong can be reached at s.dejong@geog.uu.nl.

## 3    Contributors


Mark Abrams
>ITT Industries, Ft. Wayne, IN, U.S.A.

Michael Abrams
>NASA Jet Propulsion Laboratory
>4800 Oak Grove Drive, Pasadena (CA),U.S.A.
>Mike@lithos.nasa.jpl.gov

Janet M. Anstee
>Environmental Remote Sensing Research Group
>CSIRO Land and Water
>Canberra, Australia.
>Janet.anstee@cbr.clw.csiro.au

Wim Bakker
>International Institute for Aerospace Survey and Earth Sciences (ITC)
>Remote Sensing and GIS Laboratory, P.O. Box 6, 7500 AA, Enschede, The Netherlands.
>Bakker@itc.nl

Eyal Ben-Dor
>Tel-Aviv University
>Department of Geography, Tel-Aviv, Israel.
>bendor@ccsg.tau.ac.il

Vittorio E. Brando
>Environmental Remote Sensing Research Group
>CSIRO Land and Water
>Canberra, Australia.
>Vittorio.brando@cbr.clw.csiro.au

Sabine Chabrillat
>University of Colorado
>Cooperative Institute for Research in Environmental Sciences (CIRES), Center for the Study of Earth from Space (CSES), Campus Box 216, Boulder, CO, U.S.A.
>chabril@cses.colorado.edu
>presently at:
>Observatoire Midi-Pyrenees
>CNRS/UMR 5562
>14 ave Edouard Belin
>F-31400 Toulouse, France
>Sabine.Chabrillat@cnes.fr

Jan Clevers
>Wageningen-UR
>Centre for Geo-Information, P.O. Box 47, 6700 AA Wageningen, The Netherlands.
>jan.clevers@staff.girs.wag-ur.nl

Steven de Jong
>Wageningen-UR

Centre for Geo-Information, P.O. Box 47, 6700 AA Wageningen, The Netherlands.
s.dejong@frw.ruu.nl

Arnold Dekker
CSIRO Land and Water
Environmental Remote Sensing Research Group
Canberra Laboratories
Clunies Ross Street
GPO Box 1666 Canberra, ACT 2601, Australia
arnold.dekker@cbr.clw.csiro.au

Stephen Dury
Australian National University (ANU)
Faculty of Sciences, Department of Forestry, Forestry Building 048, ACT 0200, Canberra, Australia.
Stephen.dury@anu.edu.au

Gerrit Epema
Wageningen-UR
Centre for Geo-Information, P.O. Box 47, 6700 AA Wageningen, The Netherlands.
Gerrit.epema@staff.girs.wag-ur.nl

Alexander Goetz
University of Colorado
Cooperative Institute for Research in Environmental Sciences (CIRES), Center for the Study of Earth from Space (CSES), Campus Box 216, Boulder, CO, U.S.A.
goetz@cses.colorado.edu

Erin J. Hoogenboom
National Institute for Coastal and Marine Management/RIKZ
Ministry of Transport and Waterworks
The Hague, The Netherlands
Hoogenboom@rikz.rws.minvenw.nl

Simon Hook
NASA Jet Propulsion Laboratory
4800 Oak Grove Drive, Pasadena (CA),U.S.A.
simon@lithos.nasa.jpl.gov

Raymond Jongschaap
Plant Research International
Business Unit Agrosystems Research, P.O. Box 16, 6700 AA Wageningen, The Netherlands.
R.E.E.Jongschaap@plant.wag-ur.nl

Salle Kroonenberg
Delft University of Technology,
Faculty of Civil Engineering and Geosciences, Department of Applied Earth Sciences, P.O. box 5028, 2600 GA, Delft, The Netherlands.
s.b.kroonenberg@citg.tudelft.nl

Lisa Krosley
Colorado School of Mines
Department of Engineering, Golden, CO, U.S.A.
Krosley@mines.edu

Lalit Kumar
International Institute for Aerospace Survey and Earth Sciences (ITC)
Division Agriculture, Conservation and Environment, P.O. Box 6, 7500 AA, Enschede, The Netherlands.
kumar@itc.nl

Tiit Kutser
CSIRO Marine Research
Canberra, Australia
Tiit.kutser@cbr.clw.csiro.au

Harold Lang
NASA Jet Propulsion Laboratory
4800 Oak Grove Drive, Pasadena, CA,U.S.A.
Harold@lithos.nasa.jpl.gov

Tim J.M. Malthus
Department of Geography
University of Edinburgh
Edinburgh, Scotland
Dr.T.J.Malthus@ed.ac.uk

Carsten Olbert
Freie Universität Berlin
Berlin, Germany
olbert@zedat.fu-berlin.de

Harold Olsen
Colorado School of Mines
Department of Engineering, Golden, CO, U.S.A.
Holsen@mines.edu

Reinold Pasterkamp
Institute for Environmental Studies
Vrije Universiteit
Amsterdam, The Netherlands
Reinold.pasterkamp@ivm.vu.nl

Steef Peters
Institute for Environmental Studies
Vrije Universiteit
Amsterdam, The Netherlands
steef.peters@ivm.vu.nl

Nicole Pinnel
Environmental Remote Sensing Research Group
CSIRO Land and Water
Canberra, Australia.
Nicole.pinnel@br.clw.csiro.au

Karin Schmidt
International Institute for Aerospace Survey and Earth Sciences (ITC)
Division Agriculture, Conservation and Environment, P.O. Box 6, 7500 AA, Enschede, The Netherlands.
schmidt@itc.nl

Klaas Scholte
Delft University of Technology,

XX

Faculty of Civil Engineering and Geosciences, Department of Applied Earth Sciences, P.O. box 5028, 2600 GA, Delft, The Netherlands.
k.h.scholte@citg.tudelft.nl

Andrew Skidmore

International Institute for Aerospace Survey and Earth Sciences (ITC)
Division Agriculture, Conservation and Environment, P.O. Box 6, 7500 AA, Enschede, The Netherlands.
skidmore@itc.nl

Freek Van der Meer

International Institute for Aerospace Survey and Earth Sciences (ITC)
Division of Geological Survey, P.O. box 6, 7500 AA, Enschede, The Netherlands.
vdmeer@itc.nl
Delft University of Technology,
Faculty of Civil Engineering and Geosciences, Department of Applied Earth Sciences, P.O. box 5028, 2600 GA, Delft, The Netherlands.
f.d.vandermeer@citg.tudelft.nl

Paul Van Dijk

International Institute for Aerospace Survey and Earth Sciences (ITC)
Division of Geological Survey, P.O. box 6, 7500 AA, Enschede, The Netherlands.
vandijk@itc.nl

Harald Van der Werff

International Institute for Aerospace Survey and Earth Sciences (ITC)
Division of Geological Survey, P.O. box 6, 7500 AA, Enschede, The Netherlands.
vdwerff@itc.nl

Robert Vos

Institute for Environmental Studies
Vrije Universiteit
Amsterdam, The Netherlands
Robert.vos@ivm.vu.nl

Hong Yang

Shell Exploration and Production B.V.
Rijswijk, The Netherlands.
h.yang@siep.shell.com

## 4    Introduction

Imaging spectrometers acquire images in a large number (typically over 40), narrow (typically 0.01 to 0.02 μm. in width), contiguous (i.e., adjacent and not overlapping) spectral bands to enable the extraction of reflectance spectra at a pixel scale that can be directly compared with similar spectra measured either in the field or in a laboratory. Different names have been coined to this field of remote sensing including "imaging spectrometry", "imaging spectroscopy" and "hyperspectral remote sensing". Although they have a different meaning in the sense of a direct translation of the term (i.e., spectrometry="measuring", spectroscopy="seeing", hyperspectral="too many bands"), the significance and perception to the remote sensing community is the same: "the acquisition of images in hundreds of registered, contiguous spectral bands such that for each picture element of an image it is possible to derive a complete reflectance spectrum (Goetz, 1992)".

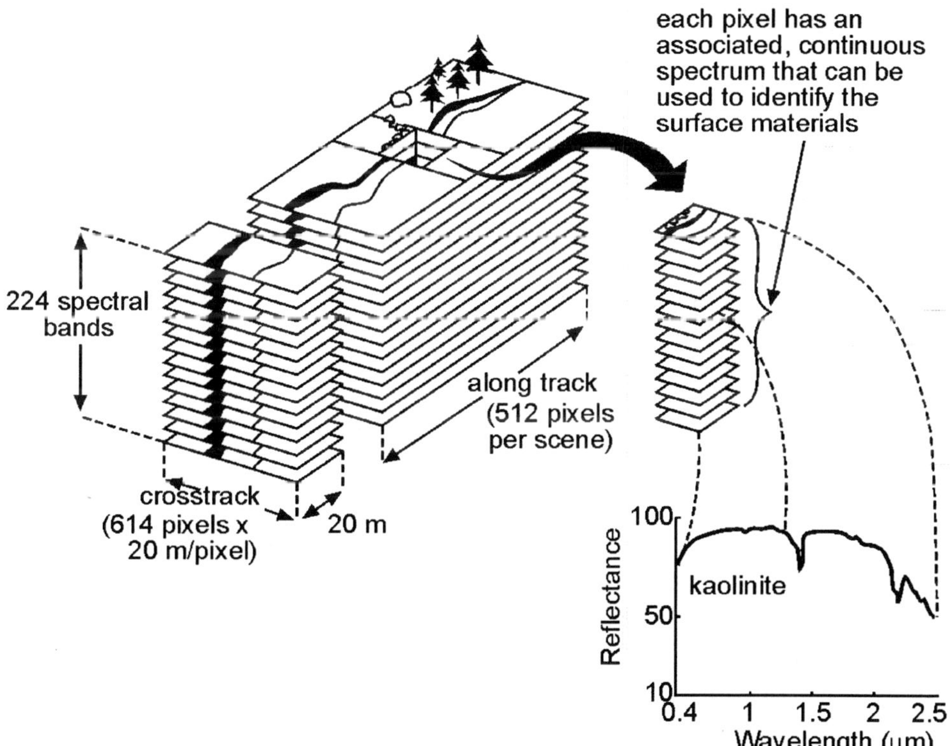


*Figure 1*. Concept of imaging spectrometry.

The objective of imaging spectrometry is to measure quantitatively the components of the Earth System from calibrated spectra acquired as images for scientific research and applications (Vane & Goetz, 1988; Figure 1). Thus we are interest in measuring physical quantities at the Earth surface such as upwelling radiance, emissivity, temperature and reflectance.

Through measurement of the solar reflected spectrum, a wide range of scientific research and application is being pursed using signatures of energy, molecules and scatterers in the spectra measured by imaging spectrometers. These fields, which are further described in subsequent Chapters to this book, include:

1. Atmosphere: water vapor, cloud properties, aerosols, absorbing gases.
2. Plant ecology: chlorophyll, leaf water, lignin, cellulose, pigments, structure, vegetation species and community maps, nonphotosynthetic constituents, etc.
3. Geology and soils: mineralogy, soil type, etc.
4. Coastal and Inland waters: chlorophyll, plankton, dissolved organics, sediments, bottom composition, bathymetry, etc.
5. Snow and Ice Hydrology: snow cover fraction, grainsize, impurities, melting, etc.
6. Biomass Burning: subpixel temperatures and extent, smoke, combustion products, etc.
7. Environmental hazards: contaminants directly and indirectly, geological substrate, etc.
8. Calibration: aircraft and satellite sensors, sensor simulation, standard validation, etc.
9. Modeling: radiative transfer model validation and constraint, etc.
10. Commercial: mineral exploration, agriculture and forest status, etc.
11. Algorithms: autonomous atmospheric correction, advanced spectra derivation, etc.
12. Other: human infrastructure, mine detection, etc.

In this book we bring together applications of imaging spectrometry and provide the analytical background and physics for understaing the signal received at the sensor. Imaging spectrometry originated largely in geology (Chapter 7, 8 and 10) and is used in an operational mode by the mineral industry (Chapter 7) for surface mineralogy mapping to aid in ore exploration. Other applications of the technology include lithological and structural mapping. The petroleum industry (Chapter 8) is developing methods for implementation of imaging spectrometry at an reconnaissance stage as well. The main targets are hydrocarbon seeps and microseeps. Other application fields include environmental geology (and related geobotany) in which currently much work is done on acid mine drainage and mine waste monitoring. Atmospheric effects resulting from geologic processes as for example the prediction and quantification of various gases in the atmosphere such as sulfates emitted from volcanoes is also an important field. In soil science (Chapters 3 and 4), much emphasis has been placed on the use of spectroscopy for soil surface properties and soil compositional analysis to aid in land degradation studies. Major elements such as iron and calcium as well as cation ion exchange capacity can be estimated from imaging spectrometry. In a more regional context, imaging spectrometry has been used to monitor agricultural areas (per-lot monitoring) and semi-natural areas. Recently, spectral identification from imaging spectrometers has been successfully applied to mapping of swelling clays minerals smectite, illite, and kaolinite in order to quantify the swelling potential of expansive soils (Chapter 4). Much research in vegetation studies (Chapters 5 and 6) has emphasized on leaf biochemistry and structure and canopy structure. Biophysical models for leaf constituents now are available as well as soil-vegetation models. Estimates of plant material and structure and biophysical parameters include: carbon balance, yield/volume, nitrogen, cellulose, chlorophyll, etc. The leaf area index and vegetation indices have been extended to the hyperspectral domain and remain important physical parameters characterizing vegetation. Ultimate goal is biomass estimates and monitoring of changes therein. Several research groups investigate the bi-directional reflectance function in relation to vegetation species analysis and floristics. Vegetation stress by water deficiency, pollution sources such as acid mine drainage, and geobotanical anomalies in relation to ore deposits or petroleum and gas seepage link vegetation analysis to exploration. Another upcoming field is precision agriculture in which imaging spectrometry aids in better agricultural practices (Chapter 6). An important factor in vegetation health status is the chlorophyll absorption and in relation to that the position of the red edge determined using the red edge index. In hydrological applications of imaging spectrometry (Chapter 11), the interaction of electromagnetic radiation with water and the inherent and apparent optical properties of water are a central issue. Very important in imaging spectrometry of water bodies is the atmospheric correction and air-water interface corrections. Water quality of freshwater aquatic environments, estuarine environments and coastal zones are of importance to national water bodies. Detection and identification of phytoplankton-biomass, suspended sediments and other matter, coloured dissolved organic matter and aquatic vegetation (i.e., macrophytes) are crucial parameters in optical models of water quality. Much emphasis has been put on the mapping and monitoring of the state and the growth or brake-down of coral reefs as these are important in the CO2 cycle. In general, many multisensor missions such as TERRA and ENVISAT are directed toward integrated approaches for global change studies (and models) and global oceanography. Atmosphere studies are important in global change models and aid in the correction of optical data for scattering and absorption due to atmospheric trace gasses. In particular the optical properties and absorption characteristics of ozone, oxygen, water vapor, and other trace gasses and scattering by molecules and aerosols are important parameters in atmosphere studies. Urban studies (Chapter 9) are a new and relative un-explored field of application of imaging spectrometry.

The accompanying CD-ROM contains the color photographs and images which, from cost perspective, were reproduced in black-and-white in this book. Furthermore the CD-ROM contains some sample data sets and ENVI viewing software (Freelook) for further studies. We hope you will enjoy reading this book.


Freek van der Meer                    Steven de Jong


July, 2001.

# PART I: BASIC PRINCIPLES OF IMAGING SPECTROMETRY

*CHAPTER 1*

# BASIC PHYSICS OF SPECTROMETRY

Freek D. VAN DER MEER
Delft University of Technology, Delft, The Netherlands
International Institute for Aerospace Surveys and Earth Sciences (ITC),
Enschede, the Netherlands

## 1   Introduction

When light interacts with a mineral or rock, light of certain wavelengths is preferentially absorbed while at other wavelengths is transmitted in the substance. Reflectance, defined as the ratio of the intensity of light reflected from a sample to the intensity of the light incident on it, is measured by reflection spectrophotometers which are composed of a light source and a prism to separate light into different wavelengths. This light beam interacts with the sample and the intensity of reflected light at various wavelengths is measured by a detector relative to a reference standard of known reflectance. Thus a continuous reflectance spectrum of the sample is obtained in the wavelength region measured.

Reflectance spectra have been used for many years to obtain compositional information of the Earth surface. Similarly, it has been shown that spectral reflectance in visible and near-infrared offers a rapid and inexpensive technique for determining the mineralogy of samples and obtaining information on chemical composition. Electronic transition and charge transfer processes (e.g., changes in energy states of electrons bound to atoms or molecules) associated with transition metal ions such as Fe, Ti, Cr, etc., determine largely the position of diagnostic absorption features in the visible- and near-infrared wavelength region of the spectra of minerals (Burns, 1970; Adams, 1974; 1975). In addition, vibrational processes in $H_2O$ and $OH^-$ (e.g., small displacements of the atoms about their resting positions) produce fundamental overtone absorptions (Hunt, 1977; Hunt & Salisbury, 1970). Electronic transitions produce broad absorption features that require higher energy levels than do vibrational processes, and therefore take place at shorter wavelengths (Hunt, 1970; Goetz, 1991). The position, shape, depth, and width of these absorption features are controlled by the particular crystal structure in which the absorbing species is contained and by the chemical structure of the mineral. Thus, variables characterizing absorption features can be directly related to the mineralogy of the sample.

3

*F.D. van der Meer and S.M. de Jong (eds.), Imaging Spectrometry, 3–16.*
© 2006 *Springer. Printed in the Netherlands.*

## 2    Radiation principles

The following discussion is based on Rees (1996). Thermal radiation is emitted by all objects at temperatures above absolute zero. Consider an area dA and radiation arriving at a direction $\theta$ to the normal of dA, but in the range of directions forming a solid angle of $d\Omega$ steradians. Radiance, L in $Wm^{-2}sr^{-1}$, is defined as

$$d\Phi = LdAd\Omega\cos\theta \tag{1}$$

The total power falling on dA from all directions is given by the integration over $2\pi$ steradians in case of no absorption nor scattering by

$$\Phi = dA \int_{2\pi} L\cos\theta d\Omega = EdA \tag{2}$$

where E is called irradiance measured in $Wm^{-2}$. In the reverse case when E measures the total radiance leaving dA we refer to exitance, M, and the total power emitted by the source is radiant intensity, I, defined as

$$\Phi_{total} = Id\Omega \tag{3}$$

We can also express these quantities spectrally by introducing an interval of wavelength $\lambda$ and integrating over $d\lambda$. Spectral radiance according to Planck's relationship is given by

$$L_\lambda = \frac{2hc^2}{\lambda^5}(e^{hc/\lambda kT} - 1)^{-1} \tag{4}$$

where h is Planck's constant, k is Boltzmann's constant ($1.38 \times 10^{-23}JK^{-1}$), c is the speed of light (in a vacuum) and T is the temperature. The total outgoing radiance of a black body of temperature T is given by

$$L = \int_0^\infty L_\lambda d\lambda = \frac{2\pi^4 k^4}{15c^2 h^3}T^4 \tag{5}$$

where $\sigma = \dfrac{2\pi^4 k^4}{15c^2 h^3}$ is known as Stefan-Boltzmann's constant ($5.67 \times 10^{-8}Wm^{-2}K^{-4}$).

Wien's displacement law gives the relation between the wavelength at which the maximum radiation is reached, $\lambda_{max}$, and the temperature of the black body as

$$\lambda_{max} = c_w / T \tag{6}$$

where $c_w$ is a constant (Wien's constant) of $2.898 \times 10^{-3}Km$.


## 3    Surface scattering properties

When radiation interacts with a surface, it is partly absorbed into the substance, and partly scattered or reflected by the object. Consider a collimated beam of radiation incident on a surface at an incidence angle $\theta_o$. The irradiance E is given by $F\cos\theta_o$ and scattered into a solid angle $d\Omega$ in a direction $\theta_1$. The outgoing radiance of the surface as a result of this illumination is $L_1$ in the direction $(\theta_1,\phi_1)$ where $\phi_1$ is the azimuthal angle. The bidirectional reflectance distribution function (BRDF) R (in $sr^{-1}$ ) is defined as

$$R = L_1 / E \tag{7}$$

R is a function of the incident and scattered directions and can thus be noted as $R(\theta_o, \phi_o, \theta_1, \phi_1)$. The reflectivity of the surface, r (also known as albedo), is the ratio of the total power scattered to the total power incident as

$$r(\Theta_o, \Phi_o) = \int\limits_{\Theta=0}^{\pi/2} \int\limits_{\phi=0}^{2\pi} R \cos\Theta_1 \sin\Theta_1 d\Theta_1 d\phi_1 \qquad (8)$$

Two extreme cases of scattering surface can be defined, the perfectly smooth surface (specular surface) and the perfectly rough surface (the Lambertian surface). A perfect Lambertian surface will scatter all the radiation incident upon it so that the radiant exitance M is equal to the irradiance E and the albedo is unity. A measure of roughness of a surface is given by the Rayleigh criterion. For a surface to be smooth according the Rayleigh criterion it should satisfy

$$\Delta h \cos\Theta_o / \lambda < 1/8 \qquad (9)$$

where $\Delta h$ is the surface irregularity of height and $\lambda$ the wavelength considered.


## 4      Reflectance spectroscopy

The following discussion is based on Schanda (1986). Radiation incident onto a material is preferentially absorbed by molecules forming the structure of the substance at wavelengths pre-determined by quantum mechanical principles. The total energy of a molecule $W_t$ is the sum of the electronic energy $W_e$ the vibrational energy $W_v$ and the rotational energy $W_r$ as

$$W_t - W_e \mid W_v + W_r \qquad (10)$$

Changes in energy states of molecules due to changes in rotational energy levels do not occur in solids and will not further be treated here, we restrict this discussion to changes in electronic states and vibrational processes. Harmonic vibration is described by

$$\omega_v = \sqrt{\frac{f(m_1 + m_2)}{m_1 m_2}} \qquad (11)$$

where f is the restoring force constant (spring constant), v the vibrational states v=0,1,2,... and $m_1$ and $m_2$ are the masses of the molecule. The possible energy states of the harmonic vibration are given by quantum mechanics as

$$W_v = (v + \frac{1}{2})h\omega_v \qquad (12)$$

where h is Planck's constant equal to $6.6 \times 10^{-34}$ W s$^2$. These define the spectral regions where absorption can occur as a result of vibrational processes.

Electrons in solid materials can occupy a discrete number of energy states q=1,2,... When photons are incident on a material they interact with the electrons and the absorption of a photon of a proper energy $h\nu$ may cause the transition of the electron to a higher state. Absorption of energy into the medium results in absorption features in reflectance spectra. The velocity of an electron is given by

$$v = \frac{qh}{2mL} \tag{13}$$

where m is the mass of the electron and L=Nl with l is the internuclear bond length and N the number of electrons. Thus the total energy of the electron is the sum of the kinetic energy $mv^2/2$ and the potential energy (often set to zero). Now the energy states of the electron can be described by the following formula

$$W_q = mv^2/2 = \frac{q^2h^2}{8mL^2} \tag{14}$$

Each molecular orbital starting from the one with the lowest energy can accommodate only two electrons with antiparallel spin orientations (i.e., the Pauli exclusion principle). Therefore N electrons can maximally occupy q=N/2 states in the lowest energy state representing an energy of $W_{N/2}$ and the lowest empty one has $W_{(N/2)+1}$. We can now calculate that absorption of a photon of proper energy $hv$ causes the transition of an electron from orbital q to q=1 corresponding to an energy difference of

$$\Delta W = W_{(N/2)=1} - W_{N/2} = \frac{h^2}{8mL^2}(N+1) \tag{15}$$

Acknowledging that $hv = hc/\lambda$ where c is the speed of light (in a vacuum) and $\lambda$ is the wavelength at which the absorption occurs we can find the longest wavelength $\lambda_{max}$ which can be absorbed by the molecule due to the first transition as

$$\lambda_{max} = \frac{8mc}{h}\frac{N^2l^2}{N+1} \tag{16}$$

To be able to interpret or predict the wavelengths at which absorptions due electronic transitions occur it is useful to know the energy level schemes of the molecules involved in the transitions. Absorption in a spectrum have two components: continuum and individual features. The continuum or background is the overall albedo of the reflectance curve which for cross comparison is often removed by Hull subtraction or division (Clark & Roush, 1984).The depth of an absorption band, D, is usually defined relative to the continuum, $R_c$

$$D = 1 - \frac{R_b}{R_c} \tag{17}$$

where $R_b$ is the reflectance at the band bottom, and $R_c$ is the reflectance of the continuum at the same wavelength as $R_b$.


## 5    Reflectance properties of materials

Isolated atoms and ions have discrete energy states. Absorption of photons of a specific wavelength causes a change from one energy state to a higher one. Emission of a photon occurs as a result of a change in an energy state to a lower one. When a photon is absorbed it is usually not emitted at the same wavelength, hence absorption features may occur at other wavelength than the frequency of the original state. Various processes take place under the umbrella of electronic processes crystal field effects, charge transfers, conduction bands and color centers. In general, electronic processes cost relatively much energy hence give rise to broad absorption features at short (VIS-NIR) wavelength. In this Chapter, spectra of some basic earth materials will be discussed. The interested reader is also referred to the Chapter 8 on petroleum geological applications for crude oil spectra and to the Chapter 9 for spectra of materials characterizing the urban environment.


## 5.1    MINERALS AND ROCKS

Reflectance spectra have been used for many years to obtain compositional information of the Earth surface. Spectral reflectance in visible and near-infrared offers a rapid and inexpensive technique for determining the mineralogy of samples and obtaining information on chemical composition. Electronic transition and charge transfer processes associated with transition metal ions determine largely the position of diagnostic absorption features in the visible- and near-infrared wavelength region of the spectra of minerals (Burns, 1970). In addition, vibrational processes in $H_2O$ and $OH^-$ produce fundamental overtone absorptions (Hunt, 1977). The position, shape, depth, width, and asymmetry of these absorption features are controlled by the particular crystal structure in which the absorbing species is contained and by the chemical structure of the mineral. Thus, variables characterizing absorption features can be directly related to the mineralogy of the sample.

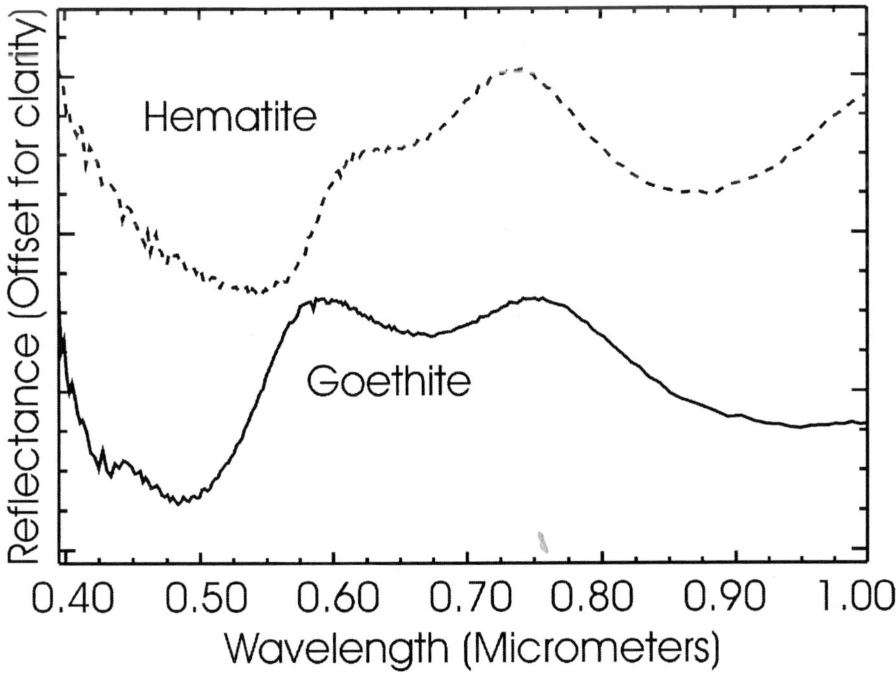


*Figure 1.* Reflectance spectra of two minerals dominated by iron absorption features.

The most common electronic process revealed in the spectra of minerals is the crystal field effect which is the result of unfilled electron shells of transition elements (e.g., Fe, Cr, Co, Ni; Figure 1). All transition elements have identical energies in an isolated ion, but the energy levels are split when the atom is located in a crystal field. This splitting of the orbital energy states enables an electron to be moved from a lower level into a higher one by absorption of a photon having an energy matching the energy difference between the states. Absorption bands can also be the result of charge transfers. An example is the charge transfer of $Fe^{2+}$ to $Fe^{3+}$. These absorption bands are diagnostic of mineralogy. Some minerals exhibit two energy levels in which electrons may reside: a higher level called the conduction band, where electrons move freely throughout the lattice, and a lower energy region called the valence band, where electrons are attached to individual atoms. Changes between these states result in absorption features in the visible portion of the electromagnetic wavelength spectrum. A few minerals show color due to absorption by color centers. Crystals have lattice defects that disturb the irradiation of the crystal. These defects can produce discrete energy levels to which electrons can be bound. This causes the color centers.

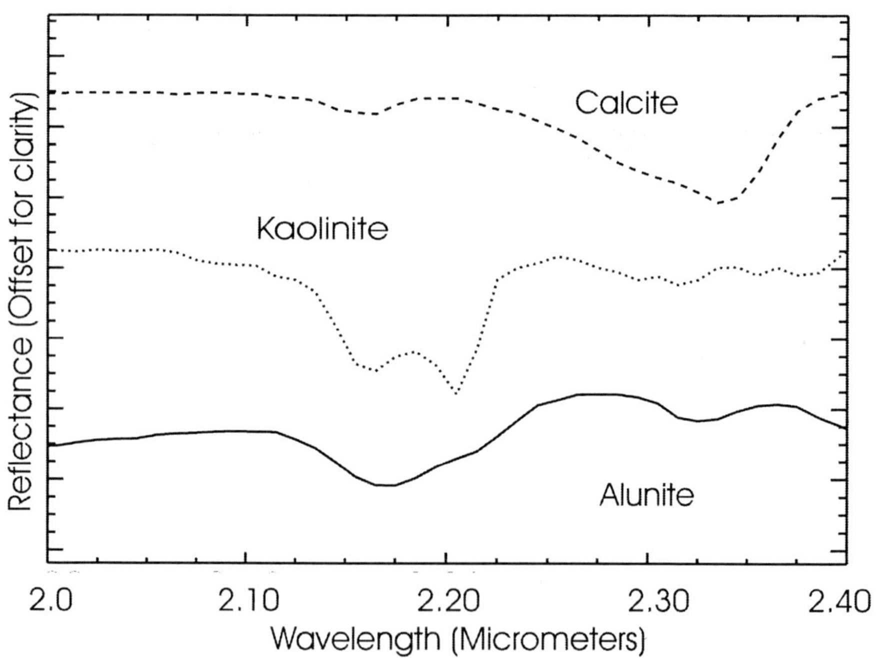


*Figure 2.* Reflectance spectra of minerals dominated by absorption features resulting from vibrational processes.

The bonds in a molecule or crystal lattice are like springs with attached weights: the whole system can vibrate. The frequency of vibration depends on the strength of each spring (the bond in a molecule) and their masses (the mass of each element in a molecule). Each vibration can also occur at roughly multiples of the original fundamental frequency referred to as overtones and combination (tones) when they involve different modes of vibrations. The frequencies of fundamental vibrations are labeled with the greek letter nu ($v$) and a subscript. If a molecule has vibration fundamentals $v_1$, $v_2$, $v_3$, then it can have overtones at approximately $2v_1$, $3v_1$, $2v_2$ and combinations at approximately $v_1+v_2$, $v_2+v_3$, $v_1+v_2+v_3$. Each higher overtone or combination is typically 30 to 100 times weaker than the last, hence the features will be more subtle and difficult to sense. Vibration costs relatively little energy and results in absorption features at high (SWIR-MIR) wavelengths with deep and narrow features.

Reflectance spectra of minerals are dominated in the visible to near-infrared wavelength range by the presence or absence of transition metal ions (e.g., Fe, Cr, Co, Ni). The presence or absence of water and hydroxyl, carbonate and sulfate determine the absorption features in the SWIR region due to so-called vibrational processes (Figure 2). The hydroxyl is generally bound to Mg or Al. The water molecule ($H_2O$) has N=3, so there are 3N-6=3 fundamental vibrations. In the isolated molecule (vapor phase) they occur at 2.738 $\mu$m ($v_1$, symmetric OH stretch), 6.270 $\mu$m ($v_2$, H-O-H bend), and 2.663 $\mu$m ($v_3$, asymmetric OH stretch). In liquid water, the frequencies shift due to hydrogen bonding: $v_1$=3.106 $\mu$m, $v_2$=6.079 $\mu$m, and $v_3$=2.903 $\mu$m. The overtones of water are seen in reflectance spectra of $H_2O$-bearing minerals. The first overtones of the OH stretches occur at about 1.4 $\mu$m and the combinations of the H-O-H bend with the OH stretches are found near 1.9 $\mu$m. The hydroxyl ion has only one stretching mode and its wavelength position is dependent on the ion to which it is attached. In spectra of OH-bearing minerals, the absorption is typically near 2.7 to 2.8 $\mu$m, but can occur anywhere in the range from about 2.67 $\mu$m to 3.45 $\mu$m. The OH commonly occurs in multiple crystallographic sites of a specific mineral and is typically attached to metal ions. Thus, there may be more than one OH feature. The metal-OH bend occurs near 10 $\mu$m (usually superimposed on the stronger Si-O fundamental in silicates). The combination metal-OH bend plus OH stretch occurs near 2.2 to 2.3 $\mu$m and is very diagnostic of mineralogy. Carbonates also show diagnostic vibrational absorption bands due to the $CO_3^{-2}$ ion. There are four vibrational modes in the free $CO_3^{-2}$ ion: the symmetric stretch, $v_1$: 1063 cm$^{-1}$ (9.407 $\mu$m); the bend, $v_2$: 879 cm$^{-1}$ (11.4 $\mu$m); the stretch, $v_3$: 1415 cm$^{-1}$ (7.067 $\mu$m); and the, $v_4$: 680 cm$^{-1}$ (14.7 $\mu$m). The $v_1$ band is not infrared active in minerals. There are actually six modes in the $CO_3^{-2}$ ion, but 2 are degenerate with the $v_3$ and $_4$ modes. In carbonate minerals, the $v_3$ and $v_4$ bands often appear as a doublet. Combination and overtone bands of the $CO_3$ fundamentals occur in the near IR. The two strongest are $v_1 + 2v_3$ at 2.50-2.55$\mu$m (4000-3900 cm$^{-1}$), and $3v_3$ at 2.30-2.35 $\mu$m (4350-4250 cm$^{-1}$). Three weaker bands occur near 2.12-2.16 $\mu$m ($v_1 + 2v_3 + v_4$ or $3v_1 + 2v_4$; 4720-4630 cm$^{-1}$), 1.97-2.00 $\mu$m ($2v_1 + 2v_3$; 5080-5000 cm$^{-1}$), and 1.85-1.87 $\mu$m ($v_1 + 3v_3$; 5400-5350 cm$^{-1}$).

In addition, the absorption band depth is related to the grain or particle size as the amount of light scattered and absorbed by a grain is dependent on grain size. A larger grain has a greater internal path where photons may be absorbed according to Beers Law. In a smaller grains there are proportionally more surface reflections compared to internal photon path lengths, if multiple scattering dominates, the reflectance decreases with increasing grain size.

The reflectance spectra of minerals are well known (Hunt, 1977; Hunt & Salisbury, 1970; 1971; Hunt *et al.*, 1971a+b; 1972; 1973a; Clark *et al.*, 1990; Crowley, 1991; Grove *et al.*, 1992) and several studies have been conducted to determine reflectance spectra of rocks (Hunt *et al.*, 1973b+c; 1974; 1975; 1976a+b). The reflectance characteristics of rocks can be simulated accurately by studying the compound effect of reflectance of minerals in a spectral mixture forming the rock. Reflectance spectra of minerals measured by different spectroradiometers with different spectral resolution are stored in spectral libraries that are available in digital format (e.g., Grove *et al.*, 1992; Clark *et al.*, 1990).

## 5.2    VEGETATION

Reflectance studies of vegetation (e.g. Gates *et al.*, 1965) generally restrict to the green leaf part of the plants giving little attention to the non-green dry vegetation components. Reflectance properties of vegetation in the visible part of the spectrum are dominated by the absorption properties of photosynthetic pigments of which chlorophyll, having absorptions at 0.66 and 0.68 $\mu m$ for chlorophyll a and b respectively, is the most important (Figure 3).

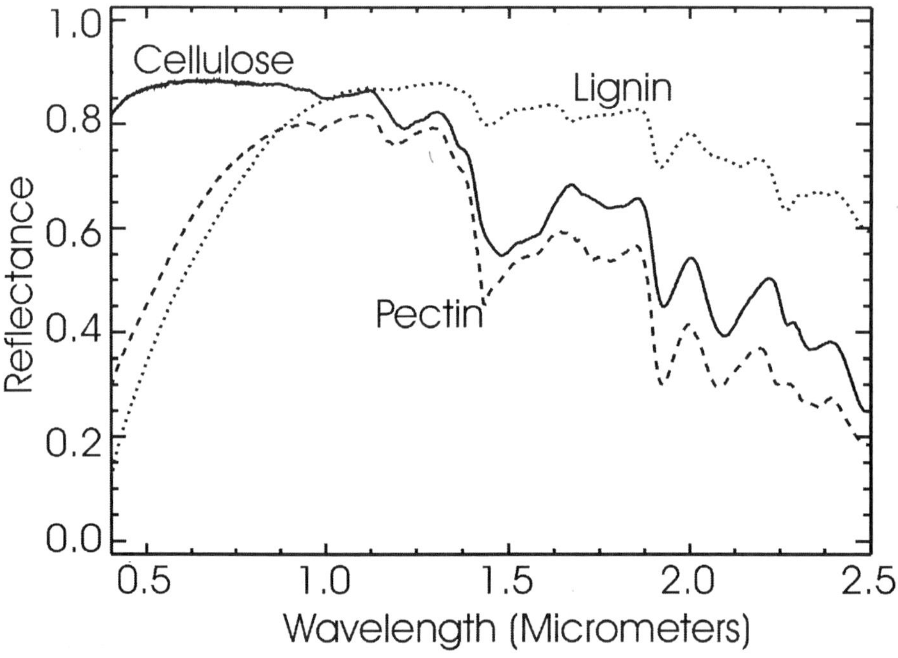


*Figure 3.* Reflectance spectra of plant constituents.

Changes in the chlorophyll concentration produce spectral shifts of the absorption edge near 0.7 $\mu m$: the red edge. This red edge shifts toward the blue part of the spectrum with loss of chlorophyll. The mid-infrared and short-wave infrared part of the vegetation spectrum is dominated by water and organic compounds of which cellulose, lignin, starch and protein (Elvidge, 1990). Absorption features due to bound and unbound water occur near 1.4 and 1.9 $\mu m$ and at 0.97, 1.20 and 1.77 $\mu m$. Cellulose has absorptions at 1.22, 1.48, 1.93, 2.28, 2.34, and 2.48 $\mu m$ while lignin has absorption features at 1.45, 1.68, 1.93, 2.05-2.14, 2.27, 2.33, 2,38, and 2.50 $\mu m$ (Elvidge, 1990).

Starch has absorption features at 0.99, 1.22, 1.45, 1.56, 1.70, 1.77, 1.93, 2.10, 1.32, and 2.48 $\mu m$ (Elvidge, 1990). The most abundant protein in leaves is a nitrogen bearing compound having absorption features at 1.50, 1.68, 1.74, 1.94, 2.05, 2.17, 2.29, 2.47 $\mu m$ (Elvidge, 1990). Dry plant materials lack the chlorophyll absorptions and intense water absorptions that are characteristic for green leaves and thus lack the intense absorption wing produced by high blue and UV absorptions.  Dry plant materials have diagnostic ligno-cellulose absorption features at 2.09 and in the 2.30 $\mu m$ region (Elvidge 1990). Examples of some vegetation spectra are shown in Figure 4.

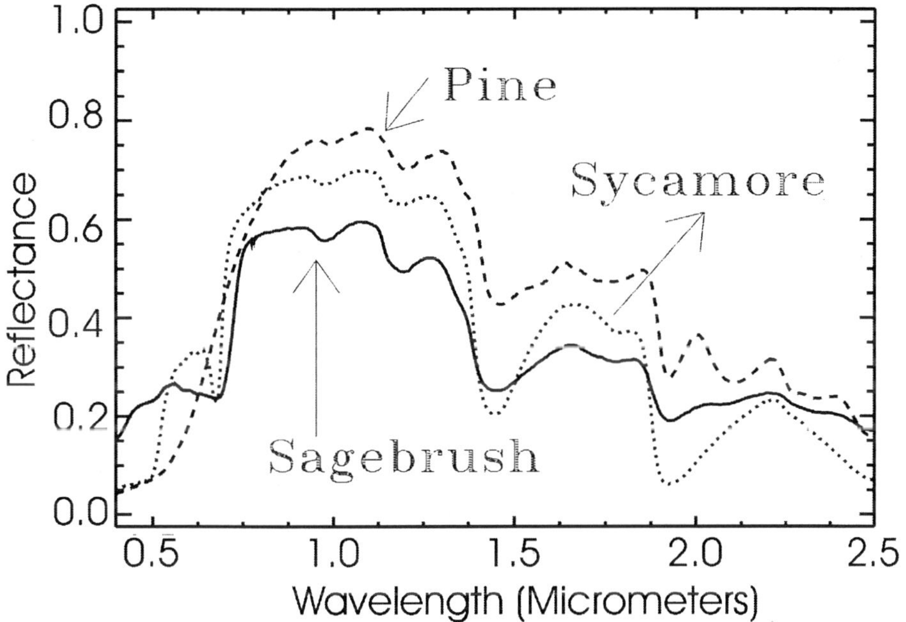


*Figure 4*. Reflectance spectra of some vegetation species.

In vegetation much variation in spectral properties result from the viewing geometry. Viewing geometry, includes the angle of incidence, angle of reflection, and the phase angle: the angle between the incident light and observer (the angle of reflection). These affect the intensity of light received. These effects are marginal for minerals, more pronounced for rocks and soils and of much importance in studying vegetation. The Bidirectional Reflectance Distribution Function (BRDF) should be constructed and analyzed. More details on reflectance properties of plant materials can be found in Elvidge (1990), Gates *et al.* (1965), and Wessman *et al.* (1988).


5.3    SOILS


Spectral reflectance characteristics of soils are the result of their physical and chemical properties and are influenced largely by the compositional nature of soils in which main components are inorganic solids, organic matter, air and water. In the visible and near-infrared regions extending to 1.0 $\mu m$, electronic transitions related to iron are the main factor determining soil spectral reflectance.

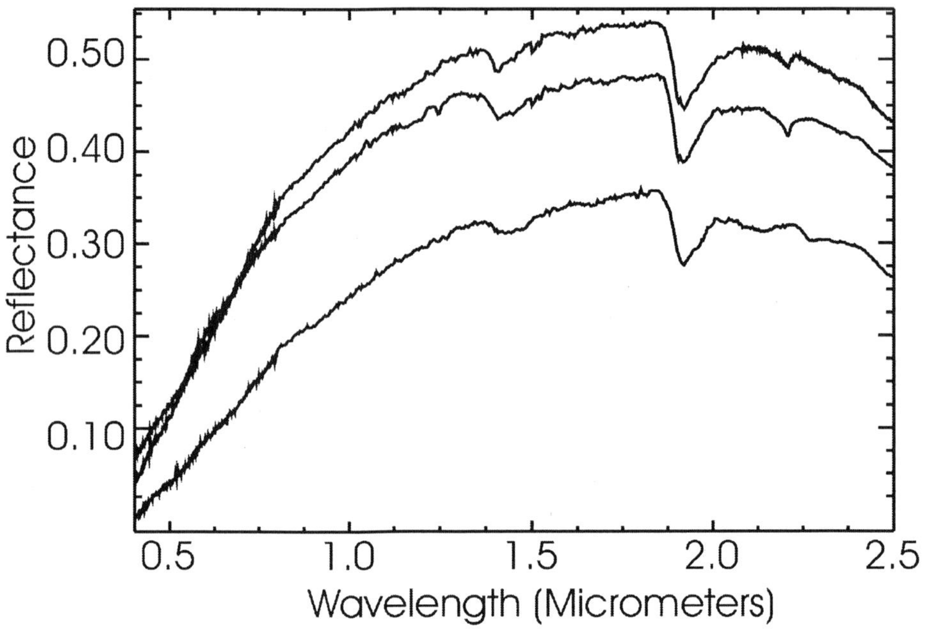

*Figure 5.* Reflectance spectra of three soils.

The majority of absorption features diagnostic for mineral composition occur in the short-wave infrared (SWIR) portion of the wavelength spectrum ranging from 2.0 to 2.5 $\mu m$. A strong fundamental OH⁻ vibration at 2.74 $\mu m$ influences the spectral signature of hyrdoxyl-bearing minerals. Furthermore, diagnostic absorption features characteristic for layered silicates such as clays and micas and also of carbonates occur in the SWIR region. Organic matter has a very important influence on the spectral reflectance properties of soils because amounts exceeding 2% are known to have a masking effect on spectral reflectance thus reducing the overall reflectivity of the soil and reducing (and sometimes completely obscuring) the diagnostic absorption features. Thus soils with a high (>20%) amount amount of organics appear dark throughout the 0.4 to 2.5 $\mu m$ range. In contrast, less decomposed soils have higher reflectance in the near-infrared region and enhanced absorption features. Prominent absorption features near 1.4 and 1.9 $\mu m$ due to bound and unbound water are typical for soil reflectance.

Less prominent water absorption features can be found at 0.97, 1.20 and 1.77 $\mu m$.

Increasing moisture content generally decreases the overall reflectance of the soil. A similar effect results from increasing the particle size resulting in a decrease in reflectivity and contrast between absorption features. Some studies on the spectral reflectance characteristics of soils and attempts to make classifications can be found in Baumgarner *et al.* (1985), Condit (1970), Stoner & Baumgardner (1981), Irons *et al.* (1989) and Singh & Sirohi (1994). Examples of some soil spectra are shown in Figure 5.

## 5.4    WATER

Another important material that can be observed in remote sensing imagery is water. Water bodies have a different response to EMR than water bound up in molecules: they do not exhibit discrete absorption features. Water has a high transmittance for all visible wavelengths, but the transmittance increases with decreasing wavelength. In deep and clear water nearly all radiation is absorbed, however suspended sediment, plankton and pigment cause increased reflectance in the visible portion of the spectrum. In the near infrared, almost all energy is absorbed by water as is also the case in the SWIR. I restrain from providing much details on the optical properties of water in the visible and near-infrared.

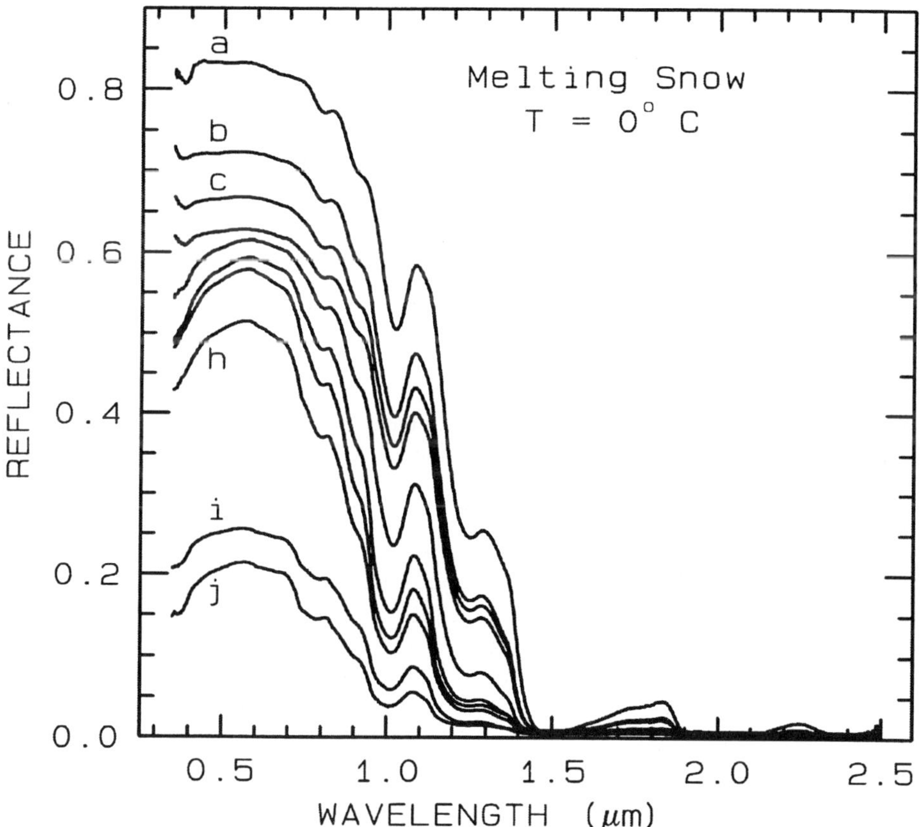



*Figure 6.* A series of reflectance spectra of melting snow. The top curve (a) is at 0° C and has only a small amount of liquid water, whereas the lowest spectrum (j) is of a puddle of about 3 cm of water on top of the snow. Note in the top spectrum, there is no 1.65-μm band as in the ice spectra in figure 22a because of the higher temperature.. The 1.65-μm feature is temperature dependent and decreases in strength with increasing temperature (see Clark, 1981). Note the increasing absorption at about 0.75 μm and in the short side of the 1-μm ice band, as more liquid water forms. The liquid water becomes spectrally detectable at about spectrum e, when the UV absorption increases (source: http://speclab.cr.usgs.gov/index.html:).

14                                    F.D. VAN DER MEER

The major water absorption features centered near $0.97\mu$m, $1.2\mu$m, $1.45\mu$m, and $1.9\mu$m result from vibrational transitions involving various overtones and combinations (Clark, 1981; Figure 6). The first overtone is a H-O-H symmetric stretch mode transition centered at $2.73\mu$m. The second is a H-O-H bending mode transition centered at $6.27\mu$m, the third is a H-O-H asymmetric stretch mode transition centered at $2.66\mu$m. The absorption feature centered near $0.97\mu$m is attributed to a 2V1 + V3 combination. That near $1.2\mu$m to a V1 + V2 + V3 combination. The feature near $1.45\mu$m is a V1 + V3 combination and the feature near $1.9\mu$m is a V2 + V3 combination.

## 5.5    MAN-MADE AND OTHER MATERIALS

It is impossible to provide a thorough review including an understanding of the physical nature of reflectance spectra of man-made structures (Figure 7 shows some examples) and other materials covering the Earth surface. There where these structures are designed of natural materials which can be mono- or multi-mineralic the spectroscopic properties are directly related to these basic materials. However for a large suite of chemical constituents no clear background information on spectral reflectance characteristics is available.

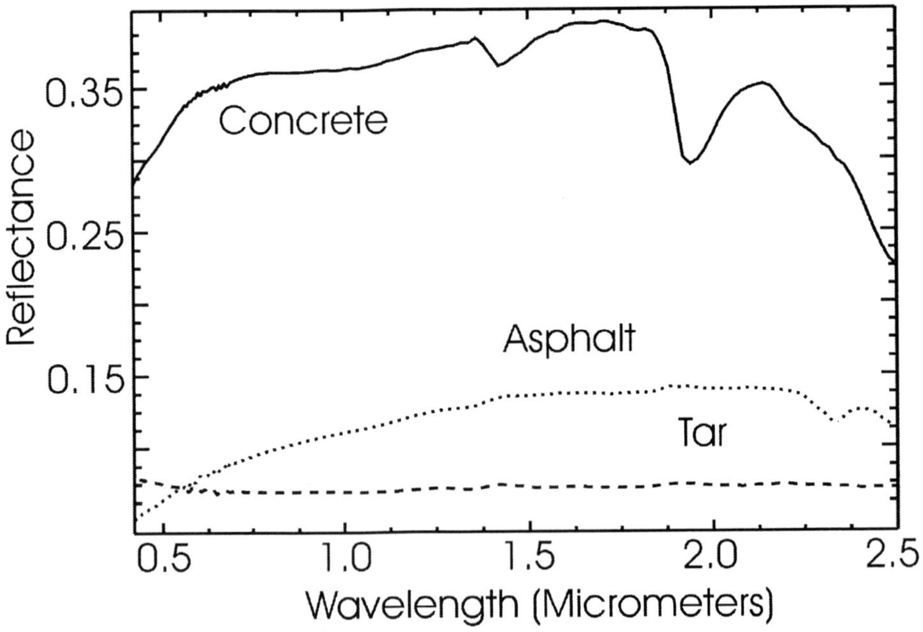


*Figure 7.* Reflectance spectra of some man-made materials.

## 5.6    THE EFFECT OF THE ATMOSPHERE

The general appearance of the solar irradiance curve shows radiance decreasing towards longer wavelengths. Various absorption bands are seen in the solar irradiance curve due to scattering and absorption by gasses in the atmosphere. The major atmospheric water vapor bands ($H_2O$) are centered approximately at 0.94 µm, 1.14 µm, 1.38 µm and 1.88 µm, the oxygen ($O_2$) band at 0.76 µm, and carbon dioxide ($CO_2$) bands near 2.01 µm and 2.08 µm. Additionally, other gasses including ozone ($O_3$), carbon monoxide (CO), nitrous oxide ($N_2O$), and methane ($CH_4$), produce noticeable absorption features in the 0.4-2.5 µm wavelength region (Figure 8).

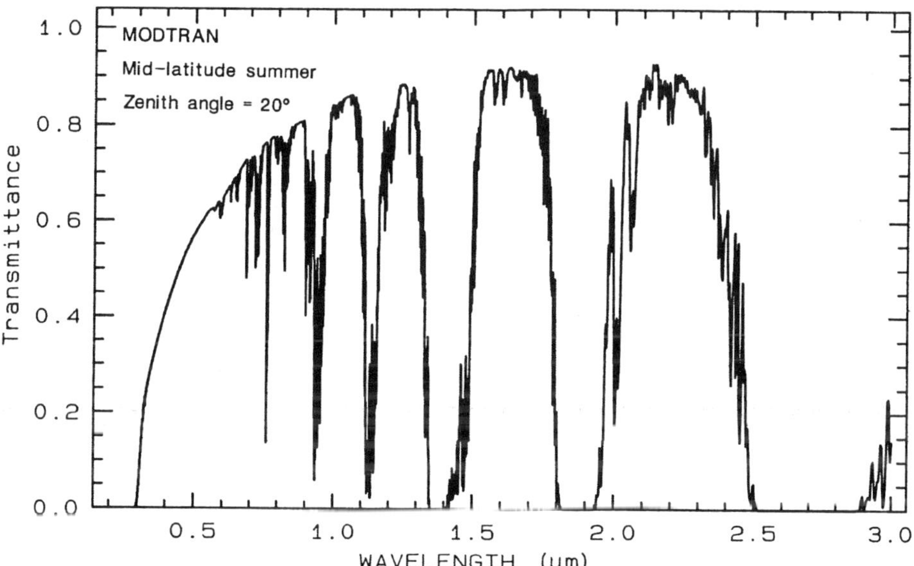


*Figure 8.* Modtran modeled atmospheric transmittance, visible to near-infrared. Most of the absorptions are due to water. Oxygen occurs at 0.76 *µm*, carbon dioxide at 2.0 and 2.06 *µm*. (source: http://speclab.cr.usgs.gov/index.html:).

## 6    Mixing problematics

Mixture modeling aims at finding the mixed reflectance from a set of pure end-member spectra. This is based on work by Hapke and Johnson *et al.* on the analysis of radiative transfer in particulate media at different albedos and reflectances by converting reflectance spectra to single-scattering albedos. Acknowledging the fact that in remote sensing spectra are measured in bidirectional reflectance, the following expression can be derived relating bidirectional reflectance R(i,e) defined as the radiant power received per unit area per solid angle viewed from a specific direction $e=\cos^{-1}\mu$ and a surface illuminated from a certain direction $i=\cos^{-1}\mu_o$ by collimated light to the mean single-scattering albedo W as

$$R(i,e) = \{w/4(\mu + \mu_o)\}\{H(\mu)H(\mu_o)\} \qquad (18)$$

where e is the viewing angle, $\mu_o=\cos(i)$, $\mu=\sin(e)$ and $H(\mu)$ is a function describing multiple scattering between particles that can be approximated by

$$H(\mu) = \frac{1 + 2\mu}{1 + 2\mu(1 - W)^{0.5}} \tag{19}$$

The mean single-scattering albedo of a mixture is a linear combination of the single-scattering albedos of the end-member components weighted by their relative fraction as

$$W(\lambda) = \sum_{j=1}^{n} W_j(\lambda) F_j \tag{20}$$

where W is the mean single-scattering albedo, $\lambda$ is the spectral waveband, n is the number of end-members and $F_j$ is the relative fraction of component j. $F_j$ is a function of the mass fraction $M_j$ the density, $\rho_j$ and the diameter $d_j$ of the end-member j as

$$F_j = \frac{M_j / \rho_j d_j}{\sum_{j=1}^{n} M_j / \rho_j d_j} \tag{21}$$

In general, there are 4 types of mixtures:

1.   Linear Mixture. The materials in the field of view are optically separated so there is no multiple scattering between components. The combined signal is simply the sum of the fractional area times the spectrum of each component. This is also called areal mixture.
2.   Intimate Mixture. An intimate mixture occurs when different materials are in intimate contact in a scattering surface, such as the mineral grains in a soil or rock. Depending on the optical properties of each component, the resulting signal is a highly non-linear combination of the end-member spectra.
3.   Coatings. Coatings occur when one material coats another. Each coating is a scattering/transmitting layer whose optical thickness varies with material properties and wavelength. Most coatings yield truly non-linear reflectance properties of materials.
4.   Molecular Mixtures. Molecular mixtures occur on a molecular level, such as two liquids, or a liquid and a solid mixed. Reflection is non-linear.

The type and nature of the mixing systematics are crucial to the understanding of the mixed signal. In many processing approaches, linearity is assumed (e.g., spectral unmixing). From the above, it can be observed that this is only true in few cases.

*CHAPTER 2*

# IMAGING SPECTROMETRY: BASIC ANALYTICAL TECHNIQUES

Freek VAN DER MEER [α,β] , Steven DE JONG [χ] & Wim BAKKER [α]

[α] International Institute for Aerospace Survey and Earth Sciences (ITC), Division of Geological Survey, Enschede, The Netherlands.

[β] Delft University of Technology, Department of Applied Earth Sciences, Delft, The Netherlands

[χ] Wageningen University and Research Center, Center for Geo-information, Wageningen, the Netherlands

## 1    Introduction

Remote sensing (e.g., the observation of a target by a device separated from it by some distance thus without physical contact) of the surface of the Earth from aircraft and from spacecraft provides information not easily acquired by surface observations. Until recently, the main limitation of remote sensing was that surface information lacked detail due to the broad bandwidth of sensors available. Work on high-spectral resolution radiometry has shown that earth surface mineralogy can be identified using spectral information from sensor data (Goetz 1991). Conventional sensors (e.g., Landsat MSS and TM, and SPOT) acquire information in a few separate spectral bands of various widths (typically in the order of 0.1-0.2 μm), thus smoothing to a large extent the reflectance characteristics of the surface (Goetz & Rowan 1981). Most terrestrial materials are characterized by spectral absorption features typically 0.02-0.04 μm in width (Hunt 1980). High-spectral resolution remotely sensed images are acquired to produce reflectance or radiance spectra for each pixel in the scene. Based upon the molecular absorptions and constituent scattering characteristics expressed in the spectrum we seek to:

- Detect and identify the surface and atmospheric constituents present
- Assess and measure the expressed constituent concentrations
- Assign proportions to constituents in mixed spatial elements
- Delineate spatial distribution of the constituents
- Monitor changes in constituents through periodic data acquisitions
- Simulate, calibrate and intercompare sensors
- Validate, constrain and improve models

New analytical processing techniques have been developed to analyze such high spectral dimensional data sets. These methods are the scope of this chapter. The pre-

17

*F.D. van der Meer and S.M. de Jong (eds.), Imaging Spectrometry,* 17–61.
© 2006 *Springer. Printed in the Netherlands.*

processing of imaging spectrometer data and the calibration of the instruments is briefly addressed. The chapter focuses on the processing of the data and new analytical approaches developed for the specific use with imaging spectrometer data. First we present a review of existing systems and design philosophy.


## 2    Imaging spectrometry: airborne systems

Imaging spectrometers have been used for many years in military applications such as the detection of camouflage from real vegetation. Due to the classified nature of the data and sensors not much can be said about the origin and applications being served. The first scanning imaging spectrometer was the Scanning Imaging Spectroradiometer (SIS) constructed in the early 1970s for NASA's Johnson Space Center. After that, civilian airborne spectrometer data were collected in 1981 using a one-dimensional profile spectrometer developed by the Geophysical Environmental Research Company which acquired data in 576 channels covering the 0.4-2.5 µm wavelength range (Chiu & Collins, 1978) followed by the Shuttle Multispectral Infrared Radiometer (SMIRR) in 1981. The first imaging device was the Fluorescence Line Imager (FLI; also known as the Programmable Line Imager, PMI) developed Canada's Department of Fisheries and Oceans (in 1981) followed by the Airborne Imaging Spectrometer (AIS), developed at the NASA Jet Propulsion Laboratory which was operational from 1983 onward acquiring 128 spectral bands in the range of 1.2-2.4 µm. The field-of-view of 3.7 degrees resulted in 32 pixels across-track. A later version of the instrument, AIS-2 (LaBaw, 1987), covered the 0.8-2.4 µm region acquiring images 64 pixels wide. Since 1987, NASA is operating the successor of the AIS systems, AVIRIS, the Airborne Visible/Infrared Imaging Spectrometer (Vane *et al.*, 1993). Since that time many private companies also started to take part in the rapid development in imaging spectrometry. Initiatives are described in a later paragraph on airborne systems.

Currently many space agencies and private companies in developed and developing countries operate there own instruments. It is impossible to describe all currently operational airborne imaging spectrometer systems in detail. Some systems will be highlighted to serve as examples rather than to provided an all-inclusive overview.

The AISA Airborne Imaging Spectrometer is a commercial hyperspectral pushbroom type imaging spectrometer system developed by SPECIM based in Finland. The spectral range in standard mode is 430 to 900 nm and a spectral sampling interval of 1.63 nm a a total of 288 channels. Spectral channel bandwidth are programmable from 1.63 to 9.8 nm. The Field of view is 21 degrees across-track and 0.055 degrees along-track resulting in typical spatial Resolutions of 360 pixels per swath, 1 m across-track resolution at an aircraft altitude of 1000 m.

The Advanced Solid-state Array Spectroradiometer (ASAS) is a hyper-spectral, multi-angle, airborne remote sensing instrument maintained and operated by the Laboratory for Terrestrial Physics at the NASA Goddard Space Flight Center. The system acquires 62 spectral channels in visible to near-infrared (404 to 1020 nm) with a spectral bandwidth of approximately 10 nm. The across-track resolution is 3.3 m (at nadir) to 6.6 m (60 deg) at 5000 m altitude, the along-track resolution is 3 m (at nadir) at 5000 m altitude.

In 1987 NASA began operating the Airborne Visible/Infrared Imaging Spectrometer (AVIRIS). AVIRIS was developed as a facility that would routinely supply well-calibrated data for many different purposes. The AVIRIS scanner collects 224 contiguous bands resulting in a complete reflectance spectrum for each 20*20 m. pixel in the 0.4 to 2.5 µm region with a sampling interval of <10 nm. The field-of-view of the AVIRIS scanner is 30 degrees resulting in a ground field-of-view of 10.5km. AVIRIS uses scanning optics and four spectrometers to image a 614 pixel swath simultaneously in 224 contiguous spectral bands over the 400 to 2500nm. wavelength range.

The Thermal Infrared Multispectral Scanner (TIMS) is a multispectral scanner that collects data in six channels. The six-element HgCdTe detector array provides six discrete channels:

- Channel 1: 8.2 - 8.6 micrometers
- Channel 2: 8.6 - 9.0 micrometers
- Channel 3: 9.0 - 9.4 micrometers
- Channel 4: 9.4 - 10.2 micrometers
- Channel 5: 10.2 - 11.2 micrometers
- Channel 6: 11.2 - 12.2 micrometers

Flown aboard NASA C-130B, NASA ER-2, and NASA Learjet aircraft, the TIMS sensor has a nominal Instantaneous Field of View of 2.5 milliradians with a ground resolution of 25 feet (7.6 meters at 10,000 feet). The sensor has a selectable scan rate (7.3, 8.7, 12, or 25 scans per second) with 698 pixels per scan. Swath width is 2.6 nautical miles (4.8 kilometers at 10,000 feet) while the scanner's Field of View equals 76.56 degrees.

In Canada, ITRES developed the Compact Airborne Spectrographic Imager (CASI) that became operational in 1989. Recently, a revised version of the instrument has been put on the market. This pushbroom sensor has 288 spectral narrow (1.9nm.) bands with 512 pixels across track in the 400-870nm. range .

The Geophysical Environmental Research Corporation (GER) based in Millbrook, U.S. develops a line of imaging spectrometers that included the GERIS (a 63 channel instrument no longer in production), the Digital Airborne Imaging Spectrometer (DAIS), and the Environmental Protection Systems (EPS) systems. The 79-channel Digital Airborne Imaging Spectrometer built by the Geophysical Environmental Research corp. (GER) is the successor of the 63-channel imaging spectrometer GERIS. This 15-bit instrument covers the spectral range from the visible to the thermal infrared wavelengths at variable spatial resolution from 3 to 20 m. depending on the carrier aircraft flight altitude. Six spectral channels in the 8 - 12 µm. region are used for temperature and emissivity of land surface objects. These and 72 narrow band channels in the atmospheric windows (e.g. those wavebands that pass relatively undiminished through the atmosphere) between 0.450 and 2.45 µm. are sensed using four spectrometers. The FOV is 32-39 degrees (depending on aircraft) and the IFOV is 3.3 mrad. (0.189 degrees) yielding a GIFOV depending on aircraft altitude of between 5 - 20 m.

The Hyperspectral Mapper (HyMAP), in the U.S. known as Probe-1, was build by Integrated Spectronics in Australia and operated by HyVISTA. The 126 channel Probe 1 (operated by ESSI in the US) is a "whiskbroom style" instrument that collects data in a cross-track direction by mechanical scanning and in an along-track direction by movement of the airborne platform. The instrument acts as an imaging spectrometer in the reflected solar region of the electromagnetic spectrum (0.4 to 2.5 $\mu$m) and collects broadband information in the MIR (3 - 5 $\mu$m) and TIR (8 - 10 $\mu$m) spectral regions. In the VNIR and SWIR, the at-sensor radiance is dispersed by four spectrographs onto four detector arrays. Spectral coverage is nearly continuous in these regions with small gaps in the middle of the 1.4 and 1.9 $\mu$m atmospheric water bands. The spatial configuration of the Probe-1 is:

- IFOV - 2.5 mrad along track, 2.0 mrad across track
- FOV - 61.3 degrees (512 pixels)
- GIFOV - 3 – 10 m (typical operational range)

The Multispectral Infrared and Visible Imaging Spectrometer (MIVIS) is a 102 channel imaging spectrometer developed by SensyTech. It has 92 channels covering the 400-2500nm region and 10 thermal channels. The IFOV is 2 mrad and the FOV is 71 degrees.

ROSIS (Reflective Optics System Imaging Spectrometer) is a compact airborne imaging spectrometer, which developed by Dornier Satellite Systems, GKSS Research Centre (Institute of Hydrophysics) and the German Aerospace Center (DLR, Institute of Optoelectronics). ROSIS is a pushbroom imaging spectrometer acquiring 32 or 84 bands in the 430 to 850nm range with a spectral halfwidth of 7.6nm. The FOV is 16 degrees and the IFOV is 0.56 mrad. For a flight altitude of 10 km the pixel size is about 5.6 m. with a swath width of 2.8 km.

TRW currently performs airborne data collection with the TRWIS III with image spatial resolutions spanning from less than 1 meter to more than 11 meters, with spectral coverage from 380 to 2450 nm. Spectral resolution is 5.25 nm in the visible/near infrared (380 - 1000 nm) and 6.25 in the short wave infrared (1000 - 2450 nm).


## 2.1    AIRBORNE SIMULATORS

Several examples can be listed of airborne simulators for spaceborne instruments. NASA has a strong airborne programme to support future and planned spaceborne imaging spectrometer missions.

The MODIS Airborne Simulator (MAS) is a multispectral scanner configured to approximate the Moderate-Resolution Imaging Spectrometer (MODIS; Justice *et al.*, 1998), an instrument orbiting  on the NASA TERRA platform. MODIS is designed to measure terrestrial and atmospheric processes. The MAS was a joint project of Daedalus Enterprises, Berkeley Camera Engineering, and Ames Research Center. Sensor/Aircraft Parameters include 50 channels acquired at 16-bit resolution, a IFOV of 2.5 mrad yielding a ground Resolution of 50 meters at 65,000 feet and a swath width of 36 km.

The MODIS/ASTER Airborne Simulator (MASTER) is similar to the MAS, with the thermal bands modified to more closely match the NASA EOS ASTER (Advanced Spaceborne Thermal Emission and Reflection Radiometer; Kahle *et al.*, 1991; Table 1) satellite instrument. It is used to study geologic and other Earth surface properties. The MASTER sensor/aircraft parameters include 50 channels acquired at 16-bit resolution, a IFOV of 2.5 mrad yielding a ground Resolution of 12-50 meters.

The Airborne Ocean Color Imager (AOCI) is a high altitude multispectral scanner built by Daedalus Enterprises, designed for oceanographic remote sensing. It provides

10-bit digitization of eight bands in the visible/near-infrared region of the spectrum, plus two 8-bit bands in the near and thermal infrared. The spatial resolution is 50 meters (from 20km) and the total field of view is 85 degrees (IFOV of 2.5 mrad).

The Multispectral Atmospheric Mapping Sensor (MAMS) is a modified Daedalus Scanner flown aboard the ER-2 aircraft. It is designed to study weather related phenomena including storm system structure, cloud-top temperatures, and upper atmospheric water vapor. The scanner retains the eight silicon-detector channels in the visible/near-infrared region found on the Daedalus Thematic Mapper Simulator, with the addition of four channels in the infrared relating to specific atmospheric features. The spatial resolution is between 50 and 100 meters with a total Field of View of 85.92 degrees and a IFOV: 2.5 or 5.0 mrad (selectable).

Currently ESA is involved in the airborne PRISM experiment. The Airborne Prism Experiment (APEX), will be a pushbroom imager with 1000 pixels across track and a swath width of 2.5 - 5 km. The spectral wavelength range covering is 450 - 2500 nm with a spectral sampling interval < 15 nm at a spectral sampling width < 1.5 times the sampling interval. The APEX instrument will have two detectors cover the specified spectral range: a VIS detector, sensitive in the spectral range 450 - 950 nm (Si CCD) and an IR detector, sensitive in the spectral range 900 - 2500 nm. (HgCdTe or InSb detectors).


## 3    Imaging spectrometry: spaceborne instruments

The first satellite imaging spectrometer to be expected was the LEWIS Hyperspectral Imager (HSI) from TRW company which was launched in 1997 but failed. It covered the 0.4-1.0 $\mu$m range with 128 bands and the 0.9 - 2.5 $\mu$m range with 256 bands of 5 nm and 6.25 nm band width respectively. Several other instruments are in the design stage. The turn of the millennium marks the onset of a new era in imaging spectrometry where we see the first satellite-based instruments being operational.

TABLE 1. ASTER band passes.

| Characteristic | VNIR | SWIR | TIR |
|---|---|---|---|
| Spectral Range | 1*: 0.52 - 0.60 $\mu$m<br>Nadir looking | 4: 1.600 - 1.700 $\mu$m | 10: 8.125 - 8.475 $\mu$m |
| | 2: 0.63 - 0.69 $\mu$m<br>Nadir looking | 5: 2.145 - 2.185 $\mu$m | 11: 8.475 - 8.825 $\mu$m |
| | 3N: 0.76 - 0.86 $\mu$m<br>Nadir looking | 6: 2.185 - 2.225 $\mu$m | 12: 8.925 - 9.275 $\mu$m |
| | 3B: 0.76 - 0.86 $\mu$m<br>Backward looking | 7: 2.235 - 2.285 $\mu$m | 13: 10.25 - 10.95 $\mu$m |
| | | 8: 2.295 - 2.365 $\mu$m<br>9: 2.360 - 2.430 $\mu$m | 14: 10.95 - 11.65 $\mu$m |
| Ground Resolution | 15 m | 30m | 90m |
| Data Rate (Mbits/sec) | 62 | 23 | 4.2 |
| Cross-track Pointing (deg.) | ±24 | ±8.55 | ±8.55 |
| Cross-track Pointing (km) | ±318 | ±116 | ±116 |
| Swath Width (km) | 60 | 60 | 60 |
| Detector Type | Si | PtSi-Si | HgCdTe |
| Quantization (bits) | 8 | 8 | 12 |

*the numbers indicate the band number.

A number of orbital imaging spectrometer missions are planned or being prepared currently. HIRIS (Goetz & Herring, 1989), the High Resolution Imaging Spectrometer is designed to acquire images in 192 spectral bands simultaneously in the 0.4-2.5 µm wavelength region with a ground resolution of 30m. and a swath width of 30km. This scanner has a spectral resolution of 9.4 nm in the 0.4-1.0 µm wavelength region and a 11.7 nm spectral resolution in the 1.0-2.5 µm wavelength range. Though an interesting design, there are no plans for the HIRIS to fly.

NASA is supporting the Japanese Advanced Spaceborne Thermal Emission and Reflectance Radiometer (ASTER; Table 1); a high spectral resolution imaging spectrometer launched on December, 18, 1999 on the EOS AM1 now known as TERRA platform. The instrument is designed with three bands in the visible and near-infrared spectral range with a 15m spatial resolution, six bands in the short wave infrared with a 30m spatial resolution, and five bands in the Thermal infrared with a 90m spatial resolution. The VNIR and SWIR bands have a spectral resolution in the order of 10 to 100nm. Simultaneously, a single band in the near-infrared will be provided along track for stereo capability. The swath width of an image will be 60km with 136km across-track and a temporal resolution of <16 days. ASTER has now been in orbit for some time and the first images are available on the ASTER web pages. Data can be ordered from the archive or users can submit data acquisition requests for any area in the world. The data will be distributed at no cost. ASTER will provide high level geophysical products including surface reflectance, radiance, temperature and emissivity. The calibration is very accurate such that temperature can be derived to an accuracy of half a degree within the dynamic range.

Also on the TERRA is the Moderate resolution imaging spectroradiometer (MODIS), which is a land remote sensing instrument with high revisting time. MODIS is mainly designed for global change research. The MODIS instrument provides high radiometric sensitivity (12 bit) in 36 spectral bands. Two bands are imaged at a nominal resolution of 250 m at nadir, with five bands at 500 m and the remaining 29 bands at 1,000 m. MODIS achieves a 2,330-km swath and provides global coverage every one to two days. The optical system consists of a two-mirror off-axis afocal telescope which directs energy to four refractive objective assemblies; one for each of the VIS, NIR, SWIR/MWIR and LWIR spectral regions covering a total spectral range of 0.4 to 14.4 $\mu$m.

Within the framework of the New Millennium Program NASA is launching the Hyperion on NMP Earth Observing (EO)-1. The Hyperion instrument (build by TRW) provides a new class of Earth observation data for improved Earth surface characterization. The Hyperion provides a science grade instrument with quality calibration based on heritage from the LEWIS Hyperspectral Imaging Instrument (HSI). The Hyperion provides a high resolution hyperspectral imager capable of resolving 220 spectral bands (from 0.4 to 2.5 $\mu$m) with a 30 meter resolution. The instrument can image a 7.5 km by 100 km land area per image and provide detailed spectral mapping across all 220 channels with high radiometric accuracy. EO-1 will be launched on a Delta-2 from Vandenberg Air Force Base on November, 16, 2000. After deployment from the third stage of the Delta, EO-1 will fly in a 705 km circular, sun-synchronous orbit at a 98.7 degree inclination. This orbit allows EO-1 to match within one minute, the Landsat 7 orbit and collect identical images for later comparison on the ground. Once or twice a day, sometimes more, both Landsat 7 and EO-1 will image the same ground areas (scenes). All three of the EO-1 land imaging instruments will view all or sub-segments of the Landsat 7 swath.

The European Space Agency (ESA) are developing two spaceborne imaging spectrometers (Rast, 1992): The Medium Resolution Imaging Spectrometer (MERIS; Rast & Bézy, 1995) and the High Resolution Imaging Spectrometer (HRIS; now renamed to PRISM (Posselt *et al.*, 1996), the Process Research by an Imaging Space Mission). MERIS, currently planned as payload for the satellite Envisat-1 to be launched in June 2001(?) by the Ariane 5, is designed mainly for oceanographic application and covers the 0.39-1.04 µm wavelength region with 1.25 nm bands at a spatial resolution of 300 m or 1200 m.

ESA's Medium Resolution Imaging Spectrometer (MERIS) is a pushbroom imager that will acquire 15 bands (despite of being fully programmable) in the 390-1040nm range of the electromagnetic spectrum with a variable spectral bandwidth between 1.25 and 30nm. The pixel sizes at nadir are 260m across-track and 300m along-track which will be re-sampled (on board) to 1040m by 1200m over open ocean. The instrument has a 68.5 degree field of view providing a swath of 1150km on the ground, where the field of view is divided between five identical cameras each having a 14 degree field of view. On-board Envisat, MERIS is put into a sun-synchronous polar orbit with a mean altitude of 800km and a time of overpass of 10:00 a.m. The radiometric requirements and radiometric accuracy of MERIS are far in excess of that of sensors currently operational and are set to allow discrimination of 30 classes of pigment concentration over open ocean each smaller or equivalent of 1 mg m$^{-3}$. On-board band-to-band calibration utilizing sun illuminated spectralon diffuser panels yields spectral calibration. PRISM (SPECTRA), currently planned for Envisat-2 to be launched around the year 2000(?), is a spectrometer similar to HIRIS which will cover the 0.4-2.4 µm wavelength range with a 10 nm contiguous sampling interval at a 32 m ground resolution.

The German Space Agency (DLR) is currently operating the MOS from the shuttle mission. The Modular Optoelectronic Scanner (MOS) is a spaceborne imaging pushbroom spectrometer in the visible and near infrared range of optical spectra (400 - 1010 nm) which was specially designed for remote sensing of the ocean-atmosphere system. MOS-PRIRODA and MOS-IRS instruments are basicaly identical providing 17 spectral channels with medium spatial resolution in the VIS/NIR. The advanced instrument built for the IRS spacecraft has one additional channel in the SWIR at 1.6$\mu$m.

Other sensor systems include ARIES, NEMO and orbiew-4. The planned Australian Resource Information and Environment Satellite (ARIES) will provide contiguous coverage from 400-1050nm (VNIR) with a minimum of 20nm band spacing and contiguous coverage from 2000-2500 (SWIR-2) with a minimum of 16nm band spacing. The pixel size is 30m. at nadir with 15km swath width. Two sets of 3 bands centred at 940 and 1140nm with minimum band spacing of 16nm for atmospheric correction. In total 105 bands will be acquired. The main aims of the system are directed toward mineral exploration and vegetation studies.

The Naval EarthMap Observer (NEMO) is a space-based remote sensing system planned for launch in mid-2000 and aimed at collecting broad-area, synoptic, and unclassified Hyperspectral Imagery (HSI) for Naval Forces and the Civil Sector. A Hyperspectral Coastal Ocean Imaging Spectrometer (COIS) will provide moderate spatial resolution with a 30/60 meter ground sample distance (GSD) at a 30 km swath width with high spectral resolution of 10 nanometers with a spectral range of 400 to 2500 nanometers.

OrbView-4's (the Warfighter) imaging instrument developed by Orbimage will provide one-meter panchromatic imagery and four-meter multispectral imagery with a swath width of 8 km as well as 200 channel hyperspectral imagery with a swath width of 5 km. The satellite will revisit each location on Earth in less than three days with an ability to turn from side-to-side up to 45 degrees from a polar orbital path. The spectral range is 400-2500nm, spatial resolution is 8m. however only 24m. data will be provided to the civil users. Note that engineering data on the system are unavailable, however judging from the specifications the hyperspectral module of orbview-4 shows resemblance to the AVIRIS scanner although the original data are acquired at 8m. resolution at nadir on the ground. However, for military purposes on 24 m. data will be delivered to non-military users hence the comparison with AVIRIS seems justified.

## 4    Spaceborne versus airborne data

Airborne observations with dedicated instruments have their own history in contributing to advances in the Geosciences and extending our general knowledge in many fields, including the numerous surveys for natural resources. These observations cannot simply be regarded as an appendix to spaceborne observations, as one might assume with virtual all the publication on the spaceborne front; rather, airborne observations add another dimension to Earth observation and provide a wide field of applications of their own. They are the most efficient means by which laboratory- and ground-based observations can be extended to regional observations. The science community needs data on the local and regional scales for calibration of its global data from spaceborne observations and for understanding overall concepts.

TABLE 2. A comparison of airborne and spaceborne remote sensing.

| | Airborne | Spaceborne |
|---|---|---|
| Development cost | ?M$ | 10-1000M$ |
| "launch" cost | 10k$-100k$ | 10-1000M$ |
| cost per image | 20$-2000$ | 0-4k$ |
| Infrastructure for platform | aircraft/airfield | Ground stations |
| Mission duration | hours | 1-15 year |
| flying height | .2-20 km | 100-40,000 km |
| flight path | flexible | Fixed to orbital path |
| ground speed | variable < Mach 1 | fixed 6-7 km/s ~ Mach 20 |
| Resolution | .1-100 m | 1-1000 m |
| Coverage | local to regional | Regional to global |
| Calibration | needed every flight | linear sensor degradation |
| Geometry | GPS/INS data needed | Earth curvature |
| data transport | anything (photo, tape, ...) | X-band downlink < 300 Mbit/s |
| main qualifications | flexible, versatile | regular revisits, global coverage |

Imaging sensors may be carried on either an airborne or spaceborne platform. Depending on the use of the prospective imagery, there are trade-offs between the two types of platforms. It is indeed a tough job for anyone to keep track of the sensor zoo. An effort is made here to do some general statements regarding differences and similarities of airborne and spaceborne sensors but that doesn't mean that there aren't any exceptions, because the spectrum of airborne and spaceborne sensors is large and increasing all the time. Table 2 provides a comparison of airborne and spaceborne systems.


## 4.1    COSTS OF A SPACEBORNE SYSTEM

The total costs of a spaceborne mission are the sum of the development of the sensor and platform, the construction and integration, the launch of the mission, the cost per image, and the cost of processing in order to get the data in the required form. See Kerekes & Landgrebe (1991) for a review of parameter trade-offs of imaging spectrometer systems.

For many systems the cost of development will either remain a mystery (because the instrument has been improved over the years and it simply unknown how much effort was spent on design and re-design), or a well-kept secret (companies in general are very secretive about the real R&D costs leading to a final product). For instance NASA built AVIRIS in 1987, but it has been improved ever since. Ask NASA about the true development cost of AVIRIS; the result will be all glazed eyes.

In general a system (platform/sensor) can be built for somewhere between 10 M$ and 1 G$. A microsatellite, roughly the size and weight of a suitcase, based on an existing satellite bus (like offered by SSTL Surrey Satellite Technologies Limited, for instance) can be produced at low cost. Large satellites, roughly the size and weight of a greyhound bus, may be very expensive. Large satellites usually carry a load of sensors and other instruments, as a consequence of which your sensor may have to share power and time with all the other instruments on board. This effectively means that your instrument of interest will not be switched on all the time.

The advantages of the microsatellites over the battlestar galactica's may be obvious: they're cheaper to build and launch, and you don't put all your eggs in one basket (if Envisat falls from the sky after launch a lot of scientists will be banging their heads against the wall). But the drawback is that these systems are still experimental (using new, i.e. not proven, technologies), and do not have the long life time as their giant cousins (life-time is, mostly, determined by the amount of fuel that can be taken onboard).

Next is the cost to launch the particular spacecraft into space. Launch cost may range from 10 M$, for a microsatellite riding piggyback on a larger satellite to a low-earth orbit (LEO), to 500 M$, for a large system launched into a geostationary (GEO) orbit. Roughly one could say that a constellation of 10 microsatellites could be built and launched for the same price as a "traditional" system. In the future we will see many more microsatellites being used for Earth observation.

In order to control a satellite, a network of ground stations may be needed for command and data relay. Typically the downlink capacity of the X-band transmitter of the satellite is the bottleneck in the data throughput of the whole system. The speed on the ground of most near-earth orbiting remote sensing systems is fixed in the order of 6-7 km/s. This means there is a trade-off between resolution, coverage, and number of bands, because the data volume generated per second by any system is more or less proportional to the coverage and the number of bands, and inversely proportional to the square of the sample spacing. Downlink speeds may range from 10 megabits per second for an old system to 2x150 megabits/s for a modern system (e.g. Landsat-7). In the future data throughput may receive a considerable boost when laser satellite-to-satellite communication systems will become more commonplace, e.g. with the European Artemis communication satellite.

It is difficult to give the cost per image in general as this highly depends on the system used, the owner of the spacecraft, desired processing level, and the reseller of the data. Prices may vary from the cost of reproduction by experimental and governmental systems (SIR-C, NOAA-14, Landsat-7) to "commercial" prices for high-resolution imagery (Ikonos), ranging roughly from 0$ to 200$ per square kilometer. Regarding the high-resolution market, prices may drop in the near future because of increasing competition in this market sector. Although, this fact still remains to be seen, because current RS systems seem to be complementary rather than competitive.


## 4.2    COSTS OF AN AIRBORNE SYSTEM

The total costs of an airborne mission are the sum of the development of the sensor and aircraft, the construction and integration, the "launch" of the mission, the cost per image, and the cost of processing in order to get the data in the required form.

Aircraft are available in a multitude of types and configuration. Earth observation can be done from as low as 100 meter in an "ultra-light" (ULV) to 30 or maybe 40 kilometers in an unmanned aircraft (UAV), spanning a range of a factor of well over 100. This, together with the fact that most missions can be "launched" from any airstrip around the globe makes aircraft a very flexible and versatile tool for remote sensing. Notwithstanding some predictions, airborne remote sensing will never be completely replaced by spaceborne observations. Only in some applications and markets, for instance the high-resolution imagery market, satellite data may take as much as 50%. But with its flexibility and some new developments (e.g. UAV's) airborne remote sensing is here to stay.

Furthermore, the field of airborne sensors can be regarded as the proving ground for future spaceborne sensors. The airborne concept permits an environment for repeated instrument access on all levels, ideally suited for experimentation and validation; test flights can manage with a lot of provisional gadgets and arrangements; limitations on instrument weight and power consumption are not as restrictive as on the intended satellite mission. Complexity can be handled in an orderly way, which reduces considerably the risk of failure. The cost of airborne experimentation along with its infrastructure is relatively low compared to spaceborne experimentation.

It makes no sense here to look at the costs of aircraft and airfields, let's just consider the cost per mission. Depending on the configuration of the aircraft and sensor(s) and the location on the Earth, for the user to have a mission "launched" he pays about 10,000$ to 100,000$ just to get the plane in the air. Next to that he will have to pay a certain amount for the data. The cost per image ranges roughly from 20$ to 200$ and more, from a traditional aerial photograph to a fully processed digital image. Usually, a lump sum will be agreed on for the delivery of the data of the entire area of interest.

Contrary to the satellite systems, airborne systems do not suffer from a data-throughput bottleneck. The aircraft can be fitted with any high-speed, high-capacity data storage system for ingesting the dataflow from a high-resolution sensor. It should also be noted that the equivalent information content of one good-old aerial photograph is about half a gigabyte worth of digital data, which is about the same as one geometrically corrected, 200-band hyperspectral image.

The sensor and data storage systems do not have to meet the requirements for a life in space, they don't have to be a compromise between capabilities on the one hand and size and weight on the other hand. A major disadvantage of the airborne scheme is that mechanical stresses on an airborne sensor are much higher than a spaceborne sensor (that is, after it did survive the stresses of the launch). In fact an airborne sensor should be calibrated again after each field campaign. Spaceborne sensors tend to degrade in a more linear way with time, requiring in-flight calibration only every once in a while.

In many ways an aircraft is more flexible compared to a satellite. For one thing it is not fixed to a particular orbit, which means it can fly at various altitudes, various speeds, and variable directions. It can also fly with various sensor configurations. On the other hand the erratic behaviour of flight parameters, like roll, pitch and yaw, make control and processing of the airborne imagery difficult (see section "Processing"). For instance, an aircraft could be flown with its tail pointing towards the Sun in order to reduce BRDF (bi-directional reflectance distribution function) effects, which are a function of the illumination angle and the look-angle. Satellites can't do that, although some wide field-of-view ocean colour imagers have a side-looking capability in order to avoid sea glint (e.g. Orbview-2).


## 4.3    COVERAGE

Coverage, the amount of ground covered by one image, is proportional to the sensor field-of-view (FOV) and the flying height. The higher the platform, the more area will be imaged on one single image. The drawback of flying higher, however, is that there will be less details visible in the image. A small study revealed that to map the city of London with Ikonos images, about 300 images would be needed. Taking into account the orbit of Ikonos and the average number of cloud-free days in London, it would take at least one full year to completely cover London. With an aircraft, these images could be acquired in one single day.

It may be noted that roughly between 20 and 100 kilometer height there is a gap where aircraft and spacecraft can not go -for aircraft because there is too little atmosphere (for propulsion), for spacecraft there is too much (drag)- this is the realm of the sounding rockets. Sounding rockets are mostly used for atmospheric research, sometimes for astronomical observations, but hardly ever for Earth observation.

Total coverage depends on the duration of the mission. Typically an aircraft can only take up fuel for a couple of hours flying, limiting one "launched" mission to at most one day. Some of the working horses in spaceborne remote sensing have been in orbit for over 15 years! (E.g. Nimbus-7, NOAA-9, Landsat-5) One of the main factors determining design life of a satellite is the amount of fuel that can be taken onboard. On one occasion a SPOT satellite had to be moved in order to avoid a possible collision with a piece of debris, which took about 400 gram of rocket fuel. On average a SPOT satellite needs 150 gram of rocket fuel to maintain its orbit. Thus the debris cost the SPOT satellite 3 years of its design lifetime!

## 4.4     RESOLUTION

This is probably the most talked about quality in any remote sensing system and at the same time probably the least understood quality as well! The confusion comes from the fact that in a well-designed system the resolution is in the same order of magnitude as the sample spacing leading people to ignore all distinctions. Apart from that, resolution may be variable within one image, which gives rise to a whole class of additional problems and misunderstandings. (For instance, the pixel spacing of the NOAA/AVHRR sensor is about 800 meter at nadir, while the resolution is about 1.1 by 1.1 km at nadir and deteriorates to about 6 by 3 km at the limb of the sensor).

Resolution determines the discriminating power of the sensor, it is closely linked to the instantaneous field-of-view (IFOV) of the sensor, and can be expressed either as a small angle, or an area on the ground. Sample spacing determines at which points a measurement (pixel) is actually taken and can also be expressed as a small angle or an area on the ground. For NOAA/AVHRR IFOV is 1.4 milliradians, while the sample step is 0.95 milliradians, which means that the AVHRR data is slightly oversampled at 1.36 samples per IFOV.

The details that can be seen in any digital image are a function of the resolution as well as the sample spacing. Furthermore it should be noted that the concepts of resolution and sample spacing can be easily understood in the spatial domain, but they are equally valid in the spectral, temporal, and radiometric domain of the data.

In this report it is assumed that systems have a balanced design and that resolution and sample space are of about the same value.

There is a linear relation between the height of the sensor and its ground resolution. This means that if you double the height then the finest details that can be discriminated on the ground are twice as big. For an airborne system it's generally easy to improve the amount of details visible in the image; just fly at lower altitudes. Of course such flexibility can not be offered by the satellite, because most satellite orbits are circular and can not be changed unless a considerable amount of fuel is used. (Note: on planetary missions the elliptical orbit is the rule making possible the acquisitions of multi-resolution images, thus serving a multitude of scientific requirements in one single spacecraft). Thus the satellite has to compensate the loss of resolution -due to the increased height- by bringing a telescope onboard. For instance, the 1-meter resolution OSA (Optical Sensor Assembly) sensor of the Ikonos satellite features a telescope with a 10-meter focal length. In order to make the telescope fit in the satellite the telescope has a complex folded optics design. Note: a significant advantage of using Synthetic Aperture Radar (SAR) is that the spatial resolution is independent of platform altitude. Thus, fine resolution can be achieved from both airborne and spaceborne platforms. In theory the ground resolution can be expressed as half the antenna length; in practice the real resolution depends on radar system electronic stability, antenna design, accuracy of navigation data, and precision of the processing; it is difficult to achieve the theoretical attainable resolution.


## 4.5     GEOMETRY

Viewing geometry and swath coverage can be greatly affected by altitude variations. At aircraft operating altitudes, an airborne sensor must image over a wide range of incidence angles, perhaps as much as 60 or 70 degrees, in order to achieve relatively wide swaths (let's say 50 to 70 km). It so happens that the incidence angle (or look angle) has a significant effect on the backscatter from surface features and on their appearance on an image. Image characteristics such as pixel displacement, reflectance, and resolution will be subject to wide variations, across a large incidence angle range. Spaceborne sensors are able to avoid some of these imaging geometry problems since they operate at altitudes up to one hundred times higher than airborne sensor. At altitudes of several hundred kilometres, spaceborne sensor can image comparable swath widths, but over a much narrower range of incidence angles, typically ranging from five to 15 degrees. This provides for more uniform illumination and reduces undesirable imaging variations across the swath due to viewing geometry.

## 4.6    PROCESSING

As with any aircraft, an airborne sensor will be susceptible to variations in velocity and other motions of the aircraft as well as to environmental (weather) conditions. In order to avoid image artifacts or geometric positioning errors due to random variations in the motion of the aircraft.

*Figure 1*. Imagery acquired by the WAAC sensor from DLR. The flight direction was from top to bottom. The left image is raw and has been supplemented by curves representing the roll (white line) and pitch (black line) behavior of the aircraft. The right image has been rectified.

The sensor system must use sophisticated navigation/positioning equipment and advanced image processing to compensate for these variations (Figure 1). Generally, this will be able to correct for all but the most severe variations in motion, such as significant air turbulence. Spaceborne sensors are not affected by motion of this type. Indeed, the geometry of their orbits is usually very stable and their positions can be accurately calculated.

However, geometric correction of imagery from spaceborne platforms must take into account other factors, such as the rotation and curvature of the Earth, to achieve proper geometric positioning of features on the surface.

## 4.7    FLEXIBILITY

Although airborne systems may be more susceptible to imaging geometry problems, they are flexible in their capability to collect data from different look angles and look directions. By optimizing the geometry for the particular terrain being imaged, or by acquiring imagery from more than one look direction, some of these effects may be reduced. Additionally, an airborne sensor is able to collect data anywhere and at any time (as long as weather and flying conditions are acceptable!). A spaceborne sensor does not have this degree of flexibility, as its viewing geometry and data acquisition schedule is controlled by the pattern of its orbit. However, satellite sensor do have the advantage of being able to collect imagery more quickly over a larger area than an airborne sensor, and provide consistent viewing geometry. The frequency of coverage may not be as often as that possible with an airborne platform, but depending on the orbit parameters, the viewing geometry flexibility, and the geographic area of interest, a spaceborne sensor may have a revisit period as short as one day.

## 4.8    COMPLEMENTARY DATA

Most airborne data collection is being provided for a number of research projects or for surveys seemingly independent of spaceborne data. However, there are two application fields of airborne data collection where airborne and spaceborne data are used side-by-side (Figure 2).

Airborne sensors can deliver complementary data for use with other remotely collected data. In particular spaceborne remote sensing is targeted at global observations, while airborne remote sensing more serves the local and regional market. Together they offer a multi-scale data set suited for serving a multitude of scientific missions.

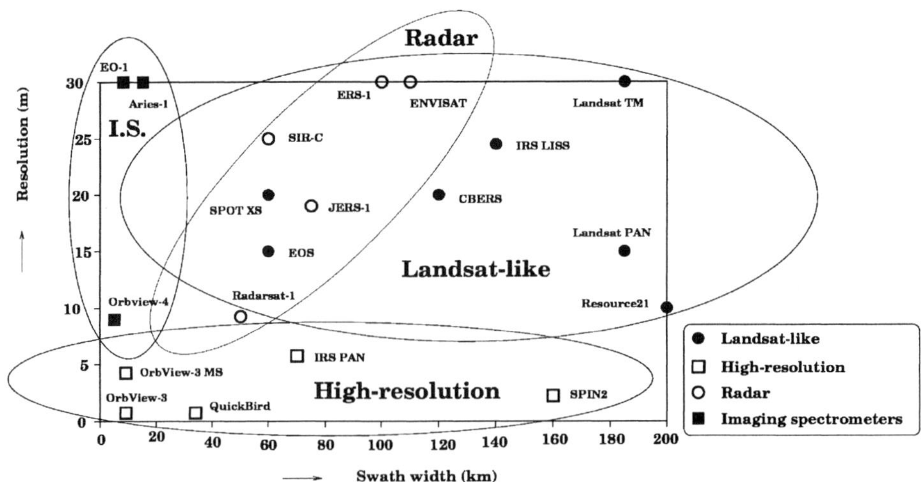


*Figure 2.* Overview of remote sensing systems that can be used in a synergetic fashion.

Yet another application field of airborne remote sensing is to use under-flight data to be used as a calibration standard for spaceborne sensors. Some agencies (e.g. CCRS) already carry out such calibration flights on a regular basis.


## 5    Pre-processing

Pre-processing generally is conducted by the manufacturers, institute/agency or commercial company that maintains and operates the instrument. In most cases the user is provided with at-sensor radiance data and the necessary instrument characteristics (e.g., spectral response functions, band passes etc.) needed for further analysis. The on-ground calibration of an VIS / TIR imaging spectrometer consists of a sequence of procedures providing the calibration data files and ensuring the radiometric, spectrometric and geometric stability of the instrument:

- measurement of noise characteristics of the sensor channels
- measurement of the dark current of the channels
- measurement of the relative spectral response function of the channels
- derivation of the effective spectral bandwidth of the channels
- definition of the spectral separation of the channels
- measurement of the absolute radiometric calibration coefficients (the "transfer functions" between absolute spectral radiance at the entrance of the aperture of the sensor and the measured radiation dependent part of the output signal of each channel)
- definition of the Noise Equivalent Radiance (NER) and the Noise Equivalent Temperature Difference (NE DeltaT) of the channels
- measurement of the Instantaneous Field of View (IFOV), the spectral resolution and the deviations in the band-to-band registration

In this Chapter we will describe the basic pre-processing and image data calibration steps that need to be performed in order for imaging spectrometer data to be used for thematic applications. Next we will discuss quality control checks that can be used to analyze the data prior to analysis. Finally, we present thematic mapping techniques that allow the derivation of surface geophysical products. The interested reader is referred to the Chapter 10 on thermal imaging spectrometry  for calibration routines of thermal infrared data. Dekker *et al.* in their chapter on hydrological applications of imaging spectrometry describe calibration at the land-water interface as well as optical models for water penetrating EM radiation. Thematic applications that make use of the analytical techniques described in this Chapter can be found in the Chapter 7 on geological applications , in the Chapter 8 on petroleum applications and in the Chapter 3 on soil and land degradation.


## 6    Laboratory set-up of a pre-processing calibration facility

A typical Laboratory Calibration Facility for VIS - TIR wide - angle video spectrometric airborne sensors as found at DLR-Oberpfaffenhofen consists of the following four main parts:
1.    Spectrometric/geometric Calibration Part (SCP)

2.   Relative Radiometric Calibration Part (RRCP)
3.   Thermal Absolute Calibration Part (TACP)
4.   Absolute Radiometric Calibration Part (ARCP)

Part 1 (SCP) comprises a mirror collimator with a flat folding mirror at the parallel-light output of the collimator. The mirror collimator has a focal length of 150 cm and a circular aperture of 30 cm in diameter. At the focus position of the mirror collimator can be installed:

- the output slit of a monochromator for the 0.4 - 2.5 μm spectral region,
- the interference filter arrangement,
- different illuminated point or slit targets (for geometric calibration)

The flat folding mirror at the collimator output allows to realize different illumination directions of the parallel light from the collimator output to the sensor entrance. Part 2 (RRCP) is mobile and included, because it is impossible to check the absolute radiometric calibration of the installed sensor by means of an integrating sphere. The RRDS consists of an optical module and power supply unit. The optical module contains 4 halogen lamps of 100 Watt and several dissipative panels made of ground glass. The dimensions of the illuminated surface are 40 cm by 55 cm. Part 3 (TACP) is mobile and consists of 2 water-cooled blackbodies, each with the dimensions of 100 x 100 $cm^2$ - one at ambient or up to 5 C below ambient temperature, the other at 25 - 30 C above ambient temperature - and the supply module. The TACP is used for the radiometric calibration of the mid-infrared and the thermal channels of the sensor in the laboratory and in the hangar. Part 4  (ARCP) is used for the absolute on-ground calibration of the sensor in the VIS-SWIR spectral region (0.4 - 2.5 μm.) by means of an Integrating Sphere (IS). The interior of the Integrating Sphere (1.65 m. in diameter) is coated with barium sulfate. The eighteen 200-Watt lamps are mounted internally. Each lamp has its own individual power supply, so that each may be lightened independently, providing 18 equal steps of radiance. The maximum radiance of the IS is 0.08 mW/sr cm² nm at the wavelength of 1.0 μm. The homogeneity of the radiance at the 40 cm x 55 cm rectangular output part of the sphere is better than 1 %, if at least 4 lamps are lightened. The IS has to be calibrated from time to time by establishing the ratio between the output of the sphere and a well known radiance standard - the so called Absolute Diffuse Source (ADS) - a separate unit of the LCF.


## 7    The spectral pre-processing chain

Data acquisition by imaging spectrometers can be done using the whiskbroom, pushbroom or staring principle. Whiskbroom imagers are electromechanical scanners. On-axis optics or telescopes with scan mirrors sweep from one edge of the swath to the other. The FOV of the scanner can be detected by a single detector or a single-line-detector. Simultaneously the movement of the satellite or airplane guarantees the sweeping scan over the earth. This means that the dwell time for each ground cell must be very short at a given IFOV because each scan line consists of multiple ground cells which will be detected. In whiskbroom scanning each pixel is viewed separately which allows a wide field of view and only one detector in each spectral band to be calibrated. The disadvantage is the short dwell time (e.g., the time that the instrument "sees" each pixel) which limits the spatial and spectral resolution as well as the ratio of signal versus noise. Furthermore, the rotating mirror often causes resonance that may be observed in the data as striping.

Pushbroom scanners are electronic scanners that use a line of detectors to scan over a two dimensional scene. The number of pixels is equal to the number of ground cells for a given swath. The motion of the aircraft or the satellite provides the scan in along-track-direction, thus, the inverse of the line frequency is equal to the pixel dwell time. By using a two dimensional detector, one dimension can represent the swath width (spatial dimension, y) and the other the spectral range. These imaging spectrometers can be subdivided in Wide Field Imagers (MERIS, ROSIS) and Narrow Field Imagers (HSI, PRISM).

Staring imagers are also electronic scanners. They detect a two dimensional FOV instantly. The IFOV along and cross track corresponds to the two dimensions of the detector area array. Two subgroups of staring imagers are Wedge Imaging Spectrometer (WIS) and Time Delay Integration Imagers (TDI). If the incoming light passes a linear wedge filter each row xn of the ground segment is seized by the detector row xn for a determined wavelength. For very high ground resolution and low sensitivity applications, xn rows of the ground can be traced by using a TDI. The light from the scene will be separated by a linear filter for spectral band definition. On the 2D-detector the signal for this line can be read out from multiple rows caused by the forward movement of the sensor. Therefore the sensitivity of a TDI with n rows is n times that of an imager using the pushbroom principle.

Incoming light or radiation from the surface is split by beam splitters into the wavelength ranges specified by the instrument. By means of a set of lenses, the incoming photons are projected onto the array containing light sensitive elements on which the charge accumulating is proportional to the integrated light intensity. For signal read-out, the charges accumulated on the detector elements are passed through a amplifier and digitizer for analogue to digital conversion. This results in a digital signal, referred here to as raw radiance, digitized in 8, 12 or 16bit. Detector materials currently used are silicon for the 0.4 -1.0 μm, lead sulfide for the 0.8 - 2.5 μm and indium antimonide for the 0.8 - 5.0 μm range. Spectral separation into the channels needed can be done through a dispersion element (grating/prism) or a filter-based system. Dispersion elements collect spectral images by using a grating or a prism. The incoming electromagnetic radiation will be separated into distinct angles. The spectrum of a single ground pixel will be dispersed and focused at different locations of one dimension of the detector array. This technique is used for both, whiskbroom and pushbroom image acquisition modes. Hyperspectral imagers are using mainly gratings as the dispersive element. A narrow band of a spectrum can be selected by applying optical bandpass filters (tunable filters, discrete filters and linear wedge filters). A linear wedge filter functions by transmitting light at a centre wavelength that depends on the spatial position of the illumination in the spectral dimension. The detector behind the device receives light at different wavelengths of the scene.

Spectral response is not homogenous when measured over the area covered by a pixel. The point spread function (PSF) describes the decline of the measured signal and is found using a monochromator covering the sensor designated wavelength coverage and different pinhole targets.

For each spectral channel that the sensor acquires, the radiance is variable. Rather than a channel sensing only photons of one particular wavelength, the channel measures radiance in a wavelength range that stretches from a few nanometers lower and a few nanometers higher wavelength than the centre wavelength of the channel.

The curve describing the decline of the radiance levels around the central channel wavelength for each channel is known as the spectral response function (SRF) which can again be deduced with monochromator measurements.

Imaging spectrometers take indirect measurements of physical parameters in the sense that the digitized signal recorded is directly proportional to the incoming photon energy but not measured in any physically meaningful unit. The relation between the raw digitized signal and a physical meaningful parameter is established after radiometric correction yielding spectral radiance measured as the photon flux power per unit solid angle per wavelength interval. During radiometric calibration the radiometric response function is derived from the relation between the signal, caused on the detectors by the bombardment by photoelectrons, and the incoming radiance. This function translates raw radiance into spectral radiance. The function is measured by mounting the sensor onto a so called integrating sphere; a half sphere reference surface coated with highly reflective barium sulfate and isolated from daylight. A set of lamps produce light of known spectral radiance in the integrating sphere which can be compared with the measurements of the sensor. These are cross calibrated using a field spectroradiometer with known standard. The radiometric response function corrected for the spectral response and geometric response as well for the temporal response (not further elaborated here) gives the at-sensor spectral radiance. Since the radiometric response function is linear it can be represented by two coefficients, the gain and offset, $C_1$ and $C_0$, respectively for each channel. Calculating at-sensor radiance, $L$, from raw recorded digital signal, DN, using given gain and offset values is done by

$$L = c_0 + c_1 * xDN = mWcm^{-2}sr^{-1}\mu m^{-1} \tag{1}$$

Often thermal sensors are calibratable using two on-board black bodies (bb), one with a low temperature, $T_{bb1}$ and a second with a high temperature $T_{bb2}$. The at-sensor radiance $L_{bb}$ can now be found as

$$L_{bb} = L_1(T_{bb1}) + \frac{L_2(T_{bb2}) - L_1(T_{bb1})}{DN_{bb2} - DN_{bb1}}(DN - DN_{bb1}) \tag{2}$$

where $DN_{bb1}$ and $DN_{bb2}$ are the digital numbers for black body 1 and 2 and $L_1(T_{bb1})$ and $L_2(T_{bb2})$ are the corresponding spectral radiances for black body 1 and 2.

Most instruments measure physical parameters indirectly by generating and recording a signal, i.e., a digital number (DN), which is related to this physical parameter, i.e., radiance. The empirical relationship between the raw signal and the desired physical parameter is done through instrument calibration. The radiometric response function defines the relation between the signal caused by $N_{photons}$ and the incoming spectral radiance, $L_\lambda$. Thus each image channel needs to be calibrated to derive its radiometric response function which translates raw signal into (at-sensor) spectral radiance through a linear relationship. $N_{photons}$ is not only a function of the incident spectral radiance but is also related to sensor characteristics of which the spectral response function and the point spread function are the two most important. Image channels are generally defined with an upper and lower wavelength defining the portion of the electromagnetic (EM) spectrum for which they are sensitive. However what actually is sensed in terms of the wavelength range of the electromagnetic spectrum and its contribution to the total signal is represented by the spectral response function; recording the relative contribution to the signal of a channel of each wavelength portion of the EM spectrum. Similarly the point spread function defines the aerial sensitivity of the instrument.

## 8 Spatial pre-processing

Aircraft data suffer from spatial distortions related to the carrier movements and the ruggedness of the terrain. Modern aircraft remote sensing campaigns are flown with onboard GPS for absolute location of the aircraft at acquisition time and with onboard gyros that record tilt of the aircraft in terms of roll, pitch and yaw. In geometric correction, for each pixel the original observation geometry is reconstructed based on the flight line, aircraft altitude, surface topography and aircraft navigational information. The result of the correction is geocoded at sensor radiance data. Geometric distortion of airborne imaging spectrometer data resulting from data recording can be characterized through four effects:

1.  Panoramic effect: Due to the scanning with constant angular scanning speed, pixels become larger from nadir to the left and right-hand sides of the scan line.
2.  Over- and undersampling: Due to non-perfect synchrony between air-speed, altitude, and scan speed in the flight direction redundant information is scanned or data holes occur.
3.  Geometric distortions due to projection: Due to movements of the aircraft, roll pitch and or yaw distortions may occur.
4.  Drift effect: Due to side wind or other effects the airplane may get out of course and as a result distortions occur in the flightpath.

In addition, topographic effects result in (1) shift of pixel locations compared to the true position and (2) affects the pixel size. In order to allow a geocoding of airborne imaging spectrometer data, on board gyroscopic measurements record the attitude of the aircraft in terms of roll, pitch and yaw and differential GPS measurements record the flight path in x, y and z-absolute coordinates. This information can be used for geocoding of the image data. Three approaches can be applied:

1.  Geocoding using control points and registration to a map base
2.  Geocoding using pixel transformations through gyroscope data
3.  Parametric geocoding using both gyroscopic data, flight line information and a digital terrain model

## 9 Quality control: signal to noise characterization

Signal in imaging spectrometry is considered to be the quantity measured by an imaging spectrometer sensor, whereas noise describes the random variability of the signal. The quantification of the noise level alone is not a very useful measure for the quality of a imaging spectrometer data set since the effect is more severe when signal is low. Therefore, in most studies the signal-to-noise ratio is used to estimated as the ratio of the signal's mean to its standard deviation. Imaging spectrometer data sets contain both periodic (coherent) sensor noise that can be removed and random noise that cannot. The signal-to-noise calculation is done on data sets with periodic noise removed. The remaining random noise can be additive noise, which is independent of the signal, and multiplicative noise which is proportional to the signal. The major part of the noise in imaging spectrometer data sets is additive and decreases sharply with both an increase in wavelength and atmospheric absorption. This random noise component consists of random sensor noise (which is image independent), intra-pixel variability (resulting from spatially heterogeneous pixel contents), and interpixel variability.

Imaging spectrometer data sets contain both periodic (coherent) sensor noise that can be removed and random noise that cannot. The signal-to-noise calculation is done on data sets with periodic noise removed. Curran & Dungan used a notch filter on fast-fourier transformed data for this purpose. The remaining random noise can be additive noise, which is independent of the signal, and multiplicative noise which is proportional to the signal. The major part of the noise in imaging spectrometer data sets is additive and decreases sharply with both an increase in wavelength and atmospheric absorption. This random noise component consists of random sensor noise (which is image independent), intra-pixel variability (resulting from spatially heterogeneous pixel contents), and interpixel variability.

Several methods have been proposed to estimate the signal-to-noise ratio of imaging spectrometer data sets which can be separated into three classes: "laboratory methods", "dark current methods", and "image methods". A typical "laboratory method" uses the signal's mean ($z$) and its standard deviation ($s$) of a bright surface to estimate the signal-to-noise ratio for a few spectral bands. The estimated signal-to-noise ratio is in most cases too high since a very bright target is used. A typical "dark current method" uses variation in the signal dark currents as a measure of noise. One simple "image method" for estimating the signal-to-noise ratio of an image is to find a homogeneous area within the scene and compute the signal mean, $z$, and its standard deviation, $s$. The ratio of the mean to the standard deviation gives an estimate for the signal-to-noise ratio of the image. The method in most cases underestimates the signal-to-noise ratio since interpixel variability contributes to the noise component, $s$, thus suppressing the signal-to-noise ratio. Another shortcoming of the approach is the difficulty in finding an homogeneous area which is representative for the entire image. Lee and Hoppel developed a method for automatic estimation of signal-to-noise ratios for imaging spectrometer data. According to their method, the image is divided into small 4x4 or 8x8 pixel blocks and the signal's mean and standard deviation is calculated for each block. From this, a scatter plot is constructed plotting the squares of the means versus the variances defining a straight line representing the noise characteristics of the image. Meer and others developed a parallel algorithm which allows to separate additive noise from multiplicative noise.

In this section, we will use three methods which will be referred to as the "homogeneous area method", the "local means and local variances method" and the "geostatistical method".


## 9.1    THE "HOMOGENEOUS AREA METHOD"

The "homogeneous area method" for estimating the signal-to-noise ratio of an image is to find a homogeneous area within the scene and compute the signal mean, $z$, and its standard deviation, $s$. The ratio of the mean to the standard deviation gives an estimate for the signal-to-noise ratio of the image.


## 9.2    THE "LOCAL MEANS AND LOCAL VARIANCES METHOD"

The "local means and local variances method" builds on dividing an image into small blocks and computing local means and variances. An image is divided into small blocks of

4x4, 5x5,...., 8x8 pixels. For each block, a local mean (denoted LM) is calculated according to

$$LM = \frac{1}{N} \sum_{i=1}^{i=N} S_i \tag{3}$$

where $S_i$ is the signal value of the $i$th pixel in the block and $N$ is the total number of pixels in the block. The local standard deviation (LSD) is calculated a

$$LSD = \sqrt{\frac{1}{(N-1)} \sum_{i=1}^{i=N} (S_i - LM)^2} \tag{4}$$

Homogeneous blocks have small LSD values and therefore provide information on the noise of the image, while inhomogeneous blocks, such as those containing edges have large LSD values. The distribution of the LSD values is used to characterize the random noise. It is proposed to set-up a number of classes of LSD values between the minimum and maximum values occurring. The LSD value corresponding to the most frequent class (e.g. that class containing most of the blocks) is the mean noise. The signal-to-noise ratio is found as the ratio of mean signal and the mean noise.

## 9.3    THE "GEOSTATISTICAL METHOD"

The "geostatistical method" allows to calculate the signal-to-noise ratio of a imaging spectrometer data set free of interpixel variability. The intrapixel variability and the random sensor noise are estimated using the semi-variogram produced from a transect of pixels. The semi-variogram, $\gamma(h)$, defines the variances of the differences between pairs of pixel signals, $z(\mathbf{x}_i)$-$z(\mathbf{x}_i+h)$, a distance $h$ apart and can be estimated by

$$\gamma(h) = \frac{1}{2n} \sum_{i=1}^{n} [z(x_i) - z(x_i + h)]^2 \tag{5}$$

where $n$ is the number of pairs of points used to calculate the lag $h$. The semi-variogram is characterized by three parameters: the range, the sill, and the nugget component. The sill is the asymptotic upper bound value which is often equal to the variance of the data set. The range is the distance or lag-spacing at which the sill is reached defining the degree of spatial correlation of the data set. The nugget component is the non-zero y-intercept of the semi-variogram implying that measuring a point twice would yield different results. This variance does not have a spatial component and is composed almost entirely of random sensor noise. It is proposed that the signal-to-noise ratio (SNR) is estimated from the signal's mean, $z$, and the nugget component $C_o$ by

$$SNR = z / \sqrt{C_o} \tag{6}$$

Difficulties with the "geostatistical method" are the estimation of the nugget component from the variograms since the semi-variogram itself at a lag of 0 pixels is not defined. Furthermore, data sets are seldom isotropic and anisotropy may result in direction-dependent nugget components.

## 9.4    NOISE ADJUSTMENT

An increase in signal to noise ratio can be obtained by reducing the noise and retaining the signal. A method for doing so is the Maximum Noise Fraction (MNF) transform developed. The MNF algorithm is a method for ordering data cubes into components of image quality using a cascaded principal components transform that selects new components in order of decreasing signal-to-noise ratio. In this approach, the grey levels are considered linear combinations of a uncorrelated signal component and a correlated noise component in which the grey level covariance is the sum of the signal and the noise covariance matrices. The difficulty in applying the MNF technique lies in finding these covariance matrices. The grey level covariance matrix can be readily derived as the sample covariance matrix of the data, however the noise covariance matrix is more complex to assess. The MNF transformation can be explained mathematically as follows.

The MNF transformation can be illustrated mathematically as follows. Consider stochastic variables

$$\mathbf{Z}(\mathbf{x}) = [Z_1(\mathbf{x})...Z_p(\mathbf{x})]^T \tag{7}$$

with expectation $E\{\mathbf{Z}\} = 0$ and dispersion $\mathbf{D}\{\mathbf{Z}\} = \Sigma$. New mutually orthogonal variables with maximum signal-to-noise ratio can be constructed as

$$Y_i = a_{i1}Z_1 + ... a_{ip}Z_p = a_i^T, i = 1,....,p \tag{8}$$

Assuming additive noise (i.e., $\mathbf{Z}(\mathbf{x}) = \mathbf{S}(\mathbf{x}) + \mathbf{N}(\mathbf{x})$ where $\mathbf{S}$ and $\mathbf{N}$ are uncorrelated signal and noise components, respectively) the response equals $\Sigma = \Sigma_S + \Sigma_N$ where $\Sigma_S$ and $\Sigma_N$ are dispersion matrices for $\mathbf{S}$ and $\mathbf{N}$, respectively. The signal-to-noise ratio (estimated from the ratio of mean and variance in a local window) for $Y_i$ that needs to be maximised can be defined as

$$\frac{VAR\{a_i^T\mathbf{S}\}}{VAR\{a_i^T\mathbf{N}\}} = \frac{a_i^T\Sigma a_i}{a_i^T\Sigma_N a_i} - 1 = \frac{1}{\lambda_i} - 1 \text{ and } \frac{a_i^T\Sigma_N a_i}{a_i^T\Sigma a_i} = \lambda_i \tag{9}$$

the factors $\lambda_i$ are the eigenvalues of $\Sigma_N$ with respect to $\Sigma$ and the factors $a_i$ are the corresponding conjugate eigenvectors. The inherent spectral dimensionality of the data set can be found by examining the eigenvalues and associated MNF images. Some of these MNF images are associated with large eigenvalues and coherent (MNF) eigenimages while the remainder if the MNF bands have near-unity eigenvalues and images dominated by noise. Thus the MNF eigenvalues and eigenimages yield the absolute number of end-members required to model the image spectral response.

A procedure known as the Minimum/Maximum Autocorrelation Factors (MAF) is used to estimate the noise covariance matrix for more complex cases. MAF exploits the fact that in most remote sensing data the signal at any pixel is strongly correlated with the signal at neighbouring pixels, while the noise is not. The original data are transformed into linear orthogonal combinations in order of increasing spatial correlation which can be subsequently treated. The common approach is to apply low-pass filtering to the low order MAF factor images which contain most of the low spatially correlated noise and a degraded signal component and removal of the high MAF factor images that contain nearly only noise. After filtering, the cleaned MAF factor images are backtransformed to the original data space.

## 10 Atmospheric correction

Raw calibrated imaging spectrometer data have the general appearance of the solar irradiance curve, with radiance decreasing towards longer wavelengths, and exhibit several absorption bands due to scattering and absorption by gasses in the atmosphere. The major atmospheric water vapor bands ($H_2O$) are centered approximately at 0.94 $\mu$m, 1.14 $\mu$m, 1.38 $\mu$m and 1.88 $\mu$m, the oxygen ($O_2$) band at 0.76 $\mu$m, and carbon dioxide ($CO_2$) bands near 2.01 $\mu$m and 2.08 $\mu$m. Additionally, other gasses including ozone ($O_3$), carbon monoxide ($CO$), nitrous oxide ($N_2O$), and methane ($CH_4$), produce noticeable absorption features in the 0.4-2.5 $\mu$m wavelength region. The effect of atmospheric calibration algorithms is to re-scale the raw radiance data provided by imaging spectrometers to reflectance by correcting for atmospheric influence thus shifting all spectra to nearly the same albedo. The result is a data set in which each pixel can be represented by a reflectance spectrum which can be directly compared to reflectance spectra of rocks and minerals acquired either in the field or in the laboratory. Reflectance data obtained can be absolute radiant energy or apparent reflectance relative to a certain standard in the scene. Calibration to reflectance can be conducted to result in absolute or relative reflectance data.

Radiation reaching the sensor can be split into four components: path radiance, reflected diffuse radiance, reflected direct radiance, and reflected radiance from neighborhood. Radiative transfer RT codes model the atmosphere's optical behavior given user defined boundary conditions. The inverse problem of atmospheric correction of imaging spectrometer data with the aim of obtaining radiance and/or reflectance at the ground surface can be achieved in three ways:

1. Empirical correction methods to obtain apparent surface reflectance
2. Use of RT codes to obtain absolute reflectance
3. In-flight calibration of airborne optical sensors

It should be noted that empirical approaches such as those listed above only approximate the highly variable processes in time and space in the atmosphere controlling the transfer of radiance to and from the Earth's surface to the sensor.

### 10.1 RELATIVE REFLECTANCE

In relative reflectance data, reflectivity is measured relative to a standard target from the scene. Correction methods currently available for this purpose include:

- Flat field correction,
- Internal average relative reflectance correction, and
- Empirical line correction.

The purpose of the flat-field correction is to reduce the atmospheric influence in the raw imaging spectrometer data and eliminate the solar irradiance drop-off, as well as any residual instrument effects. This is achieved by dividing the whole data set by the mean value of an area within the scene which is spectrally and morphologically flat, and spectrally homogeneous. The flat-field chosen should have a high albedo to avoid decreasing the signal-to-noise ratio. This can also be achieved by increasing the number of pixel spectra used to produce the flat-field spectrum. In order to select properly an flat-field target area, ground truth data is necessary to ensure that the calibration target is indeed spectrally flat. In that case, the flat-field method removes the solar irradiance curve and major gaseous absorption features as well as system induced defects.

The Internal Average Relative Reflectance (IARR) correction method allows the calibration of raw imaging spectrometer data to reflectance data when no calibration information is available. This procedure uses an "Average Reference Spectrum" (ARS) calculated as the average pixel spectrum of the entire scene. This spectrum is divided into each image radiance spectrum to produce a relative reflectance spectrum for each pixel. Care should be taken when cover types with strong absorption features are present in the scene. In such a case the IARR correction method may cause artefacts which may be wrongly interpreted as being spectral features.

Conversion of raw imaging spectrometer data to reflectance data using the Empirical Line method requires the selection and spectral characterization of two calibration targets, thus assuming a priori knowledge of each site. This empirical correction uses a constant gain and offset for each band to force a best fit between sets of field spectra and image spectra characterising the same ground areas thus removing atmospheric effects, residual instrument artefacts and viewing geometry effects.

## 10.2 ABSOLUTE REFLECTANCE

Absolute reflectance data without a-priori knowledge of surface characteristics can be obtained using atmosphere models. These models correct for scattering and absorption in the atmosphere due to water vapor and mixed gases as well as for topographic effects and different illumination conditions. The 0.94 μm and 1.1 μm water absorption bands are used to calculate water vapor in the atmosphere while transmission spectra of the mixed gases in the 0.4 - 2.5 μm wavelength region are simulated on basis of the water vapor values found and the solar and observational geometry. Scattering effects in the atmosphere are modeled using radiative transfer codes. A typical atmospheric correction algorithm using RT codes typically models the atmosphere's behavior on incident radiation through deriving transmission spectra of the main atmospheric gases and water vapor which is integrated with effects of atmospheric scattering from aerosols and molecules. User input on atmosphere condition required is the date, time and location of data take, ozone depth, aerosol type, visibility, elevation. This is derived from radiosonde data or meteorological stations in the area to be imaged. RT codes often used are LOWTRAN, MODTRAN, 5S and 6S.

In general, in atmospheric correction the reflective (VIS-SWIR) and thermal spectral regions are treated separately since the influence of the sun dominates the solar reflective region while it can be neglected in the thermal region. The basic relation defining spectral radiance in the VIS-SWIR region is given by

$$L_\lambda = L_0(\lambda) + \frac{E_g(\lambda)}{\pi}[\tau_{dir}(\lambda) + \tau_{dif}(\lambda)]\rho(\lambda) \qquad (10)$$

where $L_0(\lambda)$ is the path radiance for a black body ($\rho=0$), $E_g(\lambda)$ is the global irradiance on the ground, $\tau_{dir}(\lambda)$ is the direct atmospheric transmittance (ground to sensor), $\tau_{dif}(\lambda)$ is the diffuse atmospheric transmittance (ground to sensor), and $\rho(\lambda)$ is the reflectance of a Lambert surface. The basic relation defining spectral radiance in the thermal region is given by

$$L_\lambda = [\varepsilon_\lambda L_{bb\lambda}(T) + (1 - \varepsilon_\lambda)L_{sky\lambda}]\tau_\lambda + L_{atm\lambda} \tag{11}$$

where $L_\lambda$ is the spectral radiance for the wavelength $\lambda$, $\varepsilon_\lambda$ is the surface emissivity at wavelength $\lambda$, $L_{bb\lambda}(T)$ is spectral radiance from a blackbody at surface temperature T, $L_{sky\lambda}$ is spectral radiance incident upon the surface from the atmosphere, $\tau_\lambda$ is spectral atmospheric transmission, and $L_{atm\lambda}$ is spectral radiance from atmospheric emission and scattering that reaches the sensor. Atmospheric values in these equations are typically simulated using RT codes in conjunction with radiosonde data to derive wavelength dependent surface emissivity (TIR) or reflectance (VIS-SWIR).

For spaceborne sensor, the following relation can be demonstrated to relateground reflectance, ρ, to the DN value

$$\rho = \frac{1}{a_1}\left[\frac{\pi(c_0 + c_1 DN)}{E_s \cos\Theta_s} - a_0\right] \tag{12}$$

where $E_s$ is the extraterrestrial solar irradiance, $\Theta_s$ is the solar zenith angle, and $a_0$ and $a_1$ are atmospheric functions relating planetary albedo, $\rho_p$ to ground albedo, $\rho$, as

$$\rho_p = a_0 + a_1 x\rho \tag{13}$$

For airborne sensors the term $E_s \cos\Theta_s$ has to be replaced by $E_g$ the global downwelling flux at the sensor altitude. The adjacency effect is approximately taken into account by calculating an N x N pixel low pass filter image $\overline{\rho}^{(1)}$ of the reflectance data $\rho^{(1)}$ and by weighing the difference of $\rho^{(1)} - \overline{\rho}^{(1)}$ for each pixel with a correction function q to obtain the corrected reflectance values $\overline{\rho}^{(1)}$ as

$$\rho^{(2)} = \rho^{(1)} + q(\rho^{(1)} - \overline{\rho}^{(1)}) \tag{14}$$

where q is a function of the strength of the adjacency effect depending on the diffuse transmittance (ground to sensor) which results in the scattering of radiation of neighboring fields onto the field of view.

## 11 Re-sampling and image simulation

Spectral simulation is possible when the source sensor has a spatial and spectral resolution finer than the target simulated sensor. The procedure of simulation involves estimating the gaussian spectral response function (SRF) of each channel using the band center and width (in full-width-half-maximum). This SRF is of the type

$$R_i(\lambda) = \frac{1}{\sigma\sqrt{2\pi}}e^{-0.5(\lambda-\mu)^2/\sigma^2} \tag{15}$$

where μ is the band centre, σ the standard deviation (equivalent to the FWHM) of the channel and λ the wavelength relative to μ and $R_i(\lambda)$ the spectral response. The integral over the SRF is 1. An example is given for a laboratory spectrum in Figure 3 and for image (MERIS) data in Figure 4. In Figure 4, for each MERIS channel the spectral response curve is calculated using the gaussian approximation. The channels of the higher spectral resolution airborne imaging spectrometer data set that are within the wavelength range of the selected MERIS channel are integrated by dividing them into the derived SRF assuming that the initial SRF of the input data channels are a single bright source (i.e., a continuous input spectrum). This is done using

$$\rho_{res}(\lambda_i) = \frac{\int_{\lambda_1}^{\lambda_2} \rho(\lambda) R_i(\lambda) d\lambda}{\int_{\lambda_1}^{\lambda_2} R_i(\lambda) d\lambda} \tag{16}$$

where $\rho_{res}(\lambda_i)$ is the re-sampled spectrum using the continuous spectrum $\rho(\lambda)$ and the SRF $R_i(\lambda)$.

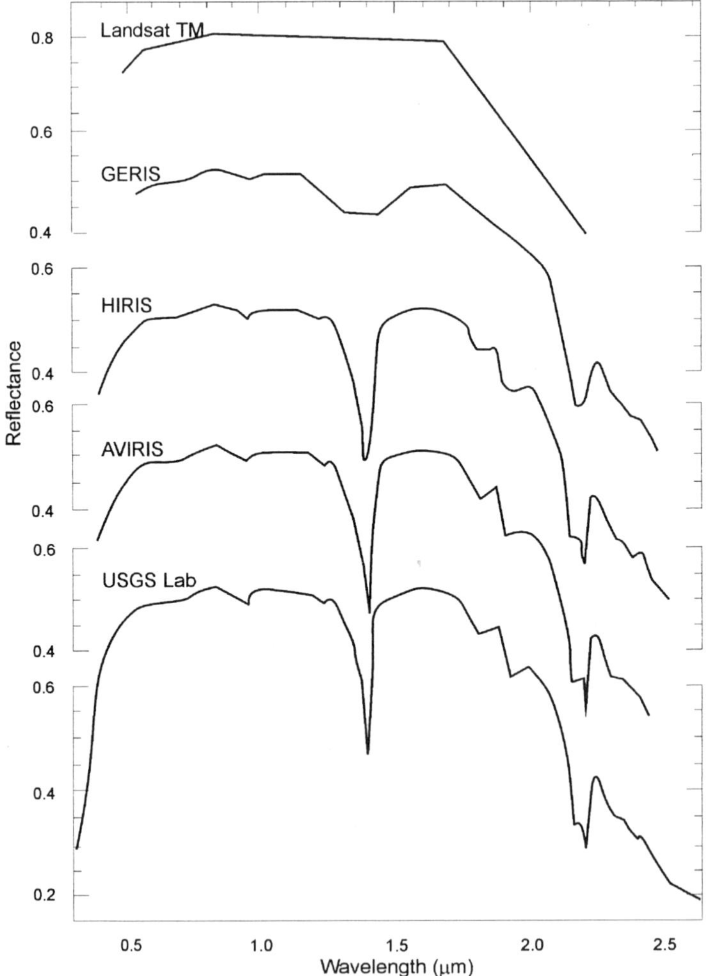


*Figure 3.* Example of the effect of spectral resolution on a kaolinite spectrum resampled to various resolutions.

This deconvolution is achieved by integrating between the lower $(\lambda_1)$ and upper $(\lambda_2)$ wavelength limit of the designated (i.e., to be simulated) image channel. Since the gaussian SRF only asymptotically approaches zero, we truncated the SRF at 1.5 times the FWHM. In case the spectral resolution of the image data to be simulated is equivalent to or less than that of the input data, band matching has to be applied. In this technique, the bands which position, in terms of wavelength of the band center, is closest to the channel to be simulated are used for the simulation process.

Spatial degradation follows two steps. The first step requires modeling the transfer function between the initial data and the desired data and deriving a spatial filter that allows simulation of the coarse spatial resolution imagery. The second step involves re-sampling to the desired pixel size.

As is the case in spectral re-sampling with the spectral response function, the sensor's Point Spread Function (PSF) gives the spatial response of the sensor. The PSF is defined as

$$e_b(x, y) = \sum_{\alpha_{\min}}^{\alpha_{\max}} \sum_{\beta_{\min}}^{\beta_{\max}} (\alpha, \beta) PSF(x - \alpha, y - \beta) d\alpha d\beta \qquad (17)$$

where $e_b(x, y)$ is the resulting electronic signal for band $b$ at location given by coordinate (x,y), $s_b(\alpha, \beta)$ is the input signal, whereas the limits of the integral determine the spatial extend in two dimensions, $\alpha$ and $\beta$, over which the physical signal is weighted.

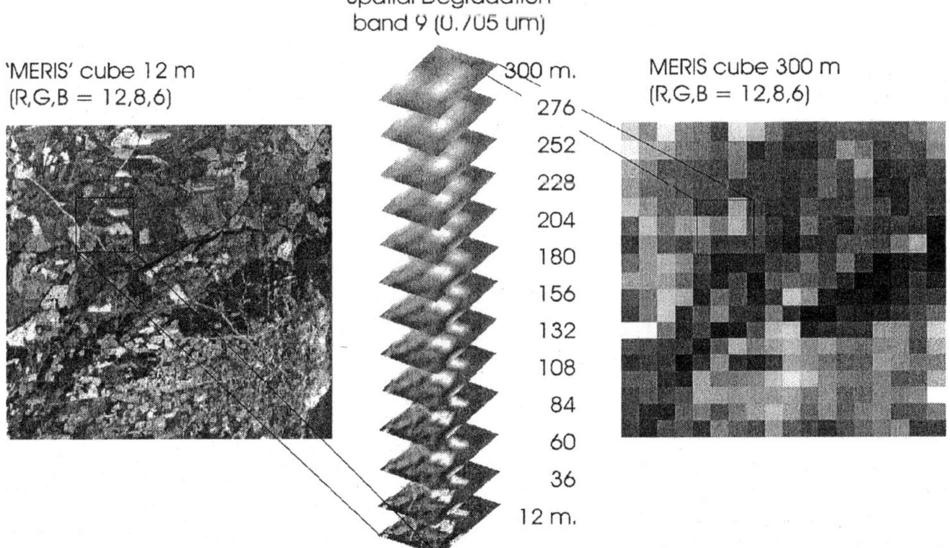


*Figure 4.* Spectral resampling of simulated MERIS data.

A common model for the PSF is given by

$$PSF(x, y) = \frac{1}{2\pi ab} e^{-x^2/2a^2 e^{-y^2/2b^2}} \tag{18}$$

where $a$ and $b$ determine the width of the optics of the PSF in the cross- and along track direction. In the absence of details on the PSF for MERIS, we used a normalized detector PSF

$$PSF(x, y) = \frac{1}{ab} \sum_{a=1}^{a} \sum_{b=1}^{b} s_b \tag{19}$$

This is equivalent to applying a square moving average filter of the desired spatial resolution.


## 12    Analytical processing techniques


### 12.1    INTRODUCTION

Once reflectance-like imaging spectrometer data is obtained, the logical next step is to map absorption features to determine surface composition. "Conventional" remote sensing techniques such as band depth mapping using ratios or differences or visual inspection of image spectra can be applied. These often provide useful information and visualizations of the data, however need experience in interpretation. Here I restrict to algorithms developed to specifically make use of or cope with the high spectral dimensionality typical for imaging spectrometer data. Such techniques include:
- Binary encoding
- Waveform characterization
- Spectral Feature Fitting (SFF)
- Spectral Angle Mapping (SAM)
- Spectral unmixing
- Linear unmixing
- Foreground-background analysis
- Iterative spectral unmixing
- Constrained energy minimization (CEM)
- Classification
- CCSM
- Geophysical inversion


### 12.2    BINARY ENCODING

A simple binary code for a reflectance spectrum can be described as (Goetz *et al.* 1985)

$$h(n) = 0 \text{ if } x(n) \geq T$$
$$= 1 \text{ if } T < x(n) \tag{20}$$

where $x(\mathbf{n})$ is the brightness value of a pixel in the nth channel, $T$ is the user specified threshold which often equals the average brightness value of the spectrum, and $h(\mathbf{n})$ is the resulting binary code for the pixel in the n[th] band. A variation to this simple encoding is to break the spectral range into a number of subregions and to code separately within these subregions following the same procedure as described above. A modification to the simple encoding has been proposed by Jia & Richards (1993) who exploit multiple thresholds. Their method consist of determining the mean brightness of a pixel vector and then setting additionally upper and lower thresholds. The binary code in two bit format can now take four different values

$$
\begin{aligned}
h(n) &= 00 \text{ if } x(n) \geq T1 \\
&= 01 \text{ if } T1 < x(n) \leq T2 \\
&= 10 \text{ if } T2 < x(n) \leq T3 \\
&= 11 \text{ if } T3 < x(n)
\end{aligned}
\tag{21}
$$

where $T_1$ is the lower threshold, $T_2$ is the mean brightness of the spectrum, and $T_3$ is the higher threshold. Binary encoding provides a simple mean of analyzing data sets for the presence of absorption features, however, with the simple 0-1 coding much depends on the threshold chosen. Comparison of 0-1 coded pixel and laboratory spectra yields a qualitative indication for the presence or absence of absorption features where the depth and thus the significance of the absorption feature is not considered.


## 12.3   WAVEFORM CHARACTERIZATION

In waveform characterization (Okada & Iwashita 1992), first the imaging spectrometer data are normalized to produce reflectance-type images. The upper convex Hull is calculated as an enveloping curve on the pixel spectra having no absorption features. Next, the Hull-quotient reflectance spectrum is derived by taking the ratio between the pixel reflectance spectrum and the enveloping upper convex Hull (Green & Graig 1985; Figure 5).

These Hull quotient spectra are used to characterize absorption features, known to be attributed to a certain mineral of interest, in terms of their position, depth, width, asymmetry, and slope of the upper convex Hull.

The absorption band position, $\lambda$, is defined as the band having the minimum reflectance value over the wavelength range of the absorption feature. The relative depth, D, of the absorption feature is defined as the reflectance value at the shoulders minus the reflectance value at the absorption band minimum. The width of the absorption feature, $W$, is given by

$$
W = A_{all} / 2D
\tag{22}
$$

where $A_{all}$ is the sum of the area left of the absorption band minimum, $A_{left}$, and the area right of the absorption band minimum, $A_{right}$, forming the total area under the convex hull enclosing the absorption feature. The symmetry factor, $S$, of the absorption feature is defined as

$$
S = 2(A_{left}/A_{all}) - 1
\tag{23}
$$

where $A_{left}$ is again the area of the absorption from starting point to maximum point. Values for $S$ range from -1.0 to 1.0 where $S$ equals 0 for a symmetric absorption feature.

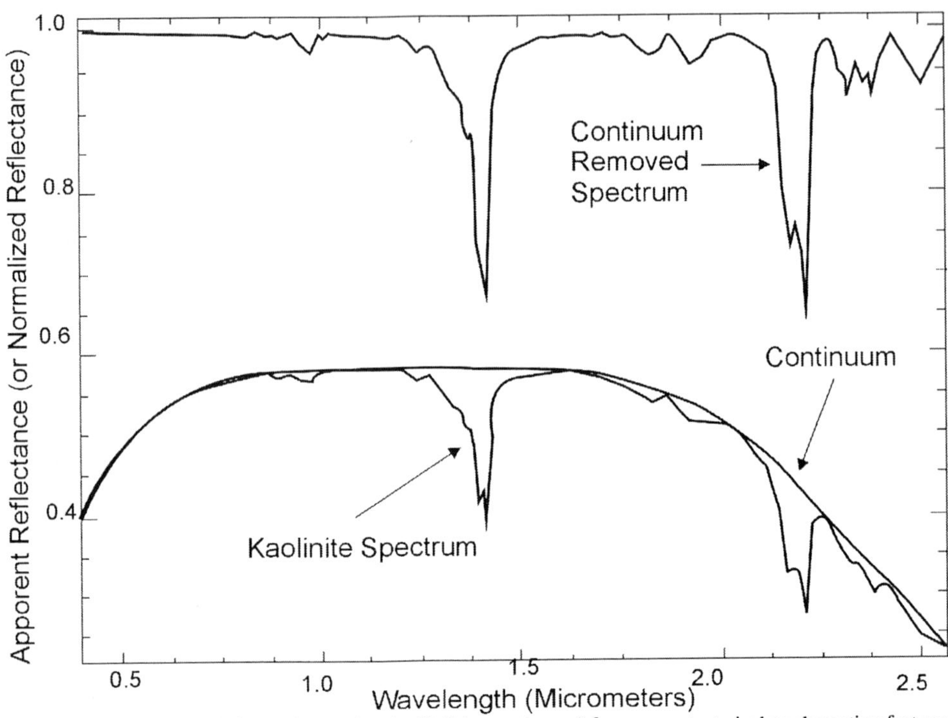


*Figure 5.* Example of deriving and removing the Hull (or continuum) from a spectra to isolate absorption features from the background signal.

The slope of the upper hull, $\phi$, characterizes the slope of the upper hull over the absorption feature and is defined as

$$\phi = \tan^{-1}\left(\frac{R_e - R_s}{\lambda_e - \lambda_s}\right) \tag{24}$$

where $R_e$ and $R_s$ are the reflectance at the ending and starting point of the absorption feature, respectively, and $\lambda_e$ and $\lambda_s$ are the wavelength at the ending and starting point of the absorption feature, respectively. As a result of the waveform characterization, five images (e.g., an position, depth, width, symmetry, and slope of upper Hull image for each absorption feature selected) can be generated defining the similarity between a pixel spectrum and an laboratory spectrum of a mineral of interest based on the presence of characteristic absorption features in both spectra. However, Okada & Iwashita (1992) give no solution to the mineral mapping from these images. Techniques need to be developed to quantitatively estimate the probability of a mineral occurrence from the waveform characteristics.

Data normalization using a single, internally derived, reference spectrum can result in spectral curves that are distorted from those observed in true reflectance spectra. To avoid this problem, Crowley *et al.* (1989) developed a method of mineral mapping from imaging spectrometer using Relative Absorption Band-Depth Images (RBD) generated directly from radiance data. In essence, RBD images provide a local continuum correction (Clark & Roush 1984) removing any small channel to channel radiometric offsets, as well as variable atmospheric absorption and solar irradiance drop off for each pixel in the data set.

To produce a RBD image, several data channels from both absorption band shoulders are summed and then divided by the sum of several channels from the absorption band minimum. The resulting absorption band-depth image gives the depth of an absorption feature relative to the local continuum which can be used to identify pixels having stronger absorption bands indicating that these may represent a certain mineral.

## 12.4   SPECTRAL FEATURE FITTING (SFF)

Spectral feature fitting, also known as the TRICORDER algorithm (Crowley *et al* 1989; Clark *et al.*1990; Crowley & Swayze 1995), builds on waveform characterization in that it is a absorption feature based method for matching image to laboratory spectra. Again continuum removed pixel spectra are used and compared to continuum reference spectra of known mineralogy (possibly derived from a spectral library). A least-squares fit is calculated band by band between each reference end member and the unknown pixel spectra. The root mean square (RMS) error of this fit is used to create an RMS image that shows pixels that are more and pixels that are less similar to the selected reference end member.

## 12.5   SPECTRAL ANGLE MAPPING (SAM)

Spectral Angle Mapping (Kruse *et al.* 1990) calculates the spectral similarity between a test reflectance spectrum and a reference reflectance spectrum assuming that the data is correctly calibrated to apparent reflectance with dark current and path radiance removed. The spectral similarity between the test (or pixel) spectrum, *t*, and the reference (or laboratory) spectrum, *r*, is expressed in terms of the average angle, $\Theta$, between the two spectra as calculated for each channel, *i*, as

$$\Theta = \cos^{-1}\left( \frac{\sum_{i=1}^{n} t_i r_i}{\sqrt{\sum_{i=1}^{n} t_i^2 \sum_{i=1}^{n} r_i^2}} \right) \tag{25}$$

In this approach, the spectra are treated as vectors in a space with dimensionality equal to the number of bands, *n*. The outcome of the spectra angle mapping for each pixel is an angular difference measured in radians ranging from zero to $\pi/2$ which gives a qualitative estimate of the presence of absorption features which can be related to mineralogy.

## 12.6   SPECTRAL UNMIXING

The most widely used method for extracting surface information from remotely sensed images is image classification. With this technique, despite the stochastic concept of the method, each pixel is assigned to one out of several known categories or classes through a statistical separation approach. Thus an image is decomposed into an image containing only thematic information of the classes previously selected as the expected image elements. In general, a training sample set of pixels is defined by the user to train the classifier. The spectral characteristics of each training set are defined through a statistical or probabilistic process from feature spaces and unknown pixel to be classified are statistically "compared" with the known classes and assigned to the class to which they mostly resemble. In this way thematic information is obtained disregarding the mostly compositional nature of surface materials. Reflected radiation from a pixel as observed in remote sensing imagery has rarely interacted with a volume composed of a single homogenous material because natural surfaces composed of a single uniform material do not exist in nature. Most often the electromagnetic radiation observed as pixel reflectance values results from the spectral mixture of a number of ground spectral classes present at the surface sensed.

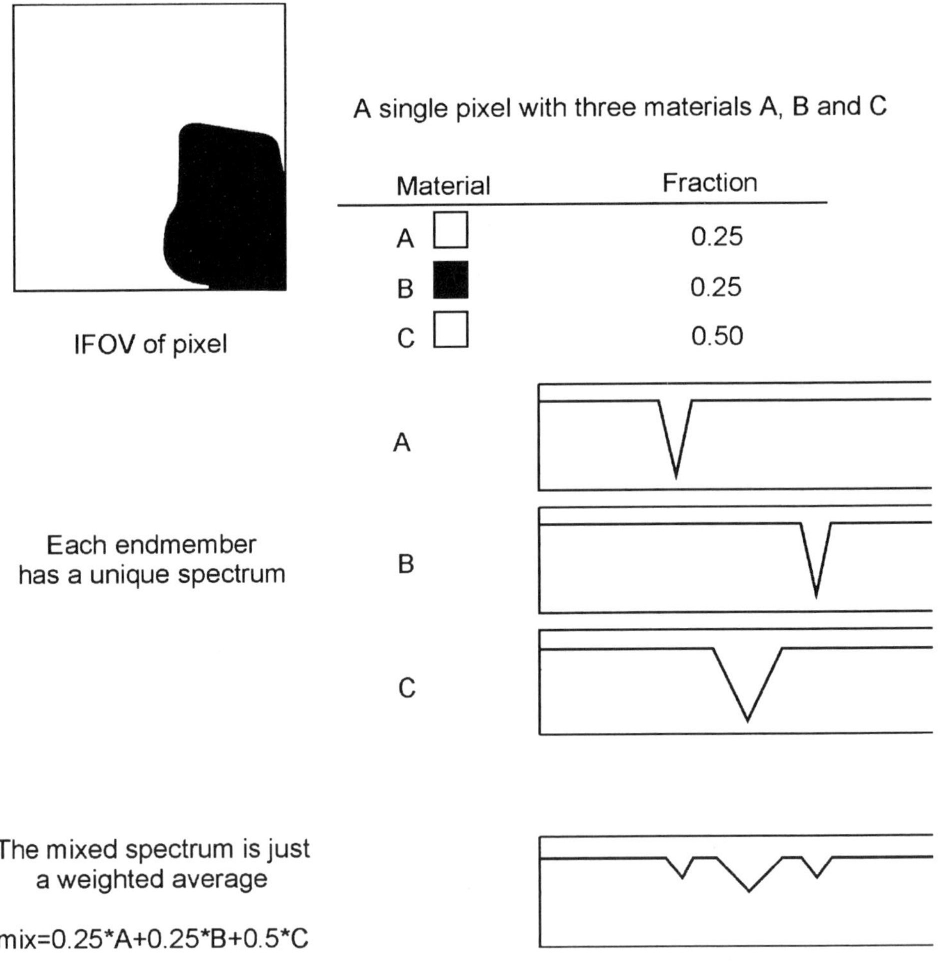


*Figure 6.* Principle of spectral mixing and unmixing.

Singer & McCord (1979) showed that if mixing scale is macroscopic such that photons interact with one material rather than with several materials, the mixing can be considered linear and the resulting pixel reflectance spectrum is the linear summation of the individual material reflectance functions multiplied by the surface fraction they constitute. Various sources contribute to spectral mixing: (1) optical imaging systems integrate reflected light from each pixel, (2) all materials present in the field of view contribute to the mixed reflectance sensed at a pixel, and (3) variable illumination conditions due to topographic effects result in spectrally mixed signals. Mixing can be considered a linear process if: (1) no interaction between materials occurs, each photon sees only one material, (2) if the scale of mixing is very large as opposed to the size of the materials, and (3) if multiple scattering does not occur.

Rather than aiming at representing the landscape in terms of a number of fixed classes, mixture modeling and spectral unmixing (Adams *et al.* 1985) acknowledge the compositional nature of natural surfaces and strive at finding the relative or absolute fractions (or abundance) of a number of spectral components or end-members that together contribute to the observed reflectance of the image (Figure 6). Therefore the outcome of such analysis is a new set of images that for each selected end-member portray the fraction of this class within the volume bound by the pixel. Mixture modeling is the forward process of deriving mixed signals from pure end-member spectra while spectral unmixing aims at doing the reverse, deriving the fractions of the pure end-members from the mixed pixel signal. A review of mixture modeling and spectral unmixing approaches can be found in Ichoku & Karnieli (1996).

A linear combination of spectral end-members is chosen to decompose the mixed reflectance spectrum of each pixel, $R_i$, into fractions $f_j$ of its end-members, $Re_{ij}$, by

$$R_i = \sum_{j=1}^{n} f_j \, Re_{ij} + \varepsilon_i \quad \text{and} \quad 0 \le \sum_{j=1}^{n} f_j \le 1 \tag{26}$$

where $R_i$ is the reflectance of the mixed spectrum in image band $i$ for each pixel, $f_j$ is the fraction of each end-member $j$ calculated band by band, $Re_{ij}$, is the reflectance of the end-member spectrum $j$ in band $i$, $i$ is the band number, $j$ is each of the $n$ image end-members and $f_j$ is the residual error or the difference between the measured and modelled DN in band $i$. A unique solution is found from this equation by minimizing the residual error, $\varepsilon_i$, in a least-squares solution. This residual error is the difference between the measured and modelled DN in each band and should in theory be equal to the instrument noise in case that only the selected end-members are present in a pixel. Residuals over all bands for each pixel in the image can be averaged to give a root-mean square (RMS) error, portrayed as an image, which is calculated from the difference of the modelled ($R_{jk}$) and measured ($R_{jk}'$) pixel spectrum as

$$RMS = \sum_{k=1}^{m} \frac{\sqrt{\sum_{j=1}^{n} (R_{jk} - R_{jk}')^2 / n}}{m} \tag{27}$$

where $n$ is the number of spectral bands and $m$ the number of pixels within the image. The solution to the unmixing is found through standard matrix inversion such as the gaussian elimination method. Boardman (1989) however suggested the use of singular value decomposition of the end-member matrix which has the advantage that singular values can be used to evaluate the orthogonality of the selected end-members. If all end-members are spectrally unique, all singular values are equal. For a degenerate set of end-members all but one singular value will be zero. Figure 7 shows a typical set of resulting images from spectral unmixing, whereas Figure 8 provides a closer look at the retained abundance estimates (fractions).

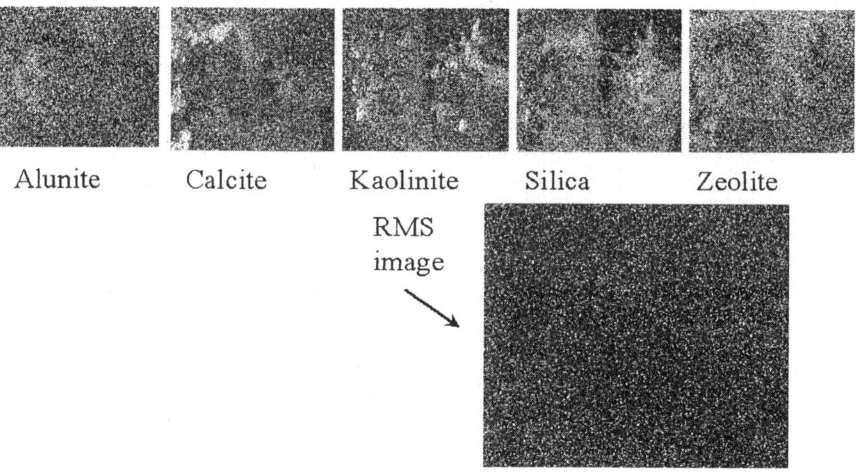


*Figure 7.* Abundance images resulting from spectral unmixing for the various endmembers and the RMS spectral goodness of fit image.

A solution to the mixing as found by a process of matrix inversion is give by Settle & Drake (1993). These authors define a vector of expected pixel signals $\mu_i = \{\mu_{i1}, \mu_{i2} \ldots\ldots \mu_{in}\}^T$ for the n ground cover classes giving an expected mixed pixel signal under strictly linear conditions as

$$f_1\mu_1 + f_2\mu_2 \ldots. f_c\mu_c = Mf \tag{28}$$

The columns of the matrix M are the vectors $\mu_i$ which are the end-member spectra. The observed signal of pure pixels will exhibit statistical fluctuations due to sensor noise characterised by a noise variance-covariance matrix $N_i$. Therefore pixels with the mixture f will exhibit fluctuations around their mean value Mf characterized by the noise covariance matrix N(f) given by

$$N(f) = f_1N_1 + f_2N_2 \ldots.+f_cN_c \tag{28}$$

If the N(f) is independent of f, thus when the noise components are un-correlated, the linear model can be defined according to Settle & Drake (1993) as

$$x = Mf + e \tag{29}$$

where e is the vector of errors satisfying

$$E(e) = 0 \text{ and } E(ee^T) = N \tag{30}$$

This signifies that the expectation of e is zero and that the expectation of noise component is close to the sensor noise component.

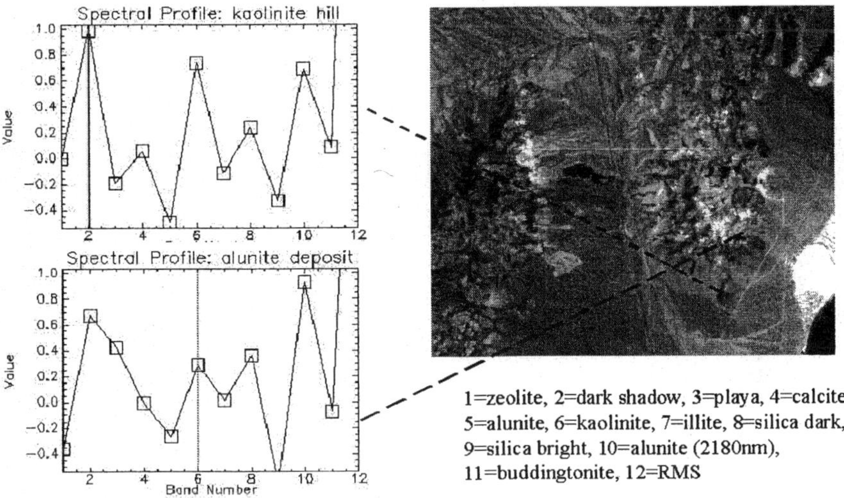



*Figure 8.* Abundance (fraction) estimates for selected targets.

The spectral unmixing model requires a number of spectrally pure end-members to be defined which cannot exceed the number of image bands minus 1 to allow a unique solution to be found that minimizes the noise or error of the model. Furthermore constraining factors are that the fractions sum to unity and that the fractions for the individual end-members vary from 0 to 100%. When adopting these assumptions, constrained unmixing is applied in other cases we refer to unconstrained unmixing.

Two versions of spectral unmixing are generally implemented: constrained and unconstrained unmixing. Constraining assumptions that can be implemented separately are that (1) the fractions are non-negative and (2) that the fractions sum to unity. These constrains are only meaningful when considering the scientific field of applications (e.g., the earth science perspective), from a statistical viewpoint or even from an image interpretation viewpoint, it is meaningless to force mixture models to constraining to the data. A comparison of SAM and spectral unmixing is given in Figure 9.

## 12.7   ITERATIVE SPECTRAL UNMIXING

Iterative spectral unmixing (ISU) is a recent extension to the above described spectral unmixing (Van der Meer, 1999; Figure 10). The iterative implementation of spectral unmixing follows 4 basic steps:

Step 1: "Deriving end-members"
The initial set of end-members is derived using the procedures outlined in the previous section.

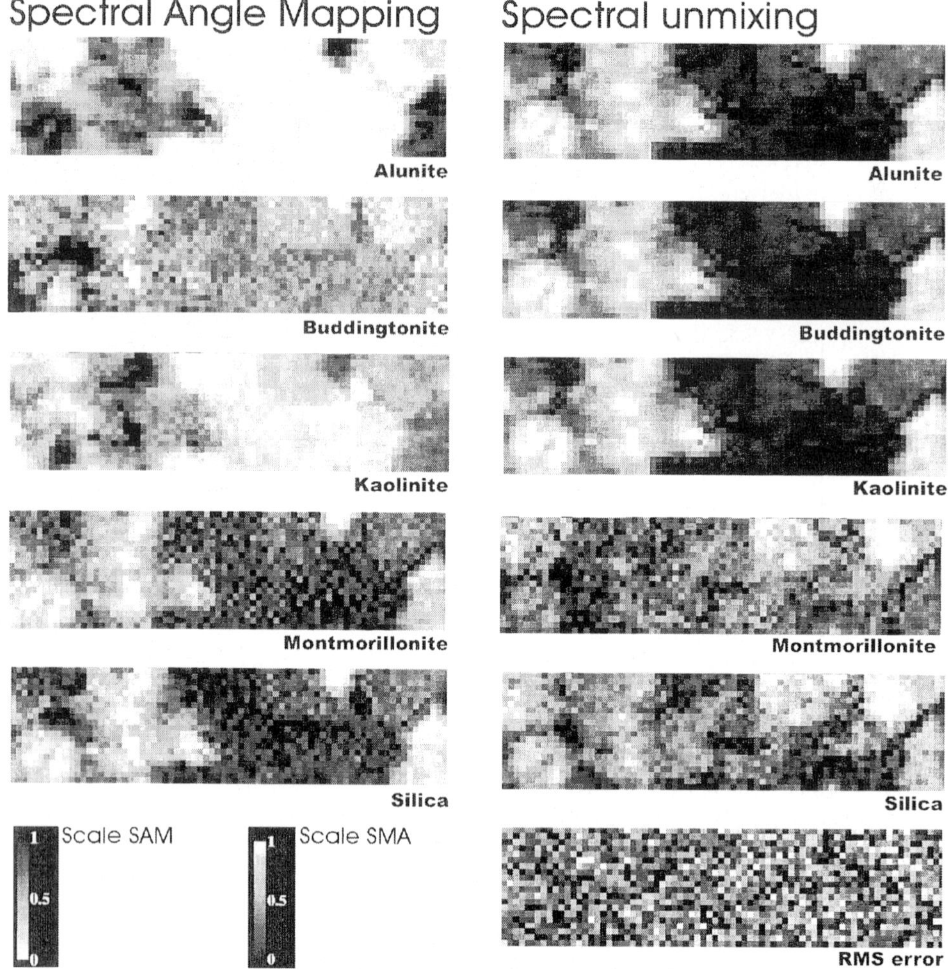


*Figure 9.*Comparison of spectral unmixing and spectral angle mapper for selected mineral endmembers for a small portion of an AVIRIS data set acquired over the Cuprite test site (see Figure 11 for location).

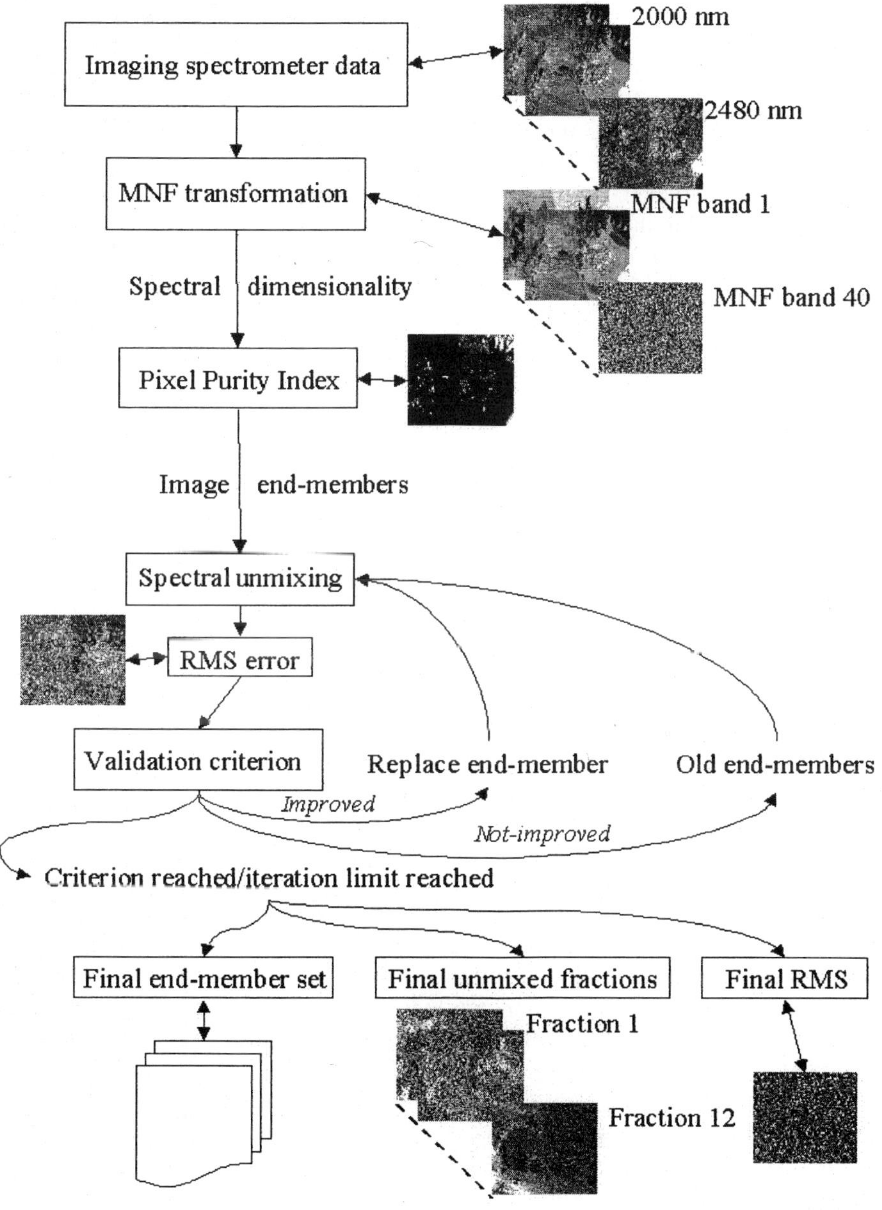



*Figure 10.* Graphical explanation of the working principle of iterative spectral unmixing (ISU; Van der Meer, 1999).

Step 2: "First iteration of spectral unmixing"
Spectral unmixing is performed using the $n$-end-members derived from the image data leading to $n$-fraction images and an initial RMS error image.

Step 3: "Calculating the optimization criterion"

An optimization criterion is defined and calculated on the RMS image. Optimization criteria that have been defined and tested in the present analysis include:

- Minimizing the mean RMS; this is calculated as the weighted average RMS value.
- Minimizing the spread of the RMS values; this is calculated as the maximum minus the minimum RMS value
- Minimizing the spatial structure in the RMS values; a measure for the spatial structure is found in the (semi) variogram. The variogram in remote sensing can be estimated from p(h) pairs of observations, $z_v(x_l)$ and $z_v(x_l + h)$ with l=1,2,...,p(h) as

$$\gamma(h) = 1/2p(h) \sum_{l=1}^{p(h)} \{z_v(x_l) - z_v(x_l + h)\}^2 \qquad (31)$$

Parameters of a fitted mathematical function (the variogram model) include a range, a nugget and a sill. The range is the distance at which the curve levels of to a constant value of semi-variance indicating the spatial scale of the pattern in the image. The nugget variance is a non-zero y-intercept of the variogram model indicative of variability at a resolution smaller than the image resolution often indicating the level of un-correlated noise in the data. The sill, i.e., the value of maximum variance, is equivalent to the variance of the image pixel values. Optimizing the RMS by minimizing the spatial structure of the RMS image strives at reaching a pure nugget variogram of which the level of the nugget variance represents the noise characteristics of the data. A measure for the spatial structure to be minimized is derived from the ratio of the total variance over the non-structurally controlled nugget variance. When no spatial structure exists in a RMS image, the total variance is equal to the nugget variance and the ratio is 1. Else the ratio is larger than 1.

- Minimizing the directional spatial structure or anisotropy; an effective way to detect anisotropy in the RMS pattern is by calculating its 2D-variogram (i.e., the variogram surface). In this surface the (semi) variance is not only related to the mere distance between two points (the lag), but also to the direction. In calculating the variogram value for pairs of points separated by a vector $(h_x, h_y)$ all pairs are grouped together whose separation is $h_x \pm \Delta x$, $h_y \pm \Delta y$. In this case $\Delta x$ and $\Delta y$ are half the support size, i.e. (in a rectangular coordinate system) the pixel size (Isaaks & Srivastava 1989). The result is a surface with for each vector a computed (semi) variance. A measure for the directional spatial structure or anisotropy is the ratio of the longer axis of the anisotropy ellipse and the shortest axis of the anisotropy ellipse. When no anisotropy exists in the data, this ratio is 1 (i.e., the spatial structure is isotropic).
- Minimizing the local variance; the local variance can be computed over a (2n+1) x (2m+1) window as (Woodcock & Strahler 1987)

$$\sigma_{ij}^2 = 1/(2n+1)(2m+1) \sum_{k=i-n}^{i+n} \sum_{l=j-m}^{j+m} \{z(x_{kl}) - u_{ij}\}^2 \qquad (32)$$

where $u_{ij}$ is the mean of the (2n+1) x (2m+1) window centred on $x_{ij}$ and $z(x_{ij})$ is the value of the pixel located at $x_{ij}$ in the $i^{th}$ row and the $j^{th}$ column of the image.

Step 4: "Iterative spectral unmixing"
The pixel underlying the end-member in which neighborhood the largest RMS errors occur as compared to the other end-members is removed and replaced by an end-member on the cluster of pixels of highest error in the RMS image. Spectral unmixing is performed, the optimization criterion (step 3) calculated, if this has improved the new end-member set is adopted else the process continues with the old set of end-members. The process of unmixing continues until the optimization criterion is satisfied or until no significant improvements in this criterion are observed.

## *12.8* CONSTRAINED ENERGY MINIMIZATION (CEM)

Constrained energy minimization (CEM) developed by Farrand & Harsanyi (1997) is an extension to spectral unmixing. CEM maximizes on a pixel by pixel basis the response of a target signature and suppresses the response of undesired background signatures. It is assumed that foreground and background signatures are mixed linearly such as is the case when each photon only interacts with one material. The CEM strives at finding a vector that suppresses the unknown background signature while enhancing the target signature. This is achieved by minimizing the total output energy of all pixels and by assuming that the energy of an individual pixel summed across the wavelength range to be 1 when applied to the target pixel spectrum. The result of CEM is a vector component image that is comparable to fraction abundance images typically obtained through unmixing.

## 12.9 FOREGROUND-BACKGROUND ANALYSIS

Spectral unmixing addresses the mixed-pixel problem, calibration, and variations in lighting geometry and displays the results in terms of proportions of endmembers. The approach works when describing a few spectral types that, in various mixtures, can account for most of the variance in an image data set. Spectral unmixing works less well when the spectral features of interest are minor components of the total variance. The disadvantage of approximating linearly the natural (non-linear) complexity of materials represented by the mixture of endmembers is the main problem of unmixing. This produces a non-unique mixing model to identify and quantify materials that occur at the sub-pixel scale.

In order to develop a directed search methodology to locate the desired robustness (analytic) property, Smith *et al.* (1994) proposed a revised SMA technique, that they termed Foreground/Background Analysis (FBA). Constrained energy minization (Harsanyi & Chang 1994) shares the properties of orthogonal space projection and a similar rationale with the FBA technique. In this technique, spectral measurements are divided in two groups of foreground and background spectra that comprise a selected subset of spectra which emphasizes the presence of a signature of interest. In defining both groups they do not include intermediate mixtures between foreground and background. In that way, FBA vectors should be sensitive to minor sources of foreground spectral variation and insensitive to background spectral variation. The goal of FBA is to project spectral variation along the most relevant axis of variance that maximizes the spectral differences between the foreground and background, while minimizing spectral variation within each group. Their FBA approach defines a weighting vector $w = (w_1, w_2, ..., w_{Nb})$, with components $w_b$ at each channel $b = 1, ..., Nb$, such that all foreground spectral vectors, $R_f = (R_{f,1}, R_{f,2}, ..., R_{f,Nb})$, are projected to 1 while background spectral vectors, $R_b$, to 0. As stated FBA is in essence another linear classifier of the spectra that can be applied to identify low and high material abundances. Pinzon *et al.* (1994) modified the FBA linear system to project a subset of spectra into relevant axis of continuous chemical variation. The FBA system to be solved is composed only of the training (calibration) samples that reflect the distribution of the feature to be detected in the whole data set.

## 12.10  CLASSIFICATION

Classification of remotely sensed imagery into groups of pixels having similar spectral reflectance characteristics is often an integral part of digital image analysis. On basis of the spectral reflectance which training pixels (e.g. pixels known to represent a ground class of interest) exhibit, pixels for which the ground cover type is unknown are classified. Classification routines aim at comparing the observed spectral reflectance of pixels with unknown composition with that of training pixels, and assign the unknown pixel to that group which resembles most their spectral reflectance characteristics. Techniques making use of training data sets are referred to as supervised classification algorithms as opposed to unsupervised classification techniques in which no foreknowledge of the existence of ground classes is required. Supervised and unsupervised image classification techniques have been widely used in the analysis of conventional remote sensing data type (e.g., Landsat MSS and TM, and SPOT) and several studies have been undertaken to develop algorithms based on classification for hyperspectral data types. Cetin *et al.*(1993) and Cetin & Levandowski (1991) use n-Dimensional Probability Density Functions based on the principle of the maximum likelihood classifier to analyse AVIRIS, TIMS, and Landsat TM images. Lee & Landgrebe (1993) give algorithms for the Minimum Distance Classifier in high-spectral resolution imagery. An approach using neural networks is given in Benediktsson (1995). An alternative classification algorithm for imaging spectrometer data based on indicator kriging is given in Van der Meer (1994;1996).

## 12.11  CROSS CORRELOGRAM SPECTRAL MATCHING (CCSM)

Cross correlogram spectral matching (CCSM; Van der Meer & Bakker 1997a+b) is a new approach toward mineral mapping from imaging spectrometer data using the cross correlogram of pixel and reference spectra. A cross correlogram is constructed by calculating the cross correlation at different match positions, $m$, between a test spectrum (i.e., a pixel spectrum) and a reference spectrum (i.e., a laboratory mineral spectrum or a pixel spectrum known to represent a mineral of interest) by shifting the reference spectrum over subsequent channel positions by

$$r_m = \frac{n\Sigma \lambda_r \lambda_t - \Sigma \lambda_r \Sigma \lambda_t}{\sqrt{[n\Sigma \lambda_r^2 - (\Sigma \lambda_r)^2][n\Sigma \lambda_t^2 - (\Sigma \lambda_t)^2]}} \tag{33}$$

where $r_m$ is the cross correlation at match position $m$, $\lambda_t$ is the test spectrum, $\lambda_r$ is the reference spectrum, $n$ is the number of overlapping positions (spectral bands), and $m$ the match position. The statistical significance of the cross correlation coefficient can be assessed by the a student's $t$-test and the skewness can be calculated as an estimator of the goodness-of-fit. The cross correlogram for a perfectly matching reference and test spectrum is a parabola around the central matching number ($m=0$) with a peak correlation of 1. Deviations from this shape indicate a different surface mineralogy. Mineral mapping on a pixel by pixel basis is achieved by extracting three parameters from the cross correlograms and combining these into a statistical estimate of the goodness of fit of the two spectra compared: the correlation coefficient at match position zero, the moment of skewness (based on the correlation differences between match numbers of equal but reversed signs, e.g., $m=4$ and $m=-4$), and the significance (based on a student $t$-test testing the validity of the correlation coefficient at $m=0$). In order to evaluate the surface mineralogy maps a root mean square error assessment procedure is proposed in Van der Meer & Bakker (1998) in which the error is calculated from the difference between the calculated pixel cross correlogram and the ideal cross correlogram calculated for the reference as

$$ RMS = \sqrt{\frac{\sum_0^M (R_M - R'_M)}{N}} \tag{34} $$

where $R_M$ is the pixel cross correlation at match position $m$, $R'_M$ is the reference cross correlation at match position $m$, $N$ is the number of match positions, $M$ is the match number. An example of this technique is shown in the Chapter 7.


## 12.12  GEOPHYSICAL INVERSION

A set of physical measurements $\{\mathbf{m}\}$ can be inverted to their resulting variables $\{\mathbf{x}\}$ if the underlying physical process is known and the following relationship is assumed

$$ \mathbf{m} = \Phi(\mathbf{x}) + \mathbf{n} \tag{35} $$

where $\mathbf{n}$ is the sensor noise. The direct inversion in the presence of (random) noise results in

$$ \mathbf{x} = \Phi(\mathbf{m})^{-1} + \mathbf{n} \tag{36} $$

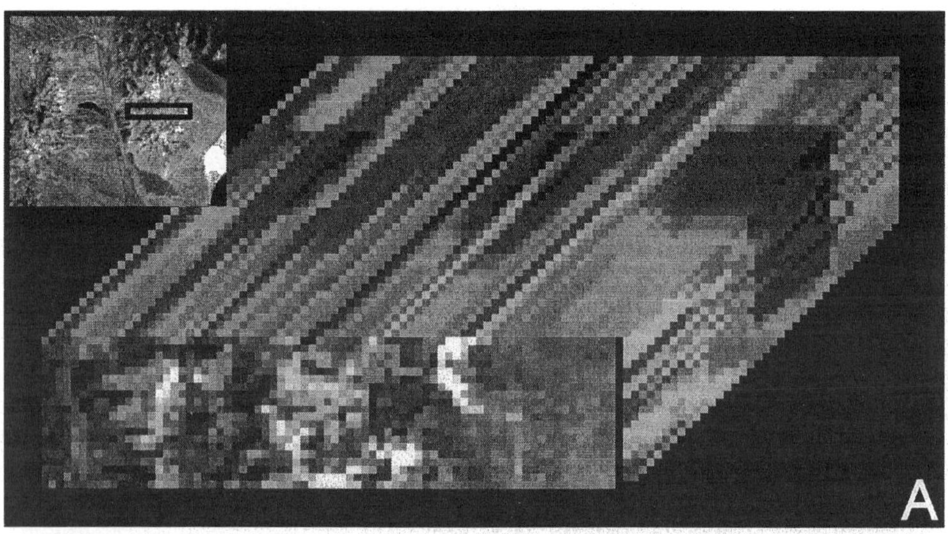


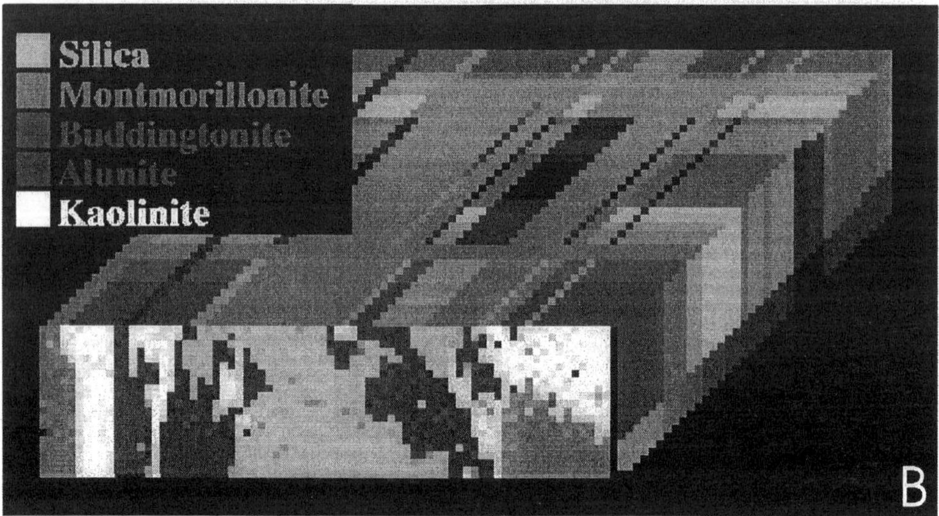


*Figure 11*. (A) Geographical subset from a AVIRIS scene at Cuprite used as sample data from the geophsyical inversion method. The results of SMA and SAM for this subset are shown in Figure 9. (B) Geological model for the selected spectral subset.

A more flexible iterative geophysical inversion can be found with the help of a neural network based on a Bayesian approach (Van der Meer, 2000). This neural network allows incorporating ground truth information and neighborhood information. We aim at finding the set of variable $\{\mathbf{x}\}$ given the set of measurements $\{\mathbf{m}\}$ that maximize the conditional posterior probability $f(\{\mathbf{x}\}|\{\mathbf{m}\})$. Bayes's basic equation states that the *joint probability* that both events A and B occur is equal to the probability that B will occur given that A has already occurred, times the probability that A will occur

$f(A, B) = f(B|A)f(B)$ where $f(B|A)$ is the *conditional probability* expressing the probability that B will occur given A has already occurred.

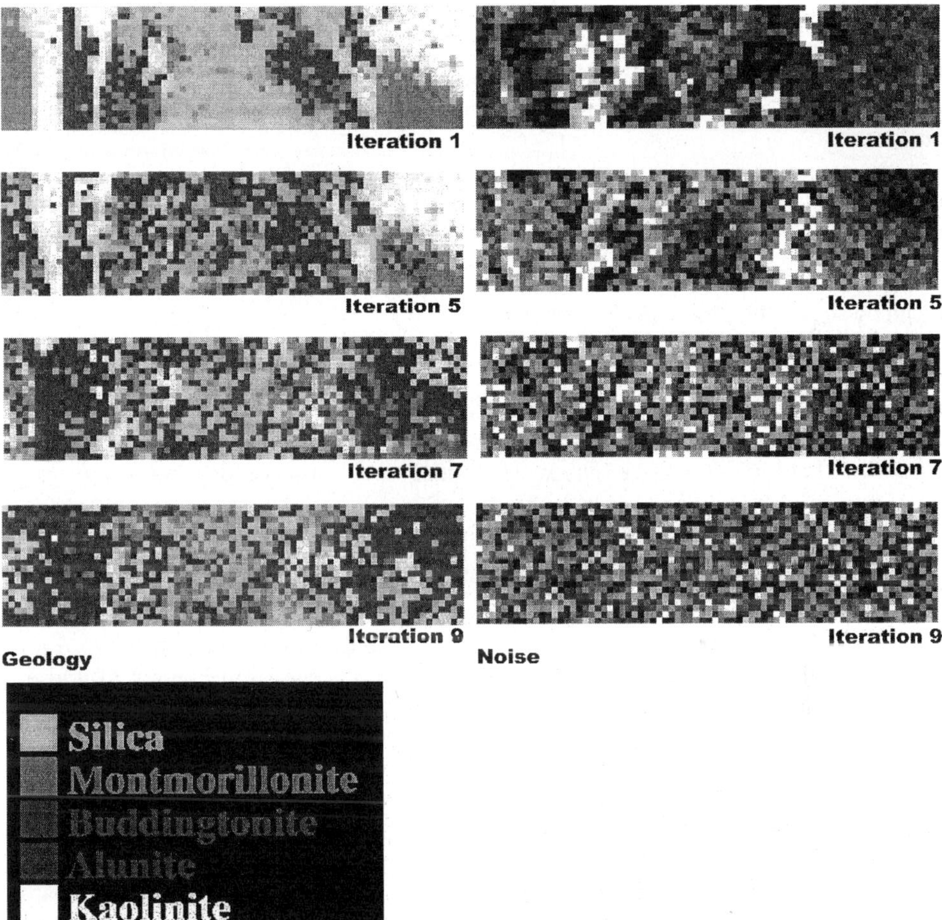


*Figure 12.* Result of geophysical inversion for 9 iteration showing the geological model and the noise component (Van der Meer, 2000).

Translating this to our specific case we can state that in case of an all-inclusive number of independent events $\mathbf{x}_i$ that are conditionally related to the measurement $\mathbf{m}$, the probability that $\mathbf{m}$ will occur is simply

$$f(\mathbf{m}) = \sum_{i=1}^{n} f(\mathbf{m}|\mathbf{x}_i)f(\mathbf{x}_i) \tag{37}$$

Finding a set of parameters that maximize the observations or measurements now yield maximization of

$$\max_{\forall \mathbf{x}_i} . f(\mathbf{x}_i | \mathbf{m}) = \frac{f(\mathbf{m}|\mathbf{x}_i)f(\mathbf{x}_i)}{\sum_{i=1}^{n} f(\mathbf{m}|\mathbf{x}_i)f(\mathbf{x}_i)} \tag{38}$$

Note that we for the time being assume one physical measurement by the sensor (e.g., radiance or reflectance). Now let us introduce $\{\mathbf{y}\}$ as the set of parameter vectors associated with neighboring pixels. This yields the following joint probability to be maximized

$$\max_{\forall \mathbf{x}_i} . f(\mathbf{x}_i | \mathbf{m}, \mathbf{y}) = \frac{f(\mathbf{m}|\mathbf{x}_i)f(\mathbf{x}_i)f(\mathbf{y}|(\mathbf{x}_i|\mathbf{m}))}{\sum_{i=1}^{n} f(\mathbf{m}|\mathbf{x}_i)f(\mathbf{x}_i)} \tag{39}$$

It can be shown that this is proportional to

$$\max_{\forall \mathbf{x}_i} . f(\mathbf{x}_i | \mathbf{m}, \mathbf{y}) = f(\mathbf{m}|\mathbf{x}_i)f(\mathbf{y}|\mathbf{x}_i)f(\mathbf{x}_i) \tag{40}$$

Thus maximization of the conditional probability of the set of variables $\{\mathbf{x}_i\}$ given the set of measurements $\{\mathbf{m}\}$ within a neighborhood $\mathbf{y}$ is equivalent to minimizing the mismatch between (geologic-physical) model and observation (allowing some mismatch due to sensor noise), $f(\mathbf{m}|\mathbf{x}_i)$, given a prior probability, $f(\mathbf{x}_i)$, for the variables, $\mathbf{x}_i$, and a neighborhood distribution, $f(\mathbf{y}|\mathbf{x}_i)$, which is arbitrarily inferred from a kernel of 8 surrounding pixels. An example of the geophysical inversion process for a portion of a AVIRIS data set acquired over the Cuprite mining district (see also the Chapter 7 on geological applications in this book) is shown in Figures 11 and 12. The corresponding SAM and SMA images are shown in Figure 7.


## 13    Endmember selection for spectral unmixing and other feature finding algorithms

Selection of spectral endmembers for unmixing or other types of analysis are crucial to understanding imaging spectrometer data. The set of end-members should describe all spectral variability for all pixels, produce unique results, and be of significance to the underlying science objectives. Selection of end-members can be achieved in two ways:
1. selecting end-members from a spectral (field or laboratory) library, and
2. deriving end-members from the purest pixels in the image.
The second method has the advantage that selected end-members were collected under similar atmospheric conditions. The end-members resulting through the first option are generally denoted as 'known' end-members while the second option results in 'derived' end-members. Identification of the purest pixels in the scene is done through compression of the data using principal component analysis (PCA) following a method developed by Smith *et al.* (1985). A very useful tool for interactive endmember selection is the 3D visualizer in ENVI (Figure 13).

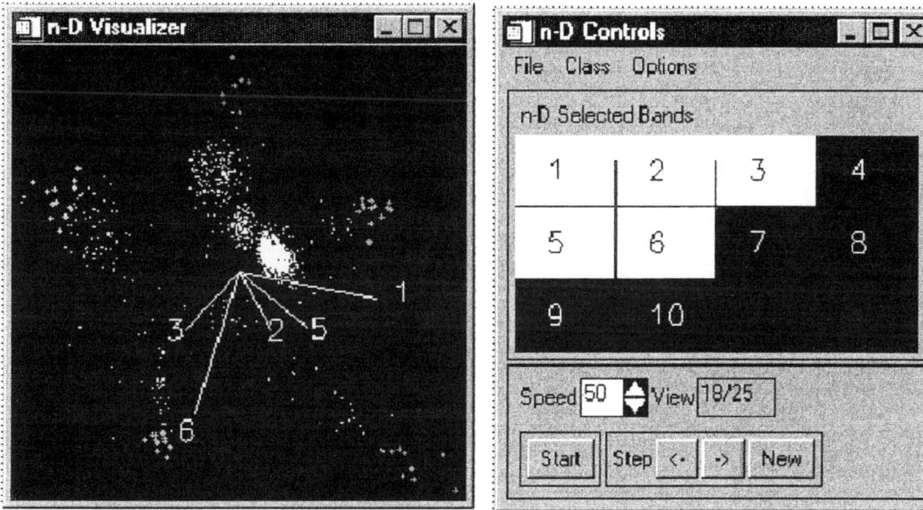

*Figure 13.* ENVI's N-dimensional visualiser that allows to interactively display image data in a multiband feature space and select image endmembers from the data (Source: ENVI tutorial).

Deriving image end-members starts by determining the number of end-members needed to optimally characterize the image data using the Minimum Noise Fraction (MNF) transformation (Green *et al.* 1988). The MNF transformation is a two-step principal component transformation where during the first step using the noise covariance matrix the noise is decorrelated to have unit variance and no band-to-band correlation. The second principal component transformation results in a data set where components are ranked in terms of noise equivalent radiance. The MNF transform is explained mathematically in a previous section.

Next, the locations of the end-member pixels need to be determined from the image data. Here the Pixel Purity Index (PPI; Boardman *et al.* 1995) based on the approach developed by Smith *et al.* (1985) is applied to the MNF transformed data. This approach regards spectra as points in an $n$-dimensional space ($n$ being the number of bands). The body (referred to as simplex) that spans the data points in a $n=2$-dimensional space is a triangle having $n+1$ facets. The purest end-members are found at the locations in this space where the sides of the triangle intersect. This principle can be extended to higher dimensions in a similar way. The PPI approach compares the pixels in a scene with the best fitting simplex and records the number of times a pixel is found at the extreme facets of the simplex. Portraying this as an image of cumulative count values allows to identify the locations in the data space of the initial set of spectrally pure end-members. Van der Meer & De Jong (2000) point out the importance of proper endmember selection as input to unmixing to avoid singularity and orthogonality problems with matrix inversion of linear systems.

# PART II PROSPECTIVE APPLICATIONS OF IMAGING SPECTROMETRY

*CHAPTER 3*

# IMAGING SPECTROMETRY FOR SURVEYING AND MODELLING LAND DEGRADATION

Steven M. DE JONG & Gerrit F. EPEMA
Wageningen University and Research Center, Center for Geo-Information,
Wageningen, The Netherlands

## 1      Introduction

In the late 1970s and the early 1980s alarming reports were published about desertification and land degradation. At that time desertification and land degradation have been described as the world's most pressing environmental issues and one of the major threats for food security of the ever-growing world population (FAO, 1983) and emotionally charged expressions such as 'man-made deserts', 'deserts on the march' and 'rape of the earth' were introduced (Rubio, 1995). Nowadays there seems to be sufficient scientific evidence that during the 1960s and 1970s various areas such as the Sahel, indeed suffered from drought and degradation due to a combination of human activities and a maximum of sun spot activity during its 11 year cycle.

Later publications put the problem of desertification and land degradation into perspective (Thomas & Middleton, 1994; Hillel, 1993). The individual processes of degradation such as soil erosion, wind erosion, salinization and desert expansion were properly described and defined. The term land degradation is generally preferred over desertification. In the late nineties our society again proved to be prone to degradation processes such as flooding in many places in western Europe and southern Africa, massive soil erosion in southern America and China and large mass movements in various places.

In many studies remote sensing techniques and aerial photographs are recommended tools for 1) assessing the spatial and temporal distribution of land degra-dation features and 2) for collecting input data for process simulation models in order to produce land cover maps, vegetative cover maps, bare soil fraction maps etc. (Morgan, 1995; Hill *et al.* 1996; Bolle 1995; Thomas & Middleton, 1994; De Jong, 1994b; Barrow, 1991; FAO, 1983). In the first case remote sensing is primarily used for surveying i.e. to assess the current status of the land in terms of ongoing degradation processes. Such a survey aims at determining the spatial variability of e.g. the status of the natural vegetation (coverage and structure), the status of the agricultural crops (coverage, performance), the surface status of the soil (sealed or crusted), the presence of erosion surface features such as gullies and rills. Next, remote

65

*F.D. van der Meer and S.M. de Jong (eds.), Imaging Spectrometry, 65–86.*
© 2006 *Springer. Printed in the Netherlands.*

sensing plays a role for monitoring changes over time e.g. the development of crop cover over a growing season as an indicator for erosion or for monitoring the long-term development of rill and gully formation in an area. In the second case process controlling variables such as rainfall interception, water canopy storage and changing agricultural land use through the growing season are derived from airborne or spaceborne images to use that information in process simulation models.

In this chapter the use of imaging spectrometry for degradation studies is discussed. We start by restricting the discussion to a limited number of degradation processes, we describe what methods are available to survey and model these processes and we wrap up by illustrating the role of imaging spectrometry.


## 2      Processes of land degradation

Land degradation is a general term name for a large number of unfavourable processes in our environment such as surface wash, rill and gully formation, salinization, soil sealing and crusting, impoverishment of the vegetation, wind erosion and desertification. Land degradation is the result of a complex interaction of many variables, which change over time and space. Land degradation results often in some adverse effects from which we would like the spatial and temporal variation. Erosion often causes a decrease in crop performance which can easily be mapped for the individual agricultural lots by remote sensing. Knowledge of the processes of land degradation and hence, the process-controlling variables, and of the adverse effects of degradation is a prerequisite to determine which variables can be derived from remotely sensed images. The importance of the factors controlling spatial variations of land degradation depends on the scale under investigation. At macro-scale (e.g. 1:1.000.000) climate is considered as a very important factor (Morgan, 1995; FAO, 1979). Going from macro to the micro-scale (1:50.000 or better) changes occur in the relevance of the variables. At micro-scale climate is fairly uniform and variation of soil properties, lithology, topography and vegetation properties becomes important. That means that investigating the degradation process at the local scale, the short distance spatial variability of process-controlling factors of land degradation becomes very important. Hence, at this level of scale it does not make sense to use vegetation and soil maps with large lumped mapping units but local variation should be captured. Remote sensing plays an important role in capturing that local variation of vegetation and soil properties.The scale-dependency of studying degradation processes is also important when remote sensing is used as surveying tools or for data collection. Working at macro-scale other type of information must be extracted from images than working at microscale (ISRIC, 1995). Although degradation of the land cannot be separated into a pure soil or a pure vegetation component, the discussion here splits the role of the two and the effects of the two into a soil section and a vegetation section.


## 2.1     SOILS AND DEGRADATION

Degradation of soils comprise processes such as soil erosion and soil surface crusting. Soil erosion refers to the removal of the fertile top layer of the soil and the incision of the land resulting in rill and gullies. The susceptibility of soils to sheet wash, rill and gully formation and surface crusting is referred to as soil erodibility. Erodibility defines the resistance of the soil to both detachment and transport. At a general level, erodibility depends primarily on the structural stability of the soil and on its ability to absorb or store water. Erodibility of the soil varies with soil texture, aggregate stability, shear strength, infiltration capacity and organic and chemical content (Morgan, 1995). Consequently, monitoring and modelling of erosion processes at a regional scale requires the assessment of the spatial distribution of the critical soil parameters or at least an assessment of the spatial distribution of soil types which are associated with these soil properties. The use of imaging spectrometry and other remote sensing techniques to contribute to degradation monitoring and modelling depends on the ability to distinguish these soil properties and soil types.

*Figure1*. Soil surface crusting generally leads to an increase of the albedo of the soil and not to a change of the spectral shape because the composition of the soil is not changed by the crusting process.

Soil crusting is the response of the soil surface to intense rainfall and can be described as the consolidation of surface particles. Generally, two types of crusts can be identified: structural crusts caused by compaction and depositional crusts built up of several layers of sedimentation. Crusts at the soil surface have several adverse effects (Valentin & Bresson, 1992; Sombroek, 1985; Hillel, 1980; Epstein & Grant, 1973):

- The infiltration capacity of the soil is reduced, this induces runoff and leads to rill and gully formation.
- Germination and the development of vegetation is hampered.
- The water retention capacity of the topsoil is reduced.
- Soil aeration is reduced, retarding root development.
- Downslope soil erosion occurs through sheetwash and rill formation.

The process of crust formation is complex and not yet completely understood. Although a considerable number of studies report on crusting, quantitative data on soil crust properties are scarce. The process of soil surface crusting does only change the arrangement of soil particles at the soil surface and does not really change the upper soil layer in terms of organic matter content or texture. This hampers the possibilities to identify crusts by imaging spectrometry. The soil aggregates (clusters of clay, silt and sand particles cemented together by organic matter, lime, moisture etc.) are destroyed during the process of crust formation and the finer soil particles form a seal or crust at the surface (Marshall & Holmes, 1981). Generally this leads to an increase of the overall reflectance (or albedo) of a soil. Figure 1 illustrates the increased albedo of a crusted soil surface. As the composition of the soil does not really change, crusting does not result in a change of absorption features in the soil reflectance spectrum but only in a overall increase of reflectivity. Consequently, it is difficult to identify crusts and seals at the soil surface by imaging spectrometry since an increase of the albedo can also be caused by other factors. An diminishing water content of a soil also causes an increase of the overall reflectance of the soil surface. The relative higher concentration of small particles at the soil surface after crusting is also difficult to detect by imaging spectrometry although a relative enrichment of clay particles may result in a stronger absorption band around 2200 nm.

## 2.2    VEGETATION AND DEGRADATION PROCESSES

Vegetation plays an important role in degradation processes, vegetation preserves the underlying soil from aggressive raindrop impact. If a soil surface is covered by a dense vegetation (figure 2), it is very unlikely that gully or rill erosion, sheetwash or crusting will occur. The effects of a vegetative cover on erosion comprise (Coppin & Richards, 1990; Sala & Calvo, 1990; Thornes, 1990a, 1990b; Morgan, 1995; Laflen & Colvin, 1981; Wishmeier & Smith, 1978; Carson & Kirkby, 1972):

- Interception. This accounts for the changes in volume and energy of the rainfall reaching the ground surface as a result of interception of rainfall by the plant cover. Its main effect is a decrease of the amount of soil detached.
- Evapotranspiration, This accounts for the transfer of moisture from the soil to the atmosphere through the plant cover. Its effect is a reduction of runoff by maintaining a drier soil and therefore greater soil moisture storage capability at the start of each storm event.
- Rooting depth. This accounts for the increase in pore spaces along the lines occupied by plant roots, which help to reduce runoff by enlarging the infiltration capacity.
- Ground cover. This accounts for the effect of a vegetation cover in decreasing the transport capacity of the runoff through reductions in flow velocity and the physical restraint of soil particle movement. Vegetation generally improves the soil's structure and it increases the organic matter content.

Because it is too time consuming and costly to measure interception, evapotranspiration and rooting depth separately to obtain an understanding of how vegetation protects the soil from erosion, many researchers use compound indices. The best known of these indices is the empirical crop cover factor USLE-C (Wischmeier & Smith, 1978), which is used in many erosion models such as the Areal Non-point Source Watershed Environment Response Simulation: ANSWERS (Beasley & Huggins, 1982), Water Erosion Prediction Project: WEPP (Foster, 1987), Soil Loss Estimator for Southern Africa: SLEMSA (Stocking, 1981), the Morgan, Morgan & Finney method: MMF (Morgan *et al.*, 1995) and theLISEm model (De Roo & Jetten, 1999). The USLE-C is defined as the ratio of soil loss from land cropped under specified conditions to the corresponding loss from clean-tilled, continuous fallow. The USLE-C accounts for the different crop stages during the growing period and for differences of land management.

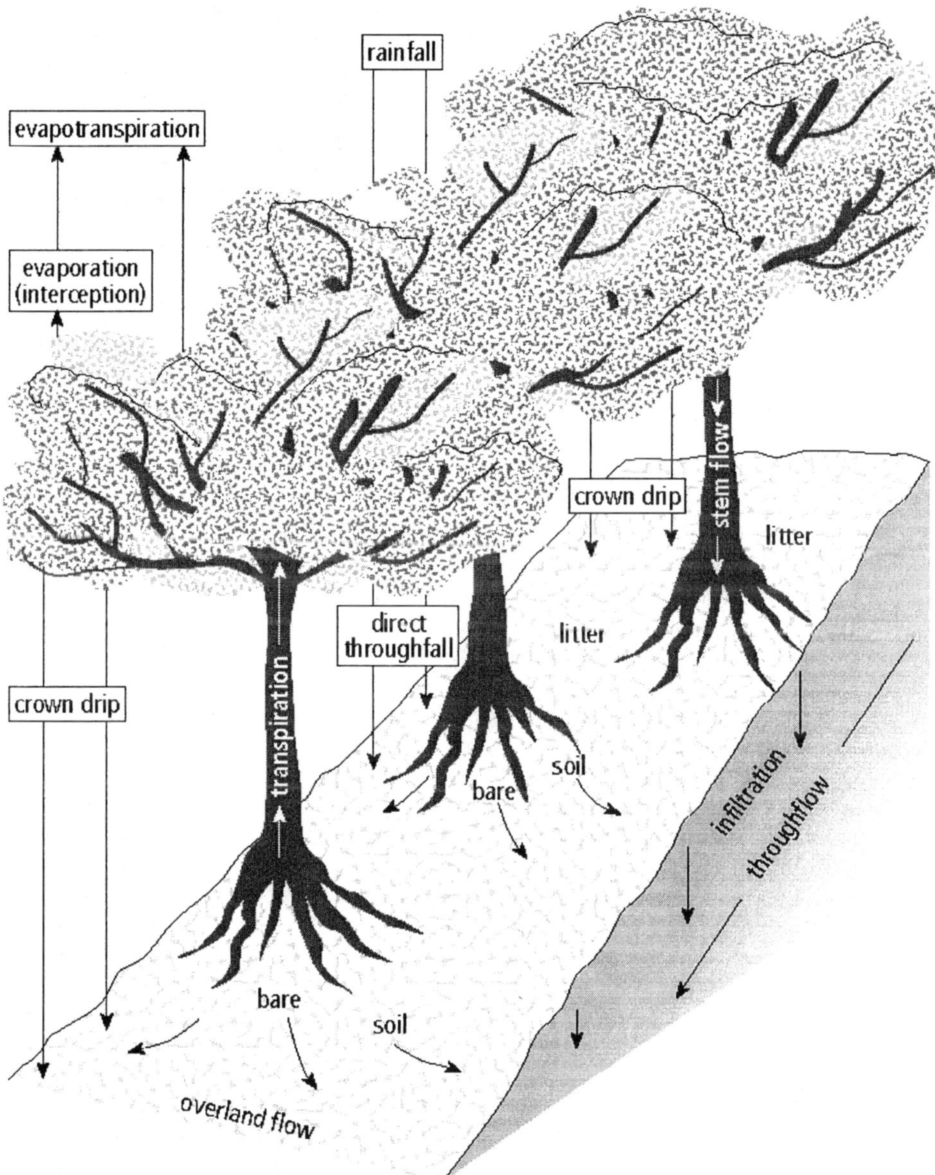


*Figure 2.* Vegetation (cover, type and density) plays an important role in the various degradation processes. It protects the underlying soils against all kind of unfavourable processes, reduces surface runoff and improves the soil top soil structure.

Most estimates of C relates to agricultural crops, there are few estimates of C for areas of natural or semi-natural vegetation. A major problem in erosion studies is the collection of data on spatial and temporal variation of cover (De Jong, 1994a). Remote sensing starts playing an important role in collecting this data. Remote sensing can also provide information on the `type of natural cover', which a) is different from mono-crop systems and b) may vary in vigour status throughout the year without necessarily affecting its physical ability to protect the soil. Next, the spatial distribution and status of the natural vegetation itself is useful as an degradation indicator. This is sometimes the case in Mediterranean landscapes, which show complex spatial patterns due to their long-term land use history of clearing, grazing, abandonment, regrowth, terracing and fire which has led to typical Mediterranean landscape types such as maquis, garrigue, rangelands and badlands. The various stages of the natural vegetation can be mapped by remote sensing and may be used as an indicator of the degradational state of the land (De Jong & Burrough, 1995).

## 3    Spectrometry for land degradation

Previous sections have described degradation processes of soil erosion, soil surface crusting and impoverishment of the natural vegetation and their controlling variables. It was concluded that remote sensing can play a role as:

1)    A surveying tool to assess the spatial (and temporal) variation of the degradation process: spatial and temporal patterns of natural vegetation and agricultural land use, condition of the soil surface, and surveying land degradation features in natural areas or on agricultural land such as badlands, gully and rill development, salinization, sedimentation patterns, surface crusts, decline or re-growth of ecosystem patterns. Various examples of surveying degradation features or processes by remote sensing can be found in literature e.g. Escadafal *et al.* (1997), Metternicht & Fermont (1998), Bonn *et al.* (1997).

2)    A tool to collect information on, again the spatial and temporal variation of the process-controlling variables such as land cover type, crop and vegetation cover, crop and vegetation status, soil surface status, soil organic matter content, soil minerals as necessary input for process simulation models. Simulation models are increasingly used to simulate degradation processes such as runoff, soil and wind erosion within the computer to improve our understanding of the process and to assess future developments taking place in the landscape (De Roo, 1999; De Jong *et al.*, 1999; Morgan 1995). These models simulating the degradation processes are moslty linked or integrated with geographical information systems (GIS). The direct integration with GIS enables us to simulate the processes in a more realistic way and it allows the simulation of exchange of fluxes between the rastercells (downtracing of surface runoff, sediment transport or grazing behaviour of ecosystems) but this integration puts increasing demand on the volume and accuracy of spatially and temporal distributed model inputs (De Roo, 1999; Jetten *et al.* 1999; Brouwer & De Jong, 1998; Burrough & MacDonnell, 1998; De Jong *et al.*, 1999). Depending on the type of model used, the input requirements of data can be very large. Distributed models incorporate data concerning the spatial distribution of variables together with computational algorithms to evaluate the influence of the spatial distribution on the simulated process. Modern geographical information systems enable us to simulate exchange of fluxes between raster cells (Burrough & McDonell, 1998). Next, the fast development in computer technology and disk space allows us to simulate the temporal behaviour of the processes, often

referred to as dynamic modelling (Burrough & McDonnell, 1998). Such models requires as input not only maps displaying the spatial distribution of process-controlling variables but also time-series of the temporal changes of rainfall and vegetative cover. The EU-MWISED project models even aims at developing spatially distributed, dynamic models that account for changes within one rainfall event of e.g. the soil surface (roughness) and vegetation canopy properties (Van de Vlag *et al.*, 2000). The role of various Earth observation techniques will in the near future significantly enhance the possibilities to collect the required information for these simulation models. Spectroscopical imagery will provide information on vegetation dynamics and soil surface conditions but one can also think of using ground based radar remote sensing systems to map the spatial distribution and intensities of rainfall within one storm.

## 4    Soils

Traditionally, existing choropleth maps of soils and geomorphological units are used in land degradation studies and soil erosion models to derive specific model input parameters such as texture and organic matter content. The sparse information on physical and chemical soil properties coming with such a soil map is used to assign these attributes to the mapping units. The mapping units are considered as homogeneous areas and `within mapping unit' variance is ignored. Mostly, the `within unit' variance is unknown and in practice that `within unit' variance can be very large (Burrough, 1986). Remote sensing might play a role in estimating the spatial variability of soil properties (De Jong & Riezebos, 1994). A limiting factor is that optical remote sensing does not penetrate into the soil and that not every pertinent soil property can be deduced from the soil surface. In this section the spectral properties of bare soil surface are presented and methods are discussed to extract soil information from imagery.

## 4.1    CLASSICAL SOIL DESCRIPTION METHODS

Pedologists have a wide variety of approaches to study the soil (USDA, 1999). In traditional mapping methods aerial photographs are used to stratify the area under investigation in a number of physiographic units. For these units, observations are done at specific locations by describing soil characteristics of auguring and soil pits (soil surface properties, soil horizons, parent material etc.). In addition to the descriptions made in the field, at a few representative locations samples are taken of the various soil horizons in the profile and brought to the laboratory for soil physical and soil chemical analyses. Based on field information, aerial photo interpretation, results of physical and chemical laboratory analyses and expert knowledge a final soil map is constructed, with a description of each unit and a description of a limited number of soil characteristics, whether or not in a digital form. For an in-depth description of a standard soil mapping procedure and the described soil qualities, please consult USDA (1999) or FAO (1990). Nowadays, geostatistical approaches are sometimes applied to quantify the spatial variability of soil properties. Variogram analysis and interpolation algorithms such as Kriging are then applied to produce spatial continuous soil maps and to assess the degree of uncertainty (Heuvelink, 1993). Furthermore, soil properties are not only described in the traditional way but currently also for specific purposes such as land degradation and soil contamination.

One of the most interesting and most widely used descriptive variables of soils is soil colour (Driessen & Dudal, 1989; FAO, 1988). Colour is used to assist the classification of soils into specific categories and next, colour is used to infer soil characteristics. Although remote sensing from airborne or spaceborne platforms, from portable field instruments or from laboratory spectrometers can measure soil reflectance over different wavelengths, soil colour plays still a major role in modern soil classification (Irons et al 1989). The Munsell soil colour charts (Munsell Color, 1975) comprise the standard to describe the color visually in the field. This system describes the colors in terms of hue, value and chroma and in standard descriptive names. Hue is the dominant spectral colour and is derived mainly from a combination of pigments in minerals and organic matter in the visible part of the spectrum. Value refers to the brightness and is besides presence of minerals and organic matter also influenced by moisture content of the soil and particle size. Chroma is the purity of the hue and is influenced by the spectral distribution of the reflected light. By inverting the colour relations, it is in fact possible to derive soil characteristics from the colour. A dark red colour is generally an indication that a soil is rich on some type of iron or manganese. The importance of soil colour for classification purposes is illustrated in figure 3.

Since the system of soil colour description is not a physically based one, approaches have been developed to convert these colours into CIE units. The CIE is an internationally accepted standard for colour measurement (Melville & Atkinson, 1985). Soil reflectance spectra were converted to CIE values by for instance Escadafal *et al.* (1989). Escadafel describes model functions between high resolution soil spectra and traditional soil colour descriptions. Such functions enable us to relate soil colour, soil spectra and soil properties. One important limitation will of course always remain: in the optical wavelengths remote sensing only allows us to study soil surface properties, it is not possible to study sub-surface properties by spectroscopical remote sensing techniques. All sub-surface properties must be derived, if possible, from surface characteristics. Various authors describe in literature fairly specific relations between soil colour and soil components such as organic matter content, iron oxides, carbonate content, clay content, cation exchange capacity and water content (Karmanova, 1981; Torrent, 1983; Escadafal *et al.*, 1989; Da Costa, 1979). In-depth overviews are given by Irons *et al.* (1989) and Ben-Dor *et al.* (1999).


## 4.2    SOIL SPECTRAL PROPERTIES

The spectral behaviour of a soil is considerably different from vegetation when soil reflectance is measured with conventional broad-band sensors (60 nm or more) a soil reflectance curve in visible and infrared wavelengths (400-2500 nm) shows less `peak and valley' variation. The factors that influence soil reflectance seems to act over less specific spectral bands than for vegetation. When soil spectra are studied in more detail i.e. at increased spectral resolution (15 nm or less), significant differences between soil spectra become apparent (Goetz, 1989; Goetz *et al.*, 1985). A soil spectrum measured at high spectral resolution shows several diagnostic absorption features. In particular the shortwave infrared spectral region from 2000 to 2500 nm contains diagnostic absorption features of soil constituents.

*Figure 3.* Soil colour is an important criterium in traditional soil classification. Top figure shows a Luvisol according to the FAO classification in the Ardèche province (France). A Luvisol is a well-developed soil and in this case relatively rich on iron. Bottom figure shows a Leptosol located at 200 m distance of the soil in 3a and strongly affected by water induced erosion processes.

The characteristics of a soil that determine its reflectance properties in the solar part of the spectrum have been described in several studies (Ben-Dor *et al.*, 1999; Irons *et al.*, 1989; Mulders, 1987; Baumgardner *et al.*, 1985; Myers, 1983; Hunt, 1980). The most important soil properties regarding optical reflection are: moisture content, organic matter content, texture, structure, iron content, mineral composition, type of clay minerals and surface conditions of the soil.

- An increasing moisture content generally decreases soil reflectance across the entire spectrum. In fact, wet soils also appear darker to the eye than dry soils. Furthermore, reflectance spectra of moist soils include prominent absorption bands for water and hydroxyl at 1.9 and 1.4 µm and some weaker absorption bands at 0.97, 1.20 and 1.77 µm (Pieters & Englert, 1993; Irons *et al.*, 1989; Baumgardner *et al.*, 1985).

- Iron is an important constituent of a soil regarding its reflective properties. Many absorption features in soil reflectance are due to the presence of iron in ferric or ferrous form. For example, the general strong decrease of reflectance of soils towards the blue and ultraviolet wavelengths is caused by iron. Furthermore, absorption features occur around 0.7 $\mu m$, 0.9 μm and 1.0-1.1 $\mu m$. Some weaker absorption features due to iron are located near 0.4 and 0.55 $\mu m$ (Pieters & Englert, 1993; Hunt & Salisbury, 1970; Baumgardner *et al.*, 1985).

- An increase of the organic matter content of a soil generally causes a decrease of reflectance over the entire spectrum that is similar to moisture. A high organic matter content and hence, a strong decrease of overall reflectance, might even mask other absorption features in the soil spectra (Irons *et al.*, 1989). However, this effect is minimal for soils having organic matter contents below 2.0 to 2.5% (Baumgardner *et al.*, 1985). Unfortunately, little knowledge of the influence of different organic soil compounds and the different forms of organic matter on the spectral behaviour of soils is available in the literature.

- The basic components of soil minerals are silicon, aluminium and oxygen. None of these soil constituents have diagnostic absorption features in the solar part of the spectrum (Burns, 1993; Irons *et al.*, 1989). Quartz for example, has a large reflectance throughout the shortwave spectrum and absorption features in shortwave quartz spectra are only due to impurities. Other minerals are less reflective. Soil minerals affect spectra in an indirect manner. Their crystal structure determines the possible contents of ions (e.g. ferrous iron or hydroxyl), which do have certain typical reflectance properties (Burns, 1993; Hunt and Salisbury, 1970). Mulders (1987) found absorption bands of gypsum at 1.8 and at 2.3 $\mu m$. Pure calcite has rather strong absorption features near 2.35 $\mu m$. Clay minerals such as illite and kaolinite have strong absorption features near 2.2 $\mu m$ due to the presence of hydroxyl-ions.

- Soil particle size influences some soil properties such as moisture content and soil structure. Both properties also influence the reflectance of soils. Consequently, it is very difficult to measure the exact effect of increasing soil particle size on reflectance. Orlov (in Myers, 1983) describes the dependence of reflection on aggregate diameter with an exponential equation. Apart from the reflectance differences which can be accounted for by differences in particle size and soil structure, size and shape of soil aggregates appear to influence the soil reflectance in varying manners (Baumgardner, 1985). Bowers & Hanks (1965) measured reflectance of pure kaolinite in size fractions from 0.022 to 2.68 mm diameter and found a rapid exponential increase in reflectance at all wavelengths between 0.4 and 1.0 $\mu m$.

Most of the studies unravel the contribution of individual soil properties to reflectance. Although the reflectance of some individual soil components is well-known it is not yet possible to model accurately the reflectance of a soil surface, which is a mixture of many components. Furthermore, outside the laboratory in the real world, reflection measurements of a soil detect a complex mixture of soil components, varying soil surface conditions and atmospheric circumstances. Changes in surface conditions complicate the remote sensing of soils. Conditions such as surface roughness, moisture content and the presence of plant residues are frequently altered by weather (or tillage). These properties show a very large temporal and spatial variability. Changes in these conditions affect soil reflectance to a large extent.

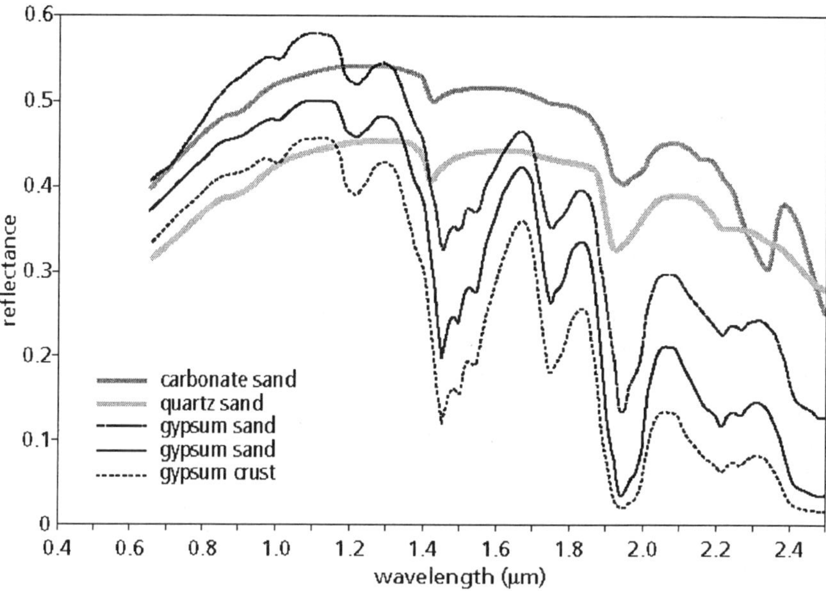


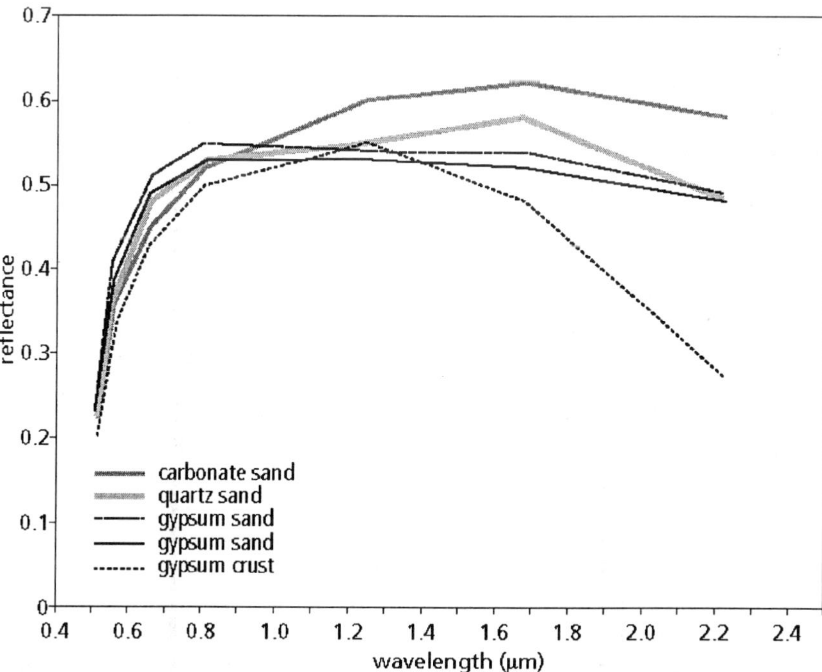


*Figure 4.* High resolution spectral curves (top) for carbonate sand (1), quartz sand (2), gypsum sand (3, 4) and gypsum crust (5) measured in southern Tunisa (Epema, 1992). Note that the overall shape and absorption features remain almost the same after crusting and that only the albedo changes significantly. Figure 4b (bottom) shows spectra of the same surface measured at low spectral resolution using only 7 broad spectral bands.

## 5 Mapping the soil degraded state by imaging spectrometry and spectral unmixing

In the previous sections processes of land degradation and spectroscopical properties of soils and vegetation were discussed. In this section an example is presented how imaging spectrometry is applied to survey the degraded state of the soils in the southern part of the Ardèche province in France. A detailed and elaborated desciption of this example is available in Hill *et al.* (1996). In the southern part of this province soil erosion is a local process and partly controlled by the transition of north-south oriented limestone plateaux and marl sediments. Erosion mainly occurs in the marls areas. The ongoing soil erosion and degradation processes result in the removal of the upper layer of top soil. Organic matter and small particles are washed downslope. As a result degraded soils can be distinguished from non-degraded soils by a change in colour of the soil surface. The degraded soils show an increased brightness of the surface. An example is shown in the field photos of figure 3a and 3b. These field photos illustrate this ongoing degradation process in the Ardèche province where at a distance of only 200 meters apart these two soil types are found: a well-developed reddish Luvisol and a degraded Leptosol. Other soils found in the area include (calcaric) Fluvisols next to outcrops of marls and limestone.

A prerequisite for proper application of spectral unmixing (see also Chapter 2 in this book) is the careful consideration of regional soil types and their degradational forms. The regional soils show in fact a sequence of ongoing erosion. Reddish coloured Cambisols and Luvisols with a clear profile development are found in the marl areas on small patches or on sites so far protected against erosion. The high resolution spectra reveal the presence of iron oxides, clay minerals and carbonates. Their presence is indicated by absorption features near 650 and 900 nm and 2200 and 2350 nm respectively. Around these areas the most common soil type is the Leptosol and Fluvisol. The Leptosol are degraded or weakly developed soil showing only a AC or AR profile. These soils are much brighter due to the absence of organic matter and iron. Where the soil is completely removed by surface wash and other erosion processes outcops of marls and limestone lay at the surface. A second prerequisite for applying spectral unmixing is that the spectral properties of the surface allows the remote identification of these soils. A digital spectral library of local soils and other surfaces was built for this case study during various field campaigns. Analysis of this local spectral library showed that the previous described soil and surface types sufficiently vary to allow remote identification.

Spectral mixture analysis was then applied to an AVIRIS image and a Landsat TM image of the Ardèche (Hill *et al.*, 1995). An endmember mixing model was constructed comprising spectra of green vegetation, limestone, marls and soils. The green vegetation spectrum was added to extract photosynthetically active vegetation from the scene. A shade spectrum was added to the unmixing model to isolate the influence of shadows in the image. This mixing model was then translated in terms of local soil degradation conditions on the basis of approximately 100 field observations. Four soil conditions stages are distinguished: fair, degraded, severely degraded and parent rock as shown in figure 5. The abundance maps of the individual endmembers were then translated into these four thematic classes of soil conditions. Figure 6 shows the result as applied to the AVIRIS image (Hill *et al.*, 1996).

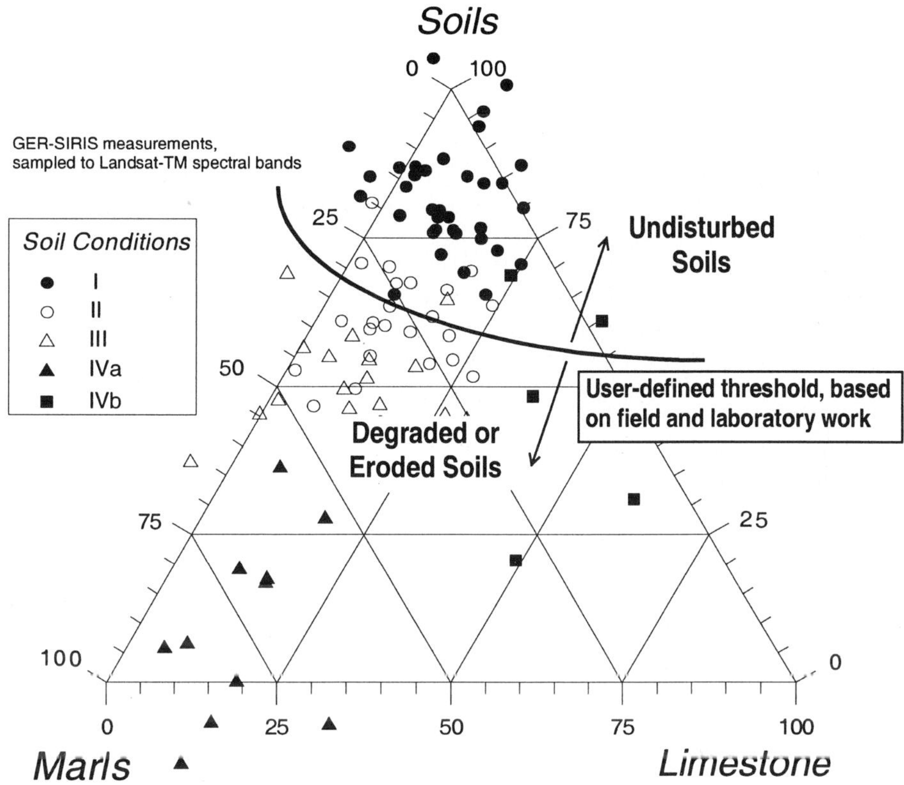


*Figure 5.* Triangle showing the basis for the spectral unmixing of the soils of the southern Ardèche into their degraded state. The symbols refer to the I) well-developed to IV) severely degraded (Source: Prof. Dr. J. Hill, Trier University, Germany).

## 5.1 MAPPING SOIL UNITS USING IMAGING SPECTROMETRY AND SPECTRAL MATCHING

Another approach (section 4.3) for mapping of soil properties and soil types was proposed by Jansen (1993; personal communication) and applied by de Jong (1994b). The most important difference compared to spectral unmixing is that less field spectra and field data are required and that spectral matching works on a pixel by pixel basis. The concept of the spectral matching approach is that a pixel spectrum of an image is matched with reference spectra for the entire spectral range (400-2500 nm) or parts of it. But again, the use of imaging spectroscopy for surveying soils or minerals is difficult in areas with a vegetation cover because radiation in visible and infrared wavelength hardly penetrates through vegetation. The Similarity Index (SI) computes a normalized Euclidian distance between two spectra in a user-defined spectral window on a pixel-by-pixel basis (Jansen, 1993)

$$SI = \sum_{i=m}^{n} \frac{(W_i - R_i)^2}{(n - m + 1)} \tag{1}$$

where $W_i$ is the spectrum of the current pixel, $R_i$ are the reference spectra and $n$ and $m$ are the band numbers of a user-defined window.

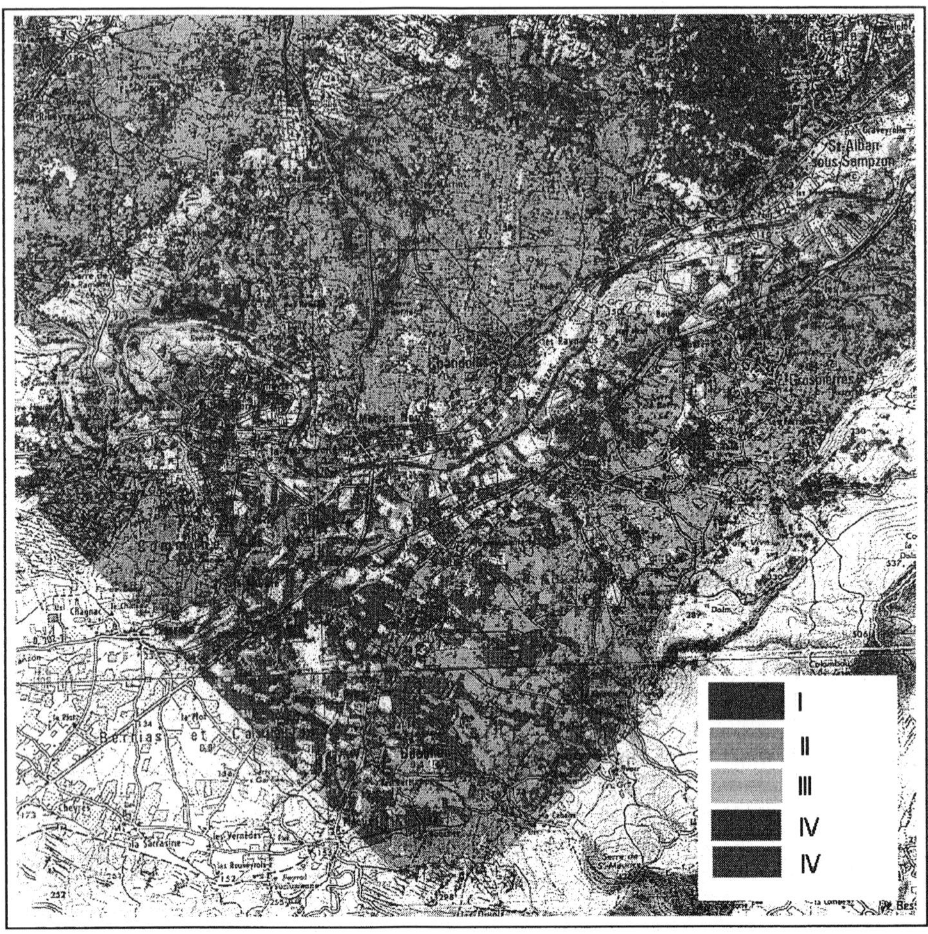


*Figure 6.* Using spectral mixture analysis of AVIRIS imagery to survey the degradational state of the soils in the Ardèche province in France. Class I) are well-developed soils, class II) and III) represent degraded and strongly degraded soils and class IV) and V) refers to parent material of marls and limestone respectively (Source: Prof. Dr. J. Hill, Trier University, Germany).

The smaller the SI-value, the better the match of the spectrum of interest and the reference spectrum. A perfect match is indicated by an SI-value of zero. Jansen (1993) suggests that SI-values below 100 are reliable matches. Using this method of spectral matching, it is also possible to normalize the reference spectrum to the image spectrum by first averaging their reflectance values. In fact the spectra are first corrected for their vertical offset and next their similarity index is computed. Such a normalisation makes it possible to match two spectra which are similar in shape but are different regarding absolute reflectance (they have an offset of absolute reflectance value). The reference spectra can be collected in the field or taken from an existing spectral library such as produced by NASA Jet Propulsion Laboratory (Grove *et al.*, 1992; http://speclib.jpl.nasa.gov) or by the CSIRO (Huntington *et al.*, 1989; http://www.csiro.au, 2000). Spectral matching was applied to

GERIS images of the same region in the Ardèche province in southern France using as reference spectra field spectra, the JPL library and the CSIRO library.

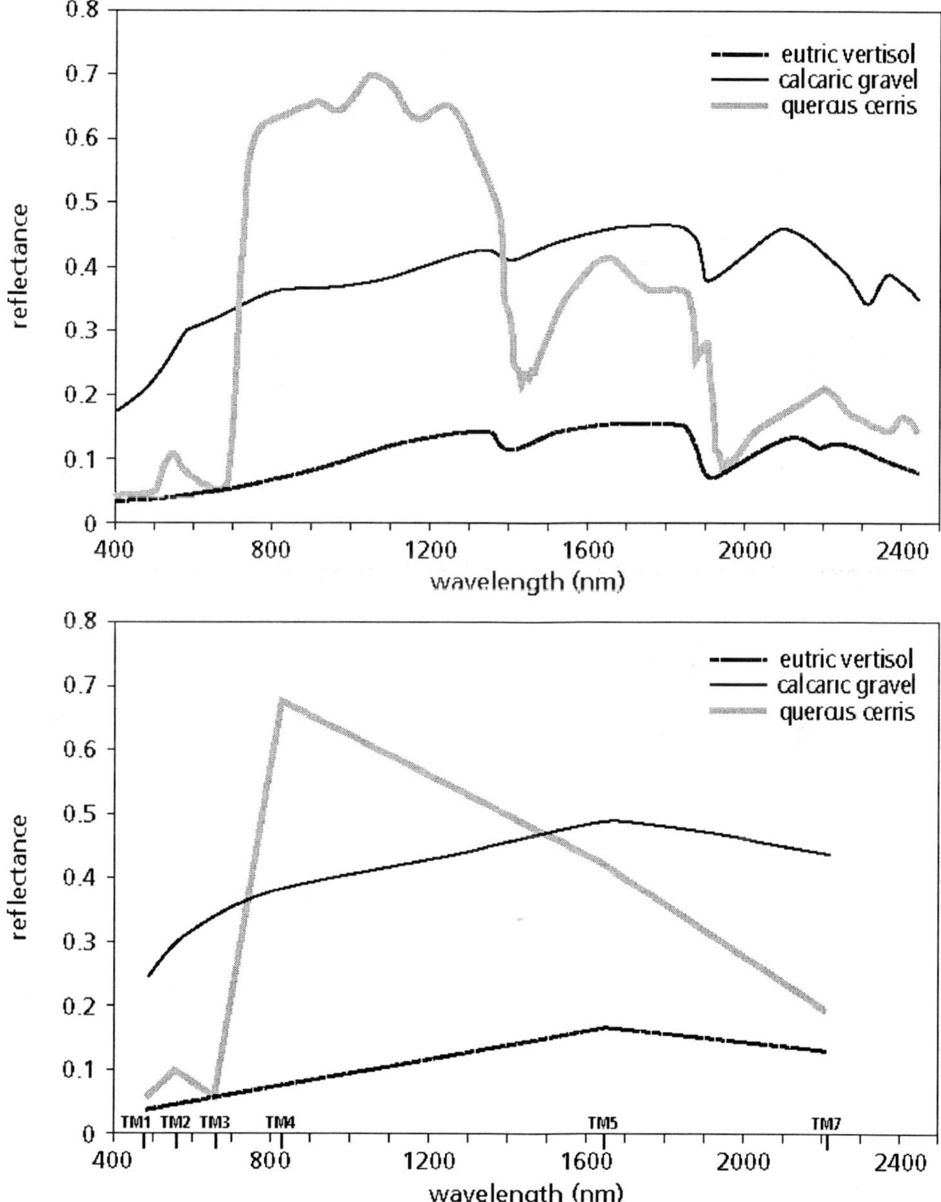


*Figure 7.* Spectral signatures of vegetation, soil and parent material measured at a high spectral resolution of 2 nm (upper graph) and at broad band resolution ranging from 60 to 270 nm corresponding to the Landsat TM sensor. Note the loss of information due to spectral band width.

Results were disappointing for the case using the spectral libraries. Both libraries contain spectra of air dry, grinded samples of pure minerals. Such spectra are not very representative for soil surfaces in the real world. Soil surfaces in the real world are generally moist, the minerals are weathered which change their spectral response and soil surfaces have often a crust or a seal which also has an impact on the spectral response. Best results for soil property survey were obtained by using field spectra for spectral matching. In the study area in the Ardèche 14 soil units were distinguished by spectral matching. These 14 images of soil units were used to refine the existing soil map of the area. The 14 images were exported from the image processing system into a geographical information systems and used to feed the soil erosion model Semmed (de Jong, 1994b; de Jong *et al.*,1999).

## 6    Vegetation

In section 3 of this chapter the important role of vegetation in the land degradation and erosion process was discussed. Vegetation protects the underlying soil against rain drop impact, it improves the structure of the soil which reduces the erodibility and enhances the infiltration and absorption rate of the soil for water. Furthermore, the status of the vegetation in terms of density, coverage and performance e.g. in terms of 'distance to the climax vegetation' is in itself an indication for the degree of degradation of an ecosystem. In the previous section it was concluded that information is required on the spatial distribution of vegetation and questions to be answered are: where is what type of vegetation growing, what is its fractional coverage, what are dynamics of ractional vegetation coverage over the seasons but also what are the long-term dynamics of the vegetation fractional coverage and what are long-term changes of vegetation composition. Before the role of imaging spectrometry for vegetation mapping and monitoring is discussed, we first look at the spectral properties of vegetation and traditional vegetation mapping methods using remote sensing approaches.

## 6.1    SPECTRAL REFLECTANCE OF VEGETATION

The optical properties of vegetation and leaves are described by various authors e.g. Baret (1994), Wessman (1989), Elvidge (1990), Goel (1989) but also chapters 5 and 6 of this book provide a detailed description. In general the reflectance of vegetation in the visible wavelengths (0.43 - 0.66 $\mu m$) is small and reflectance in near infrared (0.7 - 1.1 $\mu m$) is large (figure 7; see also the chapter 1 on reflectance spectroscopy in this book and the Chapter 6 on agricultural applications in this book). Three features of leaves have an important effect on the reflectance properties of leaves:
- Pigmentation
- Physiological structure
- Water content.

Pigments (chlorophyll a and b) absorb radiation of the visible wavelengths with strongest absorption in blue and red wavelengths. The species-specific structure causes discontinuities in the refractive indices within a leaf, which determine its near infrared reflectance (figure 7). The reflection properties of a vegetation canopy are affected by the spatial distribution of vegetated and non-vegetated areas, types of vegetation e.g. deciduous species versus pine species, leaf area index, leaf angle distribution and vegetation conditions (Goel, 1989). Some narrow absorption bands due to lignin or cellulose are present in near and shortwave infrared wavelengths (Wessman *et al.*, 1989; Vane & Goetz, 1988) but the presence of water often masks their absorption features. Furthermore, the bandwidth of the TM is too broad to detect these narrow absorption bands. Water content of the leaves, and water in the atmosphere, reduce the overall leaf reflectance and causes absorption features in the spectra: the water absorption bands. Three major water absorption bands are located near 1.4, 1.9 and 2.7 $\mu m$, two minor absorption bands occur near 0.96 and 1.1 $\mu$ m (figure 7). The combined effect of pigments and physiological structure give healthy vegetation its typical reflectance properties. Combinations of the visible and near infrared spectral bands enables us to discriminate bare soil surfaces or water bodies from vegetation. These arithmetical band combinations can be referred to as `spectral vegetation indices'.

*Figure 8.* Field picture of the Peyne area in southern France showing a logged hill side with runoff and erosion features. This logged hill slope is clearly visible (within the square) on the DAIS7915 image of figure 9.

## 6.2    CONVENTIONAL REMOTE SENSING AND VEGETATION

Traditional approaches of monitoring vegetation in natural and semi-natural areas by remote sensing comprise the use of spectral indices for assessing vegetative cover and/or bare soil abundance. Spectral indices aim at enhancing the spectral contribution of green vegetation in images while minimizing contributions from soil background, sun angle and atmosphere by combining various spectral bands e.g. near infrared reflectance and visible bands. Examples of such indices are the Normalised Difference Vegetation

Index: NDVI (Tucker, 1979), the Soil Adjusted Vegetation Index: SAVI (Huete, 1988) and the Tasselled Cap (Crist & Cicone, 1984). The statistical analyses of field measurements of green coverage and corresponding values for these spectral indices yield mathematical functions which can be used to assess green coverage for large areas, examples of such an approach is described by De Jong (1994a) and De Jong *et al.* (2000). Although the use of spectral indices is straightforward and computationally simple, a number of disadvantages can be mentioned which can partly be overcome by using spectroscopical images as discussed below. The spectral indices yield fairly reliable results for green vegetation cover estimates. Their performances however regarding senescent vegetation (which protects the soil as well against rainfall impact) is less satisfactory because the spectral response for bare soil and dead vegetation is too similar for these traditional broad band remote sensing sensors e.g. Hurcom and Harrison (1998). Furthermore, the impact of the soil background on the values of the spectral indices is significant. Studies by Jasinsky (1990) and others showed that approximately 20% of the variation of the NDVI values of various surfaces can be explained by soil background colour. Another disadvantage of these spectral indices is that they use only two or a limited number of bands available while the spectral shape of green, senescent or dead vegetation is much better expressed by the entire spectral shape in the solar part of the spectrum. Therefore, various studies focussed on spectral mixture analysis (SMA) to map in a quantitative way the abundance of vegetation e.g. Elmore *et al.* (2000), Van der Meer & De Jong (2000). Last but not least, the spectral indices work on a pixel by pixel basis while alternative approaches such as spectral mixture analysis are capable of analysing the spectral signature of a pixel at the sub-pixel level which is beneficial for ecosystems with irregular vegetation coverage and composition.

Benefits of using spectroscopical images over traditional broad-band remote sensing systems for vegetation are: 1) the full spectral shape of spectrometers allows much better the discrimination between green, senescent and dead vegetation as is shown in figure 7, 2) vegetation coverage can much better be separated from soil background and 3) the many spectral bands allows a sub-pixel analysis of the image into fraction maps or abundance maps of bare soil, vegetation, rock outcrops etc.


## 6.3    ASSESSING VEGETATION PROPERTIES FOR EROSION MODELS FROM SPECTROSCOPICAL IMAGES

Vegetation cover and vegetation types and their spatial distribution are importants factors in the soil erosion process. Soil erosion model therefore requires information on the fractional vegetation coverage and on vegetation properties to compute the percentage of rainfall interception, to assess the storage of water in the canopy i.e. the volume of water not available for surface runoff and to assess the structure of the vegetation which controls the transport capacity of the surface runoff. The latter factor is the C-factor in the Universal Soil Loss Equation. In order to fulfill these model requirements we would like to produce from a spectroscopical image (or any other remote sensing image) individual fraction maps of green vegetation, senescent vegetation and bare soil.

Figure 8 shows a field situation of a moderate steep slope in the Peyne catchment in the Hérault province in France. The Peyne area is located approximately 60 km west of the city of Montpellier. The slope is covered by a Mediterranean mixed oak forest except for one section where the forest is removed. Where the forest is removed clear indications are visible of surface runoff and rill and gully development. On the right hand side of this logged hill side the forest is re-growing but still rills and gullies are visible. In the summer of 1997 and 1998 spectroscopical images were collected over the Peyne catchment and this site by DAIS7915 (DLR: www.dlr.de/dais). The technical details of the DAIS7915 sensor are described in the introductory remarks in this book. Spectral mixture analysis (see also the chapter 3 on analytical techniques in this book) was applied to this region aiming at assessing abundance maps for green vegetation, senescent vegetation and bare soil in order to use them as input in a soil erosion model. Figure 9 illustrate some results. The upper left graph shows the original DAIS7915 image in a false colour combination. The upper right, lower left and lower right show the computed fraction or abundance maps of green vegetation, senescent or dead vegetation and bare soil respectively. Note the good correspondence with the field situation shown in figure 8. The hill side where the forest has been removed is clearly visible as a bright white spot on the bare soil abundance map and as a dark area on the green vegetation abundance map. In theory these two maps should be their opposites and they apparently are so.

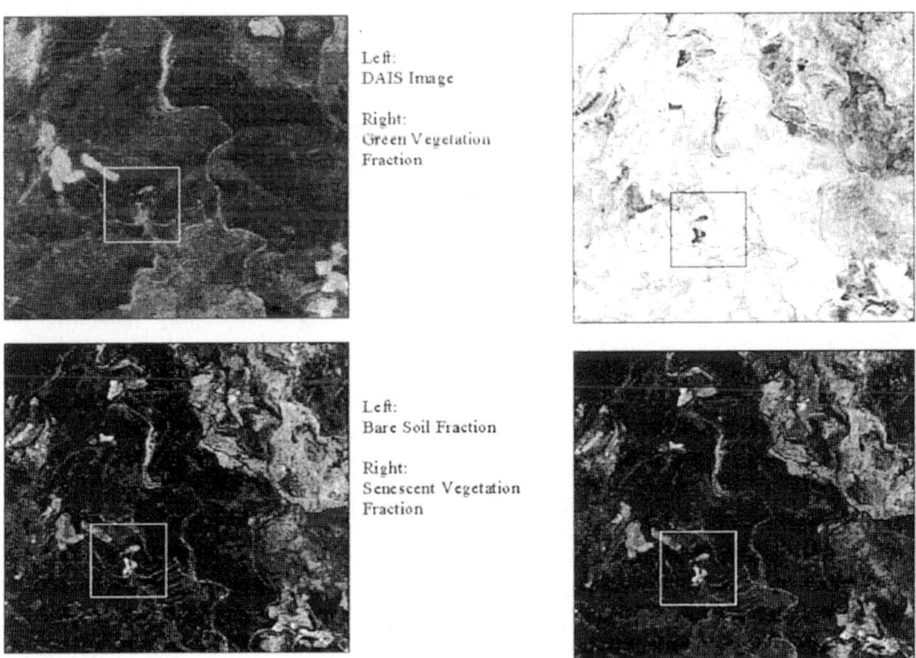


*Figure 9.* Example of the use of spectral mixture analysis on a hyperspectral DAIS7915 image. The image is unmixed into fractions of bare soil, senescent vegetation and green vegetation. The three resulting fraction maps are directly used as GIS input layers in a distributed soil erosion model.

## 7    Contextual approaches to land cover mapping

Various researchers have described contextual approaches to land cover or vegetation cover mapping using TM images or SPOT images. Better results can be achieved when spectroscopical images are used because they provide more information on the subtle differences between land cover and vegetation types and their properties. Contextual image classification methods include reflectance patterns captured by neighbouring pixels in their analysis. Such an approach is especially successful where open type of vegetation or ecosystems occur. Pixel per pixel classifiers often fail to recognise these spatially highly variable patterns of vegetation. The underlying theory of investigating spatial patterns for a landscape or for vegetation patterns is that the computed patterns express a kind of 'natural characteristic' of a spatially contiguous set of pixels for a certain ecosystem type and although the individual pixel values may vary, the pattern is distinctive. Local variability in an image can be described by computing statistics of pixels e.g. coefficient of variance in a kernel of a specific site, autocovariance or by fractals or wavelets (Atkinson & Lewis, 2000; Berberoglu *et al.*, 20000; Chica-Olma & Arbarca-Hernandez, 2000; Wallace *et al.*, 2000; De Carvalho, 2000; Hornstra & Lemmens, 1999; De Jong & Burrough, 1994; Xia, 1993; De Cola, 1989). Open types of Mediterranean vegetation often display such patterns. Open type of vegetation often occur where land degradation plays a role. Degradation processes cause a deterioration of the natural vegetation and hence the current vegetation reflects ecological conditions related to water and nutrient availability and human influence. This is especially true in European Mediterranean regions where past and present human activities have strongly affected the vegetation. These activities have caused the development of semi-arid landscapes with vegetation ranging from maquis, garrigue and rangelands to badlands (Grenon & Batisse, 1989; Le Houérou, 1992, 1981). These units are characterised by the specific spatial patterns i.e. bare soil patches and rock outcrops next to green and senescent shrubs. Broadband sensors such as Landsat TM and SPOT-HRV fail to identify these individual objects where high resolution, spectroscopical images have a definite advantage over these conventional systems as they are capable of separating bare soil surfaces from senescent vegetation. Next, the spatial patterns are so variable and complex that pixel-by-pixel classifiers do not recognise adjacent pixels as belonging to the same vegetation class. Therefore, algorithms are required that account for the entire spectral shape and for the spectral information in neighbouring pixels.

Hornstra & De Jong (2000) developed such an approach that combines spatial and spectral information in two steps and used it to classify open Mediterranean land cover types. The method is referred to as the Spatial & Spectral Classification method: SSC. Spatial information is first extracted by segmentation: spectral homogeneous regions in the image are identified on the basis of similarity of the entire spectral shape (72 DAIS bands) from visible to shortwave infrared. A type of epsilon band is computed along the spectral signature of each pixel. All neighbouring pixels having a spectral signature within this epsilon band are assigned as belonging to the homogeneous areas. The user can interactively adjust the width of the epsilon band. This first step of dividing the image into homogeneous areas is a necessary one and the results are much better if, as in this case, the entire spectral shape of the solar spectrum can be used compared to using a few broad spectral bands as available from Landsat TM. After this segmentation step, the homogeneous image parts are classified using conventional 'per-pixel' methods. The heterogeneous parts are classified using a combination of spectral and contextual information by a modified 'minimum distance to mean' algorithm. The pixel is assigned to class c using the following function (Hornstra & De Jong, 2000)

$$C = \min(\alpha \cdot d^{spectral} + (1 - \alpha) \cdot d^{spatial}) \qquad (2)$$

where d $^{\text{spectral}}$ is the Euclidean spectral distance to training polygons, d $^{\text{spatial}}$ is the distance to neighboring classified pixels and $\alpha$ is the adjustable calibration factor. The spatial distance d$^{\text{spatial}}$ is calculated by defining a growing ring of pixels surrounding the unclassified pixel. In the first run this ring exists of the 8 pixels that directly surround the pixel, in the second run the ring exist of the 16 pixels that are surrounding the first ring etc. Within every ring, the number of previously classified pixels is counted for every class and is used to determine the weight of d$^{\text{spatial}}$ The value of d$^{\text{spectral}}$ is the conventional Euclidean distance as it is calculated in the minimum distance to mean classification algorithm (Schowengerdt, 1997). The factor $\alpha$ is a ratio factor for describing the impact between the spatial and the spectral weight during pixel class assignment.

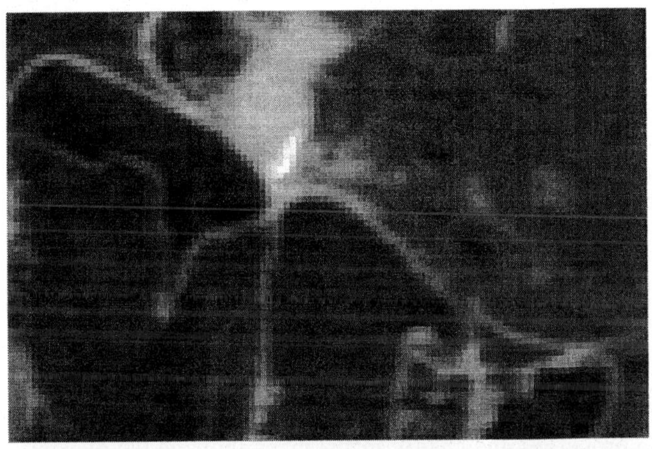

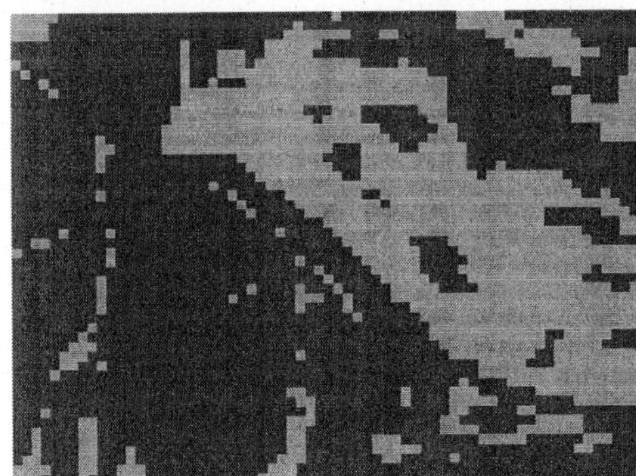
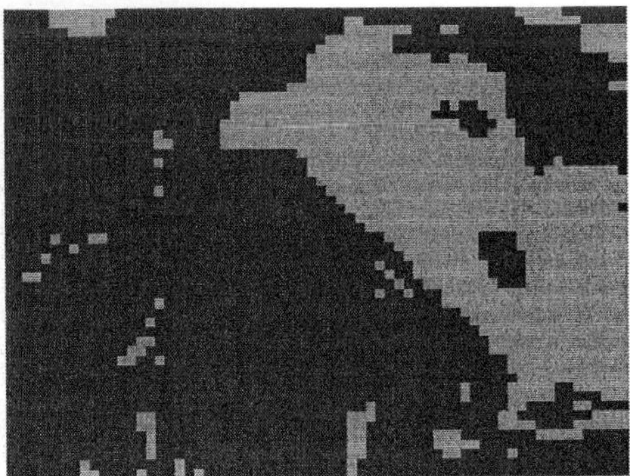

*Figure 10.* Image classification results of a DAIS7915 image of open Mediterranean vegetation types. Upper left: original DAIS7915 image with various open vegetation types. Lower left: the result of a conventional minimum distance to mean classifier. Lower right: the result of the spectral & spatial classifier SSC. For description and legend, see text.

The results of applying this algorithm to the DAIS7915 images with 72 spectral bands and 6 by 6 meter pixels are promising. In parts of the image where maximum likelihood and traditional minimum distance to mean classifiers fail to identify the open Mediterranean vegetation types, the SSC method successfully identifies classes such as open Maquis forests and open shrub ecosystems such as Garrigues. Figure 10 illustrates these results for a small test site in France. In dark green a Garrigue shrub vegetation is shown, in light green a Maquis type of forest and in between these two mapping units a bare, unpaved road and a bare agricultural lot can be seen. The left image is the result of a conventional minimum distance to mean classification (MDTM) while the right image is the result of the SSC-approach. The SSC method delineates the two vegetation cover types much better than the MDTM approach. Overall classification accuracy improves with 18% but this improvement is of course a function of the presence or absence of respectively open and closed vegetation types. The mixed pixels along the road clearly visible in the left image are also much better classified by the SSC method. The availability of many spectral bands as from an imaging spectrometer is a prerequisite to successfully delineate homogeneous image section using the previously described epsilon band.


## 8    Conclusions

Imaging spectrometry is slowly developing from an experimental research instrument towards an operational valuable mapping tool. Methods and software for pre-processing of the spectroscopical data are nowadays available enabling the user to transfer their data into well-calibrated reflectance images. At the same time we see fast developments taking place in our efforts to model and simulate landscape processes such as land degradation, soil erosion but also crop growth and surface runoff. We are not only capable of assessing much better the spatial distribution of landscape elements but we have also the tools and the computing power to simulate the dynamics of various landscape processes. These developments put large claims on the availability of geographical data: maps of soil types, maps of soil properties, maps of land cover, maps of crops and vegetation properties and change maps: maps showing the dynamics of land cover and other processes. The various remote sensing techniques must and will play an important role in making these data available.

The merit of imaging spectrometry is that it allows to
- To distinguish much better than by means of  the conventional remote sensing methods the various vegetation cover types of green, healthy vegetation, senescent, de-colouring vegetation, dead vegetation and bare soil. The spatial distribution of these four variables plays an important role in degradation models.
- Assess the spatial distribution of soil properties where the vegetative cover is sparse or absent.
- The full spectral signature captured by imaging spectrometers has benefits for contextual approaches of image analysis methods. It enables to detect much better the subtle changes between various soil surfaces, vegetation structural components and complex mixture of surfaces. So far these contextual approaches have not been given much attention by researchers although their results are encouringing especially with the availability of high spatial resolution and high spectral resolution                          image                          availability.

*CHAPTER 4*

# FIELD AND IMAGING SPECTROMETRY FOR IDENTIFICATION AND MAPPING OF EXPANSIVE SOILS

Sabine CHABRILLAT[α], Alexander F.H. GOETZ[α,β], Harold W. OLSEN[χ] & Lisa KROSLEY[χ]

[α] Center for the Study of Earth from Space/CIRES, Boulder, USA

[β] Department of Geological Sciences, University of Colorado, Boulder, USA

[χ] Colorado School of Mines, Golden, USA

## 1    Introduction

### 1.1    NATURE OF EXPANSIVE SOILS

Swelling soils are a major geologic hazard, and expansive clays and clay-shales cause extensive damage world-wide every year. Current high signal-to-noise ratio imaging spectrometers provide high spectral resolution remote sensing data that have the potential for new applications in terrestrial geology and environmental hazard. The problems associated with expansive soils are not widely appreciated outside the areas of their occurrence. The amount of damage caused by expansive soils is alarming. It has been estimated that the damage to buildings, roads, and other structures founded on expansive soils exceeds two billion dollars annually. The origin of expansive soils is related to a complex combination of conditions and processes that result in the formation of clay minerals having a particular chemical make-up which, when in contact with water, will expand. Variations on the conditions and processes mal also form other clay minerals, most of which are non-expansive. The conditions or processes which determine the clay mineralogy include composition of the parent material and degree of physical and chemical weathering to which the materials are subjected.

The problem of expansive soil is widespread throughout the five continents. Donaldson (1969) summarized the distribution of reported instances of expansive soils around the world in 1969. The countries in which expansive soils have been reported are as follows:

| | |
|---|---|
| Argentina | Iran |
| Australia | Mexico |
| Burma | Morocco |

87

*F.D. van der Meer and S.M. de Jong (eds.), Imaging Spectrometry,* 87–109.
© 2006 *Springer. Printed in the Netherlands.*

| | |
|---|---|
| Canada | Rhodesia |
| Cuba | South Africa |
| Ethiopia | Spain |
| Ghana | Turkey |
| India | U.S.A. |
| Israel | Venezuela |

Since that time, many other nations have reported significant findings of expansive soils. Among them, some of the most prominent are China, Sudan, Cyprus, Jordan and Saudi Arabia. Figure 1 shows the distribution around the world of reported instances of heaving (Chen, 1988). It indicates that the potentially expansive soils are confined to the semi-arid regions of the tropical and temperate climate zones. Expansive soils are in abundance where the annual evapotranspiration exceeds the precipitation. This follows the theory that in semi-arid zones, the lack of leaching has aided the formation of montmorillonite. Montmorillonite is the clay mineral that has the greatest swelling potential and is the culprit for most swelling soil damages. Potentially expansive soils can be found almost anywhere in the world. In the underdeveloped nations, many of the expansive soil problems may not have been recognized. It is to be expected that more expansive soil regions will be discovered each year as the amount of construction increases.

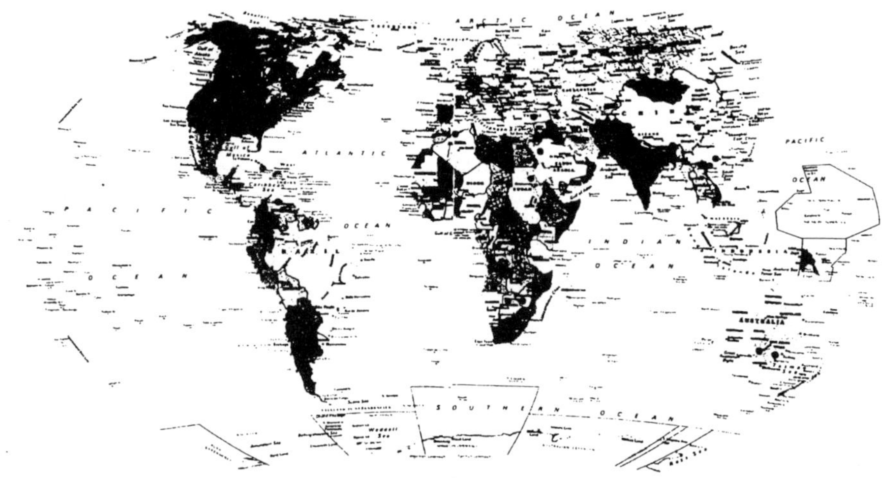

*Figure 1.* Distribution of reported instances of heaving. Reprinted from Chen (1988)

Jones & Holtz (1973) reported in the American Society of Civil Engineers (ASCE) the estimated damage attributed to expansive soil movement per construction category, e.g., single-family homes, commercial buildings, etc. Wiggins (1978) in a study on "Building losses from natural hazards" listed six major natural hazards: earthquake, landslide, expansive soils, hurricane, tornado and flood. He pointed out that expansive soils tie with hurricane wind/storm surge for second place among America's most destructive natural hazards in terms of dollar losses to buildings. Its destructive impact is currently surpassed only by that of riverine flood. This study also pointed out that few people have ever heard of expansive soils. Even fewer realize the amplitude of the damage they cause. In most cases, it takes a professional soil engineer to confirm the existence of such damage and evaluate its probable behavior. According to the study, it was projected that by the year 2000, losses due to expansive soils can exceed 4.5 billion annually as shown on Figure 2.

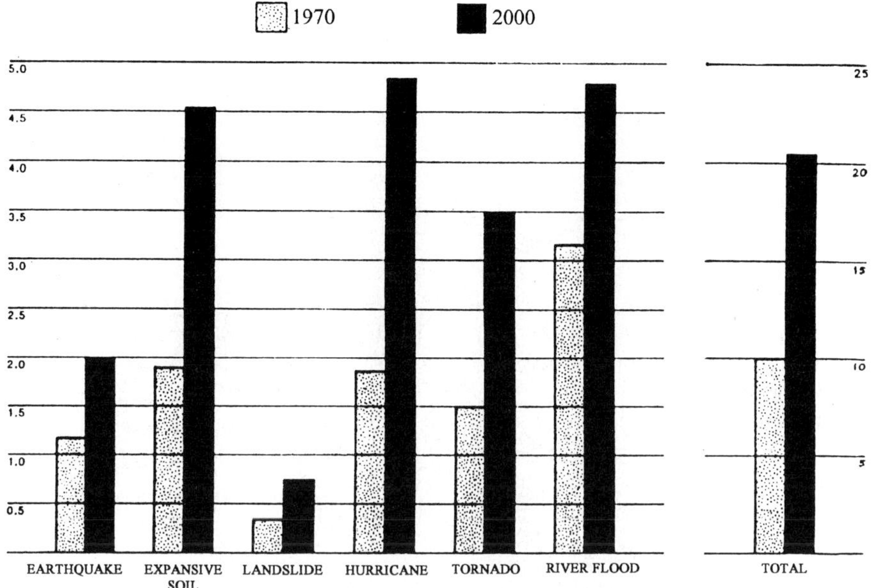


*Figure 2.* Average total annual building losses under 1970 and year 2000 conditions (1978 dollars in billions). (Wiggins, 1978)

The three most important groups of clay mineral are smectite, illite and kaolinite, which are crystalline hydrous aluminosilicates. Montmorillonite, the best-known member of the smectite group and its high alumina end-member, is the clay mineral that presents most of the expansive soil problems. The literature is confusing because the group name used to be montmorillonite. Swelling clays are also commonly referred to as bentonitic soils by laymen. Bentonite is a clay composed primarily of montmorillonite which has been formed by the chemical weathering of volcanic ash. Absorption of water by clays leads to expansion. Montmorillonite may swell up to 15 times its original volume when water is present, as shown in Figure 3, and shrink when dessicated. Natural soils, which contain considerably less than 100% smectite, can swell to more than 1 ½ times their original volume (Jones & Holtz, 1973). The swelling is caused by the chemical attraction of water. Layers of water molecules are incorporated between the flat, submicroscopic clay plates as more water is made available to the clays, and adjacent clay plates are pushed farther apart. This swelling occurs throughout the mass of soil being wetted, and causes increased volume and high swell pressures within the mass. Not all montmorillonites expand at the same potential and therein lies the crux of the problem.

The magnitude of expansion depends on several factors. From the mineralogical standpoint, the magnitude of expansion depends upon the kind and amount of clay minerals present in the soils, their exchangeable ions --sodium-rich montmorillonites swell more than calcium-rich montmorillonites--, the electrolyte content of the aquaeous phase, and the internal structure. Kaolinites have a 1:1 (silica:aluminium) layer structure. Their lack of charge unbalance and isomorphous substitutions give them low to non-swelling potential. Illites and smectites have a 2:1 (silica:aluminium) similar lattice structure (Hall, 1987). An illite differs from a montmorillonite structure as some of the silican atoms are replaced by aluminum, and in addition potassium ions are present between the tetrahedral sheet (silica) and adjacent crystals.

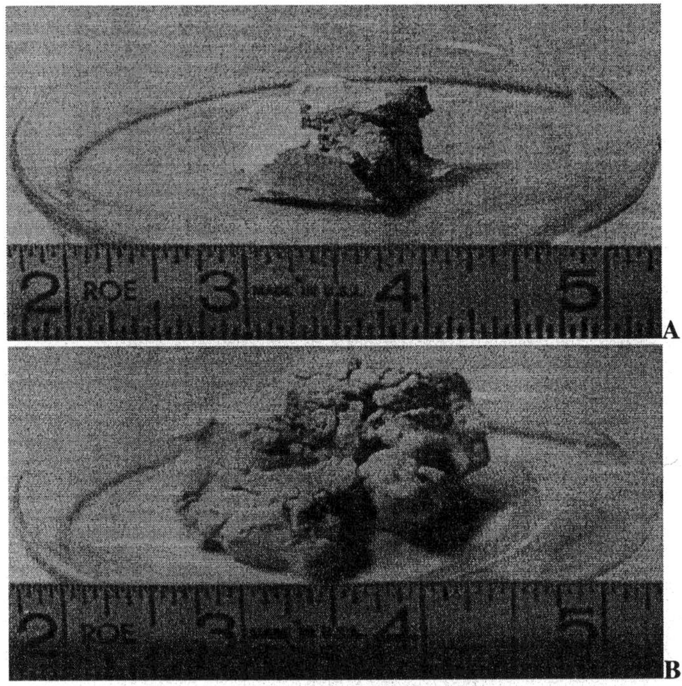


*Figure 3.* (a) Dry montmorillonite placed in a water filled dish. (b) Sample after 9 hours of standing in water. The sample has more than doubled in volume.

The illites structure is charged due to ionic substitutions, and contains interlayer non-exchangeable cations that compensate for the charge deficit. Illites have higher swell potential than kaolinite. Smectites experience a variety of isomorphous subsitutions in the octahedral layers (alumina) leading to a negative charge, which explains their readiness to exchange cations and water layers. Figure 4 shows the model of a layer of montmorillonite (Low, 1973). Previous research in understanding the nature of expansive soils has been separated into two categories. The first, from mainly theoretical approaches, involves soil mineralogy, structure, modification, effective stress, soil suction, osmotic pressure etc... little known to engineers. The second category is concerned with the field performance of expansive soils, design criteria and construction precautions for structures founded on expansive soil.

Practical approaches of combating the swelling soils problem are mostly undertaken by soils or geotechnical engineers; Therefore they must offer practical and economical solutions to their clients so that the structure will be free from damaging foundation movement. Unfortunately, present day knowledge of expansive soils has not reached a stage at which rational solutions can be assigned to the problem. It is difficult for the public to understand why the engineers cannot offer easy solutions. When the first crack appears in a structure, a lawsuit is threatened. The problem with identification and remediation of expansive soils is that engineering soil tests are expensive and therefore are used sparingly. Major difficulties with the engineering practices include the costs of subsurface exploration and soil testing to determine beforehand the likelihood of the presence of expanding soils on undeveloped land. Standard engineering tests for determination of swelling potential are time-consuming and expensive. And on the other hand, the high costs and complexity of the current methodologies, such as x-ray diffraction (XRD) analyses, to determine mineralogy, exchangeable cations, and pore-fluid chemistry of clays make it also used sparingly. Reflectance spectrometry is of interest because it has the potential of rapid identification of constituent minerals in soils. If the swelling clays and their potential can be identified spectroscopically, then new tools can be developed for identification and evaluation of expansive soils, both at specific sites such as outcrops or trench walls with field instruments, and on a regional basis with remote sensing data.

*Figure 4.* Model of a layer of montmorillonite. (Low, 1973)

## 1.2    SPECTROSCOPIC INDICATORS OF CLAY MINERALS

Previous research on synthetic monomineralic clay-water systems show that expansive potential varies directly with the amount of smectite, or montmorillonite, present in a soil (Seed *et al.*, 1962), and inversely with both the valence of the exchangeable cations and the concentration of electrolytes in the pore fluid (Mitchell, 1993). Spectroscopic indicators are of great interest because reflectance spectra show absorption bands in the visible and near-infrared (NIR) region which permit identification of smectites in natural soils. Laboratory reflectance spectra of kaolinite, illite and montmorillonite from the U.S. Geological Survey (USGS) spectral library (Clark *et al.*, 1990) are shown in Figure 5. The strong absorption bands at 1400 and 1900 nm are due to bound water, typical of montmorillonite, while strong OH bands at 1400 and 2200 nm are typical of kaolinite (Hunt and Salisbury, 1970). The presence of a 1900 nm band is key to the swelling potential as it indicates molecular water in the sample (2:1 layer clay, montmorillonite or illite), whereas its absence but the presence of a 1400 nm band indicates that only OH is present (kaolinite, 1:1 layer clay) (Kariuki, 1999).

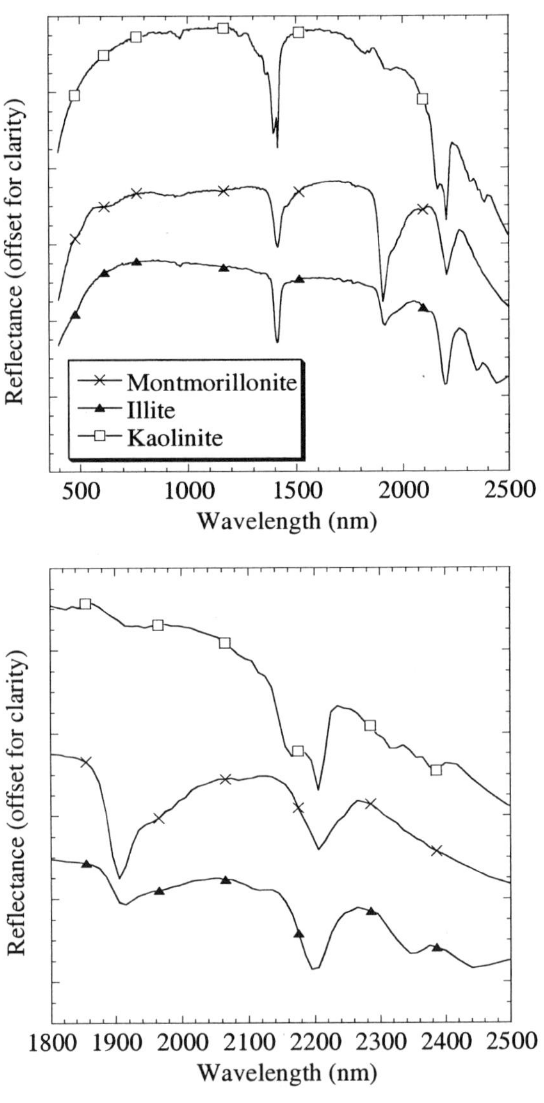



*Figure 5.* Reflectance spectra from the USGS spectral library (Clark *et al.*, 1990) of three clays: Montmorillonite (SWy-1), Illite (IL101) and Kaolinite (Kga-1); a- in the visible-NIR, and b- in 1800-2500 nm.

The Al-OH overtone-combination absorption band around 2200-2300 nm is very diagnostic of all clays. The Short Wave Infrared (SWIR) part of the spectrum, 1400-2500 nm, and particularly the SWIR 2, 1900-2500 nm, are the most useful part of the spectrum for the identification of clays. Kaolinite shows doublet absorption bands at 1400 nm and 2200 nm that are characteristic of the mineral. The doublet bands consist of a broad absorption with a sharper band at slightly longer wavelength: 1395 and 1415 nm for the OH stretch overtones and 2163 and 2208.5 nm for the combination Al-OH bend plus OH stretch. Montmorillonite and illite have similar absorption bands that do not show fine structure. Montmorillonite shows a single well-defined absorption band around 2200 nm, due to the combination Al-OH bend and OH stretch. Illite, unlike montmorillonite, shows additional absorption bands at 2340 and 2450 nm. The one at 2450 nm is poorly defined. Reflectance spectra of montmorillonites exhibit shifts in the 2200 nm absorption band, from ~2204 to 2214 nm as shown in Figure 6. There is an argument in the literature whether this shift could be correlated with an increasing Ca content, or decreasing Na content. If it is so, the position of the 2200 nm absorption band could be useful for the evaluation of swelling potential, because sodium-rich montmorillonites swell more than calcium-rich montmorillonites. The shift in the 2200 nm absorption band has been attributed to interaction with the interlayer cation field (e.g., Ryskin, 1974) although this interpretation have been disputed (Farmer, 1974). Clark *et al.* (1990) observed an apparent trend among a few samples that shows a shift in the 2200 nm band to a longward position with increasing Ca content. On the other hand, Post & Noble (1993) showed that decreasing $Al_2O_3$ content in smectite minerals shifts the 2200 nm absorption feature toward longer wavelengths.

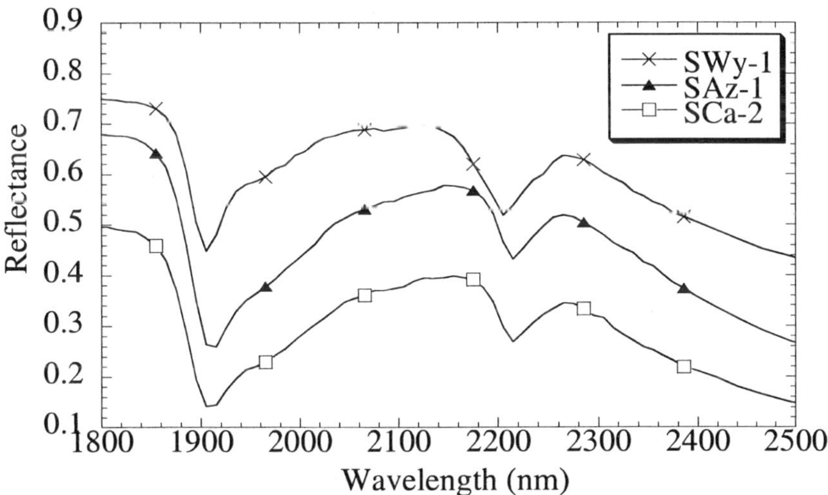


*Figure 6.* Reflectance spectra from the USGS spectral library: Na-rich montmorillonite (SWy-1), Ca-rich montmorillonite (SAz-1), and Otay bentonite, (SCa-2, Na-Ca montmorillonite).

Imaging spectrometry, also named hyperspectral imagery, has the potential of laboratory spectrometry at the remote sensing scale as it acquires simultaneously spatial images in more than 100 narrow, contiguous spectral bands (Goetz *et al.*, 1985). We thus are provided with a continuous spectrum for each picture element (pixel), which allows us to spectrally identify minerals, rocks or soils at the surface. The usefulness of imaging spectrometry for geological applications has been demonstrated in many cases. The Airborne Visible/Infrared Imaging Spectrometer (AVIRIS), developed by NASA at the Jet Propulsion Laboratory, U.S.A. (e.g., Vane *et al.*, 1993) has been flying since 1987. AVIRIS spectra have been used to identify individual minerals such as alunite, calcite, dolomite, kaolinite and muscovite, amonium minerals, crustal/mantle rocks, and enabled also the mapping of units and subunits of carbonate, clay and iron oxide minerals in sedimentary rocks (e.g., Kruse, 1988, Baugh *et al.*,

1998, Mustard & Pieters, 1987, Clark *et al.*, 1992). The best results have been obtained from areas where the rock surfaces are well exposed, little weathered, and the minerals occur in largely pure concentrations. The main limits to succesful mapping are the mixing at the sub-pixel scale, and the degree of exposure of the target at the surface. Those limits have been pushed forward over the years by the improvement of the signal-to-noise ratio (SNR) of the instruments, along with the development of new processing algorithms such as spectral mixture analysis (Adams *et al.*, 1993) or matched filtering (Boardman, 1998). The SNR of AVIRIS has evolved from less than ~50 in 1987 to more than 1000 in the visible and ~400 around 2200 nm in 1998 (Green *et al.*, 1998). As a result, hyperspectral imagery began in the recent years to be used as a new tool for environmental hazard applications, in studies such as the Environmental Protection Agency (EPA) abandoned mine lands imaging spectroscopy project (e.g., Hauff *et al.*, 1999, Rockwell *et al.*, 1999), or our case study on swelling soils.

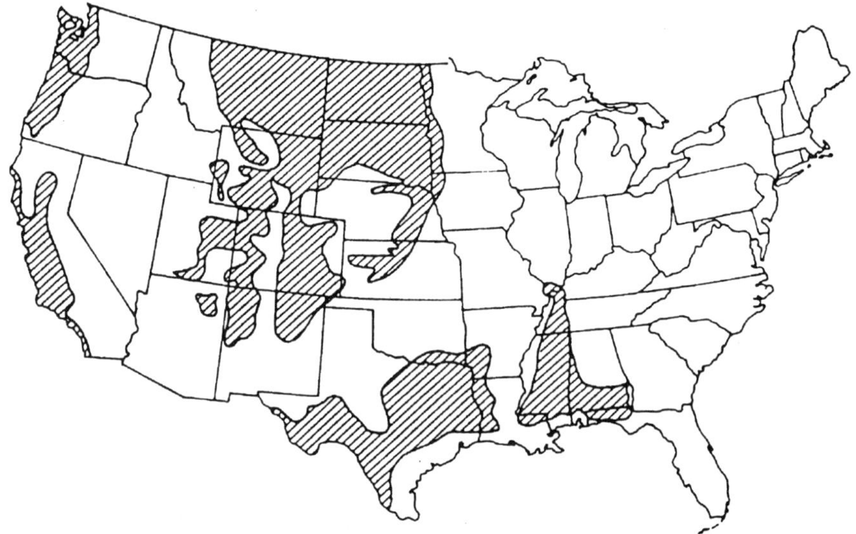

*Figure 7.* General abundance of montmorillonite in near outcrop bedrock formations in the United States. (Modified from Tourtelot, 1973)

The swelling soils project described here aims at determining the feasibility of using reflectance spectroscopy to identify and characterize the expansive clays and clay-shales along the Colorado Front Range Urban Corridor, in the field and laboratory, and with remote sensing data. It is a joint project between the University of Colorado (CSES/CIRES) in Boulder and the Colorado School of Mines in Golden. The study meant to first establish the spectral properties of swelling soils linked to their swelling potential in the field and laboratory. Then we investigated the possibilities of hyperspectral imagery in tackling the difficult operational problem of identification and mapping of expansive clays. 35% of all construction in the Front Range swelling soil corridor is affected and the remediation costs are enormous. The airborne hyperspectral sensors used in this study, the AVIRIS and the HyMap, an advanced high-quality scanner from HyVista corporation (Sydney, Australia), represent the highest performance hyperspectral systems available today, and are the precursors of what could be available tomorrow from space.

*Figure 8.* Geologic setting of the study area, steeply-dipping layers. (Gill *et al.*, 1996)


## 2    Field and laboratory analyses


### 2.1    EXPANSIVE SOILS IN THE FRONT RANGE URBAN CORRIDOR IN COLORADO

In the United States, from the Gulf of Mexico to the Canadian border and from Nebraska to the Pacific Coast, the abundance of montmorillonite is common in both clays and claystone shales. The reported problem locations are mostly in the regionally abundant montmorillonite areas indicated in Figure 7. The states that experience various degrees of expansive soil problems are listed as follows (from Chen, 1988):

| | |
|---|---|
| Severe: | Colorado, Texas, Wyoming |
| Moderate: | California, Utah, Nebraska, South Dakota, Mississippi |
| Mild: | Oregon, Montana, Arizona, Oklahoma, Kansas, Alabama |

Colorado is one of the states with the greatest swelling soil problem. One example of the problem is the current situation in the Denver Metropolitan Area. This area is underlain by Cretaceous clay-shales, including the Pierre Shale, that also underlie vast areas of the United States and Canada, including the states: Dakotas, Montana, Wyoming and Colorado. The sedimentary bedrock strata are generally flat-lying, except near the foothills of the Rocky Mountains, where they have been uplifted into steeply-dipping strata (Noe & Dodson, 1995, 1999), as illustrated in Figure 8. Such conditions underlie several cities along the Front Range Urban Corridor in Colorado that extends from Pueblo to the Wyoming border (Figure 9). This corridor is expected to provide land for much of the housing to be built for a substantial increase in Colorado's population in the next 25 years.

A high incidence of damage to roads, utilities and lightly loaded residential and commercial structures has occurred along Colorado Front Range Piedmont where steeply dipping beds of expansive claystone bedrock are encountered at shallow depth. Thompson (1992) showed that damage to residential development was substantially more extensive and severe in this outcrop belt than in areas underlain by relatively flat-lying claystone. Clay minerals dominate the constituents of the shale ranging from 50 to 75% (Schultz, 1978). The deformations causing severe damage in the Pierre Shale outcrop belt are linear ground heaves that exhibit up to 0.6 m of post-construction vertical displacement and are traceable for hundreds of meters across damaged subdivisions. Gill *et al.* (1996) showed that there is a strong correlation of these heaves with thin bentonite beds (15-30 cm thick), initially consisting of volcanic ash and altered to nearly pure smectite, that occur between much thicker silty-claystone strata consisting predominately of mixed-layer illite-smectite. The heave features are created by the differential swell between the bentonite beds that have very high swelling potential, and the surrounding mixed-layer illite-smectite soils that have a lower swelling potential. To differentiate the geological hazard responsible for this type of deformation from the general expansive soils problem, it has been termed "heaving bedrock" (Noe, 1997).

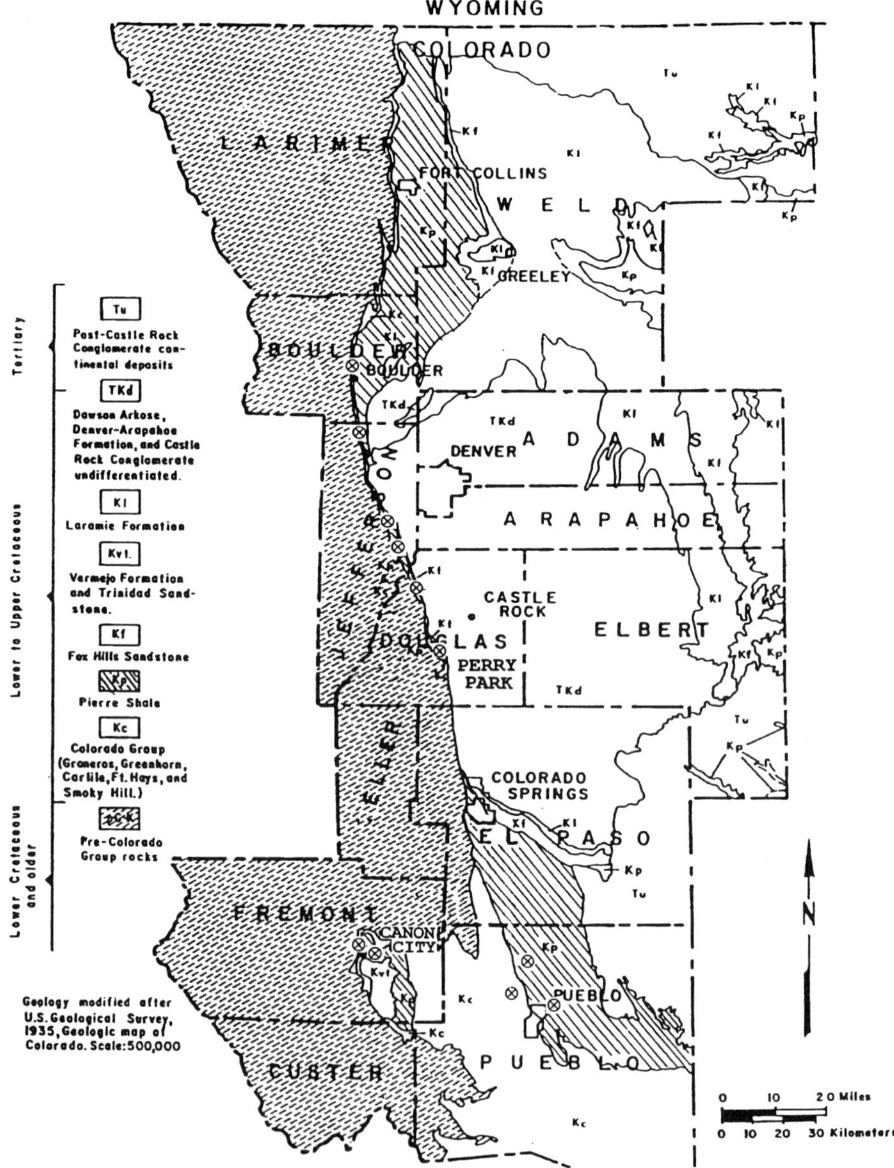


*Figure 9.* Generalized geologic map of the Colorado Front Range Urban Corridor (Modified from Hart, 1974). Dashed areas locate Pierre Shale formation. Field sampling locations are marked with a cross.

For more than four decades, the geotechnical engineering practice for mitigating damage from expansive soils in the Denver metropolitan area has involved locating potentially expansive soils in the subsurface with vertical drill holes sampled at 1.5 m intervals, characterizing the expansive potential of relatively undisturbed samples in terms of laboratory geotechnical index tests and supporting structures on drilled piers anchored in bedrock. Generally, the "percent swell" determined from the conventional swell test is used as an empirical index of the severity of potential differential vertical deformations at the site (Thompson, 1997), and as a basis for predicting when drilled piers, structural floors, and other stabilization measures are needed. In addition to the already noted problem of complexity and costs of the laboratory analyses for determination of swelling potential and mineralogy of the samples, fundamental deficiencies in this practice in the case of steeply dipping strata have become increasingly recognized. The probability of locating highly expansive beds in steeply dipping strata with vertical drill holes is low, and the design guidelines developed from experience in flat-lying strata do not take into account geologic factors in the steeply-dipping outcrop belt of the Pierre Shale. The severity of the problem in an area south of Denver that is underlain by generally flat-lying strata was reflected in two class action lawsuits in 1995 and 1996. In both cases, the developer in the area was sued successfully by the affected homeowners. The 1996 lawsuit involved 957 affected homeowners and was the biggest expansive-soil lawsuit ever filed in Colorado, and one of the largest in the U.S.A. (Denver Post, 1996).

## 2.2    FIELD SAMPLING AND LABORATORY ANALYSES

182 undisturbed samples were collected from 23 different locations along the Colorado Front Range Urban Corridor from Boulder to Pueblo. The sampling locations are shown in Figure 9. 145 samples were taken at 30-50 cm depth beneath the ground surface. 37 samples were taken at 10-20 cm depth beneath the floor of an excavation for a large building, approximately 2.5 m beneath the ground surface. The latter samples were taken along a 30-meter profile perpendicular to the strike of the underlying steeply dipping strata. Sites were selected in Cretaceous shales, most of which are uplifted into steeply dipping strata near the Foothills of the Rocky Mountains, and that are well-known to be hazardous to residential and commercial developments in this region. The sites were all located on Upper Cretaceous sediments including the Graneros, Benton, Smoky Hill, and Pierre Shales. The great majority of samples obtained were clay shales, consisting predominantly of illite-smectite interstratification with thin bentonite beds that consisted of almost pure smectite in clay fraction, and are classified as having moderate to very high swell potential.

On each sample, three types of analyses were performed: (a) X-ray diffraction (XRD) patterns were acquired on oriented samples to determine the mineralogy and mixed-layer illite-smectite expandability; (b) Reflectance spectra were acquired in the visible-NIR region; (c) Geotechnical index tests including the Atterberg limits, grain size analysis, and one-dimensional swell test were run to determine the swelling potential. For a more complete description of the methodology used for the mineralogical and geotechnical analyses, we refer to, respectively, Chabrillat *et al.* (2000b) and Olsen *et al.* (2000).

Semi-quantitative mineralogy was derived from XRD patterns of oriented preparations using a millipore filter peel transfer onto glass slides (Drever, 1973). The calculations were carried out following the equations of Schultz (1964), using the correction factors determined by Johnson (1997). The precision of the mineralogy determined with Schultz's method is generally ±10%. When possible, mixed-layer illite-smectite (I-S) expandabilities were determined by the °2θ method of Moore and Reynolds (1997). In most of the typical shales that we had, peak interferences from other phases made this determination difficult. We used the ordering of the sample and the height of the glycolated 17 Å peak above its low angle saddle to allow an educated guess of the I-S expandability. The given expandability values are usually within ±5%, but because of discrete illite interference, this error could be somewhat higher. The percent total smectite is then calculated as smectite phase + smectite in the I-S layers, and the percent total illite as illite phase + illite in I-S. The mineralogical composition determined with XRD is expressed in clay fraction, i.e. reported to 100% clays. We calculate the mineralogical composition related to the whole sample by correcting it with the percent clay of the sample determined by particle size analysis.

Bi-directional reflectance spectra were acquired in the laboratory with an ASD FieldSpec® FR spectroradiometer reproducing solar geometry conditions, off-nadir tungsten illumination. Measurements were made in the room adjacent to the humidity room where the undisturbed field samples were stored. The fiber optic was stabilized on a support. The scanning time of the instrument is 100 ms, and we averaged 10 s worth of data to increase the signal-to-noise ratio.

The geotechnical properties measured included air-dried moisture content, Atterberg limits (liquid limit, plastic limit, plasticity index) (ASTM D4318-93), particle size distribution (ASTM D 422-63), and percent swell upon inundation. They also included McKeen (1992)'s suction parameters such as the slope of the suction-water content relation dh/dw, and the slope of the suction-volume change relation dh/dV. Based on these properties, swelling potentials were then estimated using the classification systems developed by McKeen (1992), Chen (1988), Seed (1962), and the Soils Task Force of the Home Builders Association of Metropolitan Denver (Soils Task Force, 1996, Thompson, 1997). The expansive soils classification system proposed by McKeen (1992) defines categories of swelling potential as followed. Category 1: Very high swelling potential, special cases, i.e. almost pure montmorillonite. Category 2: High swelling potential. Category 3: Moderate swelling potential. Category 4: Low swelling potential. Category 5 soils are non-expansive. McKeen test was chosen as it is rapid and inexpensive, then being the only one that could be run on a sufficient number of samples to develop a significant correlation with clay mineralogy. The others swell tests were performed only on a selected suite of samples.

## 2.3    RESULTS: RELATIONSHIPS BETWEEN REFLECTANCE, MINERALOGY AND SWELLING POTENTIAL

The mineralogy in clay fraction of all the field samples ranges from pure smectite (bentonite beds) to varying degrees of illite-smectite interstratification to very low smectite (a few percent). Illite content can be up to ~80%. Kaolinite content is usually less than 15% in ~90% of the samples, but can be up to >50% in a few samples. There is almost no chlorite, less than a few percent, in the samples. The clay mineral of particular interest is smectite, because it is the one with the greatest swelling potential. From our analyses, smectite is generally the most abundant mineral in clay fraction among the field samples. The percent smectite in the clay and the percent clay vary from nearly zero to 100% (Olsen *et al.*, 2000, Chabrillat *et al.*, 2000b). The development of reflectance spectroscopy for identifying and characterizing expansive soils required the correlation of clay mineralogy with swelling potential indices determined from engineering tests, and the expansion of existing correlations of reflectance spectra with clay mineralogy determined by x-ray diffraction. Figure 10 presents the suction vs. water content data obtained on the 182 samples together with McKeen's swell potential categories, and summarizes our results. In this figure we added the schematic mineralogical composition, determined with XRD, that was associated with the samples in each category, along with characteristic spectral signatures around 2200 nm of field samples in each swelling potential category. This figure shows that the samples are widely distributed throughout all McKeen's categories, and that this is correlated with mineralogical changes. Olsen *et al.* (2000) showed that the percent smectite related to the whole sample provides a useful index of the swelling potential concept defined by Seed (correlation factor ~0.75), and correlates reasonably well with the swelling potential indices developed by Seed, Chen and McKeen (correlation factor ~0.6 to 0.75).

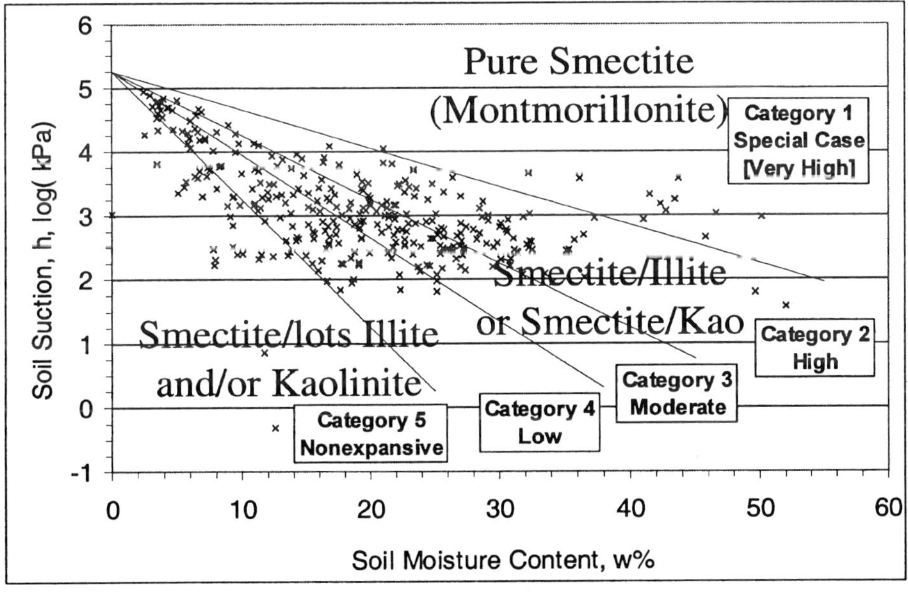


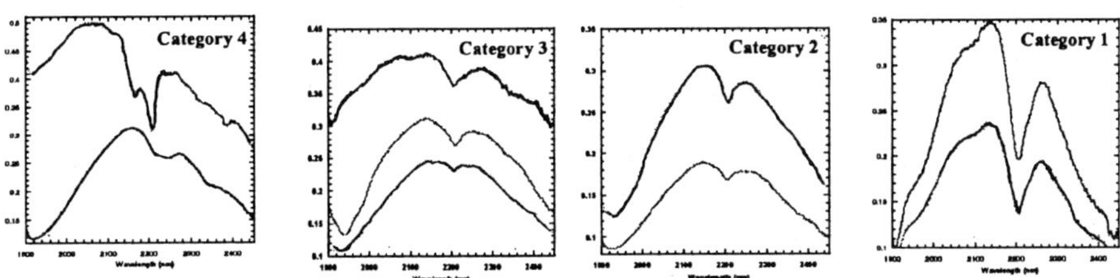


*Figure 10.* Suction vs. water content data for 182 samples together with McKeen (1992) classification categories. The schematic mineralogical composition associated with samples is added for each category. The spectra shown on the bottom of the figure are associated with field samples in each of McKeen category of swelling potential.

It was shown that the clays smectite, illite and kaolinite, present in mixed proportions throughout the field samples, can be identified spectroscopically based on their characteristic absorptions in the region 1900-2500 nm (Chabrillat *et al.*, 2000b). We can see in Figure 10 that, from spectral reflectance, we are able to discriminate among pure smectite (highest swelling potential, category 1) and mixed layer illite-smectite samples. Smectite has a characteristic single deep absorption band centered ~2200 nm. We found that the accurate location of the 2200 nm band cannot be used as an indicator of swelling potential, as a relationship with the discrimination among calcium and sodium-montmorillonite could not be observed. In mixed illite-smectite samples, the higher the smectite content, the higher the swelling potential. The absorption band ~2340 nm provides a qualitative measure of the illite content. Kaolinite is detected spectrally if above 10% in clay fraction in the sample with the appearance of an asymmetric absorption band at 2200 nm, due to the characteristic doublet feature of kaolinite. A significant amount of kaolinite (>10-15%), and/or a significant amount of illite, is indicative of low swelling potential. In the case of the spectral identification of kaolinite in the sample, but with a high smectite content (deep water absorption band at ~1900 nm), then the swell potential is not as low as the kaolinite content might suggest, according to the laboratory measurements (Chabrillat *et al.*, 1997, 1999). Spectroscopic identifications of expansive clay field samples are well correlated with mineralogical x-ray diffraction analyses and geotechnical engineering tests. Similar observations were also made by Kariuki (1999), in an equivalent study on other types of expansive soils.

It appears that the spectral determination of quantitative mineralogy and swell potential from the reflectance signature would be possible by the use of statistical multivariate analyses. The studies are currently limited as there were very few engineering-geologic studies that include clay mineralogy (XRD) and swell behavior (suction) testing and analysis. Nevertheless, using multi-dimensional statistical analyses throughout our sample set, we showed that laboratory spectral reflectance correlates reasonably well with the percent smectite of the samples in clay fraction, and also with swelling potential indices (Chabrillat *et al.*, 2000b). The correlation coefficients are of the order of 0.8 with the percent smectite, and 0.6-0.7 with swelling potential indices. However such models are presently limited, and more detailed studies are needed before they could be used in practice.


## 3    Hyperspectral image analysis


### 3.1    EXPANSIVE CLAYS IN COLORADO: REMOTE SENSING CONSIDERATIONS

Good exposures of expansive clays along the Front Range Urban Corridor are limited in size and sparse. Most of the region is covered more or less with vegetation, forests or grass, especially in the northern and Denver areas. However, south of Colorado Springs and around Pueblo and Canon City, exposures are more common and covered with desert-type vegetation. Another problem for the identification and mapping of expansive soils in Colorado from afar, is that outcrops are of variable mineralogy at a small scale (< 1m) near the mountain front where steeply dipping beds occur.

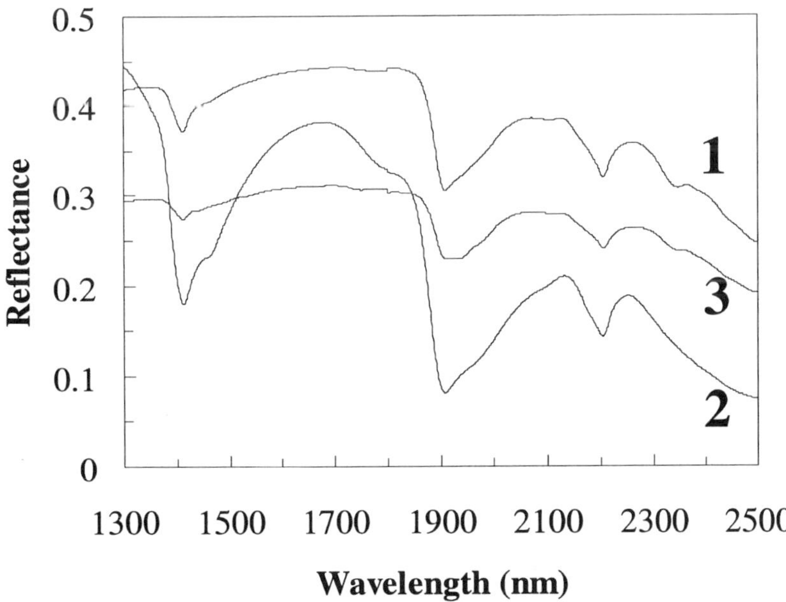


*Figure 11.* Reflectance spectra acquired in the field over a vertical exposure of steeply dipping beds of expansive clays west of Denver (Soda Lake Rd). The location of the spectra over the exposure is marked in the associated picture: 1- Grey shale, 2- Bentonite bed, 3- Grey shale.

Reflectance spectra were acquired in the field with the same portable spectrometer than the one used for laboratory studies. They were acquired over different exposures of expansive clays, associated with sampling locations and/or remote sensing data. In general, we found that smectite, illite and kaolinite can be identified spectroscopically in the field also, and the spectral discrimination between pure smectite and illite-smectite layers is possible, as shown in Figure 11. The grey shales are associated, from their spectral signature, with mixed-layer illite-smectite, and this is confirmed by the XRD analyses of field samples from these layers, unlike the bentonite beds which are associated with a characteristic smectite spectral feature. Note that for those field measurements over a vertical exposure, we used the ASD high-intensity reflectance probe that has a quartz halogen light source. In general, the main difficulties that we observed, associated with field measurements, were as followed: (a) The 1400 and 1900 nm water absorption bands are masked by the atmospheric water vapor absorption bands. The 1900 nm band, discriminative for clays, cannot be used for measurements with the sun as the light source; (b) The signal-to-noise ratio of the measurements is lower in the field than in the laboratory, because we cannot average as many data. Some shales weather to dark grey soils, and their spectra show very low in reflectance, < 10% around 2200 nm. The low reflectance level makes it difficult to identify the clay spectral feature in those shales (Chabrillat & Goetz, 1999, Chabrillat *et al.*, 2000c). Nevertheless, using the spectral region 2000-2500 nm alone, it is possible to discriminate among smectite, illite and kaolinite, of variable swelling potential. Then, provided that the clays are exposed at the surface and that exposures are good enough vs. the sensors capabilities, imaging spectroscopy should be able to help in the detection and mapping of expansive clays.

## 3.2    IMAGES ACQUISITION AND ANALYSIS

AVIRIS and HyMap hyperspectral images were acquired along the Colorado Front Range Urban Corridor in the Fall 1998 and 1999 (Table 1).

TABLE 1: AVIRIS and HyMAP hyperspectral data acquired along the Colorado Front Range Urban Corridor.

| Date | Instrument | Overflight altitude | 4    Sites |
|---|---|---|---|
| September 30th 1997 | AVIRIS | 20 km | Colorado Springs* |
| September 10th 1998 | AVIRIS | 20 km | Pueblo, Golden, Boulder |
| October    21st  1998 | AVIRIS | ~4 km | Boulder, Golden |
| September 16th 1999 | HyMap | Low altitude | Boulder, Golden, Perry Park |
| September 30th 1999 | AVIRIS | 20 km | Canon City, Pueblo, Perry Park, Golden, Boulder |
| October    1st   1999 | HyMap | Low altitude | Pueblo |

*USGS data

The overflights were requested for late summer when the amount of green vegetation cover is at a near minimum. AVIRIS measures upwelling radiance through 224 contiguous spectral channels at 10 nm intervals, from 400 nm to 2500 nm. AVIRIS spectral images are traditionally (high altitude mode) acquired from the Q-bay of a NASA ER-2 aircraft from an altitude of 20 km, except when it was mounted on-board a NOAA Twin-Otter aircraft flying at an altitude of 3.8 km for the first low altitude experiment in October 1998 (Green *et al.*, 1999b). A detailed description of AVIRIS charateristics, operations and applications can be found in Green *et al.* (1998). In 1998, AVIRIS was radiometrically calibrated at better than 96 percent accuracy (Green *et al.*, 1999a). Six flight lines were acquired from high altitude both in 1998 and 1999 for a total each year of ~300 km length. Because of the mean ~1.7 km surface elevation in the area covered, the AVIRIS pixel size is ~17m x 17m, and the swath width is ~10 km. During the low altitude overflight in 1998, AVIRIS produced a pixel size of ~2m x 2m and a swath width of ~1.2 km. AVIRIS data were distributed by JPL a few months after the overflights, in radiance, radiometrically calibrated, georectified in 1998 and non-georectified in 1999.

The HyMap images were acquired during the AIG/HyVista HyMap campaign. HyMap has 126 spectral channels, 13 to 17 nm wide (17 nm around 2200 nm), spanning the 440-2470 nm spectral region. HyMap was flown aboard a Cessna 402, at variable altitudes to achieve a 4m x 4m pixel size, thus providing images of ~2.3 km width. The SNR at 2200 nm is approximately 600 to 1. HyMap data were distributed by AIG two weeks after the overflights, both in radiance and reflectance, radiometrically calibrated and georectified.

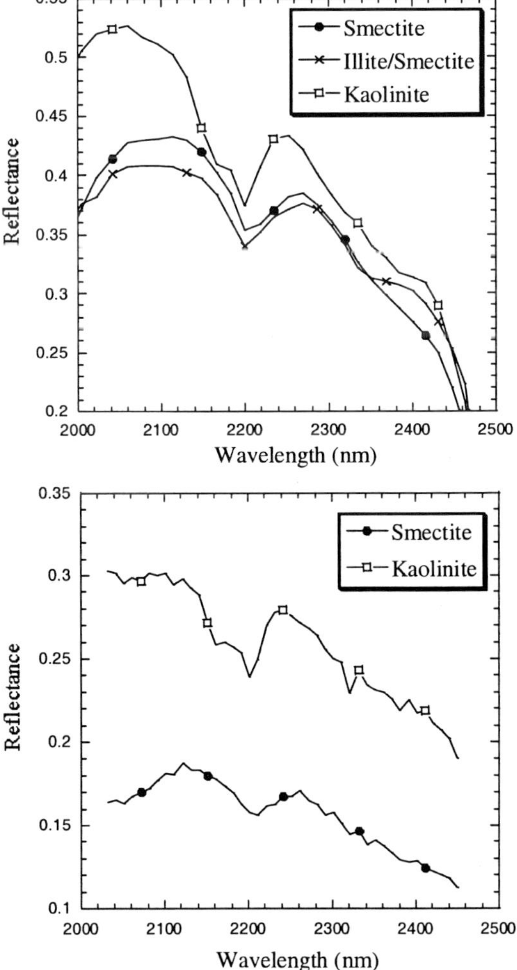


*Figure 12.* Spectra of the clay endmembers extracted from Perry Park scene from HyMap (top) and AVIRIS (bottom).

To derive surface reflectance from the radiance data, they must be corrected for solar irradiance and atmospheric effects such as two-way transmission multiple scattering and path radiance. We chose to use radiative transfer algorithms to derive reflectance from radiance. AVIRIS and HyMap data were atmospherically corrected using the ATREM algorithm (Gao *et al.*, 1993), version 3.0 and 3.1. To correct the artifacts introduced by the modeled solar irradiance, two different approaches were used, ground correction and a semi-empirical approach. AVIRIS data were ground-corrected using spectral surface measurements acquired on the day and at the time of the overflights from an almost bare field calibration target within the scene. For HyMap data, we used the reflectance data distributed by AIG, where after the ATREM correction, they used the Empirical Flat Field Optimal Reflectance Transformation (EFFORT; Boardman & Huntington, 1997). This semi-empirical approach makes the assumption that the residual spectral artifacts in the derived reflectance from ATREM are only associated with systematic errors in the model or in the spectral and radiometric calibration of the sensor.

To detect and map expansive soils along the Colorado Front Range, we applied the following processing methodology, on both AVIRIS and HyMap images, and using only the 2000-2450 nm spectral region (43 spectral channels for AVIRIS, 28 for HyMap) as characteristic clay spectral signatures are in this region:

1.  A minimum Noise Fraction (MNF) transform was performed,
2.  The Pixel Purity Index (PPI) method was used to find the most spectrally pure (extreme) pixels,
3.  The extreme pixels were visualized in the data cloud from the ten first MNF bands. One of the clusters was identified as having pixels showing characterictic spectral clay features around 2200 nm, such as kaolinite, smectite, and possibly illite,
4.  Extreme clay pixels (endmembers) were extracted from the "clay" cluster,
5.  A Mixture Tuned Matched Filtering (MTMF) method was eventually used to locate and map expansive clays outcrops. This last analysis provides an abundance map for each selected endmember associated with an "infeasibility" image to prevent detecting "false positives" in the original matched filtering algorithm (Boardman, 1998).

The advantage of this methodology, an adapted version of the AIG standardized processing methods for hyperspectral data analysis (Kruse, 1997), is that no a priori knowledge of the area is required. The difficulty associated with this type of methodology is that it needs a manual extraction of the endmembers. The use of a MNF and PPI beforehand allows to automatically subset the images spectrally and spatially, thus producing a workable dataset size to ease the determination of the image endmembers. Several other mapping tools could be used, such as the Spectral Angle Mapper algorithm (SAM). The MTMF algorithm seemed to be the most adapted to our problem, as it is well-known that the expansive soil exposures are at sub-pixel size for the airborne survey. We needed an extraction of the clay signature, and the MTMF does not need knowledge of the background signal in the pixel resulting from the mixture with other elements. Also, this algorithm is the only one that could provide in such conditions semi-quantitative information, as we expect that the clay signatures are a mixture of the three clay minerals kaolinite, montmorillonite and illite.

## 4.1    MAPPING RESULTS

We will focus here on two examples, Perry Park and the Pueblo area, for the following reasons. Perry Park is representative of the situation in northern Colorado with small sparse outcrop and heavy vegetation cover, and is close to the mountain front with variable mineralogy and many steeply dipping bentonite beds. The Pueblo area is representative of the situation in southern Colorado where vegetation is sparse and soils well exposed, and is far from the Front Range with horizontal illite-smectite layers and sometimes kaolinite in the soils. Both regions, Perry Park and Pueblo, have been overflown in 1999 with both sensors at a few weeks or even a few days interval. The results from other typical regional mapping associated, e.g., with the Colorado Springs area can be found in Chabrillat *et al.* (1999, 2000c).

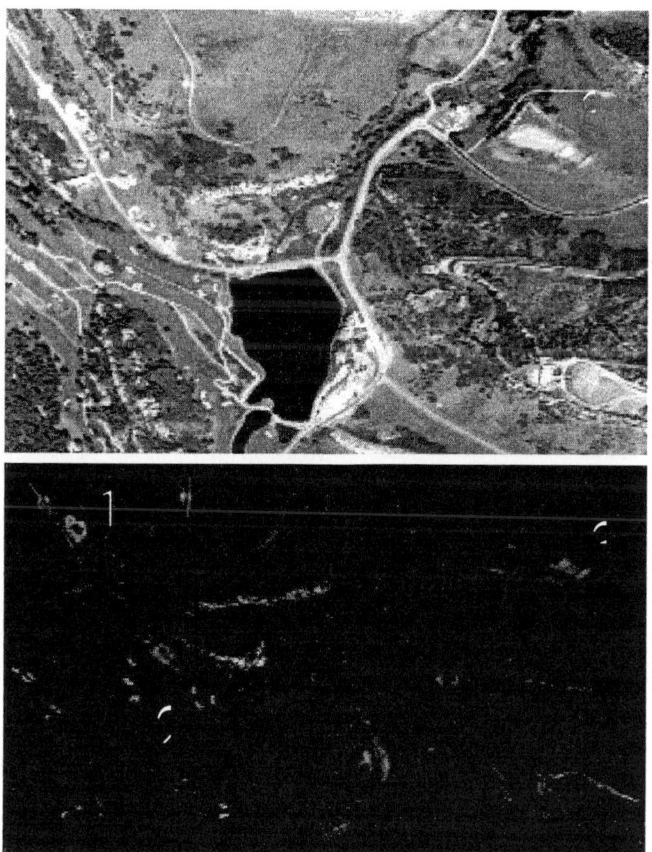

*Figure 13.* True color HyMap image over Perry Park (top) and associated clay map created from the abundance maps obtained with the Mixture Tuned Matched Filtering (MTMF) algorithm for the three endmembers shown in figure 12 (bottom).

Figure 12 presents the spectra associated with the endmembers selected in the clay cluster, for the HyMap and the AVIRIS scene over Perry Park area. Although the HyMap scanner has less spectral resolution than AVIRIS, the characteristic clay spectral features were detected. There is no illite-smectite endmember in the AVIRIS data. Figure 13 shows the true-color HyMap image, and the clay map created from the processing explained previously. The clay map shows the abundance maps obtained with the MTMF algorithm for the three endmembers smectite, kaolinite, and illite-smectite, represented as one single Red-Green-Blue (RGB) image. Each abundance map from each endmember ranges from black to a pure color, red, green or blue. Black means that the spectrum of the associated pixel does not match with the endmember one, or abundance=0%. A pure color means a perfect match between the spectrum of the pixel and the endmember, abundance=100%. Red is associated with the endmember smectite, green with the endmember kaolinite, and blue with the endmember illite-smectite. In the resulting RGB image which is a combination of the three abundance maps for the three endmembers, a black pixel means no match between the spectrum of the pixel with any endmembers.

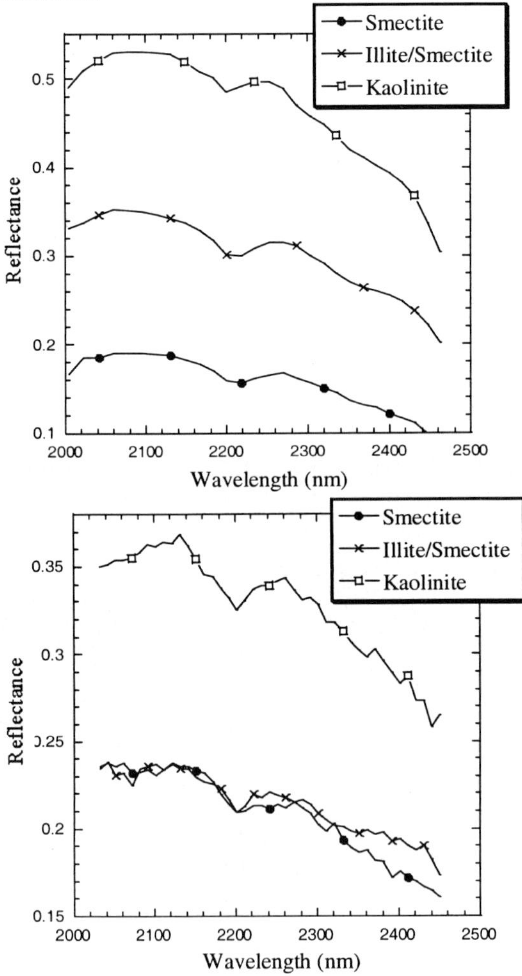



*Figure 14.* Spectra of the clay endmembers extracted from Pueblo scenes from HyMap (top) and AVIRIS (bottom).

The image presented in Figure 13b represents a map of clay materials exposed. The scattered nature of clay occurences (black pixels) is interpreted to be caused by a predominance of non-clay materials at the ground surface such as either vegetation cover or non-expansive, Quaternary-age alluvial deposits. Figure 13a shows for reference the part of the HyMap image associated with the area shown in the clay maps. By zooming on a very small area, ~1 km in extent, we can see in the abundance maps that clay materials exposed in this area are very sparse and  small in size. This agrees with our field knowledge (Chabrillat *et al.*, 2000a). Also, many areas rich in smectite were detected. This also agrees with the fact that Perry Park is known for its smectite-rich claystone exposures, along with the occurrence in the mixed illite-smectite layers of several thin bentonite beds of pure or almost pure smectite.

Several points of interest are located in Figure 13 with the labels 1, 2 and 3. The label 1 is associated with the location of the smectite endmember. This area in the field was associated with soils exposed around a house under construction (circular shape). The field samples collected around this house showed a pure smectite feature in their spectra, confirming the remote sensing identification. The same analysis on the Perry Park AVIRIS scene showed the same results, and the same location for the smectite endmember. The label 2 is associated with the location of the illite-smectite endmember in the HyMap data. This area in the field was associated with sandtraps for a golf course, a few meters in extent, surrounded by green grass. These sandtraps, and also the illite-smectite endmember, were not detected in the AVIRIS data because of the larger pixel size. They are already at a sub-pixel size for the HyMap data. The label 3 is associated with three Granero shale outcrops. Those outcrops are well exposed with almost no grass, but are small in size (10-30 m) and have a very dark grey color (low in reflectance). AVIRIS results are similar to HyMap results. They both detected one of those outcrops, brighter  than the two others, but the two darker ones were only partially detected. In this case, the detection limits in terms of SNR have been reached. This problem was already identified in Chabrillat & Goetz (1999) with AVIRIS 1998 low and high altitude data.

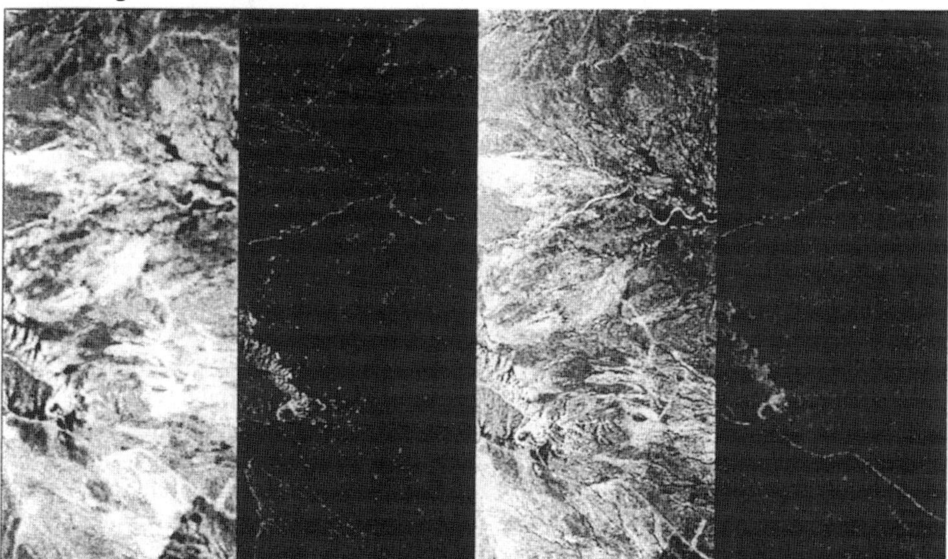

*Figure 15.* From left to right: True color AVIRIS image over Pueblo, associated clay map, HyMAP image, associated clay map for the three endmembers shown in figure 14a and 14b.

Figure 14 presents the extreme clay spectra extracted from the clay cluster for the HyMap and the AVIRIS scenes in the Pueblo area. We notice that the clay features in those extreme spectra are not as deep as, for example, the ones that could be extracted from an area more mineralogically heterogeneous like Perry Park. Indeed, the Pueblo clay endmembers are not as extreme (less "pure") than in others areas, and this is not the result of a spatial resolution effect. This is consistent with the known geological setting around Pueblo. The soils exposed contain illite-smectite layers with few bentonite beds and, in general, they are less smectite-rich than in areas against the Front Range, and less variable in mineralogy. The reduced compositional variability comes from the fact that the beds are more or less horizontal ($<30°$) as opposed to the near-vertical beds along the Front Range that expose different compositions. Figure 15 shows the HyMap and AVIRIS true-color images and the associated clay maps obtained with the MTMF algorithm from the three extreme clay spectra shown in figure 14. The colorscale coding is the same as in figure 13. The clay maps show that well-exposed illite-smectite layers, and kaolinite-rich soils are found in an area associated with the top of a large Mesa North-East of Pueblo city (Chabrillat *et al.*, 2000a). This is consistent with our field observations, i.e. laboratory spectra and analyses of field samples and field spectra from the Baculite Mesa outcrop, directly below the large Mesa. All those samples showed a mixed content in smectite and illite, with some kaolinite content. The HyMap and the AVIRIS clay maps show roughly the same results, with small differences associated with the larger pixel size.


## 5    Conclusions

The swelling soils project aimed at determining the feasibility of using reflectance spectroscopy to identify and characterize the expansive clays and clay-shales along the Colorado Front Range Urban Corridor, in the field and laboratory, and  with remote sensing data. Such studies had never been done previously. The project is still going on, and the results shown here are preliminary. However, it has been shown that there is a linear relationship between the weight percent smectite in expansive soils and the swelling potential indices developed by Seed (1962), Chen (1988) and McKeen (1992). The correlation coefficients are in the range of 0.6 to 0.75. Near-infrared (NIR) reflectance spectroscopy of swelling soil field samples shows that it is possible to discriminate among pure smectite and mixed illite-smectite layers samples. The clays smectite, illite, and kaolinite, present in different proportions in most of the field samples collected in the study area, can be identified spectroscopically throughout the suite of samples thanks to their characteristic $H_2O$ and OH-bound absorptions in the 1900-2500 nm region. Spectroscopic identifications are well correlated with mineralogical x-ray diffraction analyses and geotechnical engineering tests.

The analysis of the hyperspectral images shows that, using matched filtering algorithms, exposures of expansive clays can be detected among the other components in the images, and in the presence of significant vegetation cover. Maps of exposed clay material are produced, and among those exposures, spectral discrimination and identification of variable clay mineralogy (kaolinite, illite-smectite, smectite) related to variable swelling potential is possible. Field checks have shown that the maps of clay type derived from the imagery are accurate. The AVIRIS and HyMap sensors used in this study provide high signal-to-noise ratio data. High quality hyperspectral imagery data could be used for detection and mapping of expansive clays, at specific sites with high spatial resolution data and on a regional basis with less spatial resolution.

From additional analyses, it appears that the spectral determination of quantitative mineralogy and swell potential from the reflectance signature would be possible only by the use of statistical multivariate analyses. The studies are currently limited as there have been very few engineering-geologic studies that include clay mineralogy (XRD) and swell behavior (suction) testing and analysis. However, hyperspectral image analysis has shown to be capable of detecting and mapping expansive clays, even at a ~20m pixel size, in the presence of significant vegetation cover.

# CHAPTER 5

# IMAGING SPECTROMETRY AND VEGETATION SCIENCE

Lalit KUMAR [α], Karin SCHMIDT [α], Steve DURY [β] &
Andrew SKIDMORE [α]

[α] International Institute for Aerospace Survey and Earth Sciences (ITC)
Division Agriculture, Conservation and Environment, Enschede, The
Netherlands.

[β] Australian National University (ANU), Department of Forestry, Canberra,
Australia.

## 1 Introduction

Remote sensing is increasingly used for measurements required for accurate
determinations of the landscape and the state of agricultural and forested land. With the
deployment of early broadband sensors there was a lot of enthusiasm as data, which
was previously not feasible to obtain, was now regularly available for large areas of the
earth. For the first time vegetation mapping could be undertaken on a large (coarse)
scale and the data updated regularly. However, new technologies have shown that
while data obtained from broadband sensors have been useful in many respects, they
also have their limitations. Because of their limited number of channels and wide
bandwidths, a lot of the data about plant reflectance is lost due to averaging.
In remote sensing, the radiation values recorded by the sensor, after atmospheric
correction, are a function (f) of the location (x), time (t), wavelength ($\lambda$) and viewing
geometry ($\theta$) of the ground element, i.e.

$$R = f(x, t, \lambda, \theta) \tag{1}$$

From this it follows that sufficient change in at least one of the variables of x, t, $\lambda$ or $\theta$
has to occur and cause a detectable change in R before remote sensing can be utilized
to provide information about the environment. If we consider the wavelength factor and
look at radiation reflected from vegetation, we see that different amounts of radiation
are reflected at different wavelengths. Most natural objects have characteristic features
in the spectral signature that distinguishes them from others and many of these
characteristic features occur in very narrow wavelength regions. Hence to 'sense' these
narrow features the use of narrow band sensors is required. Broadband sensors average
the reflectance over a wide range and so the narrow spectral features are lost or masked
by other stronger features surrounding them. Thus broadband sensors, such as Landsat
MSS and TM, cannot resolve narrow diagnostic features as their spectral bandwidths
are 100-200nm wide and they are also not contiguous. For this reason hyperspectral



*F.D. van der Meer and S.M. de Jong (eds.), Imaging Spectrometry,* 111–155.
© 2006 *Springer. Printed in the Netherlands.*

remote sensing is a strong alternative for significant advancement in the understanding of the earth and its environment.

Figure 1 shows the typical spectral reflectance data of vegetation as collected by a spectrometer (GER IRIS) and a simulated model of what the resulting signal would be from Landsat TM. It is seen that considerable data are lost and that spectral fine features characteristic of vegetation is no longer discernible from broadband sensors. While the few bands of information from broadband sensors may still be useful for vegetation discrimination, it would be difficult to discriminate between species that have very similar reflectance. With the use of high spectral data, considerably more information is available which can assist in such tasks. Also high spectral data could be used for species identification, as opposed to only discrimination as is the case with broadband data.

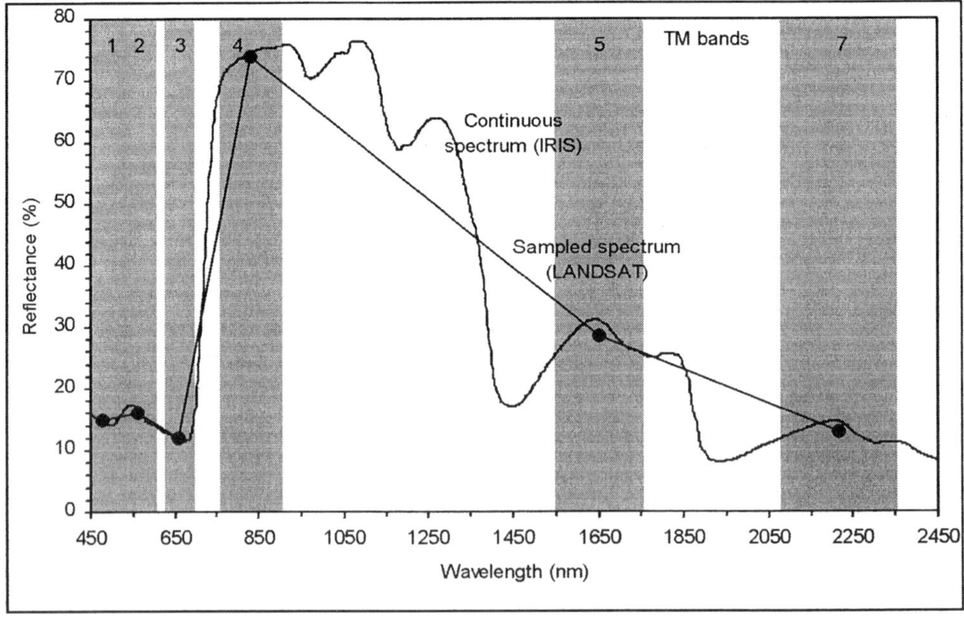


*Figure 1*. Data content of broadband (Landsat) and narrow-band (IRIS) sensors.

There is a strong optimism that with the arrival of the new generation of imaging spectrometers, significantly higher quality data will be available. Already data from imaging spectrometers have been found to yield higher quality information about vegetation health and cover than those obtained from broadband sensors (Collins *et al.* 1983, Curran *et al.* 1992, Peñuelas *et al.* 1993, Carter 1994, Carter *et al.* 1996, Kraft *et al.* 1996). However, as with any new technology, it takes time to develop new methods and algorithms to fully utilize the large information content of the hundreds of channels on imaging spectrometers. This chapter looks at a number of issues relating to the use of hyperspectral remote sensing in vegetation studies. Our discussion concentrates on the fundamental factors affecting vegetation reflectance, reflectance characteristics of vegetation in different wavelength regions (visible, shortwave-infrared, near-infrared and mid-infrared), leaf reflectance models, chemicals present in leaves and their characteristic signatures, data processing techniques and applications of hyperspectral remote sensing in vegetation studies.


## 2    Spectroscopy versus spectrometry

Spectroscopy is the branch of physics concerned with the production, transmission, measurement, and interpretation of electromagnetic spectra (Swain and Davis 1978). The basis of spectroscopy are the spectral properties of materials. The material properties which specify the response of the material to sinusoidal component waves at every frequency of wavelength are called spectral properties (Suits 1983). Radiation spectral properties are characteristic of materials, composed of molecules in gaseous, liquid, or solid form. Radiation generated from molecules is characteristic to the type of molecule, but the number of possible characteristic frequencies increases rapidly as the number of atoms in the molecule increases. Even simple molecules, such as carbon dioxide and water, have a great number of possible different characteristic frequencies. Due to the close proximity of atoms in liquids and solids, the force fields of the neighbouring atoms distort the electron orbits of each other so that a large number of different characteristic frequencies may be generated. The spectral lines of a solid and liquid are so closely spaced together in wavelength that they generally overlap and cannot be resolved at all (Suits 1983).

The term spectrometry was originally derived from spectro-photometry; spectrometry is the measure of photons as a function of wavelength. In remote sensing applications, the actual mathematical analysis of waveforms is rarely performed by numerical calculations from spectral components, instead, direct measurements are made with spectrographic devices such as airborne scanner or laboratory spectrometers (Suits 1983). Where a spectrometer is an optical instrument used to measure the apparent electromagnetic radiation emanating from a target in one or more fixed wavelength bands or sequentially through a range of wavelengths (Swain and Davis 1978).

Radiation reaching the surface of a material, is subject to one or more of several processes. It may be reflected (diffuse, specular), transmitted (with refraction), or absorbed compliant to the law of conservation of energy. This interaction of the radiation with the surface is dependent on both the properties of the radiation as well as the properties of the material (Suits 1983). The radiance spectrum of the sun differs from day to day and place to place, partly due to the changes in the sun's surface itself, but on the earth's surface mainly due to the changes in the atmosphere's composition, the atmospheric gases absorbing part of the sun's radiation at particular wavelengths. As a result of this variation of incidence spectral radiation, as well as changes in the surface characteristics, the radiance spectrum reflected from the surface of the earth also varies. Therefore, to be able to compare spectral measurements of surfaces acquired on different days and in different illumination conditions, a measure is required that is independent of illumination variation, or can somehow be calibrated for changing illumination. This measure is reflectance.

Reflectance is characteristic of any illuminated surface and is assumed to be independent of the amount of radiation reaching a surface. Some surfaces reflect most of the incident radiation and some reflect little. Reflection is defined to be the return of radiation by a surface without change of frequency of the monochromatic components of which the radiation is composed. The measure of reflectance is the dimensionless ratio of radiation reflected from a surface, to the radiation hitting that surface. Reflectance can be expressed as a percentage. However, a practical problem remains in how to measure reflectance, especially for remote sensing data. Different radiometric quantities can be ratioed to produce all kinds of reflectance measures with a multitude of names, which are sometimes confused by authors, (not even considering things like fluorescense1) who usually use many different symbols for the same quantities.

Generally the reflectance of a surface can be measured in three different ways: the bi-hemispherical reflectance which is measured with an integrating sphere mostly in a laboratory, the hemispherical-conical reflectance factor which is measured with a flat Lambertian reference panel and most commonly utilized in remote sensing research conducted in the field, and the bi-directional distribution function which is a theoretical concept and can not be measured in practice (Kimes and Kirchner 1982).

Reflectance spectra of natural surfaces are sensitive to specific chemical bonds in materials, whether solid, liquid or gas. Variation in the composition of materials causes shifts in the position and shape of absorption bands in the spectrum. Thus, the large variety of materials typically encountered in the real world, can result in complex spectral signatures that are sometimes difficult to interpret. Spectrometers are used in laboratories, in the field, in aircraft (looking both down at the Earth, and up into space), and on satellites. In fact, the human eye is a simple reflectance spectrometer: we can look at a surface and see color. But the eye can also process a spatial component, or image, for the materials within our field of view. Similarly, an image can be constructed using an imaging spectrometer that records the spectra for contiguous image pixels. Imaging spectroscopy is a new technique for obtaining a spectrum in each position of a large array of spatial positions so that any one spectral wavelength can be used to make a recognizable image. The image might be of a rock in the laboratory, a field study site from an aircraft, or a whole planet from a spacecraft or Earth-based telescope.

Every pixel in the image has a spectrum, so we are able to spatially map the presence and abundance of chemical bonds. By analyzing the spectral features, and thus specific chemical bonds in materials, one can map where those bonds occur, and thus map materials. Imaging spectroscopy has many names in the remote sensing community, including imaging spectrometry, hyperspectral, and ultraspectral imaging.

It should be noted that spectroscopy covers a wide range of topics such as the study of atomic and molecular structure and spectra (including near infrared spectroscopy (NIRS) described in section 2.6), plasma diagnostics, instrument development, and identification of materials. Spectroscopy is used in an incredibly diverse range of disciplines, for example physics and chemistry, astronomy, food science, ecology, genetic engineering, and material science. The term imaging spectroscopy occupies a small portion of this very broad field.

---

[1]Fluorescence is the conversion of absorbed radiant energy at one wavelength to emitted radiation at a longer wavelength without first converting the absorbed energy into thermal energy (Suits, 1983). Chlorophyll absorbs visible radiation and converts most of that energy into chemical energy. However, some of that absorbed energy is converted to infrared fluorescence. But the fluorescent radiation is quite small compared the the reflected infrared radiation from green leavesand thus never taken into account in passive vegetation remote sensing.

Recent successes analysing field spectrometer measurements and hyperspectral imagery using aircraft scanners, have been achieved. It should be noted that operational satellite systems have been, and are planned to be, launched (e.g., ENVISAT by ESA) The hyperspectral scanners are passive; in other words they receive radiation reflected from the earth's surface. At an altitude of 500 km, a space borne sensor receives approximately 10000 times less radiation than an aircraft at 5 km. Therefore, much less signal (information) is received by the satellite compared with the aircraft, with a lower signal to noise ratio. The signal-to-noise ratio is of prime importance in hyperspectral remote sensing. With aircraft sensors, it is possible to have small pixels (say 1-5 m) as well as narrow wavelength bands (usually around 10 nm). The advantage of aircraft scanners over satellite systems is that spatial spectral resolution is high, and these sensors are therefore ideally suited for detailed local surveys.

## 3 Fundamental factors affecting vegetation reflectance

### 3.1 LEAF OPTICAL PROPERTIES

Incoming solar radiation is the primary source of energy for the numerous biological processes taking place in plants. The interactions between solar radiation and plants can be divided into three broad categories: thermal effects, photosynthetic effects, and photomorphogenic effects of radiation. Over 70 per cent of incoming solar radiation absorbed by plants is converted into heat and used for maintaining plant temperature and for transpiration (thermal effects) (Slatyer, 1967; Gates, 1965 & 1968).

Photosynthetically active radiation (PAR) (~28% of absorbed energy) is used in photosynthesis and for conversion to high-energy organic compounds. The optical properties of leaves in the PAR region depend on a number of factors, such as conditions of radiation, species, leaf thickness, leaf surface structure, chlorophyll and carotenoid content of leaves, dry matter content per leaf unit area and leaf internal structure (Ross, 1981).

Solar radiation impinging on the leaf surface is either reflected, absorbed or transmitted. The nature and amounts of reflection, absorption and transmission depend on the wavelength of radiation, angle of incidence, surface roughness and the differences in the optical properties and biochemical contents of the leaves. The first contact of incoming radiation is with the leaf surface, which consists of the cuticle and epidermal layers. Some leaves also have wax and/or leaf hairs over the cuticle and these alter the amounts of light reflected or absorbed by the leaf. The amount of light which is absorbed or transmitted within leaves depends on its wavelength as leaf pigments absorb selectively. This is discussed in the sections that follow.

### 3.2 REFLECTANCE (400-700 NM.)

The visible region of the vegetation reflectance spectrum is characterised by low reflectance and transmittance due to strong absorptions by foliar pigments. For example, chlorophyll pigments absorb violet-blue and red light for photosynthesis. Green light is not absorbed for photosynthesis, hence most plants appear green. The reflectance spectrum of green vegetation shows absorption peaks around 420, 490 and

660nm. Most of these are caused by strong absorptions of chlorophyll. Table 1 gives the main pigments found in higher plants and their absorption maxima. Even though all pigments show strong absorption of blue light, those of chlorophyll tend to dominate the spectral response as there are 5 to 10 times as much chlorophyll as carotenoid pigments (Belward, 1991).

The energy absorbed from the visible part of the spectrum is used to synthesise the organic compounds which plants need for maintenance and growth. During photosynthesis chloroplasts absorb light energy and use this to convert carbon dioxide and water into carbohydrates;

$$CO_2 + H_2O \xrightarrow{\text{sunlight}} CH_2O + O_2 \tag{2}$$

As light enters the plant leaf it is scattered by refraction and scattering (both Rayleigh and Mie). Rayleigh scattering is due to cell organelles such as lysosomes, which are smaller than $0.1\mu m$, and hence capable of causing Rayleigh scatter. However most of the scattering is attributed to refraction due to refractive index differences between hydrated plant cells (refractive index of 1.47) and inter-cellular air (refractive index of 1.0) (Willstätter and Stoll, 1918; Allen *et al.*, 1973; Gausman, 1985a).

The mechanisms involved in the absorption of radiation by pigments in green vegetation is electron transitions ((Belward, 1991; Verdebout *et al.*, 1994). Pigments such as chlorophylls and carotenes absorb light of specific energy, causing electron transitions within the molecular structure of the pigment. The resulting energy from these electron transitions are used for the photochemical reactions. Due to light being in 'small packets' (photons) only light of certain energy can cause the electron transitions, hence plant pigments absorb light strongly at some wavelengths and not at all at others.

TABLE 1. Plant pigments and their absorption maxima.

| Type of pigment | Characteristic absorption maxima (nm) |
|---|---|
| Chlorophyll a | 420, 490, 660 |
| Chlorophyll b | 435, 643 |
| β-Carotene | 425, 450, 480 |
| α-Carotene | 420, 440, 470 |
| Xanthophyll | 425, 450, 475 |

During leaf senescence, chlorophylls degrade faster than carotenes (Sanger, 1971). This leads to a marked increase in reflectance in the red wavelengths as absorptions by chlorophylls are reduced markedly. Carotenes and xanthophylls now become the dominant chemicals in leaves, and the leaves appear yellow because both carotene and xanthophyll absorb blue light and reflect green and red light. The combination of the green and red lights gives the yellow color.

As the leaf dies, brown pigments (tannins) appear and the leaf reflectance and transmittance over the 400nm to 750nm wavelength range decreases (Boyer *et al.*, 1988).

## 3.3    THE REFLECTANCE RED-EDGE (690 - 720NM)

The red-edge, first described by Collins (1978), is a characteristic feature of the spectral response of vegetation and perhaps is the most studied feature in the spectral curve. It is characterised by the low red chlorophyll reflectance to the high reflectance around 800nm (often called the red-edge shoulder) associated with leaf internal structure and water content. Since the red-edge itself is a fairly wide feature of approximately 30nm, it is often desirable to quantify it with a single value so that this value can be compared with that of other species. For this the red-edge inflection point is used. This is the point of maximum slope on the red infrared curve.

For an accurate determination of the red-edge inflection point, a large number of spectral measurements in very narrow bands are required. For laboratory and field based studies, this is not a major obstacle as most laboratory based spectrometers record the reflectance in a large number of very narrow bands and so the derivative spectra give a fairly accurate position of the inflection point. For data where a large number of bands are not available within the red-edge region, the inflection point is approximated by fitting a curve to fewer points. One such method, described by Clevers and Büker (1991), uses a polynomial function to describe the data and then obtains the inflection point from the equation. Guyot and Baret (1988) applied a simple linear model to the red infrared slope. They used four wavelength bands, centred at 670, 700, 740 and 780nm. Reflectance measurements at 670nm and 780nm were used to estimate the inflection point reflectance (Equation 2) and a linear interpolation procedure was applied between 700nm and 740nm to estimate the wavelength of the inflection point

$$R_{red\text{-}edge} = (R_{670} + R_{780}) / 2 \tag{3}$$

$$\lambda_{red\text{-}edge} = 700 + 40 \left( (R_{red\text{-}edge} - R_{700}) / (R_{740} - R_{700}) \right) \tag{4}$$

A third method, proposed by Hare *et al.* (1986), uses an inverted Gaussian reflectance model to characterise the red-edge

$$R(W) = R_s - (R_s - R_0) \exp \frac{-(W - W_0)^2}{2S^2} \tag{5}$$

Here $R_0$ is the reflectance at the absorption maximum near 685nm, $R_s$ the reflectance at the red-edge shoulder, $W$ is the wavelength, and $S$ the Gaussian parameter which determines the red-edge slope. A comparison of the three methods by Clevers and Büker (1991) and Büker and Clevers (1992) yielded comparable results.


## 3.4    THE NEAR-INFRARED REGION (700-1300NM)

Plants generally have a high reflectance and transmittance in the near-infrared region. In contrast to light in the visible wavelengths, the energy levels of near-infrared light are not great enough for photochemical reactions and so are not absorbed by chloroplasts and other pigments. Billings and Morris (1951) and Maas and Dunlap (1989) have shown that the near-infrared reflectance characteristics of green and white (albino) leaves of the same plants are the same, thus proving that pigments do not contribute to near-infrared reflectance properties of leaves. For single citrus leaves (*Citrus sinensis*) Gausman (1985b) found that only 5 per cent of the radiation was absorbed, with 55 per cent reflected and 40 per cent transmitted. Similarly, Woolley

(1971) found that soybean leaves absorbed only 4 per cent of incoming radiation, the remaining 96 per cent being accounted for by reflectance and transmittance. The actual proportion absorbed, scattered or reflected will vary between species and depends on the internal structure of the leaves. Gates *et al.* (1965) and Sinclair *et al.* (1971) have reported that internal leaf structure is the dominant factor controlling the spectral response of plants in the near-infrared.

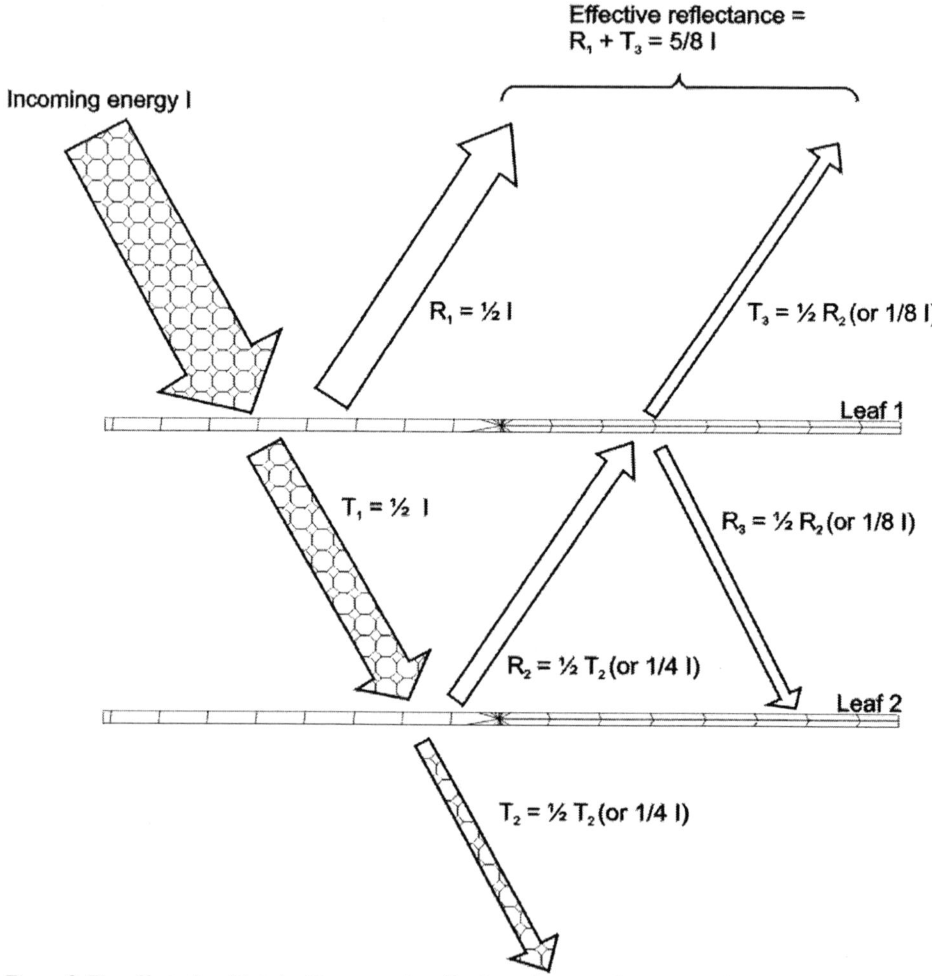



*Figure 2.* The effect of multiple leaf layers on the effective reflectance from vegetation. I = Incoming energy, T = Transmitted energy, R = Reflected energy (Adapted from Hoffer, 1978).

The distribution of air spaces and the arrangement, size and shape of cells influence the passage of light in the leaves. For example, leaves having a compact mesophyll layer will have fewer air spaces and so allow more transmission and less scattering of radiation. On the other hand, leaves having a spongy mesophyll layer have more air spaces and thus more air-water boundaries, and so induce more scattering and less transmission of radiation. Gausman *et al.* (1970) have shown that the near-infrared reflectance is strongly correlated with the volume of intercellular air spaces in the mesophyll layer of leaves. Gates *et al.* (1965) and Knipling (1970) infiltrated leaves with water and reported large decreases in reflectance in the near-infrared region. As the water fills the air gaps, the refractive index discontinuities are reduced and there is a resulting decrease in multiple scattering. This increases transmittance and reduces reflectance. The near-infrared reflectance spectra of leaves also change during development, growth and senescence.

In vegetation canopies, near-infrared reflectance is much higher than that for single leaves. As most of the radiation at near-infrared wavelengths pass through single leaves, the multiple leaf layers of a canopy have an additive effect on reflectance (Belward, 1991). Part of the radiation transmitted by the first leaf layer is reflected back by subsequent layers (Hoffer, 1978), as shown in Figure 2. For cotton leaves Myers (1970) observed that near-infrared reflectance increased from 50 per cent for one leaf to 84 per cent for 6 leaves. It should be noted that after a certain number of leaf layers, addition of extra layers does not increase near-infrared reflectance. This point is referred to as the near-infrared infinite reflectance (Belward, 1991).


## 3.5    THE MID-INFRARED REGION (1300 - 2500NM)

The mid-infrared domain is characterised by strong water absorptions and minor absorption features of other foliar biochemical contents. The reflectance in this region is much lower than in the NIR.

The main water absorption bands are centred at 2660, 2730 and 6270nm, and overtones are observed at 1200, 1450, 1940 and 2500nm. As the water absorptions in the mid-infrared are fairly strong, they have a carry-over effect such that the regions between major water absorption bands are also affected. Therefore increased water contents of leaves will not only decrease reflectance in the water absorption bands, but they will also cause a decrease in reflectance in other regions as well. Unlike pigments, where absorptions are caused by electron transitions, water absorptions are caused by transitions in the vibrational and rotational states of the water molecules (Belward, 1991).

Leaf biochemicals which absorb in the mid-infrared region include lignin, cellulose, starch, proteins and nitrogen. Specific absorption bands for each of these are discussed later in this chapter. The absorptions of these chemicals is not very strong and so are generally masked by water absorptions in fresh leaves. They are much more clearly distinguishable in dry leaf spectra.


## 3.6    BRDF

Vegetation canopies are anisotropic scatterers. That means they do not reflect radiance equally in all directions. The spectral variability of hyperspectral bi-directional reflectance distribution function (BRDF) data is related to spectral and structural characteristics of vegetation canopies and allows the derivation of their overall structure from remote sensing data (Sandmeier and Deering 1999).
Several scientists have ventured into approximating BRDF over canopies, using goiniometer measurements and modelling. The BRDF is a conceptual function, defined as the relation of that part of the total spectral radiance $dL_r\Phi(\theta_r,\phi_r)$ reflected into the direction $\theta_r$, $\phi_r$ which originates from the direction of incidence $\theta_i$, $\phi_i$, to the total

spectral irradiance $L_i(\theta_i,\phi_i)d\Omega_i$ impinging on a surface from the direction $\theta_i$, $\phi_i$ (Kimes, Smith and Ranson 1980), so that

$$\rho^B(\theta_r,\phi_r;\theta_i,\phi_i) = \frac{dL_r\Phi(\theta_r,\phi_r)}{L_i(\theta_i,\phi_i)d\Omega_i} \qquad (6)$$

The BRDF is difficult to measure, because in practice it is impossible to measure radiances of infinitesimally small solid angles, and therefore the BRDF is approximated by the bi-directional reflectance factor (BRF) and normalised to the anisotropy factor (ANIF) for comparison between surfaces. The BRF is the ratio of the reflected radiance from a target surface in a specific direction within a field of view smaller than $20°$ to the reflectance of a Lambertian reference standard using identical illumination and viewing geometry, multiplied by the reference standard's calibration coefficients, so that (Robinson and Biehl 1979):

$$R_s(\theta_i,\phi_i;\theta_r,\phi_r) = (V_s/V_r)R_r(\theta_i,\phi_i;\theta_r,\phi_r) \qquad (7)$$

where $R_r$ = bidirectional reflectance factor of the reference target,
$\theta_i, \theta_r$ = zenith angles of incidence and reflection,
$\phi_i$, $\phi_r$ = azimuth angles of incident and reflected rays,
$V_s$ = measurement of response of the instrument viewing the subject, and
$V_r$ = measurement of response of the instrument viewing the reference surface.
The ANIF is the BRF measurements normalised to the nadir reflectance, so that

$$ANIF(\lambda,\theta_i,\phi_i,\theta_r,\phi_r) = \frac{R(\lambda,\theta_i,\phi_i,\theta_r,\phi_r)}{R_o(\lambda,\theta_i,\phi_i)} \qquad (8)$$

where $R$ is the BRF, $R_0$ is the nadir reflectance factor, $\lambda$ the wavelength, $\theta$ the zenith angel, $\phi$ the azimuth angle of illumination direction, $i$, and viewing direction, $r$ (Sandmeier and Deering 1999).

Over a vegetation canopy the BRDF has certain characteristics, which broadly apply to all canopies. As a combination of gap and backshadow effect (Kimes 1983), densely vegetated erectrophile canopies produce a minimum reflectance near nadir in the forward scatter direction, a reflectance maximum (hot spot) in the backscatter direction, and variously distinct bowl shapes in BRDF characteristics (Coulson 1966, Kimes 1983, Jensen and Schill 2000, Middleton 1993, Gerstl and Simmer 1986, Ni, Woodcock and Jupp 1999, Sandmeier, Müller, Hosgood and Andreoli 1998). This effect is due to the shadowing effect of tree crowns, leaves, and other canopy elements, where from forward scattering to backward scattering, there is a decrease in the observed shadows until no shadows can be seen when the sun and the viewer direction coincide (hot spot) (Kimes 1983, Ni *et al.* 1999). In addition the BRF pattern in the visible spectrum shows a stronger and sharper hot spot than in the near infrared, and the overall reflectance values in the near infrared are higher than in the visible (Sandmeier and Deering 1999, Ni *et al.* 1999). This is because the hot spot effect occurs only for single scattering, the higher absorption in the visible wavelength over vegetation canopies results in less averaging out by multiple random scattering as occurs in the near infrared over vegetation canopies (Sandmeier *et al.* 1998, Ni *et al.* 1999). In addition, in most vegetated canopies some specular reflectance occurs and increases the forward scattering compound, this being more pronounced in planophile canopies were the leaves are mainly horizontally oriented (Sandmeier *et al.* 1998).

BRDF effects are most pronounced in erectrophile canopies (e.g. a grass canopy) if soil/background influences are negligible, and are reduced in planophile surfaces (e.g. a watercress canopy) (Sandmeier *et al.* 1998).


## 4      Vegetation reflectance curve

Leaf optical properties are influenced by the concentration of chlorophyll and other biochemicals, water content, and leaf structure (Mestre 1935; Gates, Keegan, Schleter and Weidner 1965; Allen, Gausman, Richardson and Thomas 1969, Gates 1970; Knipling 1970; Woolley 1971; Hoffer 1978; Curran, Dungan, Macler, Plummer and Peterson 1992; de Boer 1993; Fourty, Baret, Jacquemoud, Schmuck and Verdebout 1996). These leaf characteristics are all very variable and therefore also the reflectance of vegetation is a result of a very complex changing process within the leaves, the canopy and the stand. Physical consideration suggests that the most useful three broad spectral intervals for plant discrimination purposes are those associated with chlorophyll, water, and a third region where both chlorophyll and water are transparent (Gausman, Allen, Gardenas and Richardson 1973). Specifically, intervals centred around 680, 850, 1650, and 2200 nm wavelength appear to be most useful to discriminate among different kinds of leaves (Gausman *et al.* 1973). The first interval is in the visible region, and the others correspond to peaks of the atmospheric windows.


## 4.1    PIGMENTS

Vegetation, which has evolved under the selective umbrella of the earth's atmosphere of variable gaseous composition, is sensitive to ultraviolet and infrared radiation, and specialised to tap energy from the sun's visible radiation (roughly between 400 and 700 nm). Wavelengths below 400 nm are absorbed in a non-specific fashion by common biological constituents (Leopold and Kriedemann 1975), and since energy level per quantum increases with decreasing wavelength, these short wavelengths dump rather heavy packages of energy onto cellular components, but the atmosphere protects by absorbing most of the harmful ultraviolet radiation (Figure 3). With wavelengths beyond the visible region energy absorption again becomes damaging because cellular water absorbs infrared radiation to produce thermal energy (Leopold and Kriedemann 1975).

Chemical bonding energy is required for photochemical activity, and it is known that most reactions involving chemical bonds require energies of greater than a quantum of light of wavelength 950nm (Gates 1980), and efficiency of photosynthesis depending on the illumination wavelength has a longwave limit before 700nm (Emerson, Chalmers and Cederstrand 1957). Between the harmful ultraviolet and infrared wavelengths are the action spectra for the principle light-regulated systems of biology (Leopold and Kriedemann 1975). Inside a leaf cell are biochemicals which are involved in the fixation of atmospheric carbon dioxide and tapping the sun's radiative energy for the reduction of carbon (photosynthesis) (Fosket 1994). Wavelengths of the visible spectrum are selectively absorbed by unsaturated carbon skeletons, like the carotenoids; carbon associated with nitrogen, as in flavins; and carbon associated with nitrogen and metals, as in chlorophyll (Leopold and Kriedemann 1975; Figure 4). Strict selectivity in light absorption then promotes selective translation of its energy into organised biochemical systems (Leopold and Kriedemann 1975).

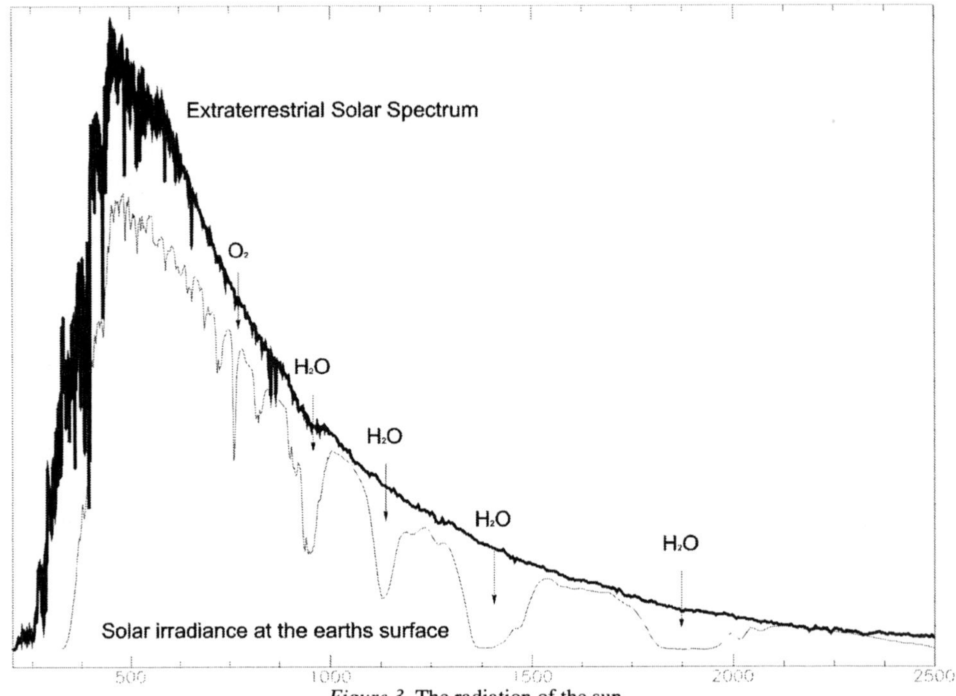



*Figure 3*. The radiation of the sun.

Radiant energy absorbed ($E_a$) by the leaf can be apportioned as

$$E_a = E_{rad} + E_{H_2O} + E_p \tag{9}$$

where $E_{rad}$ represents energy which heats the leaf and becomes re-radiated, $E_{H2O}$ signifies energy used to vaporise water (transpiration), and $E_p$ is energy used in photosynthesis (Leopold and Kriedemann 1975, Gates 1980). If a pigment absorbs light energy, one of three things will occur: energy is dissipated as heat; the energy may be emitted immediately as a longer wavelength (a phenomenon known as fluorescence; or energy may trigger a chemical reaction, as in photosynthesis. In the visible part of the spectrum most of the radiation is absorbed by chlorophylls, and other pigments, especially in the red and blue (Gates *et al.* 1965, Hoffer 1978) and converted to chemical energy. When such a pigment absorbs radiant energy, there is a displacement of pi electrons in a resonance system through the pigment molecule. This activated state may last for only a fraction of a second, and then the energy corresponding to its displacement is given up (Leopold and Kriedemann 1975). Such activation energy can be yielded in several ways: by re-emission of light (fluorescence) or of heat energy, by transmission of the activated state to another molecule, or by consumption of this energy in some biochemical event (Leopold and Kriedemann 1975). Light-activated plant pigments employ all these methods in dissimilating activation energy (Leopold and Kriedemann 1975). The chemical differences between the various forms of chlorophylls equip them with different light-absorbing properties (Leopold and

Kriedemann 1975). Chlorophyll *a* and *b*, for example, differ in the nature of their participation in the two photosystems.

All plants have chlorophyll *a*, and most plants also have chlorophyll *b*, *c*, or *d*. Chlorophyll *a* and chlorophyll *b* are most frequent in higher plants, but altogether about 10 forms have been isolated, each with a unique absorption spectrum (Leopold and Kriedemann 1975). Chlorophyll *a* exists in more than one form, and the variant known as P700 (named for its absorption peak) plays a key role in photochemical reactions (Leopold and Kriedemann 1975). P700 is very rapidly bleached and represents only 0.1% of the total Chlorophyll *a* present. A second form of Chlorophyll *a* (P690) is also fundamental to the light reaction of photosynthesis (Leopold and Kriedemann 1975). These two forms of chlorophyll are key pigments in the photochemical reactions of photosynthesis and represent the chief energy traps for photosystems I and II, respectively. The ratio between chlorophyll *a* and *b* varies according to both plant and environmental conditions. Also the contents of P700 and Chlorophyll *a* and *b* in C4 and C3 plants differs for the two different photosynthetic carbon dioxide fixation cycles (Black and Mayne 1970, Chang and Troughton 1972). Chlorophyll is housed in the chloroplast, a highly functional organelle which houses the complete machinery of photosynthesis (Leopold and Kriedemann 1975, Fosket 1994). Operative pigments in chloroplasts include chlorophyll (65%), carotenes (6%), and xanthophylls (29%); percentage distribution is of course highly variable (Gates *et al.* 1965). In the spectrum of the yellow leaf (grown in the dark), the contribution of carotenoid absorption between 400 and 500nm which is also present in green leaves can be seen (Buschmann and Nagel 1993).

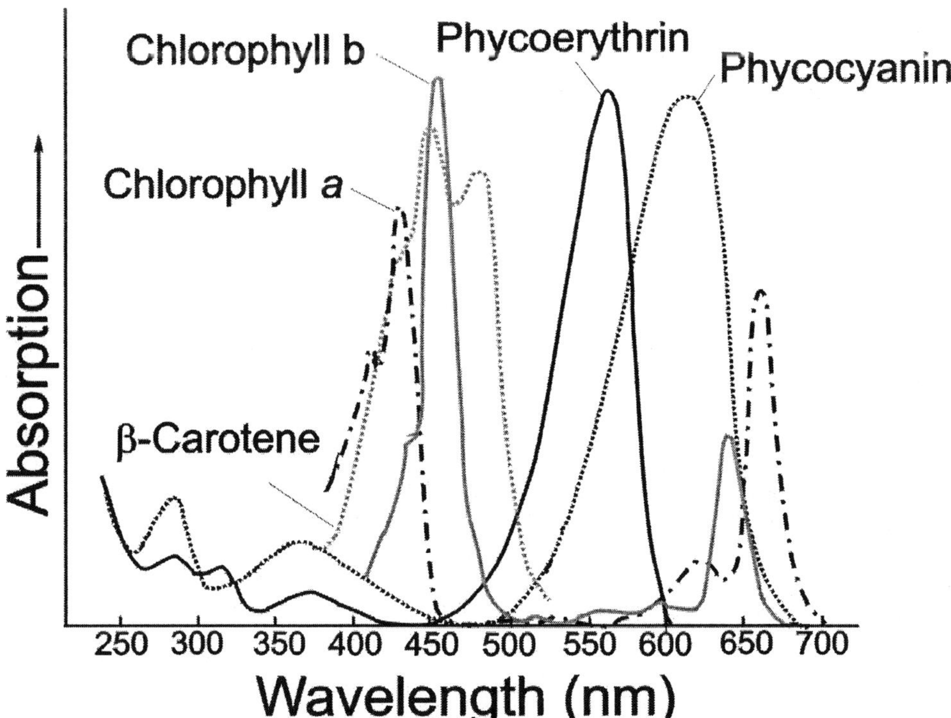


*Figure 4.* Absorption curves of plant pigments (source: Purves *et al.*, 1998)).

Excitation energy from incident light has an upper limit of actually being used for photosynthesis, because of the rate of energy transfer through, for many chlorophyll molecules, common reduction centres (photosynthesis I and II) in the plastides. These organelles show peak efficiency in weak light (Leopold and Kriedemann 1975). The relative lack of absorption in the wavelength between the two chlorophyll absorption pits allows a reflectance peak to occur in the green part of the spectrum and that is why most healthy living vegetation looks green to the human eye. This can be demonstrated by measuring the reflectance of leaves that lack the usual pigments and reflect much of the visible spectrum just as they reflect the near infrared (Knipling 1970, Woolley 1971, Hoffer 1978). In Photosynthesis, as in other light-driven reactions, absorption and action spectra do not show precise coincidence (Leopold and Kriedemann 1975). The action spectrum is not directly proportional to absorbance (Balegh and Biddulph 1970). The discrepancy between chlorophyll absorption and photosynthesis action spectra is largely due to contributing actions of accessory pigments such as carotenoids, and on oversupply of chlorophyll using common reduction centres (Leopold and Kriedemann 1975). This general broadening in an action spectrum bears some resemblance to light absorption by the intact leaf (Leopold and Kriedemann 1975). A high chlorophyll concentration guarantees a high light absorption, however, it does not necessarily mean a high photosynthetic activity and high biomass production. Different plants and even different leaves from the same plant are known to be characterised by different efficiency of photosynthesis and plant growth (Buschmann and Nagel 1993). That means, if one measures the photosynthetic rate of vegetation that is not directly proportional to the reflectance, since mostly more energy is absorbed, than is used for photosynthesis. Elvidge (1990) studied absorption features and spectra of biochemicals (cellulose, lignin, xylan, arabinogalactan, starch, pectins, waxes). Within the intact leaf the light interacts with a very complex and changing arrangement of these and other biochemicals.


## 4.2    GREEN LEAF STRUCTURE

Although plant leaves present numerous anatomical structures, the basic elements are the same, and the variability of the leaf optical properties only results from their arrangement inside the leaf (Verdebout, Jacquemoud and Schmuck 1994). A typical angiosperm leaf is a thin, flattened structure that may be only a few cells thick (Fosket 1994). Structural elements, like the cell walls and specialised cells support the leaf. The typical leaf structure is presented in Figure 5. The plant leaf consist of an outer cuticle, cells and intercellular air spaces. The internal leaf structure of C3 and C4 plants is slightly different (See Figure 5). C3 leaves have a layer of palisade mesophyll cells, which is absent on the C4 leaf, where the bundle sheath cells are more pronounced. Elvidge (1990) measured reflectance curves of pure biochemicals, but their characteristic and pronounced absorption features fade substantially in the reflectance curve of a fresh leaf, since they are bound into complex organic molecules. In addition the molecules and pigments are part of a complex structure of cells in the leaf, this has two effects on the reflectance of a leaf.

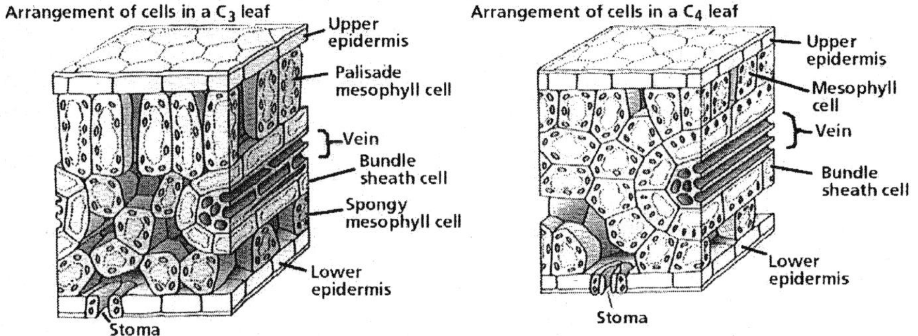


*Figure 5.* The typical structure of C3 and C4 plant leaves (after: (Purves, *et al.*, 1998)).

Cellular structures in a leaf are large compared to wavelengths of visible radiation. Typical cell dimensions of say 15000 by 20000 by 60000 nm for palisade cells are orders of magnitude greater than wavelengths in the visible spectrum (Leopold and Kriedemann 1975), but organelles in these cells present a different picture. Each photosynthetic cell may contain between 50 and 150 chloroplasts (Leopold and Kriedemann 1975). These organelles are about 5000 to 8000 nm in diameter and roughly 1000 nm thick (Leopold and Kriedemann 1975). Granal stacks in these palisades are still smaller, amounting to a length of 500 nm and thickness around 50 nm. These dimensions approach the wavelength range of visible radiation and may produce a lively optical interaction in illuminated leaves because scattering or diffraction takes place when light encounters structures whose dimensions are comparable to its wavelength (Leopold and Kriedemann 1975). Cellular dimensions are too large, but chloroplasts and grana dimensions are conducive to scattering. Compared to the absorbance spectrum of the extract, the *in vivo* spectrum of the leaf has broader absorbance maxima, and the red absorbance maximum is shifted towards longer wavelengths (Buschmann, Nagel, Szabo and Kocsanyi 1994). The reflectance spectrum of leaf extract is very different to leaf reflectance curve, in that the visible region between 500 and 650nm has a higher reflectance, and the absorption minima (the main one around 680nm) are more sharp, than that for the intact leaf (Buschmann and Nagel 1993). The sharp absorption peaks which characterise organic solutions of pigments are not evident in intact leaves for a variety of reasons (Leopold and Kriedemann 1975, Buschmann and Nagel 1993). Attenuation of light within a leaf occurs over a much broader spectral region due to scattering by organelles in the leaf, e.g. mitochondria, ribosomes, nuclei, starch grains, and other plastids (Leopold and Kriedemann 1975, Buschmann and Nagel 1993). Due to formation of pigment-protein-complexes the absorption maximum of chlorophyll is shifted towards longer wavelengths (Buschmann and Nagel 1993). Therefore the red edge is better than NDVI for chlorophyll content determination. The inflection point of the red edge is well correlated with the chlorophyll content of the leaf.

Studies indicated that reflectance at wavelengths with high absorption coefficients should be more sensitive to low concentrations of Chlorophyll *a*, while spectral regions with low absorption should be more sensitive to higher Chlorophyll *a* concentrations (Blackburn 1999). The chloroplast is motile in the cytoplasm and generally capable of altering its position in response to light. Under low light intensities chloroplasts arrange themselves into positions which maximise light interception, and under high intensities they arrange themselves in positions which minimise light interception (Leopold and Kriedemann 1975).

Rate of increase or decrease of absorption varies with plant species (Zurzycki 1953). Some plants adopt their chloroplast structure in response to the light intensity while they grow (Leopold and Kriedemann 1975, Reger and Krauss 1970). Plants that are adapted to intense light conditions have chloroplasts with fewer grana (in direct contrast to chloroplasts from shade-adapted leaves), but instead they contain a high proportion of stroma lamellae (Leopold and Kriedemann 1975). Light entering a leaf is subject to multiple reflections; at the cell walls within the leaf, from scattering within the epidermis and chloroplasts of the palisade cells, and from absorption due to plant pigments and leaf water (Leopold and Kriedemann 1975). The multiple reflections and scattering ensure that light reflected from leaves is approximately Lambertian in character (Woodgate 1985). In contrast the leaf surface reflectance is a combination of diffuse and specular reflectance, and is not Lambertian (Verdebout *et al.* 1994). The leaf can be compared to a Lambertian scatterer only for normal incidence (Verdebout *et al.* 1994). All leaf surfaces polarise light (Vanderbilt and Grant 1986). Different leaf surfaces exist for different plant strategies to excess or deficient light conditions. In the near infrared, healthy vegetation is characterised by very high transmittance and very low absorbance, as compared to the visible wavelength (Knipling 1970). For most types of vegetation the near infrared reflectance is also high - approximately 45% to 50% (Hoffer 1978, Knipling 1970), almost all the rest is transmitted. The high infrared reflectivity is caused by the intercellular structure of the vegetation (Mestre 1935). An important parameter determining the level of reflectance is the number of total area of air-wall interfaces and not the volume of air space (Knipling 1970, Buschmann and Nagel 1993). This is proven by infiltrating leaves with water, which leads to a decrease in reflectance in the NIR and to a lesser degree in the green peak (Buschmann and Nagel 1993). For the same thickness, monocotyledons whose mesophyll is compact have a lower NIR reflectance than dicotyledons which present a palisade and spongy mesophyll; on the other hand their transmittance is higher (Verdebout *et al.* 1994). During senescence of the leaf the NIR first increases the reflectance as adjacent cell walls split apart and as living cell contents shrink away from interior cell walls, thereby increasing the area of the air-wall interface, before it finally decreases as cell walls break down or deteriorate (Knipling 1970). Though, during the plant growth the NIR reflectance of a given species is almost constant (genetic determinism); the most important changes appear during maturation and senescence (Verdebout *et al.* 1994). Gausman and Allen (1973) found correlation of leaf reflectance at seven wavelengths with leaf thickness, water thickness, and internal leaf structure of 30 plant species. The variation in all the measured parameters was quite large, and that might be the reason for the good results, as opposed to other studies like for instance (Knapp and Carter 1998). The spectral composition of incident and transmitted solar radiation demonstrates how plants have become magnificently adapted to their radiation environment (Leopold and Kriedemann 1975). They absorb efficiently in those regions of the spectrum where energy is readily usable and poorly in the near infrared, which serves to minimise heat load, but to absorb far infrared to become efficient radiators. Further anatomical subtleties of leaves enhance their utilisation of absorbed sunlight (Leopold and Kriedemann 1975). Leaves are optically heterogeneous with interfaces between cells walls and air spaces offering ideal sites for internal reflection. A ray of light is therefore likely to encounter several internal reflections before transmission or reflection back outside. Photosynthetic prospects for the single leaf are therefore improved.


## 5    Leaf optical models


### 5.1    INTRODUCTION

Light passing through leaves undergo either reflection, absorption or transmission, or a combination of the three. Which one is the dominant process is determined by the optical properties of the leaf. Optical properties of leaves are determined by the following conditions (Ross, 1981):

- *Surface roughness and the refractive index of the cuticular wax of the upper epidermis.* These control the reflectance from the upper surface of the leaf.
- *Composition, amount and distribution of pigments.* These determine the absorption of radiation in the ultraviolet and visible ranges.
- *Internal leaf structure.* The internal scattering of incident radiation is dependent on the arrangement of cells and tissues. The optical path length determines leaf absorptance, reflectance and transmittance.
- *Water content.* Leaf water content determines the absorption of infrared radiation.

As evident from the preceding section, the passage of light through plant leaves is a complex process. To explain the passage of light through plant leaves, a number of models of varying complexities have been suggested. The models are either *descriptive* or *physically-based*. The *descriptive* models, such as the ray-tracing and stochastic models, are mainly for theoretical studies and cannot be inverted to extract information about leaf chemical content or structure. The *physically-based* models, such as those based on the Kubelka-Munk theory and plate-based, describe the radiative transfer within leaves by using absorption and scattering coefficients and can be easily inverted to extract leaf characteristics from reflectance and transmittance data. The theory behind some of these models are discussed below, while practical applications are discussed later in this chapter.


### 5.2    RAY TRACING MODELS

Willstätter and Stoll (1918) were the first to explain the interaction between solar radiation and leaves. According to their theory, the incident solar radiation is treated in two parts. One part of the incident solar radiation is normal to the plate-like epidermal cells and is therefore reflected. The other part enters the leaf and is subject to multiple scattering due to the refractive index discontinuities at the cell wall and air, and cell membrane and liquid boundaries. A portion of this scattered radiation would escape through the lower epidermis and is designated as transmitted radiation. The remainder either undergoes further scattering within the leaf or escapes through the upper epidermis, thus forming part of the reflected radiation.

Sinclair *et al.* (1973) showed that the high levels of reflectance in the near-infrared region could not be explained by reflection and transmission laws alone. Allen *et al.* (1973) used the W-S theory to calculate reflectance and transmission from leaves described by 100 circular arcs. The leaf was considered as an optical system consisting of only two media: air and a second medium having similar refraction as leaf tissue. The ray tracing calculations of reflectance and transmittance led to an overestimation of transmittance values and an underestimation of reflectance values.

Kumar and Silva (1973) proposed a model based on ray tracing but since it did not consider absorption of radiation, it was applicable in the near-infrared region. They also used four optical media, as opposed to two by Allen *et al.* (1973), in the leaf cross section. The media used were cell wall, chloroplasts, cell sap and air. Fresnel equations and geometrical optics were used to describe the light ray traversing through the leaf cross section, although Rayleigh and Mie scattering were neglected. The model was tested on soybean leaves and good agreement was obtained for both reflection and transmission values when the ray tracing method was compared to actual spectrometer readings.

The ray tracing models are mainly used to test assumptions about the way light interacts with other medium within the leaf. These models cannot be inverted to retrieve information about leaf constituents and so are not very useful. They also require long computational times for accurate predictions (Jacquemoud and Baret, 1990). To overcome these problems other simpler models which can be easily inverted to retrieve leaf constituent information have been developed.


## 5.3    MODELS BASED ON THE KUBELKA-MUNK (K-M) THEORY

The Kubelka-Munk theory considers the leaf as being a slab of diffusing and absorbing material. The internal radiant energy flow is described by a pair of linear, first-order differential equations and absorption and scattering coefficients. This radiative transfer model considers only the diffuse downward flux and diffuse upward flux. According to Slater (1980) the K-M model is exact only for perfectly diffuse incident flux and an ideal diffusing medium. It also works well for the case of a stack of leaves irradiated by a nearly collimated beam in a spectrophotometer. This is because the first leaf in the stack converts most of the direct radiation to diffuse radiation. The most serious shortcoming of the K-M model is that it does not account for variations in canopy radiance due to changes in sun angle. Allen and Richardson (1968) used the K-M theory to develop a model of diffuse reflectance in plant canopies. The model was then used to determine the influence of the number of canopy layers on diffuse upward reflectance. This was accomplished by measuring the reflectance and transmittance of a stack of mature cotton leaves in the 0.5 to 2.5μm range. The model-predicted and laboratory spectrometer-measured values for reflectance and transmittance showed good agreement. The model also satisfactorily explained the multiple scattering in the near-infrared region.

Allen *et al.* (1970) extended this two-flux approach by adding a third flux for direct solar radiation. The direct beam radiation was included because it influences the overall canopy reflectance and also the intercepted beam radiation represents sources of diffuse upward and downward fluxes. The resulting three-flux model was described by differential equations, called Duntley equations (Welles and Norman, 1991), and five coefficients. A major drawback of the Allen *et al.* (1970) model was that it did not address the canopy structural and optical parameters. The effect of foliage orientation, multiple layers, and the view direction were not considered. The 'phenomenological character' of the coefficients used in the transfer equations and the assumption of isotropic diffuse flux leaving the canopy were also of concern (Bunnik, 1984).

Suit (1972) proposed a model which addressed the shortcomings of the Allen *et al.* (1970) model. Canopy layers containing mixed components, such as leaves, flowers and stalks, could be included in the model. Also, laboratory measured canopy component reflectances could be used as direct input. The effect of foliage orientation, multiple layers and reflectance direction were also considered. Verhoef (1984) noted that since the Suits model assumed that all leaves were a mixture of vertical and horizontal projections, the angular response of the model was artificial. Verhoef (1984) corrected this anomaly and suggested the SAIL (Scattering by Arbitrarily Inclined Leaves) model, in which the foliage could be of any inclination distribution. Reflectance predictions based on the SAIL model were observed to be much smoother with view angle than the Suits model.

Yamada and Fujimara (1988, 1991) used the K-M theory to develop a more sophisticated model in which the leaf was described as being composed of four layers: upper cuticle, palisade parenchyma, spongy mesophyll and lower cuticle. Each layer was modelled with different parameters and the solutions coupled with suitable boundary conditions to yield the leaf reflectance and transmittance values. The calculated and modelled values of reflectance and transmittance showed good agreement. The model was also used to perform non-destructive measurements of chlorophyll concentrations.

A number of extensions of the above models have been proposed to take into account the fact that many canopies of interest are agronomic and planted systematically, so the effects of rows and clumping have to be considered. Some of these models have been described by Verhoef and Bunnik (1975), Goel and Grier (1986, 1987), Whitfield and Conner (1980) and Whitfield (1986).


## 5.4    THE PLATE MODELS

The plate model treats leaf internal structures as sheets or plates and calculates multiple reflections of diffuse radiation between these interfaces. The air cavities are excluded from all calculations. The first plate model, introduced by Allen *et al.* (1969), represented the leaf as a non-diffusing but absorbing plate with rough surfaces, hence giving it Lambertian reflectance properties. The radiation process within the single layered leaf was described using an effective index of refraction (n) and an effective absorption coefficient (k). Although this model excluded air spaces from calculations, it was still able to successfully reproduce the reflectance spectrum of corn leaves.

The plate model of Allen *et al.* (1969) was extended by Allen *et al.* (1970) and Gausman *et al.* (1970) to the case of non-compact leaves. The leaf was thus regarded as being made up of a stack of N plates separated by N-1 air spaces, rather than being a single layer as in the original model. Unlike the first model, this model also considered the effects of intercellular air spaces. Jacquemoud and Baret (1990) improved this model to develop the PROSPECT model. Variables such as water content, leaf chlorophyll concentration and leaf mesophyll structural parameter were used to simulate leaf optical properties in the 0.4 to 2.5μm range. Measurements obtained from specially prepared leaves (such as albino, dried, and etiolated leaves) were used to obtain specific absorption coefficients of water, non-photosynthetic pigments, and chlorophyll by making adjustments to the model. The main advantage of this model was that it was easily invertible and so could be used for the extraction of parameters from remotely sensed data. Jacquemoud *et al.* (1996) improved this model further to include leaf biochemical coefficients. This later model could simulate the reflectance and transmittance of fresh as well as dry leaves.


## 5.5    STOCHASTIC MODELS

Stochastic models take into consideration the different physical and physiological properties of dicotyledon leaves. Tucker and Garrat (1977) proposed a stochastic model which explicitly accounted for the palisade parenchyma and spongy mesophyll cells. Each of these cell types was treated as a separate layer, each having its own thickness and other parameters. Radiation transfer within leaves was modelled using Markov chains, which uses compartments to represent radiation states. Both scattering and absorption are treated as discrete time processes with a finite number of states. Steady state reflectance and transmittance at each wavelength are calculated by iteratively applying the one-step transition matrix and utilising prior probabilities of radiation flow within leaves. Leaf scattering is taken to be proportional to the cellular density in the palisade parenchyma and spongy mesophyll. The model was successfully used to simulate the spectrum of maple leaves (Verdebout *et al.*, 1994).


## 6    Vegetation biochemistry


## 6.1    CHEMICAL COMPOUNDS IN PLANTS AND METHODS OF ESTIMATION

The reflectance characteristics of forest canopies, and particularly that of leaves, are related to the absorption features of the compounds present in them (Table 2). Organic compounds absorb infrared radiation at fundamental stretching frequencies, both in middle and thermal infrared regions. However, the organic compound related absorption features noticed in the visible and near-infrared regions are generally due to harmonics and overtones of the fundamental stretching frequencies of C-H, N-H and O-H bonds (Hergbert, 1973).

The main biochemicals found in plants are cellulose, hemicellulose, lignin, starch and nitrogen based compounds such as proteins and chlorophyll. Chlorophyll absorptions are due to electron transitions and located in the visible region of the spectrum. The wavelengths associated with chlorophyll absorptions are 430nm and 660nm (for chlorophyll *a*) and 460nm and 640nm (for chlorophyll *b*). Proteins, mainly in the form of *D*-ribulose 1-5-diphosphate carboxylase, are the most abundant nitrogen bearing compounds in green leaves. They account for between 30 to 50 per cent of the nitrogen of green leaves (Elvidge, 1990). The main near-infrared absorptions of proteins are at 1500, 1680, 1740, 1940, 2050, 2170, 2290 and 2470nm.

Cellulose, which is a D-glucose polymer, is found in the cell walls of all plants and makes up one-third to one-half of the dry weight of most plants (Colvin, 1980). Its main function is to protect and strengthen plant structures. Since it is insoluble in water and is highly resistant to degradation, it is the main substance found in plant litter.

Cellulose has near-infrared absorptions at wavelengths of 1220, 1480, 1930, 2100, 2280, 2340 and 2480nm. As cellulose has its main absorption in the thermal infrared (Elvidge, 1990) the above absorptions are mainly due to harmonics and overtones of the peak absorptions.

TABLE 2  Absorption features related to particular plant compounds*.

| Wavelength (nm) | Absorbing Compounds | Absorption Mechanism |
| --- | --- | --- |
| 430 | Chlorophyll a | Electron transition |
| 460 | Chlorophyll b | Electron transition |
| 640 | Chlorophyll b | Electron transition |
| 660 | Chlorophyll a | Electron transition |
| 910 | Protein | C-H stretch, 3rd overtone |
| 930 | Oil | C-H stretch, 3rd overtone |
| 970 | Water, starch | O-H bend, 1st overtone |
| 990 | Starch | O-H stretch, 2nd overtone |
| 1020 | Protein | N-H stretch |
| 1040 | Oil | C-H stretch, C-H deformation |
| 1120 | Lignin | C-H stretch, 2nd overtone |
| 1200 | Water, cellulose, starch, lignin | O-H bend, 1st overtone |
| 1400 | Water | O-H bend, 1st overtone |
| 1420 | Lignin | C-H stretch, C-H deformation |
| 1450 | Starch, sugar, water, lignin | O-H stretch, 1st overtone C-H stretch, C-H deformation |
| 1490 | Cellulose, sugar | O-H stretch, 1st overtone |
| 1510 | Protein, nitrogen | N-H stretch, 1st overtone |
| 1530 | Starch | O-H stretch, 1st overtone |
| 1540 | Starch, cellulose | O-H stretch, 1st overtone |
| 1580 | Starch, sugar | O-H stretch, 1st overtone |
| 1690 | Lignin, starch, protein | C-H stretch, 1st overtone |
| 1730 | Protein | C-H stretch |
| 1736 | Cellulose | O-H stretch |
| 1780 | Cellulose, sugar, starch | C-H stretch, 1st overtone O-H stretch, H-O-H deformation |
| 1820 | Cellulose | O-H stretch, C-O stretch |
| 1900 | Starch | O-H stretch, C-O stretch |
| 1924 | Cellulose | O-H stretch, O-H deformation |
| 1940 | Water, protein, lignin, cellulose, starch, nitrogen | O-H stretch, O-H deformation |
| 1960 | Starch, sugar | O-H stretch, O-H rotation |
| 1980 | Protein | N-H asymmetry |
| 2000 | Starch | O-H deformation, C-O deformation |
| 2060 | Protein, nitrogen | N-H stretch, N=H rotation |
| 2080 | Starch, sugar | O-H stretch, O-H deformation |
| 2100 | Starch, cellulose | O-H rotation, O-H deformation, C-O-C stretch |
| 2130 | Protein | N-H stretch |
| 2180 | Protein, nitrogen | N-H rotation, C-H stretch, C-O stretch, C=O stretch |
| 2240 | Protein | C-H stretch |
| 2250 | Starch | O-H stretch, O-H deformation |
| 2270 | Cellulose, sugar, starch | C-H stretch, O-H stretch, C-H rotation, $CH_2$ rotation |
| 2280 | Starch, cellulose | C-H stretch, $CH_2$ deformation |
| 2300 | Protein, nitrogen | C-H rotation, C=O stretch, N-H stretch |
| 2310 | Oil | C-H bend, 2nd overtone |
| 2320 | Starch | C-H stretch, $CH_2$ deformation |
| 2340 | Cellulose | C-H stretch, O-H deformation |
| 2350 | Cellulose, nitrogen, protein | $CH_2$ rotation, C-H deformation |

*Compiled from: Curran (1989), Himmelsbach *et al.* (1988), Elvidge (1987, 1990) and Williams and Norris (1987).

Lignin, a polymer of phenylpropanoid, makes up 10 to 35 per cent of dry weight of plants (Crawford, 1981) and acts as a barrier to the decomposition of cellulose and hemicellulose (Elvidge, 1990). Lignin has absorptions at 1450, 1680, 1930, 2270, 2330, 2380, 2500nm and a broad region from 2050nm to 2140nm (Elvidge, 1990).

Starch, a polysaccharide of D-glucose, is the main food storage molecule of plants. It has NIR absorptions at 990, 1220, 1450, 1560, 1700, 1770, 1930, 2100, 2320 and 2480nm.

Techniques used for biochemical estimation commonly employ a linear pre-treatment of reflectance data, using Beer-Lambert's (BL) law, to model the bulk or global absorption coefficient of an organic sample (Murray and Williams, 1987, Bolster et al, 1996). The intensity of absorption can be described in terms of transmittance

$$T = I/I_o \qquad (10)$$

Where I is the intensity of energy emerging from and Io the energy incident on the sample. For absorption spectra, the intensity I can be expressed more precisely by the BL law

$$\mathrm{Log}_{10}(I_o/I) = \mathrm{log}_{10}(1/T) = kcl = A \qquad (11)$$

A is a parameter called absorbance; k is a proportionality constant called the molecular absorption (or extinction) coefficient that is a characteristic of each molecular species; c is the concentration of absorbing molecules and l is the path length of the irradiating energy through the sample. It follows that, for monochromatic light traversing a fixed path length, the absorbance is directly proportional to concentration of the absorbing species. A linear relationship exists between absorbance and concentration, provided Beer's law is upheld, and no association occurs between absorbing molecules.

Information on canopy biochemistry is of great importance for the study of nutrient cycling, productivity, vegetation stress and for input to ecosystem simulation models (Curran 1994, Wessman 1994).

In the field of NIR spectroscopy/spectrometry, reflectance (R) is analogous to transmittance; equation 7.9 can thus be expressed in terms of reflectance as:

$$\mathrm{Log}_{10}(1/R) = kcl = A \qquad (12)$$

Or, more familiarly

$$A = \mathrm{Log}_{10}(1/R) \qquad (13)$$

Log (1/R) is used instead of percentage reflectance because of the almost linear relation between the concentration of an absorbing component and its contribution to the log (1/R) value at the wavelength absorbed (Hruschka, 1987).

Five major absorption features characterise the reflectance spectra of all leaves in visible and near infrared wavelengths. These broad absorption features are responsible for the overall similarity of green vegetation spectra (Curran et al 1992). Electron transitions in photosynthetic pigments (chlorophyll, xanthophylls, and carotene) cause absorption in the 400-700 nm range and bending and stretching of the O-H bond in water and other molecules cause absorption centred at wavelengths of 970 nm, 1145 nm, 1400 nm, and 1940 nm (Danks et al 1984, Curran 1989). In ultraviolet and middle infrared wavelengths, leaves absorb strongly, primarily as a result of the stretching and bending vibrations of the C-H, N-H, C-O, and O-H bonds within organic compounds. Harmonics and overtones of these strong fundamental absorptions cause minor absorption features in visible and near infrared wavelengths (Murray and Williams 1987, Curran 1989).

Much work has focused on trying to correlate these minor absorption features with the amounts of the organic compounds containing the C-H, N-H, C-O, and O-H bonds. Laboratory-based near-infrared spectroscopy (NIRS) studies have identified approximately 42 minor absorption features in the visible and near-infrared portions of the spectrum and the corresponding chemical bond vibrations for each of the important biochemical components: lignin, proteins, cellulose, starch and oils (Norris et al 1976, Shenk et al 1981, Winch and Major 1981, Marten et al 1989).

NIRS has been developed and used for determining biochemical contents from dried, ground agricultural and food products (Williams and Norris 1987, Card et al 1988, Marten et al 1989, Wessman 1990). A regression is established between the reflectance of the dried powdered material and the biochemical content, using the results of the wet chemical analysis performed on a restricted number of samples. Once this calibration is determined, NIRS is then used to analyse the bulk of the samples with the advantage of being as accurate as the wet chemical techniques, and much faster (Williams & Norris, 1987).

NIRS studies have identified four principal zones containing biochemical absorption features (Peterson and Hubbard 1992). The zones are:

- 1100-1300 nm    Dominated by C-H stretch 2nd overtones, features common to all biochemical constituents.
- 1500-1600 nm    Contains major features for lignin (C=O), cellulose (O-H), and protein (N-H).
- 1670-1820 nm    Contains fourteen C-H bond features; three major features for lignin.
- 1980-2380 nm    Contains a variety of major absorption features due to N-H, O-H, C-O, C-H and C=C bonds.

However, it is not possible to use reflectance measurement at a single absorption feature to estimate a particular foliar chemical concentration, for the following reasons (outlined by Curran, 1989):

Absorption features are broadened by multiple scattering and often interfere with one another. For instance, the first overtones of the N-H and O-H stretches overlap for most of their width. Organic compounds absorb in similar wavebands, so that a wavelength is never uniquely related to a chemical. For instance, the strong O-H bond is a component of the absorption spectra of water, cellulose, sugar, starch, and lignin.

Each wavelength has its own measurement error, which increases toward both water absorption wavebands and longer wavelengths. For instances, many of the second overtones are hidden by sensor noise when field and airborne sensors are used.

Reflectance at the centre of an absorption feature will reach an asymptote when saturation is reached, and this can be at relatively low chemical concentrations. This explains why reflectance at an off-centre wavelength of $0.7\mu m$ is often a more accurate estimator of chlorophyll concentration than is reflectance at the centre wavelength of $0.66\mu m$. In addition to the compounds discussed earlier, leaves contain significant concentrations of phospholipids, oils, tannins, and accessory and secondary pigments. Furthermore, these compounds share many of the same chemical bonds, producing strongly overlapped spectral absorption features. Many researchers have therefore adopted an empirical, multivariate approach for leaf biochemical estimation.


## 6.2    EMPIRICAL APPROACH

This approach assumes that a foliar spectrum is the sum of each chemical, weighted by its concentration. Stepwise multiple linear regression (MLR) is commonly used to develop a calibration equation by selecting a small number of narrow band reflectances that explain a large proportion of the variation in biochemical content (Williams and Norris, 1987, Marten et al 1989). Reflectance in these wavebands can then be used to estimate chemical concentration in additional samples. Derivative transformations (see below) frequently provide the best explanation of the variation (Hruschka 1987). These measurements show a high level of reliability and reproducibility when used under carefully controlled conditions. In fact, NIRS methods are now used by many labs in place of analytical chemical techniques to determine chemical composition of dried ground samples (Williams and Norris, 1987, Marten et al 1998, Curran, 1989).

Four main assumptions underlie the multivariate approach (Curran 1989): 1) the relationships between reflectance and chemical concentration are near-linear; 2) we can extract the vegetation spectra of interest from the spectrometry data; 3) the relationship between spectra and chemical composition is not confounded by other factors, such as leaf angle and area distribution, variations in solar incidence angle and sensor view angle, and plant factors such as phenology or canopy geometry and 4) the chemical concentrations have been accurately measured.

The first assumption is reasonable whilst the second assumption is dependent on the success of spectral unmixing approaches (see below). The third assumption poses particular problems at the canopy level. Various canopy modelling approaches have been used to understand the sensitivity to these confounding factors (see below). There are two sets of problems for the fourth assumption. Firstly, the *in vivo* state of the biochemicals may significantly differ from their *in vitro* state. Isolated biochemicals may be physically altered by isolation processes such as oxidation, hydrolysis, or denaturation. Consistency in explicitly defined assay techniques is required to permit comparisons and reliable predictions (Peterson and Hubbard 1992). Secondly, chemical structure can be highly variable both within a leaf; and each chemical has many forms, for example, an "average" lignin molecule is difficult to define, and protein in chloroplasts is not necessarily the same as protein in cytoplasm. In addition, biochemistry may be highly variable within single plants, complicating the statistics of spatial sampling of plants for analysis.

Many studies have indicated that empirical estimates of canopy chemistry based on remote spectroscopic measurements may be possible. These studies fall into two categories: the assessment of reflectance at leaf scales and those at canopy scales. Studies at the leaf scale involve the use of foliage harvested and assessed through the use of either laboratory or field spectrometers (e.g. Card et al 1988, Peterson et al 1988, Curran et al 1991, Gastellu-Etchegory et al 1995, Zagolski et al 1994, Kupiec and Curran 1995, Jacquemoud et al 1995, Asner 1998). Alternatively, spectrometer measurements taken from aircraft or satellites are used to assess the characteristics at the canopy scale (e.g. Peterson et al 1988, Wessman et al 1988, Johnson et al 1994, Matson et al 1994, Gastellu-Etchegory et al 1995, Kupiec and Curran 1995, LaCapra et al 1996, Zagolski et al 1996, Martin & Aber 1993, 1997)

In these studies multiple stepwise regressions between reflectance spectra and biochemical (protein, lignin, cellulose, sugar, starch) concentration have been established (specific examples are discussed later). Fresh leaf tissue creates more problems than dried (or dried/ground) tissue for spectral analysis. This is primarily because the dominant effect of absorption by water largely masks the signatures of the biochemical components beyond 1.0um (Gates 1970, Tucker and Garratt 1977, Curran 1989, Elvidge 1990). In addition to water being a strong absorber of infrared radiation, the cell structure of fresh plant material scatters light as it passes through multiple air and water surfaces with different refractive indices, as discussed earlier. These phenomena may obscure the subtle biochemical absorption features. Furthermore, the distribution of chemical constituents in fresh leaves is not uniform because of the organisation of cells and organelles. Most proteins and all chlorophylls are packed into chloroplasts that migrate and clump, depending on the light environment, and how much lignin is in the cell walls. Non-uniformity results in micro-differential absorbance and reflectance across a leaf surface, just as non-uniform vegetation results in variation in optical properties across a landscape (Yoder & Pettigrew-Crosby, 1995). The waxy cuticle of many leaves can also cause high specular reflectance (Vanderbilt et al 1985). Calibration equations derived from dry material cannot therefore be applied directly to fresh leaves and new wavelengths have to be selected.

Although it has been used successfully, MLR has the following limitations:

- 'overfitting': the large number of wavelengths available for inclusion in calibration equations compared with the number of samples and the number of major constituents (Lindberg et al 1983).
- the nonadditive behaviour of spectra of pure constituents versus a mixture of the same constituents (Aber et al 1994, Lindberg et al 1983, Sjostrom et al 1983).
- the extensive spectral overlap of individual chemical constituents, and the problem of multicollinearity where individual biochemical concentrations may covary (Lindberg et al 1983, Martens and Jensen 1983, Naes and Martins 1984, Wold et al 1984, Otto and Wegscheider 1985, Lorber et al 1987).
- The loss of information and resulting increase in signal to noise ratio when reducing all the available data to a few selected wavelengths for the calibration equation (Martens and Jensen 1983).
- The potential that the best fitting wavelength combination fits the random errors as well as the model (Westerhaus 1989).

This concern about the suitability of MLR has been identified and acknowledged in a number of studies (Wessman et al 1988, Curran 1989, Curran et al 1992, Peterson and Hubbard 1992, Martin and Aber 1994, Kupiec 1994, Grossman et al 1996).

Grossman et al (1996) found that although MLR explained large amounts of the variation in chemical data in both fresh and dry leaf datasets containing a broad range of plant species, the bands selected were not related to known absorption bands, varied among datasets and expression basis for the chemical [concentration (g g$^{-1}$) or content (g m$^{-2}$)], did not correspond to bands selected in other studies, and were sensitive to the samples entered into the regression model. Furthermore, coefficients of determination ($R^2$'s) for correctly-paired nitrogen data and first and second derivative log(1/R) spectra only exceeded the ($R^2$'s) for artificially constructed randomised datasets by 0.02-0.42.

An alternative technique, partial least squares (PLS) regression, overcomes problems due to multicollinearity between wavelengths and the reduction of relevant information into a few available wavelengths selected as MLR regression factors (Martens and Naes 1987). PLS regression uses data compression to reduce the number of independent variables, followed by a calibration regression stage consisting of a least-squares fit of the chemical constituent to the obtained regression factors. Bolster et al (1996) showed that PLS performed better than MLR on dry, ground, green foliage samples of native woody plants. Dury et al (2000a,b) used PLS regression to estimate nitrogen and sideroxylonal (a phenolic plant "defensive" compound) concentration from laboratory spectrometer measurements of fresh eucalypt foliage, and found absorption features common to both fresh leaf and oven-dried leaf spectra, as well as spectra obtained from freeze-dried, ground leaves through near-infrared spectroscopy.

In the transition from the controlled conditions of the laboratory to airborne studies, a number of perturbing effects are introduced, including variable solar illumination intensity and angle, viewing geometry, atmospheric conditions, vegetation canopy architecture [LAI and canopy cover] and understorey. Gastellu-Etchegorry (1995) weighted chemical concentrations from pine forest foliage with local standing biomass and LAI values in an attempt to account for the influence of canopy geometry and biomass on AVIRIS spectral data, however no improvement in prediction accuracy for nitrogen, lignin or cellulose was obtained Although a number of studies have correlated plant canopy chemistry to imaging spectrometer data (Wessman et al 1988, Peterson and Running 1989, Johnson et al 1994, LaCapra et al 1996, Martin and Aber 1997), the results at leaf and canopy scales are inconsistent, and the derived regression equations are not reliable predictors for other remotely sensed data. The suggested reasons for this include inaccurate atmospheric correction, variability in the magnitudes of nitrogen levels between data sets, canopy architecture effects, sensor limitations, and background vegetation and soil influences.

Although some previous studies have concluded that nitrogen containing compounds may have different spectral signatures from plant to plant (LaCapra et al 1996 and Jacquemoud et al 1995), Kokaly & Clark (1999) achieved success at cross-species prediction of nitrogen across seven sites using normalised band depths calculated from continuum-removed spectra together with stepwise multiple linear regression. Kokaly (2000) further demonstrated that increases in nitrogen concentration have a consistent influence on the overall shape of the 2.1 µm absorption feature due to protein absorptions at 2.054 µm and 2.172 µm, thus supporting the current hypothesis that nitrogen-containing protein absorptions represent a sound basis for estimating nitrogen concentration using reflectance spectra of dried and ground leaves.

Johnson et al (1994) recommended that spectral unmixing approaches need to be developed which isolate the canopy spectrum with subtle biochemical features intact. Although spectral mixture analysis (SMA) has become a well established procedure for analysing imaging spectrometry data (Roberts et al 1990, Sabol et al 1990, Gamon et al 1993, Ustin et al 1994), the technique is relatively insensitive to subtle absorption features, and it produces significant quantification errors due to endmember variability from linear and non-linear mixtures (e.g. from scattering and lighting geometry) in a pixel (Pinzon et al 1998). Therefore SMA is not appropriate for detecting minor sources of spectral variation such as variations in canopy chemistry (Smith et al 1994, ACCP 1994). Smith et al (1994) proposed a revised SMA technique, termed Foreground/Background Analysis (FBA). However Pinzon et al (1995) found the method lacked robustness, particularly when using samples with different types of tissues or with concentrations beyond the range of variation in the training data set. Pinzon et al (1998) developed an iterative hierarchical application of a modified FBA technique (HFBA) that was found to be more robust, although the selection of a training set that explains the mixing at different spatial scales remains critical.

Niemann et al (1999): developed an unmixing procedure to isolate scene components when they are not spectrally dissimilar e.g. separation of forest canopy from other 'background' green vegetation.

## 7    Reflectance models for foliar biochemical estimations

Since statistical methods lack robustness and portability, this has led to the advancement of analytical techniques for estimating foliar biochemical content from canopy reflectance spectra, such as radiative transfer modelling (Allen & Richardson 1968, Conel *et al* 1993, Fourty *et al* 1996, 1998, Jacquemoud *et al* 1996, Verdebout *et al* 1995, Asner 1998, Ganapol *et al* 1998) or spectral matching techniques (Gao & Goetz 1995, Wessman *et al* 1993). Other studies have concentrated on development of spectral indices that are insensitive to understorey or atmospheric effects or both (e.g. Huete 1988, Verstraete & Pinty 1996) or spectral analogies between leaves and canopies (Baret *et al* 1994, Kupiec & Curran 1995).

## 7.1    RADIATIVE TRANSFER MODELLING

Radiative transfer models can adapt to various surfaces and experimental conditions because they account for the physical processes that explain the influence of biochemical information on the spectral measurements. They can be used to define and to test predictive equations, or to determine surface biochemistry directly through inversion procedures. These models have mainly been used to estimate chlorophyll and/or water contents as input parameters (Jacquemoud and Baret 1990, Fukshansky *et al* 1991, Yamada and Fujimura 1991, Martinez v Remisowsky *et al* 1992). The influence of protein, cellulose lignin and starch on leaf reflectance was introduced by Conel *et al* (1993), who proposed a two stream Kubelka-Munk model. Radiative transfer models comprise both canopy reflectance models, such as SAIL (Verhoef 1984, 1985) and THREEVEG (Myneni and Ross 1991) and leaf radiative models such as PROSPECT (Jacquemoud and Baret 1990) and LIBERTY (Dawson *et al* 1998). Such models can be used to test the apparent relationship between biochemical content and spectral derivatives at the leaf and canopy levels, since biochemical concentration and other critical factors, such as leaf thickness and canopy architecture characterisation, can be controlled by the investigator.

## 7.2    LEAF RADIATIVE MODELS

The LIBERTY leaf model (Leaf Incorporating Biochemistry Exhibiting Reflectance and Transmittance Yields) was designed to characterise and model conifer (particularly pine) needles (Dawson *et al* 1998). It can be parameterised using data from laboratory leaf optical measurements, and can be coupled with a suitable canopy model for investigations of forest canopy reflectance. The model performs a linear summation of the individual absorption coefficients of the major constituent leaf biochemicals (chlorophyll, water, cellulose, lignin, protein) according to their content per unit area of leaf. The resultant "global" absorption coefficient together with three structural parameters (mean cell diameter, leaf thickness, and an intercellular air-gap determinant) provides accurate reflectance and transmittance spectra of both stacked and individual needles (Dawson *et al* 1998). Initial inversion studies have demonstrated that significant improvements can be made to LIBERTY by utilising *in vivo* absorption coefficients which have been determined by the inversion process.

PROSPECT is a simple but effective radiative transfer model that calculates the leaf optical properties with a limited number of input parameters: a structure parameter and the leaf biochemistry (Jacquemoud *et al* 1996). In the inversion, however, it was found necessary to group some leaf components in order to estimate leaf biochemistry with reasonable accuracy. In fresh leaves, there is no sensitivity for protein, and cellulose and lignin (or other carbon combinations) is poorly estimated (Jacquemoud *et al* 1996). Fourty *et al* (1996) also found that, even in the simple case of dry leaves, inversion of the leaf model did not provide accurate estimates of the content of biochemical per unit leaf area. This was explained by the weakness and lack of specificity of the absorption features. For fresh leaves the retrieval from model inversion was again not efficient for deriving the detailed biochemical composition – the only characteristics that could be retrieved accurately were leaf water and dry matter contents per unit leaf area (Baret and Fourty 1996). Inversion of PROSPECT by Demarez *et al* (1999) revealed good agreement between measured and predicted leaf chlorophyll concentrations for hornbeam and sun oak leaves, but higher predicted values for shade oak and beech (sun and shade) leaves.


## 7.3    CANOPY REFLECTANCE MODELS

A number of approaches have involved the coupling of canopy and leaf models, such as PROSPECT and SAIL (Jacquemoud, 1993, Jacquemoud *et al*, 1995, Verdebout *et al*, 1995, Hobson and Barnsley, 1996). Turbid medium radiative transfer models, however, require a large number of input parameters to describe the canopy and there is still a long way to go before such relationships can be modelled and inverted reliably for the estimation of canopy biochemical concentrations (Jacquemoud 1993, Johnson *et al* 1994, Wessman 1994, Danson *et al* 1995). Another major limitation of turbid medium models is that they do not account for some canopy architecture variables such as tree crown closure, tree density, tree heights, and shapes and dimensions of crowns, which may lead to incorrect estimation of forest reflectances (Gastellu-Etchegorry *et al*, 1999).

Geometric-optical models are more sophisticated radiative transfer models that allow one to account for the influence of canopy architecture on reflectance, and have been shown to be invertible (Li *et al*, 1995, Gastellu-Etchegorry *et al*, 1996). These have also been coupled with leaf models to determine vegetation biochemistry. Bruniquel-Pinel and Gastellu-Etchegorry (1999) and Demarez and Gastellu-Etchegorry (2000) coupled the DART canopy model (Gastellu-Etchegorry *et al*, 1996) with PROSPECT to study the biochemical contents (lignin, nitrogen and cellulose) of a pine plantation, and the chlorophyll content of a beech forest in France, respectively, and Dawson *et al* (1997) coupled the RSADU canopy model (North, 1996) with the LIBERTY leaf model for studying a boreal pine forest. Both studies showed that spectral information associated with leaf biochemistry could be reliably detected at the canopy level. However, the disturbing influence of canopy variables such as canopy LAI and (to a lesser extent) understorey reflectance on predictive equations was demonstrated and both Dawson *et al* (1997) and Demarez and Gastellu-Etchegorry (2000) advocate an *a priori* knowledge of understorey reflectance to estimate leaf chlorophyll/biochemical concentrations.

Dawson *et al* (1999) coupled FLIGHT (forest light - a hybrid geometric optical/radiative transfer BRDF model for conifer forests; North, 1996) with

LIBERTY, and by varying LAI, canopy closure, understorey reflectance, foliar water and lignin-cellulose content, evaluated the degree to which *canopy* spectra remain sensitive to variation in the *foliar* biochemical content. The study identified absorption features or wavelength regions related to both water and lignin-cellulose that were persistent at both the leaf and canopy scales.

Ganapol *et al* (1999) developed a coupled leaf/canopy radiative transfer model called LCM2 (Leaf/Canopy Model version 2), to generate vegetation canopy reflectance as a function of leaf biochemistry, leaf morphology (as represented by leaf scattering properties), leaf thickness, soil reflectance, and canopy architecture. LCM2 couples a within-leaf radiative transfer model, LEAFMOD (Leaf Experimental Absorptivity Feasibility MODel) (Ganapol *et al* 1998) with a dense canopy radiative transfer model, CANMOD (Ganapol and Myeni 1992). The model was used to reconstruct fresh-leaf reflectance and transmittance profiles for 38 different monocot and dicot leaf species, producing RMSE's (vs. measured spectral data) of 0.5-2.7%. Spectral matching techniques.
The linear fitting technique of Gao and Goetz (1995) yielded an equivalent correlation coefficient for lignin as that derived through MLR by Martin and Aber (1994), leading the authors to propose that this technique may be more readily transferable to different sites without extensive ground calibration.


## 7.4    SPECTRAL ANALOGIES BETWEEN LEAVES AND CANOPIES

Kupiec & Curran (1995) compared canopy and leaf reflectance to determine if canopy effects (composite of LAI, biomass, structure, multiple scattering and shadow) could alter the leaf biochemical information in canopy reflectance spectra, and concluded that biochemical information was transmitted virtually unchanged from the leaf to the canopy in near-infrared wavelengths, but canopy influenced leaf reflectance substantially at wavelengths beyond 1400 nm.

In contrast to studies focused on scaling leaf biochemical properties to the canopy level (Jacquemoud *et al* 1995), Asner (1998) approached the scaling problem from the observed variability in tissue optical properties resulting from an established range of biochemical conditions, and then examined the importance of this tissue versus canopy structural variability at landscape scales using a restructured version of SAIL modified for use with hyperspectral data (Verhoef 1984). This was used to quantify the relative contribution of leaf, stem and litter optical properties (incorporating known variation in foliar biochemical properties) and canopy structural attributes to nadir-viewed vegetation reflectance data. It was concluded that leaf optical properties, and thus the biochemistry of leaf material, are expressed most directly at the canopy level within the NIR spectral region (in agreement with the findings of Kupiec & Curran, 1995). However LAI and leaf angle control the strength of this link. Leaf optical properties were generally found to be under-represented at canopy scales unless LAI is quite high. However, high LAI canopies do allow the weak leaf-level biochemical information to be enhanced at the canopy scale via multiple scattering (Baret *et al* 1994). Departure from horizontally oriented leaves results in decreased expression of leaf optical properties.

Fourty and Baret (1997), in comparing empirical and modelling approaches, concluded that the lack of robustness of empirical approaches is partly due to the lack of physical basis. Further, the sampling within the space generated by the variables is obviously limited to what was present during the experiment and is generally poor. On the other hand, the inversion of physical models is difficult because of the number of unknowns compared with the amount of independent information embedded in the spectral signature. The estimation of canopy biophysical or biochemical variables from inversion of radiative transfer models generates ambiguities in the solution, partly because of compensation between several variables that could affect canopy reflectance the same way.

## 8 Applications: Field spectroscopy for vegetation studies

Spectroradiometers can be both imaging and non-imaging. Imaging spectroradiometers, used on remote sensing platforms, provide images similar to conventional multispectral scanners but with a much higher spectral resolution. Such systems, having the potential to provide high resolution spectral signatures of vegetation, provide the means for the identification of classes or species types from air or space. Non-imaging spectrometers also provide information with high spectral resolution but are used for ground measurements. In this case the objects are pre-identified and their spectral responses are recorded for comparative purposes or for correlation with other leaf parameters such as biochemical contents. A large number of laboratory and field based studies have been undertaken to investigate the feasibility of utilising high spectral data for vegetation studies. Most of these studies have concentrated on linking leaf biochemical contents, water contents, plant stress or phenological changes to spectral response of plants. A brief overview of some of the field applications are presented in the following sections.

### 8.1 PHENOLOGICAL STUDIES

Boyer *et al.* (1988) studied the senescence and spectral reflectance in leaves of northern pine oak (*Quercus Palustris M.*). Six sequential stages of leaf senescence were identified and their reflectance properties were related to chlorophyll content and changes in leaf structure. It was found that the interpretation of the process of senescence as explored anatomically was compatible with the derived spectral information.

Miller *et al.* (1991) obtained leaf spectral measurements from ten deciduous and coniferous trees to determine how seasonal changes affected visible and red-edge reflectance characteristics. Parameters studied included the green peak reflectance at 550nm, the chlorophyll-well absorption and wavelength position, the red-edge inflection point, and the near-infrared shoulder reflectance. Phenological changes were studied using both the derivative technique and the inverted Gaussian model. Of the two red-edge position parameters, the inverted Gaussian model parameter provided a better descriptor of the red-edge position over the long term.

Rock *et al.* (1993) studied the impact of phenologic change on the spectral characteristics of needles attached to branch segments of *red spruce* and *eastern hemlock* growing under normal conditions. The spectral characteristics were also related to the leaf and canopy biophysical properties such as canopy water status and needle chlorophyll content. The results indicated that the seasonal change in conifer species over a growing season had a dramatic influence on the spectral fine features, such as green peak reflectance, red-edge parameters and near-infrared plateau reflectance. The importance of these findings were that attention has to be given to the time of the year the hyperspectral data is acquired, especially if it is used for comparison with previously acquired data or broad-band data sets.

The spectral reflectance changes associated with autumn senescence and the relationship of the spectral features to chlorophyll content was investigated by Gitelson and Merzlyak (1994). The experiments were carried out on two deciduous species (maple (*Acer platanoides L.*) and chestnut (*Aesculus hippocastanum L.*)) and the results were to serve as a basis for developing indices sensitive to pigment variation. Large differences in the spectral signature of leaves undergoing senescence were observed and a number of algorithms were proposed for correlating the changes in chlorophyll content to spectral signature of the leaves.

Maple (*Acer platanoides L.*) and horse chestnut (*Aesculus hippocastanum L.*) leaves were also used by Gitelson *et al.* (1996) to detect red-edge positions and chlorophyll contents by reflectance measurements near 700nm. Spectral measurements and leaf pigment contents were determined for both the species in spring, summer and autumn, and these were correlated. The position and magnitude of the first derivative peak at 685-706nm gave a high correlation with leaf chlorophyll concentration. The reflectance near 700nm was found to be a very sensitive indicator of the red-edge position and chlorophyll concentration.

Other studies which have attempted to characterise the spectral fine features associated with phenologic change include those by Moss and Rock (1991) and Lauten and Rock (1992).


## 8.2   PIGMENT CORRELATIONS

Considerable research has been carried out to relate leaf reflectance to leaf biochemical contents. The idea behind this is to find the wavebands which give the best correlation with different chemical compounds present in the leaves. Apart from straight correlations between specific wavelengths and chemicals, different algorithms have been proposed which utilise ratios and combinations of different wavelengths to predict chemical compositions.

Yoder and Pettigrew-Crosby (1995) conducted research to determine whether chlorophyll and nitrogen concentrations could be predicted from laboratory acquired spectra of fresh maple leaves and to extend this to canopy level spectra. Wavebands which correlated best with leaf chlorophyll and nitrogen contents were selected using stepwise regression and these were compared with those selected on the canopy scale. For chlorophyll content, the best correlations were observed at wavelengths of 550nm and 730nm, while for nitrogen content these were at 560nm and 734nm. Higher correlations were obtained by using the first difference of log (1/R).

Non-destructive determination of chlorophyll content of leaves of an aurea mutant of tobacco by reflectance measurements was carried out by Lichtenthaler *et al.* (1996). The wavelength position of the red reflectance minimum in the main red absorption bands was found to be independent of chlorophyll content of leaves and thus was not suitable for chlorophyll determination. The wavelengths from 530nm to 630nm and those near 700nm were found to have the most sensitivity to variations in chlorophyll content. The inflection point of the red-edge was also found to be closely correlated to chlorophyll content. However, the best correlations with chlorophyll content were obtained by using ratios of reflectances at 750nm and 700nm (R750/R700), and those at 750nm and 550nm (R750/R550). For both these ratios, correlation coefficients ($r^2$) were greater than 0.93. It was also found that these new ratios were very sensitive to chlorophyll content at low, medium and high chlorophyll contents, unlike NDVI which does not show good correlations at high chlorophyll values, and thus can be used for remote sensing of chlorophyll content of terrestrial vegetation.

Chappelle *et al.* (1992) also used ratio analysis method to determine chlorophyll *a*, chlorophyll *b* and carotenoid content of soybean leaves. The reflectance spectra of soybean was divided by an arbitrarily selected reference soybean reflectance spectra to obtain a ratio spectra which was then correlated to chlorophyll and carotenoid contents of leaves. Using the ratio spectra, absorption bands could be distinctly seen and their wavelengths defined. A number of algorithms for the determination of chlorophyll and carotenoid content of leaves were also developed. Similar work was also carried out by Buschmann and Nagel (1993), who related leaf reflectance to chlorophyll content and tested a number of vegetation and ratio based indices. The best correlation was found for the logarithm of the ratio of the reflection signals at 800nm and 550nm. The normalised difference vegetation index was found to exhibit a poor correlation with chlorophyll content.

Curran *et al.* (1992) used reflectance spectroscopy of fresh whole leaves to determine the concentrations of chlorophyll, amaranthin, starch and water. First derivative spectra of the reflectance spectra were obtained and stepwise regression was used to generate equations for the estimation of chemical concentrations. These estimated concentrations were then compared with the measured concentrations and coefficients of determination ($r^2$) of greater than 0.82 were obtained for all comparisons.

The red-edge inflection point is probably the best known and most widely used parameter in relating leaf chlorophyll content to its spectral reflectance. A number of authors have demonstrated the relationship between the red-edge and leaf chlorophyll content; these include Thomas and Gausman (1977), Horler *et al.* (1980), Horler *et al.* (1983), Demetriades-Shah and Steven (1988), Curran *et al.* (1990), Boochs *et al.* (1990), Curran *et al.* (1991) Peñuelas *et al.* (1993) Vogelmann *et al.* (1993) and Filella and Peñuelas (1994). The red-edge has generally been found to be very valuable for assessment of plant chlorophyll concentration and leaf area index.

Other research on estimation of leaf biochemical contents from green leaf spectrum include those of Goetz *et al.* (1990), Buschmann and Nagel (1991), Babani *et al.* (1996), Babani and Lichtenthaler (1996), Fourty *et al.* (1996), Gitelson and Merzlyak (1996), McMurtrey *et al.* (1996), Price *et al.* (1996) and Yang and Prince (1997).

## 8.3    WATER STATUS

Water availability is a critical factor in plant survival and development and water stress is one of the most common limitations of primary productivity (Boyer, 1982). The ability to assess water stress symptoms in vegetation using spectral reflectance measurements is an important goal for remote sensing research (Jackson, 1986). In agricultural crops it is important to be able to detect the onset of water stress as soon as possible so that preventive measures such as irrigation can be undertaken. In forest areas severe water stress is a strong indication of increased fuel loadings and higher probabilities of bushfires. While vegetation presence and vigour are generally computed from digital images using the broad band spectral sensors of airborne and spaceborne platforms, these bands are usually misplaced spectrally or are too broad to measure exclusively in the water absorption or stress sensitive bands (Carter, 1994). A number of recent studies have looked at the relationship between spectral reflectance and leaf water content using hyperspectral data with the view of utilising it on imaging spectrometer platforms. However, most plant moisture stress studies have been in the form of laboratory investigations. Some of these are discussed below.

Thomas *et al.* (1966, 1971) allowed fully turgid cotton leaves to dry and collected relative turgidity and leaf reflectance data in the 400nm to 2500nm ranges. It was found that the relative reflectance increased as the turgidity decreased. However the increase in reflectance was not uniform. The visible part of the spectrum showed the least change with decreasing water content. The regression equations showed that relative turgidity had to be less than 70 per cent to cause predictable changes in leaf reflectance. Olson (1967) reported that there was a linear relationship between leaf water content and leaf reflectance at 1550, 2050 and 2500nm. Increases in leaf reflectance with decreasing water content were also observed by Rohde and Olson (1971) and Sinclair *et al.* (1971). The spectral reflectance in the middle infrared was found to be more closely related to absolute water content than to relative water content (Bauer *et al.*, 1981) and the mid-infrared was seen to be asymptotic at relatively low water contents. Holben *et al.* (1983) found the near-infrared (760 - 900nm) to be a better indicator of plant water stress than either of red (630 - 690nm) or mid-infrared (1550 - 1750nm). A similar result was reported by Jackson and Ezra (1985).

Ripple (1986) used snapbean leaves (*Phaselus vulgaris L.*) to study the changes in leaf reflectance caused by changes in leaf cover, relative water content of leaves, and leaf water potential. Results showed that the reflectance in the red band (630 690nm) was sensitive to both relative water content and leaf cover, the near-infrared reflectance (760-900nm) was strongly correlated to leaf cover but weakly to relative water content, and the mid-infrared (2080-2350nm) was highly correlated to both leaf cover and relative water content. Leaf water potential was found to be strongly correlated to both mid-infrared and red reflectances. Ripple (1986) attributed this to covariances with relative water content and leaf cover.

Danson *et al.* (1992) used reflectance as well as derivative spectroscopy to relate leaf signatures to water content. Linear correlation analysis showed that a statistically significant relationship (95 per cent confidence interval) existed between the specific water density and leaf reflectance at 1450, 1650 and 2250nm. The first derivative of the reflectance spectrum at the slopes of the major water absorption bands was also found to be highly correlated with leaf water content. Danson *et al.* (1992) also noted that the first derivative of leaf reflectance was better at predicting leaf water status than the original reflectance spectrum when variations in leaf internal structure were present.

Peñuelas *et al.* (1993) found the reflectance in the 950-970nm region to be an important indicator of plant water status. The reflectance ratio of 970nm and 900nm (R970/R900), the first derivative minimum reflectance spectra in the above region, and the wavelength position of the minimum were found to be closely related to relative water content, leaf water potential, stomatal conductance and foliage-air temperature differences. Strong correlations were observed when the relative water content was less than 80 per cent and the water stress was well developed. Similar results were also reported by Hunt *et al.* (1987).

Other work relating leaf water content to leaf reflectance have been reported by Bowman (1989), Ammer *et al.* (1991), Carter (1991), Shibayama *et al.* (1993) and Schepers *et al.* (1996).


## 8.4    PLANT STRESS

Carter (1994) presented a number of ratios which could be used as indicators of plant stress. The ratios were obtained from narrow band (2nm) spectral signatures of vegetation under different stress conditions. The stress agents included competition, herbicide, pathogen, ozone, mycorrhizae, senescence and dehydration, and six different plant species were tested. It was found that the reflectance ratios of R695/R420, R605/R760, R695/R760 and R710/R760 were significantly greater ($p \leq 0.05$) in stressed compared with non-stressed leaves for all stress agents. The ratios R695/R420 and R695/R760 showed the highest correlations.

Dockter *et al.* (1988), Collins *et al.* (1980, 1983), Chang and Collins (1983) and Boochs *et al.* (1990) have related the red-edge of leaf reflectance to plant vitality and an indicator of nitrogen supply. Dockter *et al.* (1988) and Boochs *et al.* (1990) observed a distinctive shift in the red-edge of reflectance spectra of sugar beet due to differences in leaf vitality caused by differences in time of sowing. Nitrogen supply also affected the red-edge position. Collins *et al.* (1983) used airborne reflectance data for detecting stress levels in forest trees and reported that the shifts in the red-edge position could be used as an indicator of stress. A similar conclusion was reported by Chang and Collins (1983) based on laboratory reflectance spectra obtained from leaves subjected to metal-induced stress.

Carter *et al.* (1996) tested the effect of the herbicide diuron on the reflectance characteristics of loblolly pine (*Pinus taeda* L.) and slash pine (*Pinus elliottii* Engelm). Reflectance increases in the 690 to 700nm band were detected at least 16 days before the first visible signs of damage appeared on the leaves. It was also noted that thermal imagery showing canopy temperature in the 800 to 1200nm region did not differ significantly between herbicide-treated and control plots. As a result of this the authors concluded that high spectral imagery in the 690 to 700nm region was far superior to thermal imagery for the early and pre-visual detection of stress in pine.

Kraft *et al.* (1996) utilised high spectral reflectance data for detecting ozone damage to crops. Several species of plants, including spring wheat, white clover and maize, were exposed to high levels of ozone for 20-30 days and their spectral reflectance measurements were taken using a CCD camera and wavelength filters with 11 wavebands ranging from 450nm to 950nm. Leaves and canopies of all the species showed an increase in visible light reflectance after ozone treatment, however the near-infrared reflectance either decreased (white clover and maize) or remained the same (spring wheat). The normalised difference vegetation index and the inflection point of the red-edge were found to be strongly correlated to ozone damage. Williams and Ashenden (1992) also showed that ozone treatment of white clover led to a decreased vegetation index for plant canopies. A simple vegetation index utilising the ratio of near-infrared (760 to 1100nm) to red (600 to 700nm) was used.

Lorenzen and Jensen (1989) studied the changes in leaf spectral properties produced by powdery mildew on several varieties of spring barley. Significant increases in reflectance in the visible region were observed 6 days after inoculation.

Malthus and Madeira (1993) used high resolution spectroradiometry to investigate the reflectance of field bean (*Vicia faba*) leaves infected by *Botrytis fabae* fungus. A general flattening of the reflectance in the visible region and a decrease in near-infrared shoulder reflectance (800nm) were observed. These changes were attributed to a collapse of leaf cell structure as the fungus spread. It was also found that the first derivative reflectance spectra gave better correlations between reflectance in the visible region and percentage infection.

A number of researchers have investigated the effects of heavy metal stress on the absorbance and reflectance spectra of plants. A general trend for vegetation growing over mineralisations is to have higher reflectance in visible wavelengths than background vegetation (Horler *et al.*, 1980). Howard *et al.* (1971) found that the reflectance at 810nm increased for plants (*Pinus ponderosa*) growing in soils with increased copper contents. Birnie and Dykstra (1978) found that lodgepole pine (*Pinus contorta*) growing on a copper-molybdenum mineralisation zone showed increased reflectance at around 670nm, but the differences were least around 790nm. Press (1974) found a general increase in the reflectance in the wavelength range 680nm - 900nm for bean plants grown in soils containing increased amounts of lead and zinc.

Koch *et al.* (1990) investigated the effects of chlorosis, defoliation, water stress, phenological development and age on the spectral reflectance of conifers. Spectral responses of leaves were measured in the laboratory and in the field using a crane mounted spectroradiometer. It was found that defoliation caused a decrease in spectral reflectance across the spectrum due to an increase in shadowing and background within the field of view. Chlorosis produced an increase in visible reflectance due to reduced chlorophyll content. Changes in leaf reflectance due to water stress were more noticeable in the laboratory spectra than the field spectra.

Merton (1999) also found that multi-temporal second-derivative plots centred at the 733 nm band from AVIRIS data responded to patterns of environment-induced vegetation stress.

Spectral ratios tend to eliminate differences that are due only to illumination variations and provide a more stable measure of the vegetation type as well as a more reliable indicator of stress. However Demetriades-Shah *et al* (1990) noted that, while spectral ratios (e.g. VI and NDVI) reduce the soil contribution in canopy spectral measures, they do not eliminate it. The same can be said for the position of the red edge, which is determined from first derivative spectra. Demetriades-Shah *et al* (1990) proposed the use of second derivative spectra in canopy measurements since these essentially eliminate the effects of soil background, while first derivative spectra do not.

Adams et al (1999) suggest using a yellowness index (YI) as a measure for chlorosis of leaves in stressed plants, based on leaf-level measurements using a spectrometer. TheYI is an approximation of a spectral second derivative, and provides a measure of the change in shape of the reflectance spectra between the maximum near 0.55um and the minimum near 0.65um.

Lelong et al (1998) used a different approach for stress mapping of an agricultural crop (wheat), using MIVIS hyperspectral data. Firstly, two endmembers of wheat, corresponding to well-developed and stressed plants, were identified using principal component analysis; these were used in a SMA, leading to the decomposition of the detected spectrum into several constituents, enabling the soil and shade contributions to be discarded from the vegetation spectrum. The resulting spectra are interpreted in terms of crop vitality (level of green biomass) in relation to stress presence and compared to field knowledge.

Other research related to plant stress have been by *Rock et al.* (1988), Banninger (1990), Curtiss and Maecher (1991) and Matson *et al.* (1994).


## 9    Applications: Airborne imaging spectroscopy for vegetation studies

There have been a number of studies to extend the relationships of laboratory acquired leaf spectra and leaf biochemistry to airborne imagery. Information relating to canopy chemistry is of great importance for the study of nutrient cycling, productivity, vegetation stress, and for input to ecosystem simulation models (Curran, 1994a,b), therefore it is important that methods which can reliably invert (upscale) laboratory based models to the canopy level be developed. This section discusses some of the research undertaken to upscale spectral signatures and field spectroscopy results to larger scale vegetation using imaging spectrometers.


### 9.1    EXTRACTING BIOPHYSICAL VARIABLES (E.G. LAI, $F_{APAR}$, COVER)

Recent studies have shown that narrow bands may be crucial for providing additional information with significant improvements over broad bands in quantifying biophysical characteristics of vegetation. The hyperspectral data for these studies have been gathered using, for example: (1) field spectroradiometers, conducted for a) chlorophyll content of plants (Blackburn 1998, 1999, Curran *et al* 1990); b) semiarid bushland and bracken canopy LAI (Blackburn 1998, Blackburn and Pitman 1999); c) rice yield (Shibayama and Akiyama 1991); d) pinyon pine canopy LAI (Elvidge and Chen 1995); e) photosynthesis and stomatal conductance in pine canopies (Carter 1998,); f) quantifying sparse vegetation cover (Hurcom and Harrison 1998); g) savanna landscape-level $f_{APAR}$ (Asner *et al* 1998a); (2) CASI, for coniferous forest LAI (Gong *et al* 1995, Lucas *et al* 2000), and (3) AVIRIS a) quantifying sparse vegetation cover (Elvidge *et al* 1993, Chen *et al* 1995) b) LAI of arid land shrub and saltgrass communities (Chen 1998)

Leaf area index (LAI) and fraction of absorbed photosynthetically active radiation ($f_{APAR}$) are key variables in most ecosystem productivity models, and in global models of climate, hydrology, biogeochemistry and ecology which attempt to describe the exchange of fluxes of energy, mass (e.g. water and $CO_2$), and momentum between the surface and the atmosphere. LAI is one of the main drivers of canopy primary production processes because it represents the size of the interface between the plant and the atmosphere for energy and mass exchanges. It is thus of prime interest for evapotranspiration, primary productivity, and crop yield models. Similarly, $f_{APAR}$ provides a way to estimate biomass production.

Several ways to estimate biophysical variables from reflectance data have been elaborated. They can be grouped into two main classes: 1) empirical relationships based on vegetation indices and 2) physically based methods (model inversion).

LAI and $f_{APAR}$ have been related directly to spectral vegetation indices using data from visible and near-infrared wavelengths, such as NDVI, by theoretical canopy modelling and field studies, and both have been used extensively as satellite-derived parameters for calculation of plant carbon uptake or net primary productivity (NPP), which feeds into soil carbon, nutrient and water algorithms that calculate a wide range of biogeochemical fluxes (Running *et al* 1994, Field *et al*, 1995).

Currently, most biogeochemical/hydrological models are driven off biophysical parameters estimated using the NDVI. However a broad-band two-channel index fails to fully utilise the potential of spectrally rich hyperspectral data sets. Because of the numerous variables governing canopy reflectance as compared to the small amount of independent information available from the two-channel broad-band red and near-infrared reflectance measurements, few biophysical variables are actually reliably estimated (Baret *et al*, 1995). Asner *et al* (2000) argue that, particularly in arid and semiarid environments, high spectral resolution data are required to resolve subtle but important variation in vegetation condition, where a wide range of biophysical and abiotic surfaces are prominent, in order to make quantitative biophysical assessments.

Guyot *et al* (1992) showed that, at the canopy level, movement of the red edge is controlled by canopy LAI, leaf chlorophyll content and leaf inclination angle. They suggested that the wavelength of the inflexion point of the red edge (REP) is determined by the level of red and near infrared reflectance, the variation of which is dominated by LAI. Consequently, the REP should provide a useful tool for LAI estimation with high spectral resolution data providing a large number of spectral measurements in narrow bands in this region. Clevers (1994) demonstrated that the red edge index seems to be independent of soil reflectance, and the solar zenith angle seems to have only a minor influence on the position of the red edge.

Elvidge *et al* (1993) showed that the improved spectral resolution of hyperspectral remote sensing data in the region of the red edge may improve the detection of changes in sparse amounts of vegetative cover.

Using a field spectrometer, Blackburn and Pitman (1999) found that red edge position had a good correlation with (bracken) canopy LAI, and percentage reflectance at the red edge position was even more closely related to LAI. Moreover it was largely insensitive to view angle, whereas NDVI and SR (simple ratio vegetation index), although well correlated to LAI and percentage canopy cover, showed considerable changes in their angular distribution as the canopy developed. Blackburn and Steele (1999) found that for spectrometer measurements of semiarid bushland canopies, LAI and percent cover were related to ratios of reflectance in narrow bands on the near-infrared plateau and red edge features of canopy reflectance spectra, as well as the amplitude of the first derivative in the red edge and visible regions respectively. No relationship was found between broadband NDVI/SR and LAI/ percent canopy cover.

Lucas *et al* (2000) found a strong linear correlation between LAI and the REP for a conifer forest in Wales using CASI data, and the relationship was used to obtain spatial estimates of LAI, which in turn was used to provide spatial estimates of leaf nitrogen concentration (LNC). LAI and LNC, along with meteorological data, were key inputs to a general forest ecosystem model (FOREST-BGC (Bio-Geo-Chemical Cycles), Running and Coughlan 1988), to estimate stem carbon production (SCP). Estimates of SCP compared favourably with estimates derived from tree cores.

Gong *et al* (1995) evaluated the potential of estimating coniferous forest LAI from CASI data using three types of modelling techniques: univariate regression, multiple regression, and vegetation-index (VI) based LAI estimation. All three methods resulted in LAIs with reasonably low root-mean-squared errors (RMSEs). The use of the NDVI produced more accurate LAI estimates than did the use of channel ratio for the univariate regression and the VI-based LAI prediction methods. For the univariate regression, a non-linear hyperbola relationship between the LAI and the NDVI was the most appropriate for LAI estimation.

High spectral resolution derivative-based green vegetation indices derived from field spectrometer data were shown to be effective at minimising background impacts of soils on the estimation of LAI and percent green cover of pinyon pine and big sagebrush (Chen 1995, Elvidge and Chen, 1995). The DGVIs outperformed numerous vegetation indices formed with spectral data from discrete red and NIR bands, and have been successfully applied to AVIRIS data of a single-species Monterey pine plantation to map spatial variability in percent green cover levels (Chen 1995), and for measuring LAI of arid land shrub and saltgrass communities (Chen *et al* 1998). Chen *et al* (1998) conclude that the DGVI concept has potential for monitoring ecosystems in arid and semiarid lands where the influence of exposed rock-soil backgrounds reduces the effectiveness of broadband red-vs.-NIR vegetation indices.

Thenkabail *et al* (2000) investigated the optimum number of hyperspectral bands, centres and widths for establishing relationships with crop biophysical characteristics, and showed that the widely used red and NIR band combinations are not necessarily the best for two-band NDVI type indices for estimating crop variables. A rigorous procedure to identify the best narrow band NDVI type models showed that the best two-band combinations often involve: (1) a very narrow red band and a narrow or a broad NIR band, or (2) a very narrow red band and a very narrow or a narrow green band, or (3) a very narrow red band and a narrow band in the moisture-sensitive portion of the NIR, or (4) a very narrow red-edge band and a narrow or broad NIR band. Even when red and NIR bands are involved, they often appear in unique narrow portions of the spectrum that are not selected by any existing satellite sensor. In multiple band models most of the variability was explained using the first two to four narrow bands with addition of further bands adding only small (often, statistically insignificant) incremental increases in $R^2$ values. Four-variable multiple-band models performed only marginally better than two-band NDVI type models, and "overfitting" of multiple-band models was found to be a problem where the ratio of narrow bands to field samples exceeded 0.15-0.20. Narrow band transformed soil-adjusted vegetation index models are considered the best where site specific soil lines are available, but did not provide significant improvements over narrow band NDVI models. The best derivative indices performed significantly poorer than nearly all the narrow band NDVI and multipe-band models, and even with most broad band NDVI models. The study recommended 12 specific narrow bands, centres and widths, which provide optimal crop information in the 350 nm to 1050 nm range.

McGwire (2000) compared multispectral and hyperspectral remote sensing to measure small differences in percent green vegetative cover using NDVI, and compared the results to two narrow-band vegetation indices, soil-adjusted vegetation index (SAVI) and linear mixture modelling. The use of linear mixture modelling with hyperspectral data provided significantly better results than the standard vegetation indices that were tested. Results from broad-band and narrow-band NDVI were similar. Somewhat surprisingly the incorporation of a soil adjustment factor (SAVI) resulted in a significant reduction in performance, and use of a modified SAVI (MSAVI) further incorporating a variable function based on the weighted difference vegetation index (WDVI: Clevers and Verhouf, 1993) and the slope of the soil line provided an insignificant improvement over NDVI. The mixture modelling directly incorporates multiple soil endmembers into the solution, and the interaction of these with the shadow endmember goes beyond the single soil adjustment of MSAVI to compensate for multiple soil lines in the image.

The advantage in extracting biophysical information from spectral indices over individual wavebands is that they can be designed to minimise the effects of variations in extraneous factors including soil brightness (Huete, 1988) and atmospheric effects (Pinty and Verstraete, 1992); however it appears unlikely that indices can provide accurate information on a routine basis, due to the sensitivity of indices to extraneous factors (McDonald *et al*, 1998).

Asner *et al* (1999) cautioned that observed changes in the NDVI cannot resolve differences in the relative importance of: (1) canopy chlorophyll and water content that occurs via changes in foliar properties and LAI, (2) vegetation structure which occurs via changes in horizontal cover and canopy architectural development, and (3) background albedo (e.g. soil and litter). For regional ecological and biogeochemical studies, additional information, beyond that which can be provided by the NDVI, is required to quantify the relative impact of the different factors  on remotely sensed data.

The influence of sun-surface-sensor geometry has yet to be incorporated in vegetation index development (Qi *et al*, 2000). Since the optical and structural attributes of vegetation and soils vary in three dimensional space and in time (Asner 1998), a method that accounts for the scale dependence of tissue optical and canopy structural attributes is therefore needed.


## 9.2    PHYSICALLY-BASED METHODS

An alternative to empirical relationships is a modelling approach based on a set of radiative transfer equations or models. In this approach the inversion of a vegetation reflectance model may be used to estimate the biophysical characteristics of the canopy, providing sufficient information can be obtained from the combined remote sensing and ancillary data. Inversion involves adjusting model parameters until the model reflectance best matches the measured reflectance (Goel 1988, Privette *et al* 1994).

Inversion of a canopy radiative transfer model is usually achieved numerically by minimising the difference between measured canopy reflectance samples and modelled values using an optimisation routine (Goel 1988, Privette *et al* 1994). Canopy radiative transfer model inversions are a robust approach to access canopy structural information using remotely sensed data, yet are limited by the potential lack of reflectance information needed to successfully execute the model inversion (Asner *et al*, 1998c,d). However hyperspectral data has been shown to provide sufficient reflectance information from which canopy attributes can be estimated via inverse modelling. Once the canopy variables are estimated, these can be used to calculate plant functional properties such as energy absorption (using a photon transport model; Asner and Wessman, 1997) and ecological processes such as carbon uptake and evapotranspiration (using an ecosystem or land-surface model; Field *et al*, 1995; Sellers *et al*, 1997). Asner *et al* (1999) contend that hyperspectral data provide the best means to estimate most of the canopy variables of interest to ecological and biogeochemical research efforts (with the exception of gap fraction, which can be assessed using multi-angle data).

The spectral, angular, spatial and temporal dependence of ecosystem (pixel level) reflectance ($\rho_{pixel}$) can be summarised in the relationship:

$$\rho_{pixel} = f(\text{GEOMETRY, STRUCTURE, BIOCHEMISTRY, GEOCHEMISTRY}) \quad (14)$$

GEOMETRY represents the solar and sensor viewing orientation. STRUCTURE includes canopy materials and architectural features such as the horizontal distribution of vegetation types, canopy height, canopy leaf and non-photosynthetic vegetation (NPV) area, leaf and NPV angular distributions, and foliage clumping. BIOCHEMISTRY represents the chemical make-up of the tissues such as carbon, nitrogen and water content, while GEOCHEMISTRY includes the mineral and moisture content of surface soils (Asner *et al* 1999).

## 10    Hyperspectral-BRDF inverse modelling

### 10.1   INTRODUCTION

Most commonly used models are bidirectional reflectance distribution (BRDF) models. A BRDF model usually consists of a set of equations that relate surface physical properties to the observed signals as a function of wavelength. The physical properties may include soil reflectances, canopy architectures and optical properties/geometric configuration of the sensing systems, as well as the illumination sources. Some models may require other input parameters such as leaf inclination distribution and anisotropic properties of both canopy and soil substrates (Jacquemoud 1993, Qi *et al*, 1995, 2000, Running *et al* 1996). Studies using multispectral remote sensing data suggest that, in practice, such models can retrieve useful bio-physical information (Hall *et al*, 1995, Woodcock *et al*, 1997, Gemmell, 1998).

The models provide a physically consistent means to extend a field-level understanding of vegetation properties to a broader scale. They also provide a means to improve remotely sensed measurements of vegetation by allowing the incorporation of *a priori* knowledge to constrain and/or validate the physical interpretation of spectral, angular, temporal and spatial signatures. For example biogeophysical information, such as soil optical variability or tissue reflectance relationships between spectral bands (Asner *et al* 1998c), can provide constraint of the model while combinations of unknown factors (e.g. leaf area, live:dead fraction, bare soil extent) are simulated.

Huemmrich and Goward (1997) examined the effect of varying leaf, twig, and background optical properties on the determination of $f_{APAR}$ from NDVI for ten different forest species, using the SAIL radiative transfer model, and a Spectron SE590 spectrometer for measurement of the optical properties of the canopy components. The simulations indicated that at low values of LAI, the background reflectance had a significant effect on the canopy reflectance, although little effect on PAR absorption. At higher values of LAI, leaf optical properties were the factors that dominated canopy reflectance and NDVI, with some species having quite different NDVI-$f_{APAR}$ relations from most other vegetation types, which might result in significant errors in the estimation of $f_{APAR}$ unless the image is classified beforehand accordingly. Asner *et al* (1998a,b) used a combination of radiative transfer inverse modelling and imaging spectrometry to quantify the structural and biophysical attributes of plant canopies and land cover types in a South Texas savanna. The studies resulted in plant area estimates

(LAI + NPVAI) well within the error and variance limits of field measurements, although it was demonstrated that the presence of nonphotosynthetic vegetation (e.g. stem and litter) had a significant effect on canopy $f_{APAR}$. In trees with a LAI <3.0, stem surfaces increased canopy $f_{APAR}$ by 10-40%. Failure to account for the absorption of PAR by nonphotosynthetic plant components (through modelling) might lead to overestimates of primary production, especially in woodlands, savannas, and shrublands dominated by species with optically thin canopies and in grasslands that accumulate senescent material. Asner (1998) extended the Texas data set to include a wider range of climate conditions, plant functional and structural groups, and foliar chemistry. LAI and leaf angle distribution (LAD) were found to be the dominant controls on canopy reflectance data with the exception of soil reflectance and vegetation cover in sparse canopies. Stem material played a small but significant role in determining canopy reflectance in woody plant canopies, especially those with LAI <5.0. The effects of solar and viewing geometry were not addressed in these studies.

Variation in canopy gap fraction, leaf angle distribution and canopy height result in significant non-Lambertian behaviour of vegetation reflectance signatures (e.g. Li et al, 1995). Asner et al (1999) applied a photon transport inverse modelling approach (PTIM) (which physically accounts for reflectance anisotropy by using the known solar and viewing geometry in providing the canopy reflectance estimates), with AVIRIS data in two case studies to estimate PAI (leaf and non-photosynthetic vegetation area index, i.e. LAI + NPVAI) in an arid/semi-arid ecosystem, and LAI and live:dead fraction of pasture in the Amazon Basin, with high accuracy. The authors argue that this is a robust approach for simulating remotely sensed signatures of vegetation and soils, and that it provides a means to scale, in three-dimensional space, from leaf to landscape level. It is predicted that this approach will be applicable to any ecosystem, but the constraints, and thus the breadth of parameters that can be retrieved, will be ecosystem dependent. Two weaknesses are explicitly acknowledged with this approach. Firstly, the model may not adequately account for all of the physical processes involved, and thus an accurate solution via model inversion may not be obtainable, and secondly, there may be insufficient data to execute the inversion e.g. when the model is insufficiently sensitive to data characteristics (e.g. certain spectral wavelengths or viewing angles). These problems still require addressing via an integrated field and modelling approach.

The use of physical models, at both leaf and canopy scales, improves our understanding of the absorption and scattering of solar radiation in vegetation canopies. The direct problem, or the problem of modelling of the canopy wavelength, can be resolved to a high degree of accuracy by using a series of coupled models or complex functions in which there are a large number of variables (Dawson et al 1999). A canopy model used only for estimating the reflectance values can have any number of variables describing the characteristics of the canopy scene. However the inverse problem, or the problem of estimating canopy variables from the reflectance spectra, usually requires a simplified approximation of the canopy characteristics because the greater number of variables makes the inversion of the model harder. This constraint on invertibility, where only a limited number of variables can be reasonably estimated with any degree of accuracy, precludes the inversion of the complex canopy models (Dawson et al 1999).

Although model inversion is theoretically more objective than empirical schemes and is capable of accounting for the bidirectional effects of surface reflectance, the development of adequate inversion algorithms is far from complete (Qui et al, 1998).

## 10.2   BIOMASS/YIELD

Improvements in the retrieval of biophysical characteristics from using hyperspectral data should lead to further progress in the synergetic potential of coupling crop process models and satellite observations for the estimation of crop yield/biomass.

Crop growth can be monitored using crop growth models. However, estimates of crop growth are often inaccurate for nonoptimal growing conditions. Remote sensing can provide information on the actual status of agricultural crops, thus calibrating the growth model for actual growing conditions (Maas, 1988, Delecolle et al, 1992, Clevers et al, 1994).

Crop biomass cannot be measured directly by remote sensing. Although relationships between VIs and $f_{APAR}$ and LAI have been established both theoretically and experimentally, the relationships between cumulated VIs and dry biomass are empirical and only have a local value (Moulin et al, 1998). To estimate production in any conditions, it is then necessary to describe how photosynthetically active energy is absorbed, converted into dry biomass and partitioned to harvestable organs. More mechanistic or physiologically sound models are therefore necessary to assimilate remote sensing data and predict production of major agricultural crops. Moulin et al (1998) list a number of ways of combining a crop model with radiometric observations, with examples given for each method. These are:

- *the direct use of a driving variable* estimated from remote sensing information in the model;

- *the updating of a state variable* of the model (for example, LAI) derived from remote sensing

- *the re-initialization* of the model, i.e. the adjustment of an initial condition to obtain a simulation in agreement with the remotely-sensed derived observations;

- *the re-calibration* of the model i.e. the adjustment of model parameters to obtain a simulation in agreement with LAI derived from the observations.

Re-initialization/re-calibration consists of minimising the difference between a derived state variable or the radiometric signal and its simulation by the re-initialization and/or re-calibration of the crop production model. Clevers and van Leeuwen (1996) used LAI measurements derived from AVIRIS data to calibrate a crop yield model, SUCROS (simplified and universal crop growth simulator). Results for sugar beet indicated that when periodic (about every ten days) optical remotely sensed data is available throughout most of the growing season, LAI can be adequately monitored and a good estimate of sugar beet yield at the end of the season is possible by using the calibrated crop growth model. Clevers (1994) reported that for nine out of ten fields used in the study, the simulated yield agreed more closely with actual obtained yields with the calibrated (compared to non-calibrated) model, with the simulation error of fresh beet yield reduced from 6.6t/ha (8.6%) to 4.0 t/ha (5.2%). Since the calibration procedure mainly concerned the calibration of the simulated LAI, the results indicated the importance of LAI for accurate growth simulation. Although re-initialization/re-calibration using a remotely-sensed derived state variable has been shown to improve estimation of biomass time profile and final yield compared to the forcing (updating) method, Moulin et al (1998) concluded that the main drawback with this method is the empirical relation used to derive LAI from satellite data (e.g. the asymptotic nature of the relationship between canopy reflectance and LAI). However this is likely to be less of a problem with improved estimates of LAI derived from hyperspectral data, as noted above (see: *a) Empirically based methods*). An alternative approach is the direct use of radiometric information to re-initialise/ re-calibrate a crop model. This assumes that the temporal behaviour of canopy surface reflectances, as they can be observed from a sensor, can be reproduced by coupling a radiative transfer model to the crop production model (see *b) Physically-based methods*, for examples of the use of radiative transfer models). The minimisation of differences between the simulated and observed reflectances is carried out by adjusting initial conditions or model parameters. However, this approach requires multiangular measurements to retrieve canopy properties (LAI, LAD and leaf optical properties) and, as discussed in the next section, current and planned off-nadir looking satellite-based instruments are not high spectral resolution.


## 11    Other applications


## 11.1    EXTRACTING BIOCHEMICAL VARIABLES

With a large number of bands it becomes possible to more accurately identify and estimate chemical properties of plants. Quantitative estimates of the biochemical parameters of plant species can provide an index to underlying ecological processes (Peterson & Hubbard, 1992). Many of the biogeochemical processes, such as photosynthesis, net primary production, evapotranspiration, and decomposition are related to the content of chlorophyll, nitrogen, water, lignin, and cellulose in leaves (Goetz and Prince, 1996, Running, 1990, Running and Coughlan, 1988). The description and modelling of ecosystem processes such as photosynthesis, carbon and nitrogen cycling require measurements of canopy chemistry (Vitousek 1982). Curran (1994) highlighted the current emphasis in the use of hyperspectral remote sensing with plant canopies on the estimation of both nitrogen and lignin concentration because these two chemicals are indicators of the rate of carbon fixation by canopies and the rate of litter decomposition. Attempts to drive an ecosystem simulation model with meteorological and remotely sensed data have indicated that the accuracy with which such models can estimate net primary productivity can be increased markedly if estimates of foliar biochemical concentration (e.g. nitrogen) are used as one of the inputs (Curran 1994, Green *et al* 1996).

Martin *et al* (1999), report on the preliminary results of a project known as MAPBGC (Mapping and Analysis of Productivity and BioGeochemical cycles), involving the combined use of hyperspectral remote sensing (AVIRIS), intensive field data collection, historical land use records and modelling. Assessment of canopy-soil-stand interactions from a mixed hardwood/conifer forest suggest that canopy chemistry may be used as a direct scalar of ecosystem properties e.g. Aboveground net primary production (ANPP) as a function of foliar N concentration.

AVIRIS-predicted nitrogen and lignin from mixed forest stands have been used as input data for two models of ecosystem processes (Martin and Aber 1997). Canopy nitrogen concentrations were used to drive a carbon-balance model (Aber and Federer, 1992), resulting in an estimate of net ecosystem productivity, and canopy lignin concentration was used to drive a simple linear model of nitrogen mineralization. To increase the accuracy of the model predictions, Martin *et al* (1998) used the relationship between AVIRIS spectral data and foliar chemistry, in addition to species specific chemical characteristics, as a basis for classifying species composition, which was achieved with a classification accuracy of 75%.

The drawbacks that apply to using empirical methods for extracting biophysical variables also apply to the extraction of biochemical variation. Thus although a given biochemical may result in a well-defined absorption feature at a given wavelength (or series of wavelengths), remote sensing measures canopy or pixel-level absorption at the given wavelength(s), which occurs through a combination of the foliar biochemical concentration, canopy LAI, leaf orientation and canopy architectural development. Although Curran et al (1997) reported significant relationships between canopy chemistry and AVIRIS reflectance data in a slash pine stand, variation in leaf angle distribution was probably minimized through using only the one species, thereby allowing variation in foliar chemistry to translate most effectively to canopy-level reflectance. Asner *et al* (1997, 2000) and Asner (1998) showed that in most ecosystems, variation in fully-developed, green-leaf optical properties plays a secondary role to structural attributes in driving pixel-level reflectance level variation. (see also: section on canopy reflectance models, Vegetation biochemistry.

## 11.2  CARBON FLUXES

Modelling of carbon fluxes is an increasingly important application in ecosystem modelling. BRDF models have shown that some important carbon-balance parameters, such as $f_{APAR}$ are well estimated by empirical measures such as the NDVI. However, NDVI varies with look angle, so that a single look is not likely to characterise properly the behaviour of the plant cover (Myneni *et al*, 1992). Yet if the BRDF can be derived from multiangle measurements, the biophysical parameters can be properly summarised, either empirically or through more specific biophysical process models (Strahler, 1997). Multiangular satellite data, by providing constraints on the reflectance model parameters, will thus lead to more robust modelling. However, current and planned off-nadir looking satellite-based instruments are not high spectral resolution.

Since biophysical canopy-level parameters (e.g. LAI, $f_{APAR}$) are often of primary interest to remote sensing scientists, biosphere-atmosphere modellers, and ecologists, leaf optical properties tend not to be a direct goal of model retrievals (Asner *et al* 1998c). This, of course, does not hold true in efforts to retrieve canopy chemical characteristics from remotely sensed hyperspectral data. However, as previously mentioned, current and planned off-nadir looking instruments employable for BRDF inversion methods do not have high spectral resolution, which is necessary for canopy chemistry estimates. Although AVHRR, MODIS and MISR can acquire off-nadir imaging radiance measurements with adequate repeatability for BRDF model inversions (Asner *et al* 1998c) they are not of sufficiently high spectral resolution for canopy chemistry estimates.

## 11.3  COVER

Roberts *et al* (1999) report on the contribution of AVIRIS to the Boreal Ecosystem Atmosphere Study (BOREAS), aimed at modelling and predicting fluxes of energy, mass and gases (including carbon dynamics) in the boreal forest biome. The two major advantages in using AVIRIS are predicted to be: 1) direct assessment of biochemical/ biophysical information and 2) improved land cover classification. The authors describe preliminary results suggesting that an improved land-cover classification developed using spectral mixture analysis and Multiple Endmember Spectral Mixture Analysis (MESMA) of AVIRIS data is indeed possible.

Chewings *et al* (2000) applied a decision-tree approach to successfully predict broad groups of arid-zone woody vegetation using HYMAP data, and concluded that airborne hyperspectral imagery appears to be able to discriminate the main soil and vegetation components in a complex arid environment more readily than multispectral data.

# CHAPTER 6

# IMAGING SPECTROMETRY FOR AGRICULTURAL APPLICATIONS

Jan P.G.W. CLEVERS [α] & Raymond JONGSCHAAP [β]

[α] Centre for Geo-Information Wageningen-UR, The Netherlands

[β] Plant Research International, Business Unit Agrosystems Research, The Netherlands

## 1    Introduction

Currently, the use of remote sensing for agricultural applications is one of the main application fields of remote sensing techniques. The possibilities of applying remote sensing in agriculture has been demonstrated, for instance, with regard to the estimation of crop characteristics such as soil cover and leaf area index (LAI). LAI is regarded as a very important plant characteristic because photosynthesis takes place in the green plant parts. The LAI is also a main driving variable in many crop growth models, designed for yield prediction (Bouman, 1991; Clevers et al., 1994; Delécolle et al., 1992; Maas, 1988). Crop growth models describe the relationship between physiological processes in plants and environmental factors such as solar radiation, temperature and water and nutrient availability. Estimates of crop growth may be inaccurate for sub-optimal growing conditions, such as stress conditions that are not included or wrongly interpreted in the model descriptions. In this respect, information concerning leaf or crop nitrogen status currently is a key item. Remote sensing may yield information about the actual status of a crop, resulting in an improvement of crop growth modelling.

The green, red and (near-)infrared reflectances may be used as variables for estimating LAI. Much research has been aimed at establishing combinations of the reflectances in different wavelength bands, to minimise the undesirable disturbances of differences in soil background or atmospheric conditions. Such combinations are called vegetation indices (see, for instance, Rouse et al., 1973, 1974; Kauth & Thomas, 1976; Richardson & Wiegand, 1977; Huete, 1988; Baret et al., 1989; Clevers, 1988, 1989). However, when using some combination of reflectances, one should be careful not to lose sensitivity to variations in LAI after complete soil cover has been reached. This also means that the near-infrared (NIR) reflectance should play a dominant role in such a combination.

In order to ascertain such vegetation indices, one can measure vegetation reflectance in rather broad spectral bands (20 - 50 nm). Laboratory spectral measurements using spectrometers showed that specific absorption features of



*F.D. van der Meer and S.M. de Jong (eds.), Imaging Spectrometry,* 157–199.
© 2006 *Springer. Printed in the Netherlands.*

individual dried, ground leaves may be found when the spectral resolution is high (band width in the order of 10 nm or smaller). In this way, in addition to the main absorption features caused by pigments and water, a large number of minor absorption features were found (Curran, 1989). These minor features are correlated to concentrations of leaf organic compounds, such as cellulose, lignin, protein, sugar and starch. Absorption is most pronounced below 400 nm and above 2400 nm. The absorption features of these organic compounds are quite weak in the range 400-2400 nm.

Potentially, the use of spectrometers to measure the reflected radiation of vegetation offers new opportunities to estimate important carbohydrates of plants (Elvidge, 1990). Specific absorption features caused by these compounds may also be found when moving such a spectrometer into an aeroplane (or even satellite) and using it as an imaging remote sensing technique (Goetz, 1991). However, up to now the remote sensing of foliar chemical concentrations (other than chlorophyll and water) has not been very successful. The presence of water in living leaf tissue almost completely masks these biochemical absorption features (Vane & Goetz, 1988). A number of airborne spectrometers have been developed, operating in the 400-2400 nm spectral range. In the near future some spaceborne systems will become available. These instruments do not operate in the region where the absorption features of leaf chemicals are most pronounced. Below 400 nm wavelength atmospheric influence disturbs the remote recording of such features. Above 2400 nm not enough solar radiation reaches the earth's surface to allow recording from a remote platform in narrow spectral bands.

In this chapter, the possible role of imaging spectroscopy in agriculture will be studied (section 2). Subsequently, emphasis will be put on the red-edge index, its definition, and a sensitivity analysis will be performed (section 3). In section 4, ground-based observations are used for studying the relationship between the red-edge index and the nitrogen status of agricultural crops. Subsequently, in section 5 airborne observations are used for describing a framework for crop yield prediction including information from the red-edge index as an important component. Next, section 6 describes the potential of a spaceborne instrument in deriving the red-edge index. Finally, the main conclusions are listed in section 7. In Chapter 5 of this book, leaf models are also addressed.


## 2    Role of imaging spectroscopy in agriculture


### 2.1    INTRODUCTION

As stated before, growth of agricultural crops may be sub-optimal as a result of stresses, such as fertiliser deficiency, pest and disease incidence, drought or frost. Vegetation response to stress varies with both the type and the degree of stress. On the one hand, stress may cause biochemical changes at the cellular and leaf level, which have an influence, e.g., on pigment systems and the canopy moisture content. On the other hand, stress may cause biophysical changes in canopy structure, coverage, LAI or biomass.

Up to now most promising results for detecting the occurrence of plant stress are obtained by studying the sharp rise in reflectance of green vegetation between 670 and 780 nm (Horler et al., 1983). This region is called the red-edge. Both the position and the slope of the red-edge change under stress conditions, resulting into a blue shift of the red-edge position. The position of the red-edge is defined as the position of the main inflexion point of the red-NIR slope. This is called the red-edge index. Reliable detection of this index requires sampling at about 10 nm intervals or less, requiring high-resolution spectral measurements.

In this section, the potential of imaging spectroscopy for agricultural applications is evaluated by studying the information contained in imaging spectrometer data. This will result into a selection of the most significant parts of the electromagnetic spectrum for agricultural studies. It will also show whether information about plant biochemistry can be obtained and whether the red-edge is a significant region. For this study, data of the airborne visible-infrared imaging spectrometer (AVIRIS) obtained during the MAC Europe 1991 campaign in The Netherlands are used. The optimal set of spectral bands is being selected by applying a principle component analysis (factor analysis) to the available AVIRIS data.


## 2.2    SPECTRAL ANALYSIS (PCA)

In order to select the optimal set of spectral bands from a large number of bands as in imaging spectroscopy, a principal component analysis (a simplified form of factor analysis) is performed first. Factor analysis is a statistical technique used to identify a relatively small number of factors that can be used to represent relationships among sets of many interrelated variables (see, e.g., Finn, 1974; Harman, 1968; Kim & Mueller, 1978). In the case of imaging spectroscopy of vegetation the (observed) variables are the responses in the individual spectral bands, whereas (unobservable) factors could be common sources of variation like leaf chlorophyll content, leaf structure, water content, biochemistry, LAI or leaf angle distribution (LAD) at canopy level.

In a principal component analysis (used for the factor extraction), linear combinations of the observed variables are formed (Mather, 1976). The first principal component is the combination that accounts for the largest amount of variance in the sample. The second principal component is uncorrelated with the first one and accounts for the next largest amount of variance. Successive components explain progressively smaller portions of the total sample variance, and all are uncorrelated with each other. To help us decide how many principle components (factors) we need to represent the data, it is helpful to examine the percentage of total variance explained by each.

Although the factor matrix obtained in the principal component analysis indicates the relationship between the factors and the individual variables, it is usually difficult to identify meaningful factors based on this matrix. Often the variables and factors do not appear correlated in any interpretable pattern. Most factors are correlated with many variables. Since one of the goals of factor analysis is to identify factors that are substantially meaningful, a factor rotation attempts to transform the initial factor matrix (from the principal component analysis) into one that is easier to interpret. If each factor would have high loadings for only some of the variables, this would help the interpretation. Moreover, if many variables would have a high loading on only one factor, the factors could be differentiated from each other. Most rotation procedures (e.g. the varimax procedure) try to realise such a simple structure. It should be noted that the explained variance is redistributed over the individual factors, while the total variance explained by the chosen number of factors does not change.

Finally, to identify the factors, an interpretation has to be given to groups of variables that have large loadings for the same factor.

## 2.3    CASE STUDY

### 2.3.1    *MAC Europe campaign*

The potential of using imaging spectroscopy for agricultural applications was tested in a case study using data of the MAC Europe 1991 campaign from the Flevoland test site in The Netherlands. In the MAC Europe campaign, initiated by the National Aeronautics and Space Administration (NASA) and the Jet Propulsion Laboratory (JPL), both radar and optical airborne measurements were made over selected test sites during the growing season of 1991. In the optical remote sensing domain, NASA executed one overflight with the AVIRIS scanner (for system description, see Vane et al., 1984). An extensive description of the collected ground truth and of the airborne optical data during the 1991 season over Flevoland is provided by Büker et al. (1992a,b).

### 2.3.2 *Test site*

The test site was located in Flevoland in The Netherlands, an agricultural area with very homogeneous soils reclaimed from the lake "IJsselmeer" in 1966. The test site comprised ten different agricultural farms, 45 to 60 ha in extension. Main crops were sugar beet, potato and winter wheat. By considering a range of agricultural crops and bare soils it is made sure that considerable biochemical variations (referring to contents of components such as chlorophyll, cellulose, lignin and water) are present within the data set used.

### 2.3.3 *AVIRIS*

The ER-2 aircraft of NASA, carrying AVIRIS, performed a successful flight over the Flevoland test site on July 5th, 1991 (being the middle of the growing season). AVIRIS acquires 224 contiguous spectral bands from 410 to 2450 nm. Both the spectral resolution and the spectral sampling interval are about 10 nm. However, because during the recording of the Flevoland test site the fourth spectrometer in the SWIR range yielded only noise data, spectral information was available only in the 400 nm to 1860 nm wavelength range. The ground resolution is 20 m as it is flown at 20 km altitude. For this study, no atmospheric correction was applied. Assuming that atmospheric influence for all spectral bands individually is a linear function of the measured radiances or DNs as confirmed by AVIRIS calibrations using a special version of the LOWTRAN model (Van den Bosch & Alley, 1990), the calculation of factor loadings in the next section yields the same results for either DNs or reflectances.

## 2.4    RESULTS

A principal component analysis and factor rotation was applied to the AVIRIS data of July 5th 1991. As stated in section 2.3.3, the fourth spectrometer (from 1830 nm onwards) was not functioning. Moreover, measurements near the water absorption bands yielded only noisy data. As a result, spectral bands from 410 nm to 1350 nm and from 1480 nm to 1800 nm were used in the analysis (a total of 135 spectral bands). A selection of the spectral signatures of 101 pixels within the test site was made. All crops and bare soil were included in the data set, whereas for each object type pixels were randomly selected.

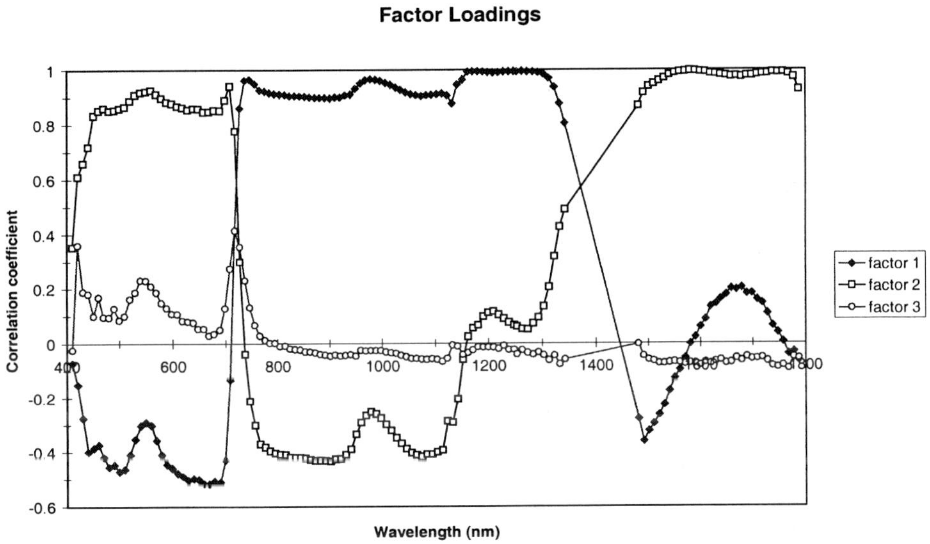

**Factor Loadings**



*Figure 1*. Factor loadings (correlation coefficients) for the main factors resulting from a principal component analysis and factor rotation for an agricultural data set based on spectral bands of AVIRIS spectrometers 1, 2 and 3. Flevoland test site, July 5th 1991.

Applying the usual criteria, the principal component analysis resulted in three factors explaining 96.8% of the total variance in the selected data set (Table 1). Subsequently, a factor rotation was performed. Figure 1 illustrates the relationship between the initial spectral bands and the three rotated factors. Depicted are the factor loadings which equal the correlation coefficients between the spectral bands and the respective factors. Figure 1 shows that factor 1 is highly correlated to the NIR region (from 730 nm up to about 1350 nm) and little to all other bands. This Figure also shows that this factor 1 may be described as one broad band in this NIR region. Factor 2 appears to be highly correlated with the visible region (from about 500 nm up to 700 nm) and with the SWIR region (from about 1500 nm onwards). As a result, factor 2 may be described as a combination of two broad bands, one in the VIS region and one in the SWIR region (up to 1800 nm). Finally, factor 3 does not exhibit high correlations with any spectral band at all. However, it should be noticed that factor 3 shows the highest correlation with a few spectral bands around 717 nm (see Figure 1). This is exactly the region of the red-edge, not covered by factors 1 or 2. All other factors did not yield any significant results or clear correlation with specific spectral bands.

It may be concluded that the principal component analysis on AVIRIS data confirms that the investigated data set can best be described by one broad spectral band in the NIR region, one broad band in the VIS region, and one broad band in the SWIR region between the two main water absorption features. However, the bands in the VIS and in the SWIR appear to be highly correlated. This may be attributed to a strong correlation between chlorophyll content and leaf water content as often observed for agricultural crops, but there was no ground truth available in this study to confirm this. The results also indicate that some extra information may be provided by spectral measurements around 717 nm (the red-edge region), not covered by the information provided by a combination of an NIR and a VIS broad spectral band. Concerning high spectral resolution data it seems to be most promising to pay particular attention to this red-edge region.

In judging the factors resulting from this analysis, factor 1 may be related to the leaf mesophyll structure and the LAI (NIR reflectance). Factor 2 may be related to the leaf chlorophyll content and the LAI (VIS reflectance). Moreover, factor 2 may be related to the leaf water content (SWIR reflectance). It is noticeable that the VIS and SWIR reflectances are highly correlated for the analysed AVIRIS data. The effects of LAI on VIS and NIR reflectances may explain the negative correlation between factor 1 and factor 2. Generally, NIR reflectance increases with increasing LAI, whereas VIS reflectance decreases.

TABLE 1: Results of the principal component analysis on AVIRIS data for the first ten factors.

| Factor | Eigenvalue | % of total variance | % of total variance cumulative |
|---|---|---|---|
| 1 | 87.779 | 64.5 | 64.5 |
| 2 | 42.655 | 31.4 | 95.9 |
| 3 | 1.248 | 0.9 | 96.8 |
| 4 | 0.975 | 0.7 | 97.5 |
| 5 | 0.610 | 0.4 | 98.0 |
| 6 | 0.539 | 0.4 | 98.4 |
| 7 | 0.467 | 0.3 | 98.7 |
| 8 | 0.255 | 0.2 | 98.9 |
| 9 | 0.178 | 0.1 | 99.0 |
| 10 | 0.174 | 0.1 | 99.2 |

## 2.5    CONCLUSIONS

As a conclusion, it must be stated that the AVIRIS data set of the Flevoland test site did not provide information on leaf biochemistry, except for leaf chlorophyll content, because the spectra were obtained from living vegetation. The influence of cell water and cell structure will obscure the effect of single cell biochemical components in the SWIR region. The results indicate that some additional information may be provided by spectral measurements at the red-edge region, not covered by the information provided by a combination of an NIR and a VIS broad spectral band. Concerning high spectral resolution data, this seems to be the major contribution to applications in agriculture. Combining high-resolution spectral measurements with broad-band measurements will yield information on LAI, LAD and leaf chlorophyll content, which is important for an accurate monitoring of crop growth and prediction of crop yield (see section 5 of this chapter). This is of importance for monitoring of agricultural production at a national and a regional level. In addition to a red and NIR spectral band this requires a few narrow bands at the red-edge slope. This may be considered as a minimal requirement for future sensor systems when they have to be applied for monitoring agricultural crops. It should be mentioned that some additional narrow spectral bands at specific wavelengths might be useful for atmospheric correction procedures (Conel et al., 1988; Gao & Goetz, 1990; Green et al., 1991). The latter topic was not part of this study. Future research should be focused at a further improvement of growth models using the information on the leaf nitrogen status as related to leaf chlorophyll content. This seems to offer the best opportunities for using imaging spectroscopy in agriculture.


## 3    Red-edge index


## 3.1    INTRODUCTION

From literature (see Clevers & Büker, 1991) it may be concluded that the red-edge shift is mainly related to the leaf chlorophyll content and to the LAI. This means that this red-edge index and a vegetation index for estimating LAI contain complementary information. In order to use the red-edge index for estimating the leaf chlorophyll content (or even the leaf nitrogen status), first (or simultaneously) the LAI has to be estimated. A framework for this will be described in case study I and II of this chapter (section 4 and 5). In this section, methods for deriving the red-edge index will be described (section 3.2). Subsequently, the variables that influence the red-edge index will be studied by performing a sensitivity analysis (section 3.4) using radiative transfer models (section 3.3). At the end the main conclusions are listed (section 3.5).


## 3.2    DEFINITION RED-EDGE INDEX

Since the position of the red-edge mostly is defined as the inflexion point (or maximum slope) of the red-NIR slope, an accurate determination requires a large number of spectral measurements in narrow bands in this region. Subsequently, the red-edge position is defined by the maximum first derivative of the reflectance spectrum in the region of the red-edge. High-order curve fitting techniques are employed to fit a continuous function to the derivative spectrum (Horler et al., 1983; Demetriades-Shah & Steven, 1988; Demetriades-Shah et al., 1990). However, existing curve fitting techniques are quite complex and computationally demanding. Recently, Dawson & Curran (1998) presented a technique based upon the three-point Lagrangian interpolation technique for location of the red-edge position accurately in spectra that have been sampled coarsely. A problem may arise when the reflectance spectrum exhibits more than one maximum in its first derivative. Horler et al. (1983) identified two components in the first derivative spectrum with peaks around 700 and 725 nm. The red-edge index then represented whichever component was dominant, and therefore a red-edge shift could involve a jump between the two components, creating a discontinuity in the red-edge index. They found that the second component moves to longer wavelengths and becomes more pronounced as the LAI increases.

For practical reasons, fitting a curve to just a few measurements in the red edge region often is applied for approximating the inflexion point. First, a polynomial function may be fitted to the data (Clevers & Büker, 1991). Secondly, a so-called inverted Gaussian fit to the red-NIR slope may be applied (Bonham-Carter, 1988). Finally, Guyot & Baret (1988) applied a simple linear model to the red-NIR slope. A comparison of the three methods yielded comparable results (Clevers and Büker, 1991; Büker & Clevers, 1992). Since the method of Guyot & Baret (1988) uses only four wavelength bands, this method is most likely to be applicable to future sensors. Moreover, Clevers et al. (2000) found for various data sets that the latter method is much more robust than procedures based on the first derivative spectrum because the first derivative spectrum did not have one unique maximum. Therefore it was chosen to use the method of Guyot and Baret for this study.

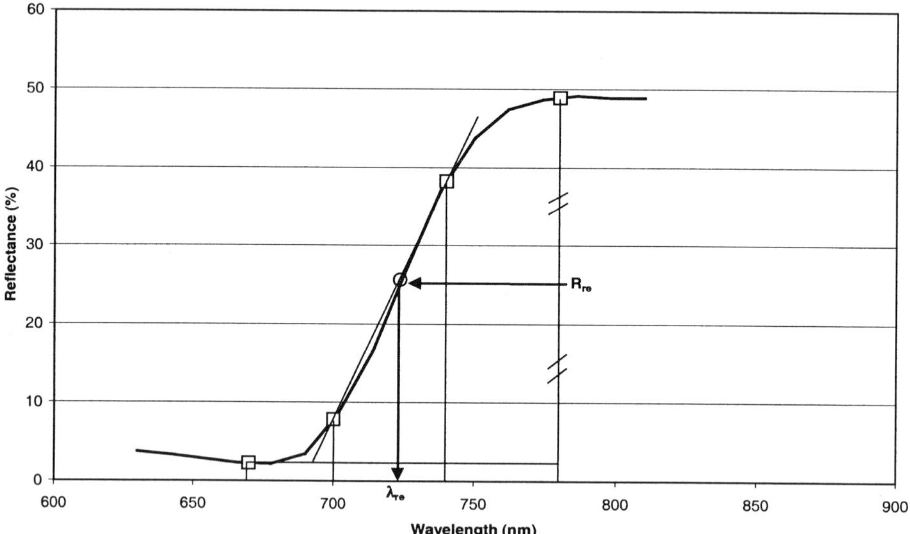


*Figure 2.* Illustration of the "linear method" (Guyot & Baret, 1988).

The linear interpolation as described by Guyot & Baret (1988) assumes the reflectance red-edge can be simplified to a straight line centred around a midpoint between the reflectance in the NIR at 780 nm and the reflectance minimum of the chlorophyll absorption feature at about 670 nm. First of all, they estimated the reflectance value at the inflexion point (Figure 2). Secondly, they applied a linear interpolation procedure between the measurements at 700 and 740 nm for estimating the wavelength corresponding to the estimated reflectance value at the inflexion point. From Figure 2 it may be concluded that such a linear interpolation will offer a good approximation. This method, which will be called the "linear method", can be described in the following way.

1) Calculation of the reflectance at the inflexion point ($R_{re}$)

$$R_{re} = (R_{670} + R_{780})/2 \tag{1}$$

2) Calculation of the red-edge wavelength ($\lambda_{re}$):

$$\lambda_{re} = 700 + 40((R_{re} - R_{700})/(R_{740} - R_{700})) \tag{2}$$

$R_{re}$ is the estimated reflectance value at the main inflexion point.
$R_{670}$, $R_{700}$, $R_{740}$ and $R_{780}$ are the reflectance values at 670, 700, 740 and 780 nm wavelength, respectively.

The value 700 in Eq. (2) refers to the wavelength position belonging to $R_{700}$. This position may deviate from 700 nm a bit as long as it is at the bottom linear part of the slope. The actual position should always be used in Eq. (2). The value 40 in Eq. (2) refers to the length of the wavelength interval between $R_{700}$ and $R_{740}$. Also, the position belonging to $R_{740}$ may deviate from 740 nm a bit as long as it is at the top linear part of the slope. The actual wavelength interval should always be used in Eq. (2).

## 3.3    SIMULATIONS USING RADIATIVE TRANSFER MODELS

In order to perform a theoretical study towards the possibilities of imaging spectroscopy for agricultural applications, a leaf reflectance model (PROSPECT) and a canopy reflectance model (SAIL) are used. Both models are linked to one another. Since it is concluded from section 2 that the information provided by the red-edge index (only obtainable from narrow spectral bands in the red-NIR region) is most promising for agricultural applications, in addition to traditional broad-band spectral information, emphasis is put on this red-edge index.

### 3.3.1    SAIL model

The one-layer SAIL radiative transfer model (Verhoef, 1984) simulates canopy reflectance as a function of canopy variables (leaf reflectance and transmittance, LAI and LAD), soil reflectance, ratio diffuse/direct irradiation and solar/view geometry (solar zenith angle, zenith view angle and sun-view azimuth angle). The SAIL model has been extended with the hot spot effect (Looyen et al., 1991). Leaf inclination distribution functions used with the SAIL model are given by Bunnik (1978) and Verhoef & Bunnik (1981). The SAIL model has been used in many studies and validated with various data sets (e.g. Goel, 1989).

### 3.3.2    PROSPECT model

Jacquemoud & Baret (1990) developed a leaf model that simulates leaf reflectance and leaf transmittance as a function of leaf properties: the PROSPECT model. The PROSPECT model is a radiative transfer model for individual leaves. It is based on the generalized "plate model" of Allen et al. (1969, 1970), which considers a compact theoretical plant leaf (without air cavities) as a transparent plate with rough plane parallel surfaces. An actual leaf is assumed to be composed of a pile of N homogeneous compact layers separated by N-1 air spaces. The compact leaf (N = 1) has no intercellular air spaces or the intercellular air spaces of the mesophyll have been infiltrated with water. The discrete approach can be extended to a continuous one where N need not be an integer. PROSPECT allows to compute the 400-2500 nm reflectance and transmittance spectra of very different leaves using only three input variables: leaf mesophyll structure parameter N, pigment content and water content. All three are independent of the selected wavelength. The output of the PROSPECT model can be used directly as input into the SAIL model. As a result, these models can be combined into one combined model.

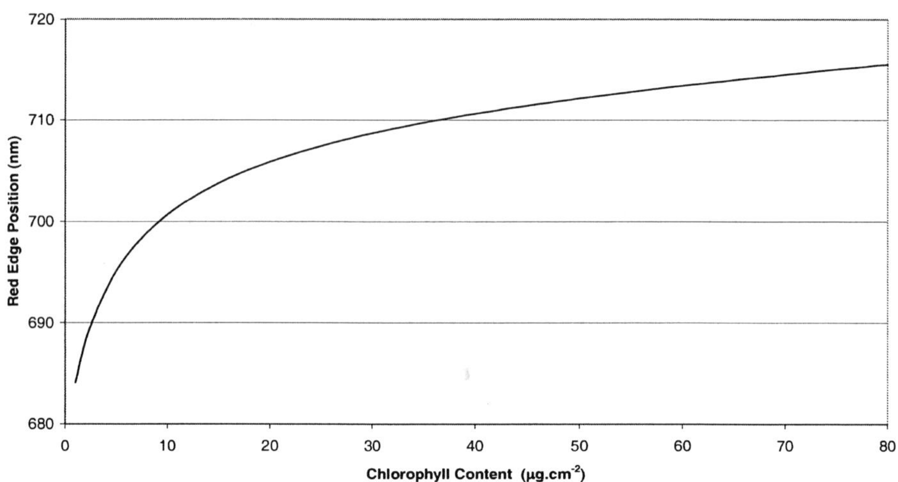


*Figure 3.* Simulated leaf red-edge values as a function of leaf chlorophyll content (N parameter of 1.8320). PROSPECT simulation.

## 3.4    SENSITIVITY ANALYSIS

One of the interesting features of the red-edge index is that it seems to be independent of soil reflectance. Moreover, the atmosphere seems to have only a minor influence on the position of the red-edge. Both the soil background and the atmospheric influence hamper the use of solely one spectral band in the visible part of the spectrum for estimating leaf chlorophyll content (leaf nitrogen). In this section a sensitivity analysis is described using theoretical leaf and canopy reflectance models in order to study the influence of leaf and canopy variables and of external variables on the relationship between red-edge index and leaf chlorophyll content. The atmospheric influence on the red-edge index will be given attention in an empirical way.

### 3.4.1    Simulation results at the leaf level

*Introduction* By using the PROSPECT model, one can simulate the spectral reflectance of individual leaves as a function of leaf chlorophyll content and mesophyll structure. Subsequently, it is possible to calculate the position of the red-edge for simulated leaves. The "linear method" is used as a simple method to calculate the position of the red-edge inflexion point. One variable will be varied at a time, whilst the other is kept constant.

*Effect of chlorophyll content on leaf red-edge* Figure 3 illustrates the red-edge index as a function of the leaf chlorophyll content. This figure clearly shows the influence of the chlorophyll content on the position of the red-edge at the leaf level.

**PROSPECT Simulation**

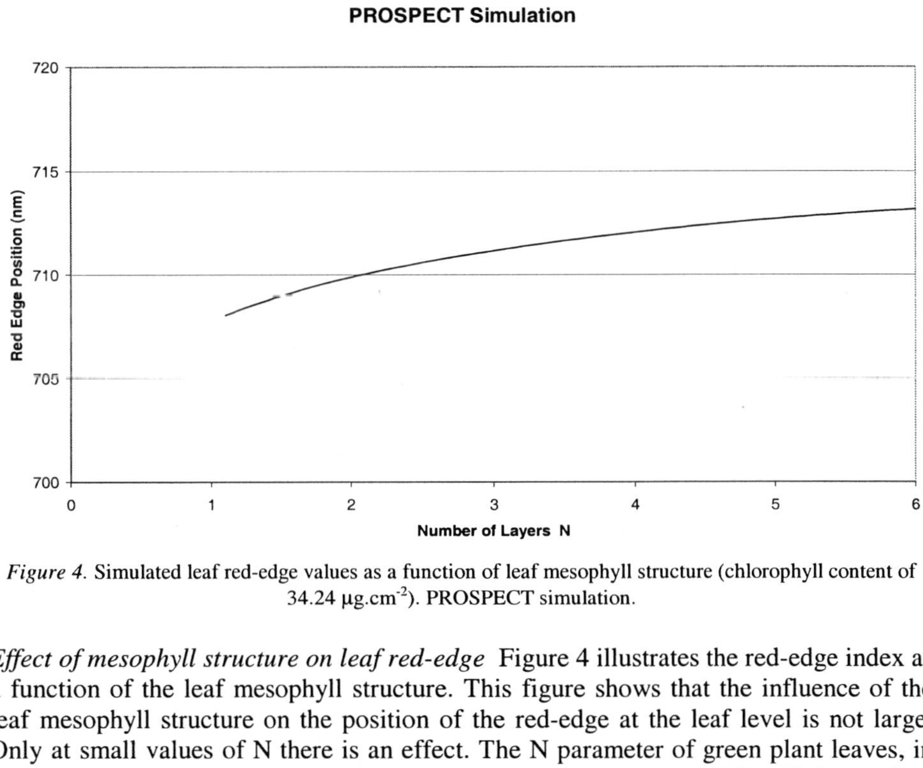

*Figure 4.* Simulated leaf red-edge values as a function of leaf mesophyll structure (chlorophyll content of 34.24 $\mu$g.cm$^{-2}$). PROSPECT simulation.

*Effect of mesophyll structure on leaf red-edge*  Figure 4 illustrates the red-edge index as a function of the leaf mesophyll structure. This figure shows that the influence of the leaf mesophyll structure on the position of the red-edge at the leaf level is not large. Only at small values of N there is an effect. The N parameter of green plant leaves, in general, ranges between 1.0 and 2.5 (Jacquemoud & Baret, 1990). Higher values represent senescent leaves with a disorganized internal structure.

### 3.4.2    Simulation results at the canopy level

In this study, simulations are performed for a standard crop under standard irradiation and viewing conditions unless otherwise indicated. The inputs for this standard crop are:

*chlorophyll content of 34.24 $\mu$g.cm$^{-2}$*     *example of a dicotyledonous plant as*
*N parameter of 1.8320*     *given by Jacquemoud and Baret*

*water content of 0.0137 cm*                    *(1990)*
*leaf area index (LAI) of 4*
*spherical leaf angle distribution (LAD)*
*hot-spot size parameter of 0*
*soil reflectance of 20%*
*only direct solar irradiation*
*solar zenith angle of 45°*
*nadir viewing.*

The effect of changing one input variable at a time is studied. All simulations will be restricted to vertical viewing. The influence of the observation geometry on red and NIR reflectances was studied by Clevers & Verhoef (1993). They concluded that off-nadir viewing effects become significant at angles larger than about 7 degrees (and most pronounced in the direction of the hot spot). So, in this study we did not investigate the influence of off-nadir viewing on the relationship between red-edge index and chlorophyll content.

The simulated position of the red-edge as a function of the leaf chlorophyll content for various LAI values is illustrated in Figure 5. As expected, the red-edge shifts to longer wavelengths with increasing chlorophyll contents. Largest effects occur at the lower chlorophyll contents. This confirms the assumption that the red-edge index can be used as a measure for estimating the leaf chlorophyll content.

**PROSPECT-SAIL simulation**

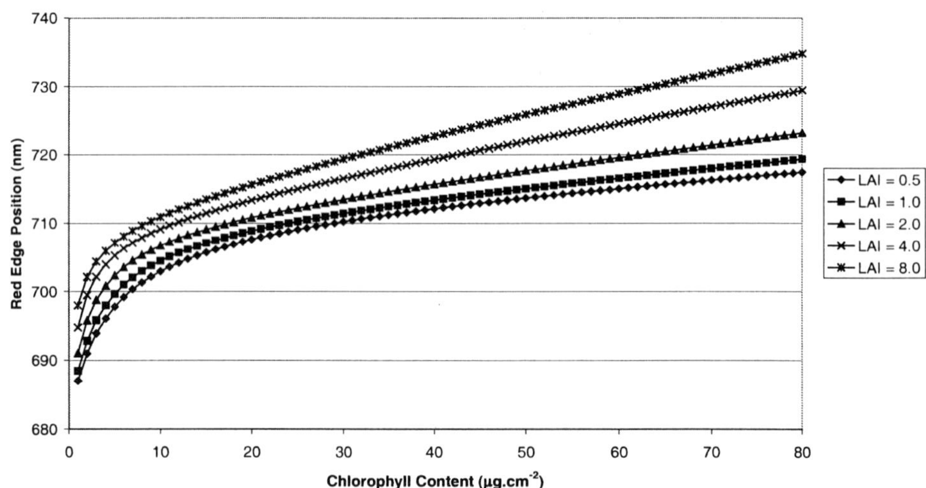


*Figure 5.* Influence of leaf chlorophyll content on simulated red-edge position for several LAI values of the standard crop. PROSPECT-SAIL simulation.

The effect of changing mesophyll structure (N parameter) on the position of the red-edge for various leaf chlorophyll contents is illustrated in Figure 6. At low chlorophyll contents an increasing N parameter (up to N=3) causes a shift of the red-edge position to longer wavelengths whereas higher N values do not change the red-edge position. At high chlorophyll contents the red-edge position changes up to N values of 4. So, the influence of variations in the N parameter will be largest at low values of N.

Figure 5 already showed that the LAI had a clear effect on the position of the red-edge index. The effect of varying LAI on simulated red-edge index for various leaf chlorophyll contents is presented in Figure 7. The LAI has a significant influence on the red-edge position. There is a distinct shift of the red-edge to longer wavelength positions with increasing LAI. This shift is most pronounced at the lower LAI values.

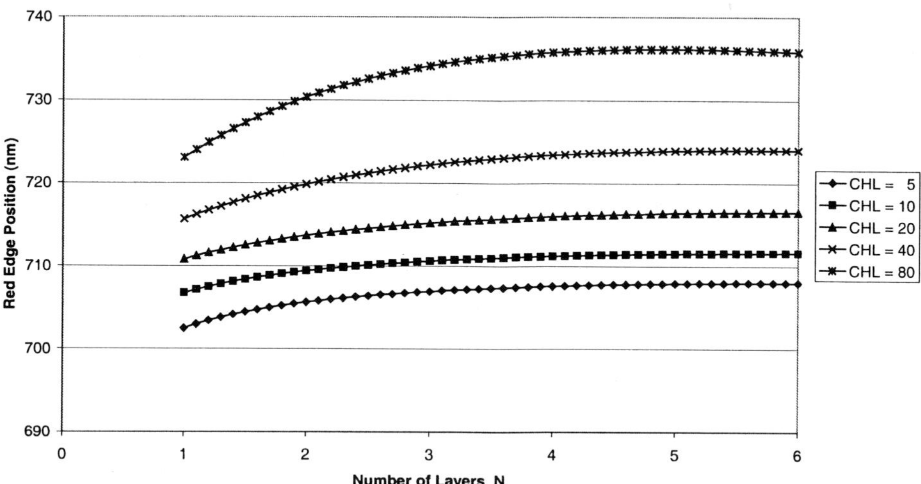


*Figure 6.* Influence of leaf mesophyll structure on simulated red-edge position for various leaf chlorophyll contents (CHL in µg/cm²) of the standard crop. PROSPECT-SAIL simulation.

The effect of varying leaf inclination angle on simulated red-edge index is presented in Figure 8. To perform these calculations a discrete distribution of leaf inclination angles is assumed (Table 2). Since the leaf inclination angle is defined as the angle with the horizontal plane, an erectophile LAD coincides with large leaf angles. The leaf inclination angle has only a small influence on the red-edge position. A small shift of the red-edge index to longer wavelength positions is found with increasing leaf inclination angle from planophile to more erectophile leaves. Whereas hardly any effect at low chlorophyll contents is measured, the influence is more pronounced at high chlorophyll contents.

TABLE 2: Leaf angle distribution functions (LAD) used for the SAIL simulations in this section (percentages per angle are given).

| LAD | Leaf angle | | | | | | | |
|-----|-----|-----|-----|-----|-----|-----|-----|-----|
| no | 5 | 15 | 25 | 35 | 45 | 55 | 65 | 75 | 85 |
| 1 | 100 | 0 | 0 | 0 | 0 | 0 | 0 | 0 | 0 |
| 2 | 0 | 100 | 0 | 0 | 0 | 0 | 0 | 0 | 0 |
| 3 | 0 | 0 | 100 | 0 | 0 | 0 | 0 | 0 | 0 |
| 4 | 0 | 0 | 0 | 100 | 0 | 0 | 0 | 0 | 0 |
| 5 | 0 | 0 | 0 | 0 | 100 | 0 | 0 | 0 | 0 |
| 6 | 0 | 0 | 0 | 0 | 0 | 100 | 0 | 0 | 0 |
| 7 | 0 | 0 | 0 | 0 | 0 | 0 | 100 | 0 | 0 |
| 8 | 0 | 0 | 0 | 0 | 0 | 0 | 0 | 100 | 0 |
| 9 | 0 | 0 | 0 | 0 | 0 | 0 | 0 | 0 | 100 |

**PROSPECT-SAIL simulation**

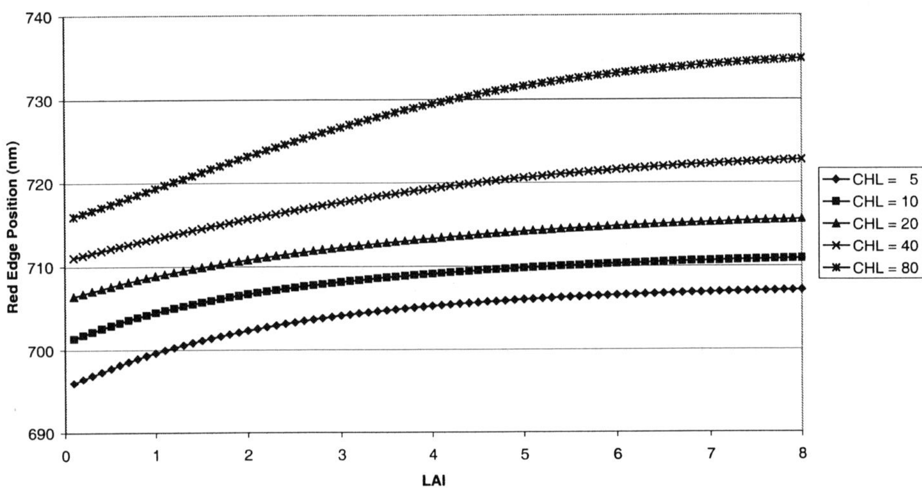

*Figure 7.* Influence of LAI on simulated red-edge position for various leaf chlorophyll contents (CHL in µg/cm²) of the standard crop. PROSPECT-SAIL simulation.

**PROSPECT-SAIL simulation**

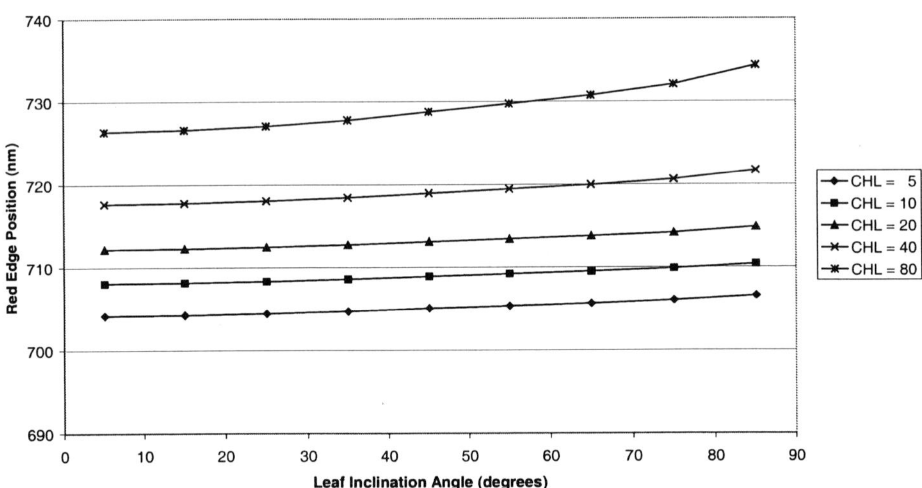

*Figure 8.* Influence of the leaf inclination angle on simulated red-edge position for various leaf chlorophyll contents (CHL in µg/cm²) of the standard crop. PROSPECT-SAIL simulation.

The effect of the size-parameter is illustrated in Figure 9. This confirms the statement that the effect of the size-parameter is only minor.

As concluded before, one of the main advantages of working with the position of the red-edge is its insensitivity to soil background. Figure 10 illustrates the effect of soil reflectance on simulated values of the red-edge for varying leaf chlorophyll contents. Under the condition of our simulations, the position of the red-edge is not very sensitive to soil reflectance (spectrally constant soil reflectance) for a given chlorophyll content. A small shift to longer wavelengths occurs with increasing soil reflectance. Since the influence of soil background on canopy reflectance is most pronounced at low LAI values, Figure 11 presents the effect of soil reflectance on simulated red-edge values for varying LAI. Even at low LAI the soil background hardly has an influence on the position of the red-edge.

**PROSPECT-SAIL simulation**

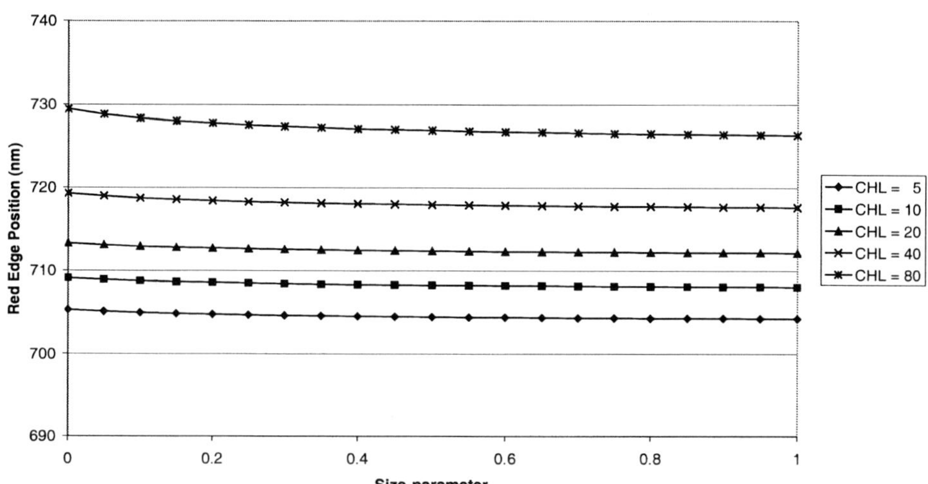

*Figure 9.* Influence of the hot spot size-parameter on simulated red-edge position for various leaf chlorophyll contents (CHL in μg/cm²) of the standard crop. PROSPECT-SAIL simulation.

**PROSPECT-SAIL simulation**

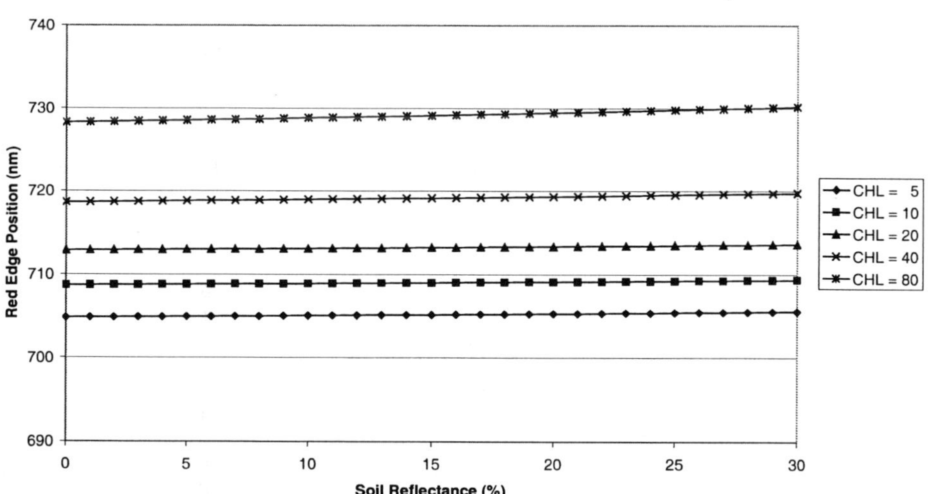

*Figure 10.* Influence of the soil reflectance on simulated red-edge position for various leaf chlorophyll contents (CHL in μg/cm²) of the standard crop. PROSPECT-SAIL simulation.

Figure 12 shows that the solar zenith angle has only a small influence on the position of the red-edge. The largest influence occurs at large chlorophyll contents at large solar zenith angles (larger than 60 degrees). These results do coincide with results reported by Guyot et al. (1988).

**PROSPECT-SAIL simulation**

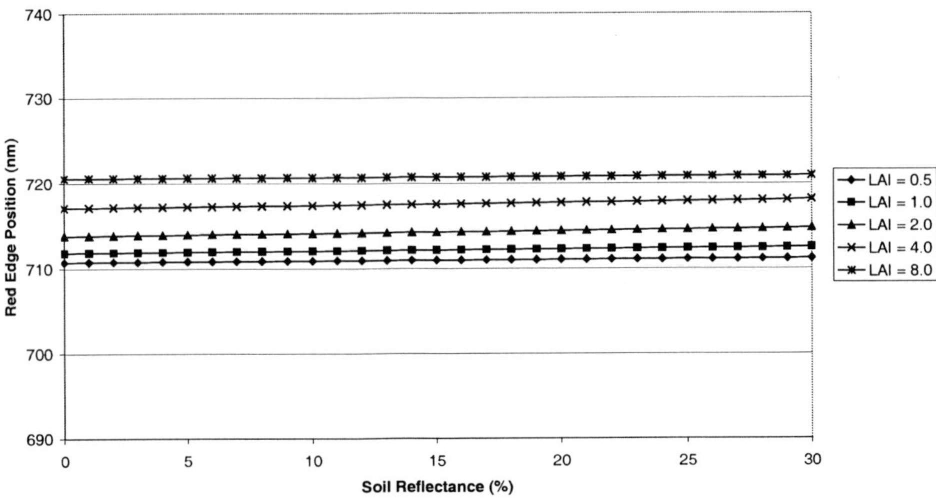

Figure 11. Influence of the soil reflectance on simulated red-edge position for various LAI values of the standard crop. PROSPECT-SAIL simulation.

**PROSPECT-SAIL simulation**

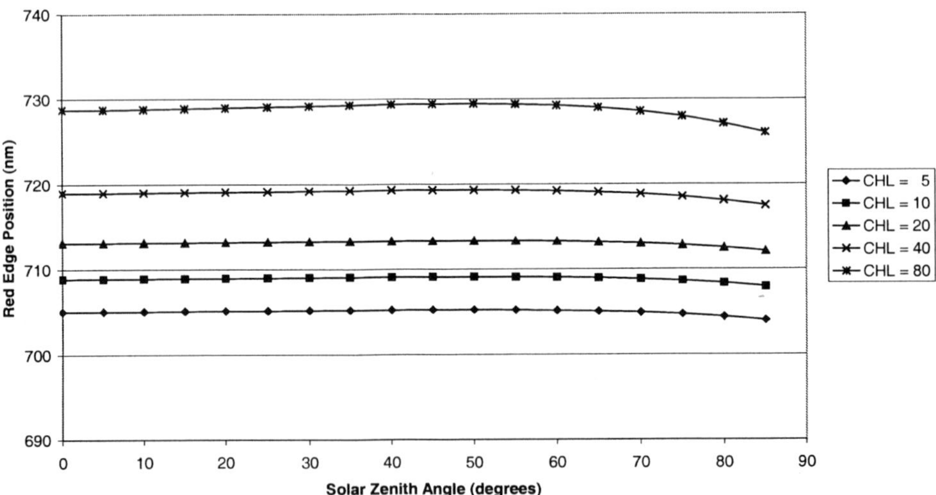

Figure 12. Influence of the solar zenith angle on simulated red-edge position for various leaf chlorophyll contents (CHL in µg/cm²) of the standard crop. PROSPECT-SAIL simulation.

Figure 13 shows that the ratio of diffuse to direct irradiation has only a minor influence on the position of the red-edge.

### 3.4.3    *Atmospheric influence*

Guyot et al. (1988) concluded from simulation studies that the position of the red-edge was unaffected by atmospheric conditions. This would be a second important characteristic of the red-edge (its independence of soil reflectance being the first one). Systematic errors in the atmospheric calibration of some spectral bands may have effects on the position of the red-edge. Moreover, bias in the reflectance measurements may cause false blue shifts that might be confused with real shifts because of vegetation stress (Bonham-Carter, 1988).

**PROSPECT-SAIL simulation**

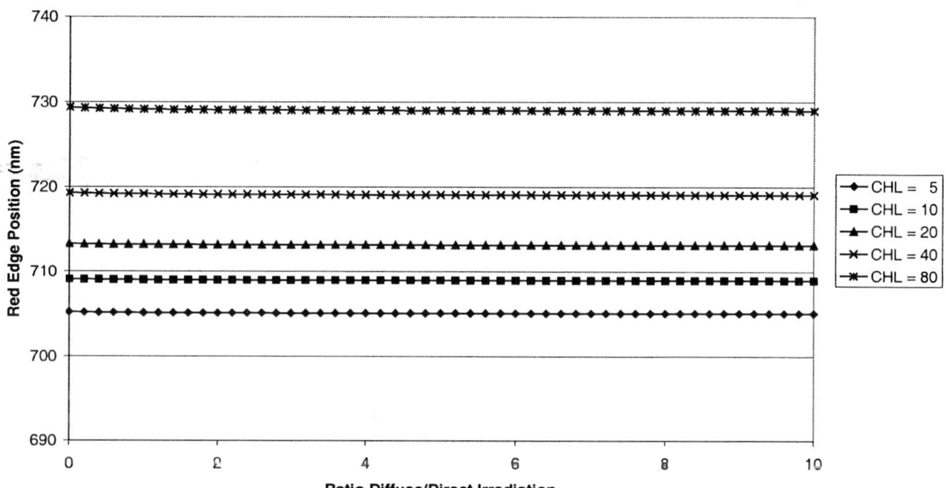


*Figure 13.* Influence of the ratio diffuse/direct irradiation on simulated red-edge position for various leaf chlorophyll contents (CHL in µg/cm²) of the standard crop. PROSPECT-SAIL simulation.

For calibration errors that are constant for a whole scene, systematic (but not constant) shifts of the position of the red-edge will be introduced, but local anomalies will be detectable. However, for multidate or interscene comparisons, calibration errors could produce misleading results.

The atmospheric influence in airborne or spaceborne measurements around the red-NIR slope is captured within the transformation from radiance values at the sensor (or digital numbers) to reflectances at ground level. Clevers & Buiten (1991) describe an atmospheric model based on the model of Richards (1986). However, they extended the model of Richards by including diffuse and direct upward fluxes in addition to the direct and diffuse downward fluxes. Moreover, they included multiple reflections between the earth's surface and the atmosphere. Results of the atmospheric correction procedure using this extended model in combination with parameterization methods of Iqbal (1983) yielded good results for the Flevoland study area in The Netherlands. Similar results were obtained using the model of Verhoef (1985a, 1985b, 1998).

The above atmospheric model was used for studying the influence of the atmosphere on the calculation of the red-edge index. The visibility is used as an indicator for the state of the atmosphere. Figure 14 depicts the influence of the atmosphere on the position of the red-edge for various chlorophyll contents. As expected, the influence of the atmosphere is only minor.

**Atmospheric Influence**

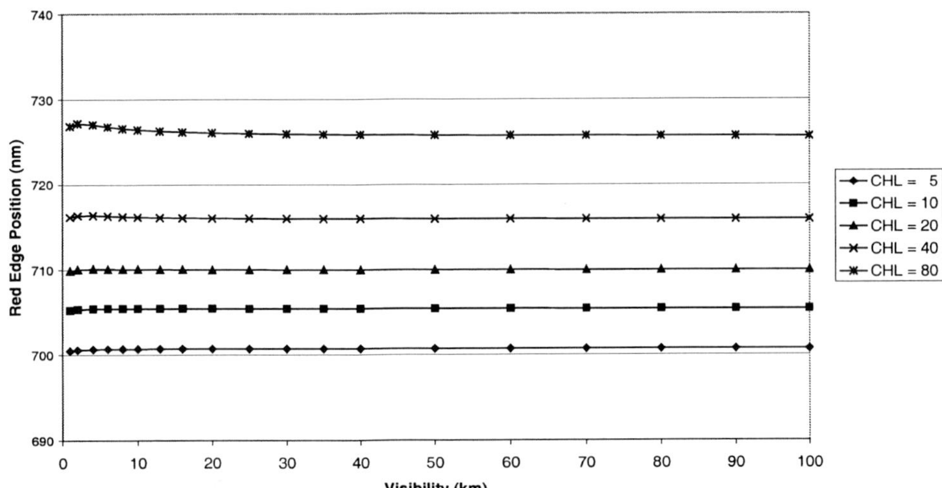


*Figure 14*. Influence of the atmospheric influence (in terms of visibility) on simulated red-edge position for various leaf chlorophyll contents (CHL in $\mu$g/cm$^2$) of the standard crop. PROSPECT-SAIL simulation.


## 3.5    DISCUSSION AND CONCLUSIONS

The merit of high-resolution spectral data for agricultural applications is determined by the possibility of estimating the red-edge index. The LAI and the leaf chlorophyll content, and the LAD to a lesser extent, are the main variables determining the value of this red-edge index as shown by a sensitivity analysis. So, after estimation of LAI and LAD, the red-edge index can be applied for estimating the chlorophyll content. Clevers et al. (1994) applied this within a framework for monitoring crop growth by combining directional and spectral remote sensing information (see section 5).

The red-edge index may be considered as a particular "vegetation index" of interest because of its low sensitivity to disturbing factors such as atmospheric conditions or soil brightness and its high sensitivity to canopy characteristics such as chorophyll content or LAI. Results of the simulation of the effects of the canopy properties and external factors on the spectral shift between red and NIR show the complexity of the phenomena involved. The principal factors are the leaf chlorophyll concentration, the LAI and the LAD. One can therefore imagine that the information provided by the red-edge shift is very close to that contained in the classical wide spectral bands. However, the relative independence for sun zenith angle and soil optical properties makes it a useful tool. Moreover, atmospheric influence on the position of the red-edge is minor.

This section has shown that the position of the red-edge offers valuable additional information for agricultural research. Four narrow spectral bands at about 670, 700, 740 and 780 nm can be used for determining this red-edge information. As a conclusion, it is stated that optimal bands in the optical region for agricultural applications are:

- about 670, 700, 740, 780 nm (5-10 nm width), and
- about 870 nm (broadband).


# 4    Case study I – Red-edge index and crop nitrogen status


## 4.1    INTRODUCTION

The green colour of a crop canopy is closely linked to crop production parameters for photosynthesis, like chlorophyll contents and organic nitrogen status. Nitrogen shortage is expressed by lower chlorophyll contents and leads to reduced photosynthesis rates. The relation between leaf chlorophyll concentration (mg g$^{-1}$), leaf nitrogen concentration (mg g$^{-1}$) and nitrate concentration in petiole sap (mg l$^{-1}$) is strong, and can be described as linear (Vos & Bom, 1993). Optical measurements at *leaf* level as described in earlier paragraphs, do not necessarily provide detailed information on *plant* and *crop* performance, especially not in stress situations. Apart from exogenous factors like climatic conditions and soil fertility, in-plant nitrogen distribution, canopy structure and green biomass will influence photosynthetic rates as well (Dreccer, 1999). Optical observations on the leaf level may provide interesting information, but the transformation into crop management practices at the field level is hardly feasible. The integration of reflection signatures over a whole plot may disregard the spatial variability in a field, which is the essence of modern techniques in agriculture like precision farming. In the presented case study (I), sub-field level observations on a potato crop are explored to demonstrate the link between leaf observations on the one hand, and field crop performance on the other hand.


## 4.2    ESTIMATION OF NITROGEN STATUS

In this case study, the red-edge index is derived from CropScan$^{TM}$ measurements (cf. section 5.3.5) using the "linear method" (Eq. 2). The main inflexion point of the red-infrared slope provides information on chlorophyll and nitrogen status of the object under study (Baret & Guyot, 1988; Clevers & Büker, 1991). For other methods using multispectral reflection patterns for nitrogen fertilisation of potatoes, reference is made of Uenk and Booij (2000), who designed and applied their method in actual crop management in The Netherlands.

Reflectance signatures can only be related directly to above-ground biomass and nitrogen contents of a crop. Nitrogen status of a crop, however, is related to the whole crop. So, it is necessary to relate the above-ground characteristic to the whole crop. An important feature of a potato crop is that the ratios of above-ground and below-ground dry matter ($R_{dm}$) and of above-ground and below-ground nitrogen contents ($R_{nc}$) decrease during the growing season, as tubers develop in the ridges below-ground and foliar biomass ceases to grow above-ground. By using a dynamic simulation model or an empirical approach, $R_{dm}$ and $R_{nc}$ can be estimated and used to determine the nitrogen concentration of the whole potato crop (Neeteson, 1989; Jongschaap, 2000). Crops that merely have their storage organs above ground need a different approach.

### 4.2.1    Estimation of nitrogen deficiency

In order to determine the actual nitrogen deficiency, the observed crop nitrogen concentration $C_{nc}$ (found by remote sensing and in combination with the distribution function for nitrogen) is evaluated against the crop nitrogen concentration needed for the maximum growth rate $C_m$. To calculate this parameter, the total crop biomass and nitrogen content need to be known. The maximum growth rate depends on the total biomass W (kg ha$^{-1}$) (Eq. 3 after Greenwood et al., 1990), which can be retrieved from WDVI measurements (Clevers, 1988, 1989; cf. section 5.2.2) and $R_{dm}$.

$$C_m = (0.0135 + 0.0403e^{-0.0026W}) \bullet 100\% \tag{3}$$

The logarithmic relationship between WDVI and green above-ground biomass (dry matter) or LAI for potato (Bouman et al., 1992) tends to become very sensitive between WDVI values of 50-60%. WDVI values of 50% are concurrent with 2500 kg ha$^{-1}$ dry matter and an LAI of about 4 m$^2$ m$^{-2}$ for a potato crop. After the 50%-value for WDVI, more accurate estimates of green above-ground biomass can be obtained by assuming that the fraction intercepted radiation (fPAR) equals 1 and using a simulation model to estimate the daily increase in total dry matter. The daily increase in dry matter is then related to the Light Use Efficiency coefficient (LUE, g MJ$^{-1}$) and the incoming radiation, as shown by Casanova et al. (1998).

### 4.2.2    Managing nitrogen stress

If nitrogen deficiency occurs ($C_{nc} < C_m$), the fertiliser amount that is necessary to overcome the nitrogen stress can be calculated by the difference between $C_m$ and $C_{nc}$. Using hand-held equipment such as the CropScan$^{TM}$ equipment allows to monitor different spots in a field to estimate the actual nitrogen status. The use of a high resolution remote sensing image to determine $C_{nc}$ as presented in this study, may provide the spatial distribution of the nitrogen deficiency in a field. This is an important feature for precision agriculture, where spatial and temporal management of growing conditions is aimed to overcome the in-field variability of crop performance.

## 4.3    EXPERIMENTAL DATA

### 4.3.1    Set-up of the potato trials

In 1997 and 1998, 2 field experiments were conducted at the Droevendaal experimental station of Plant Research International in The Netherlands (51° 58' N, 5° 40' E). In both years potato tubers (*Solanum tuberosum* L., cv. Bintje) of 35-45 mm were planted at 0.75 m distance between rows and 0.30 m between plants, resulting in a plant density of 44,444 plants ha$^{-1}$. Different nitrogen quantities were applied ranging from 0 to 300 kg ha$^{-1}$. Various applications schemes were used to ensure different crop nitrogen levels during the growing season. All treatments were applied in 3 replicates, resulting in 72 experimental plots for each year. The 1997 experiment was used for the calibration of nitrogen relations with CropScan$^{TM}$ (CropScan$^{TM}$, 1993; cf. section 5.3.5) reflectance measurements and the 1998 experiment served as a validation test.

### *4.3.2 Crop measurements*

In both years 10 periodical yield measurements were performed by analysing 12 plants per plot at each harvest (30 plants at final yield). Yields were effectuated at predetermined soil cover stages (25, 50, 75 and 100 %) and at fixed time-intervals after the 100 % cover date. Crop analysis comprised the determination of fresh and dry weight of leaves (young and fully-grown), stems, tubers and roots and the determination of leaf area (young and fully-grown). Leaves, stems, tubers and roots were chemically analysed on total nitrogen and organic nitrogen contents. At each harvest date, 3 reflectance measurements per plot were made with a CropScan™ instrument (8 spectral bands of 12-20 nm bandwidth, at 460, 510, 560, 610, 660, 710, 760, and 810 nm). The CropScan™ was positioned horizontally at ca. 1.5 m above the canopy.

**Potato Experiment 1997 - Calibration**

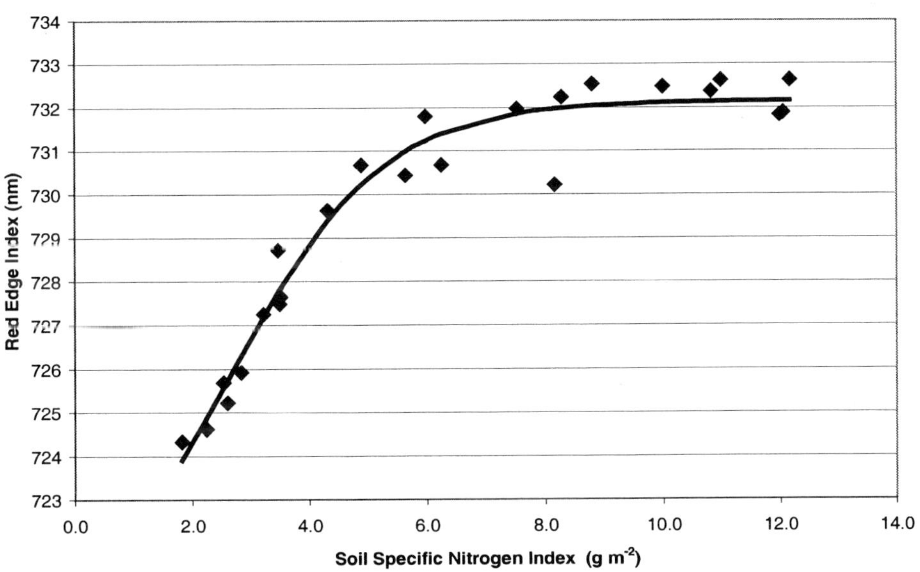

*Figure 15.* Red edge index as function of the Soil Specific Nitrogen Index for a potato crop (cv. Bintje). Data represent average values over 3 replicates per treatment. Data (♦) come from the whole growing season from June 6th to August 13th, 1997. The model (solid line) represents a sigmoidal fit ($r^2$=0.96). Source: Plant Research International, Wageningen, The Netherlands.

## 4.4    RESULTS AND DISCUSSION

The red-edge values were averaged per treatment, and plotted against the above-ground organic nitrogen content per square meter soil surface (SSN, g m$^{-2}$). Figure 15 shows the data for 4 nitrogen treatments on 8 observation dates throughout the growing season. The use of the averaged treatment values for the red-edge and for the organic nitrogen content improved the $r^2$ considerably (from 0.76 to 0.96). The relationship was validated on the 1998 trials and yielded an $r^2$ of 0.89 (Figure 16). It is clear that the useful part of the relationship is situated between 2.0 <= SSN <= 6.0 g m$^{-2}$. Values below 2.0 g m$^{-2}$ are disregarded, as the red-edge index cannot be determined accurately at low nitrogen availability and low above-ground biomass, which is the case at these values. The relationship becomes insensitive above SSN-values of 6.0 g m$^{-2}$ but were never found to be limiting in the fertiliser trials (Jongschaap and Booij, 2000).

**Potato Experiment 1998 - Validation**

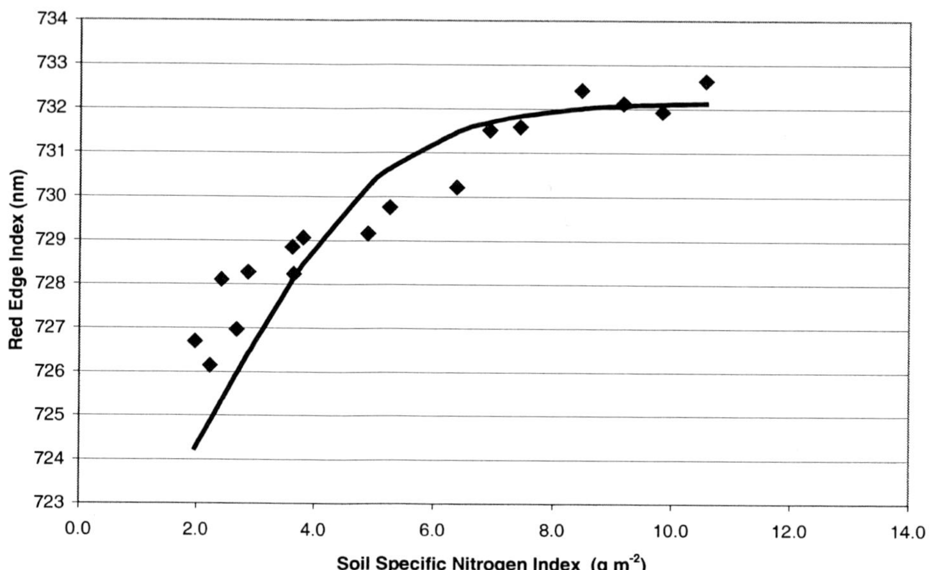


*Figure 16.* Red edge index as function of the Soil Specific Nitrogen Index for a potato crop (cv. Bintje). Data (♦) come from the whole growing season of 1998. The model (solid line) represents the sigmoidal fit on the 1997 data (cf. Figure 15). Source: Plant Research International, Wageningen, The Netherlands.

## 4.5    CONCLUSIONS

- Above-ground nitrogen availability can be estimated by the use of the red-edge index.
- Above-ground nitrogen concentration can be estimated by the use of the red-edge index in combination with the use of a vegetation index to estimate above-ground biomass.
- The combination of the red-edge index method as presented in this study and a vegetation index to estimate the nitrogen concentration of a crop canopy, together with a model to estimate optimum canopy nitrogen concentrations, can be used to determine the crop nitrogen status or nitrogen deficiency.
- In a potato crop, the red-edge index alone cannot be used as an indicator for nitrogen limited growth conditions. Additional information is needed on dry matter partitioning between above-ground and below-ground crop organs and nitrogen partitioning between above-ground and below-ground dry matter. A simulation model or an empirical approach may provide such information.
- In potato, nitrogen fertilisation practices can be quantified by the use of spectral information. The crop nitrogen status can be determined by the use of vegetation

indices derived from spectral information (red-edge index and WDVI). However, additional information is needed to determine the degree of growth reduction and the quantity of fertiliser that is needed to overcome the growth reduction.


## 5    Case study II – A framework for crop growth monitoring


### 5.1    INTRODUCTION

In agricultural market economies knowledge of crop production at an early stage is very important at both national and regional level. The two constituents of crop production are crop acreage and crop yield. In order to estimate or predict crop yield, best results are obtained if the growth of the crops is being monitored during the growing season. Using crop growth models can monitor crop growth (section 1). The leaf area index (LAI) is important as a measure for crop growth, because it is a main driving variable in many crop growth models. Remote sensing can provide information on the actual status of agricultural crops. Best results are obtained by using (reflective) optical remote sensing data (e.g. some vegetation index) in estimating the LAI regularly during the growing season and subsequently calibrating the growth model on time-series of estimated LAIs (Clevers & van Leeuwen, 1994; Clevers et al., 1994). The LAD appears the most important crop variable influencing this relationship between vegetation index and LAI.

LAI and LAD determine the amount of light interception. Leaf chlorophyll content influences the fraction of absorbed photosynthetically active radiation (APAR) and the maximum (potential) rate of photosynthesis of the leaves. A framework is described for integrating optical remote sensing data from various sources in order to estimate the mentioned variables. These concepts for crop growth estimation are elucidated and illustrated using ground-based and airborne data obtained during the MAC Europe 1991 campaign. Quantitative information concerning both LAI and LAD is obtained by measurements at two viewing angles (using data from the CAESAR scanner in dual-look mode). The red-edge index is used for estimating the leaf optical properties (using AVIRIS data). Finally, a crop growth model (SUCROS) is calibrated on time-series of optical reflectance measurements to improve the estimation of crop yield.


### 5.2    FRAMEWORK FOR YIELD PREDICTION


#### 5.2.1    Crop growth models

Since the 19th century, agricultural researchers have used modelling as a tool to describe the relationship between physiological processes in plants and environmental factors such as solar irradiation, temperature and water and nutrient availability (de Wit, 1965; Penning de Vries & van Laar, 1982; Spitters et al., 1989). The models compute the daily growth and development rate of a crop, simulating the dry matter production from emergence till maturity. Finally, a simulation of yield at harvest time is obtained. The basis for the calculations of dry matter production is the rate of gross

$CO_2$ assimilation of the canopy. Input data requirements concern crop physiological characteristics, site characteristics, environmental characteristics and the initial conditions defined by the date at which the crop emerges (or planting date).

The main driving force for crop growth in these models is absorbed solar radiation, and a lot of emphasis is given to the modelling of the solar radiation budget in the canopy. Incoming photosynthetically active radiation (PAR $\approx$ 400-700 nm) is first partly reflected by the top layer of the canopy. The direct reflectance of the canopy is a function of solar elevation, leaf area index (LAI), LAD and optical properties of the leaves. The complementary fraction is potentially available for absorption by the canopy. Subsequently, the absorptance by the canopy is a function of LAI, scattering coefficient (which may be derived from the direct reflectance) and extinction coefficient. The extinction coefficient is a function of solar elevation, LAD and scattering coefficient. The product of the amount of incoming photosynthetically active radiation (PAR) and the absorptance yields the amount of absorbed photosynthetically active radiation (APAR). The rate of $CO_2$ assimilation (photosynthesis) is calculated from the APAR and the photosynthesis-light response of individual leaves. The maximum rate of photosynthesis at light saturation is highly correlated to the leaf nitrogen content (van Keulen & Seligman, 1987). The assimilated $CO_2$ is then reduced to carbohydrates, which can be used by the plant for growth.

SUCROS (Simplified and Universal Crop Growth Simulator, Spitters et al., 1989) is a mechanistic crop growth model that describes the potential growth of a crop from irradiation, air temperature and crop characteristics. Potential growth means the accumulation of dry matter under ample supply of water and nutrients, in an environment that is free from pests and diseases. The light profile within a crop canopy is computed on the basis of the LAI and the extinction coefficient. At selected times during the day and at selected depths within the canopy, photosynthesis is calculated from the photosynthesis-light response of individual leaves. Integration over the canopy layers and over time within the day gives the daily assimilation rate of the crop (partly from Spitters et al., 1989). The excess assimilates over that needed to maintain the present biomass (maintenance respiration) is converted into new, structural plant matter (with loss due to growth respiration). The newly formed dry matter is partitioned to the various plant organs, through partitioning factors introduced as a function of the phenological development stage of the crop. An important variable that is similated is the LAI, since the increase in leaf area contributes to next day's light interception and hence to next day's rate of assimilation.

When applied to operational uses such as yield estimation, models such as SUCROS often appear to fail when growing conditions are non-optimal (e.g. fertiliser deficiency, pest and disease incidence, severe drought, frost damage). Therefore, for yield estimation, it is necessary to 'check' modelling results with some sort of information on the actual status of the crop throughout the growing season. Remote sensing can provide such information (Bouman, 1991, 1994).

Optical remote sensing can provide actual information on the status of crops for checking modelling results obtained with crop growth models. There are three 'key-factors' useful in crop growth models that may be derived from optical remote sensing data: (a) LAI, (b) LAD and (c) leaf optical properties in the PAR region.

### 5.2.2   *Estimating LAI*

The LAI during the growing season is an important state variable in crop growth modelling. Moreover, the LAI is a major factor determining crop reflectance and is often used in crop reflectance modelling (e.g., Suits, 1972; Bunnik, 1978; Verhoef, 1984). The estimation of LAI from remote sensing measurements has received much attention. Much research has been aimed at determining combinations of reflectances, so-called vegetation indices, to correct for the effect of disturbing factors on the relationship between crop reflectance and crop characteristics such as LAI (Tucker, 1979; Richardson & Wiegand, 1977; Clevers, 1988, 1989). A sensitivity analysis using radiative transfer models revealed that the main factor influencing the relationship between many vegetation indices and green LAI is the LAD (e.g., Clevers & Verhoef, 1993).

A simplified, semi-empirical reflectance model for estimating LAI of a green canopy was introduced by Clevers (1988, 1989). It is called the CLAIR model. In this model, first, the WDVI (= weighted difference vegetation index) is ascertained as a weighted difference between the measured NIR and red reflectances, assuming that the ratio of NIR and red reflectances of bare soil is constant (the weighting factor). In this way a correction for the influence of soil background is performed. Subsequently, this WDVI is used for estimating LAI according to the inverse of an exponential function

$$LAI = -1/\alpha \cdot \ln(1 - WDVI / WDVI_\infty) \qquad (4)$$

with $\alpha$ as a combination of extinction and scattering coefficients describing the rate with which the function of Eq. (4) runs to its asymptotic value, and $WDVI_\infty$ as the asymptotic limiting value for the WDVI.

The exponential relationship between WDVI and LAI means that LAI estimations will be less accurate when approximating the asymptotic value of WDVI ($WDVI_\infty$). In other words: the accuracy of LAI estimation will decrease with increasing LAI value. It is expected that the accuracy of WDVI measurement will not depend on the absolute value of the WDVI itself. However, the accuracy of LAI estimation cannot simply be expressed in terms of accuracy of the individual WDVI values, because LAI is a non-linear function of WDVI

$$LAI = f(WDVI) \qquad (5)$$

A first order approximation of the variance of a non-linear function $f(x)$ is

$$VAR[f(x)] = (\delta f / \delta x)^2 \cdot VAR(x) \qquad (6)$$

For the relationship given in Eq. (4) this gives

$$VAR(LAI) = (\alpha \cdot WDVI_\infty)^{-2} \cdot (1 - WDVI / WDVI_\infty)^{-2} \cdot VAR(WDVI) \qquad (7)$$

Combination of Eq. (7) and Eq. (4) yields for the standard deviation of LAI estimation

$$\sigma(LAI) = \exp(\alpha \cdot LAI - \ln(\alpha \cdot WDVI_\infty)) \cdot \sigma(WDVI) \qquad (8)$$

The validation of the CLAIR model for sugar beet was performed by Bouman et al. (1992). They found for sugar beet empirically for $\alpha$ an estimate of 0.485 and for $WDVI_\infty$ an estimate of 48.4, whereby the WDVI was based on green reflectance instead of red reflectance. The residual mean square for the calibration set was 4.1 (in terms of squared WDVI units). This value may be used as an estimate of the variance of the individual WDVI measurements. The resulting estimate for the WDVI standard deviation ($\sigma$[WDVI] in Eq. 8) is 2.0. Figure 17 plots the estimated LAI using the CLAIR model against the measured LAI (ground measurements) for the calibration set used by Bouman et al. (1992). In addition, the lines defining 2 standard deviations from the measured LAI are shown.

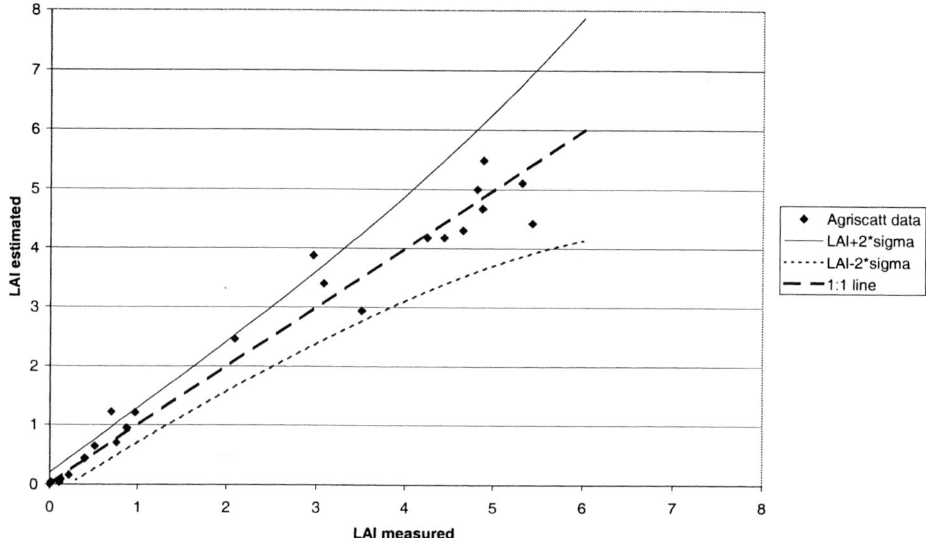


*Figure 17.* Relationship between estimated LAI using the CLAIR model and measured LAI for sugar beet. Flevoland test site, AGRISCATT campaigns 1987 and 1988.

### 5.2.3    *Estimating leaf angle distribution (LAD)*

LAD affects the process of crop growth because it has an effect on the interception of APAR by the canopy (e.g. Clevers et al., 1992). Moreover, it is the main variable influencing the relationship between vegetation index and LAI. With optical remote sensing it is more difficult to obtain quantitative information on LAD than on LAI. A solution may be found by performing measurements at different viewing angles. Using both radiative transfer models and experimental data, Goel & Deering (1985) have shown that measurements at two viewing angles for fixed solar zenith and view azimuth angles are enough to allow estimation of LAI and the LAD by the NIR reflectance. In Looyen et al. (1991) the possibilities of acquiring information on both LAI and LAD by means of a so-called dual look (two viewing angles) concept using the Dutch CAESAR scanner are illustrated.

### 5.2.4    *Estimating leaf optical properties in the PAR region*

Leaf optical properties are important in the process of crop growth because (1) they influence the fraction of absorbed PAR, and, (2) they can be indicative for the nitrogen status (or chlorophyll content) of leaves which affects the maximum rate of photosynthesis. Leaf optical properties in the PAR region may be ascertained by spectral measurements in the visible region (VIS) of the electromagnetic (EM) spectrum. Chlorophyll content is the main factor determining these optical properties in the VIS. However, at low soil cover the measured signal will be confounded by soil influence. At complete coverage spectral measurements in the VIS offer information only on leaf colour. However, since the signal in VIS at complete coverage is relatively low, it may be heavily confounded by atmospheric effects for which must be corrected.

Another measure of chlorophyll content (or leaf optical properties) may be offered by the red-edge index. A decrease in leaf chlorophyll content results into a shift of the red-edge towards the blue (see section 3).

In practice it will be very difficult to ascertain leaf optical properties unless leaves are analysed in the laboratory. A more practical measure may be offered by the red-edge index. However, in section 3 it was shown that this index is determined by both LAI and leaf chlorophyll content (related to leaf optical properties). Two (independent) measurements are therefore needed, one more related to LAI (like WDVI) and one more related to chlorophyll content (like red-edge index). In this section, the "linear method" of Guyot & Baret (1988) for determining the position of the red-edge was applied. Measurements of the WDVI and the position of the red-edge may be combined for ascertaining the leaf chlorophyll content.

### 5.2.5    *Linking optical remote sensing with crop growth models*

Two methods can be distinguished to link remote sensing data with crop growth models. In the first method, called 'model initialization', crop variables are estimated from remote sensing and 'fed' into a growth model as input or forcing function. Most crop variables that have been used successfully so far are measures for the fractional light interception by the canopy, namely LAI and soil cover (Steven et al., 1983; Kanemasu et al., 1984; Maas, 1988; Bouman & Goudriaan, 1989). However, variables like LAD and leaf optical properties can also be used as input in more complex growth models.

In the second method, called 'model calibration', crop growth models are calibrated on time series of remote sensing measurements. Maas (1988) presented a method in which crop growth model parameters were adjusted in such a way that simulated values of LAI by the growth model matched LAI values that were estimated from reflectance measurements. Bouman (1992) developed a procedure in which remote sensing models (a.o. optical reflectance) were linked to crop growth models so that canopy reflectance was simulated together with crop growth. The growth model was then calibrated to match simulated values of canopy reflectance (from a combined crop growth and reflectance model) to measured values of reflectance. The calibration in this procedure is governed by the variables which link the crop growth model and the remote sensing model.

Most crop parameters in SUCROS do not have one specific value but are characterized by a 'biological plausible range'. This variation in parameter values allows for a range in simulation results from SUCROS. The combined growth/reflectance model (Bouman, 1992) simulates crop variables such as biomass and LAI, together with canopy reflectance and WDVI during the growing season. This model now may be calibrated within the biological plausible ranges to fit the simulated remote sensing signals (WDVI) to the measured ones. Calibration of the starting conditions (e.g. sowing date) of SUCROS is called initialization. Initialization and calibration of SUCROS was based on a controlled random search procedure as developed by Price (1979) and extended by Klepper (1989). This procedure yielded per simulation condition 'optimum' initialization conditions and crop parameter values. Optimum refers to parameter values that resulted in the smallest (absolute) difference between simulated and measured remote sensing signals, averaged over the whole growing season. With optimum parameter values, the simulated time course of remote sensing signals best fitted the measured signals.

A sensitivity analysis was used to select the initialization conditions and crop species parameters to be calibrated. The growing season was divided into three distinct growth periods: initialization, exponential growth, and linear growth. Per growth period, parameters were selected that had a large effect on both the remote sensing signal and the canopy biomass and/or the LAI. Because of the redundancy in the effect of parameter changes, the selection of parameters was kept to a minimum. With optical data 'sowing date', 'relative growth rate', 'light use efficiency' and 'maximum leaf area' were selected.

Clevers et al. (1994), van Leeuwen & Clevers (1994) and van Leeuwen (1996) described a method for calibrating crop growth models on periodic remote sensing measurements by using the remote sensing models in an inverse way. The SUCROS crop growth model was initialized and calibrated to fit simulated LAI values to estimated LAI values obtained from remote sensing measurements.. Thus, first the CLAIR model was applied for obtaining LAI estimates from the remote sensing measurements. Subsequently, the SUCROS model was calibrated on these LAI estimates. Since we have seen that the accuracy of the LAI estimates depends on the absolute value of the LAI, the reciprocal of the standard deviation of LAI estimation is used as a weighting factor for each individual LAI estimate used in the optimization procedure. To obtain LAI estimates from optical measurements Eq. (4) is used. In addition, parameter estimates obtained for sugar beet during the calibration of the CLAIR model are used in this Eq..

## 5.3     EXPERIMENTAL DATA

### 5.3.1    *Introduction*

Some of the principles for linking remote sensing with crop growth models will be illustrated with results from the European multisensor airborne campaign MAC Europe in 1991. A description will be given of the MAC Europe campaign in the Dutch test site Flevoland and of the collected remote sensing and ground truth data.

In the optical remote sensing domain, NASA executed one overflight with the AVIRIS scanner. Some information concerning AVIRIS and the test site was already given in section 2.3. The preprocessing which includes correction for dark current, vignetting and radiometric response was done by the Jet Propulsion Laboratory (JPL). Atmospheric correction was also performed by JPL with a new version of the LOWTRAN 7 atmospheric model, yielding reflectances (Van den Bosch & Alley, 1990). The AVIRIS data set is available in radiance as well as in reflectance. Both are considered in this study. In addition, the Dutch experimenters flew three flights with the Dutch CAESAR scanner (for system description, see Looyen et al., 1991).

Bouman (1992) has described a version of the growth model SUCROS for sugar beet that has been parameterized for Flevoland. This version is called SBFLEVO and was used in this study.

### 5.3.2    Ground truth

Crop parameters concerning acreage, variety, planting date, emergence date, fertilization, harvest date, yield and occurring anomalities were collected for the main crops. During the growing season, additional parameters were measured in the field approximately every ten days. The selected parameters were the estimated soil cover by the canopy, the mean crop height, row distance, plants per $m^2$, the soil moisture condition and comments about plant development stage.

### 5.3.3    Meteorological data

Daily meteorological data are needed as input for crop growth simulation models. For the 1991 growing season these were obtained from the Royal Dutch Meteorological Service (KNMI) for the station Lelystad. Data consisted of daily minimum and maximum temperature, daily global irradiation and daily precipitation.

### 5.3.4    Spectra of single leaves

Leaf optical properties were investigated with a LI-COR laboratory spectroradiometer at the Centre for Agrobiological Research (CABO) in Wageningen. The reflectance or transmittance signature of the upper and lower surface of several leaves was recorded continuously from 400 to 1100 nm wavelength in 5 nm steps. The instrument was calibrated with a white barium sulphate plate.

### 5.3.5    CropScan$^{TM}$ ground-based reflectances

Field reflectance measurements were obtained during the 1991 growing season with a ground-based CropScan$^{TM}$ radiometer approximately every ten days. Eight narrow-band interference filters with photodiodes were oriented upwards to detect hemispherical incident radiation and a matched set of interference filters with photodiodes were oriented downwards to detect reflected radiation. Spectral bands were located at 490, 550, 670, 700, 740, 780, 870 and 1090 nm with a bandwidth of 10 nm. The sensor head of the radiometer was mounted on top of a long metal pole and positioned three metres above the ground surface. The distance to the crop was 2.5 to 1.5 m depending on the crop height. As the diameter of the field of view (FOV 28°) was half the distance between sensor and measured surface, the field of view varied from 1.23 $m^2$ to 0.44 $m^2$.

### 5.3.6    CAESAR

The CAESAR (CCD Airborne Experimental Scanner for Applications in Remote Sensing) applies linear CCD arrays as detectors. It has a modular set-up and it combines the possiblities of a high spectral resolution with a high spatial resolution. For land applications three spectral bands are available in the green, red and NIR part of the EM spectrum. One of the special options of CAESAR is the capability of acquiring data according to the so-called dual look concept. This dual look concept consists of measurements performed when looking nadir and under the oblique angle of 52°. Combining these measurements provides information on the directional reflectance properties of objects (Looyen et al., 1991). Successful overflights over the test site were carried out on July 4th, July 23rd and August 29th, 1991.

## 5.4    RESULTS

Results of a case study for sugar beet will be presented.

### 5.4.1    Measurements of leaf optical properties

During July and August 1991 individual sugar beet leaves were measured in the laboratory with a LI-COR LI-1800 portable spectroradiometer. During this period leaf properties were rather constant. The measurements yielded for a NIR band (at 870 nm) an average reflectance of 46.0% and an average transmittance of 48.4%. These values were respectively 7.3% and 0.6% for a red (at 670 nm) and 15.8% and 13.8% for a green band (at 550 nm). The average scattering coefficient was 0.144 for the whole PAR region.

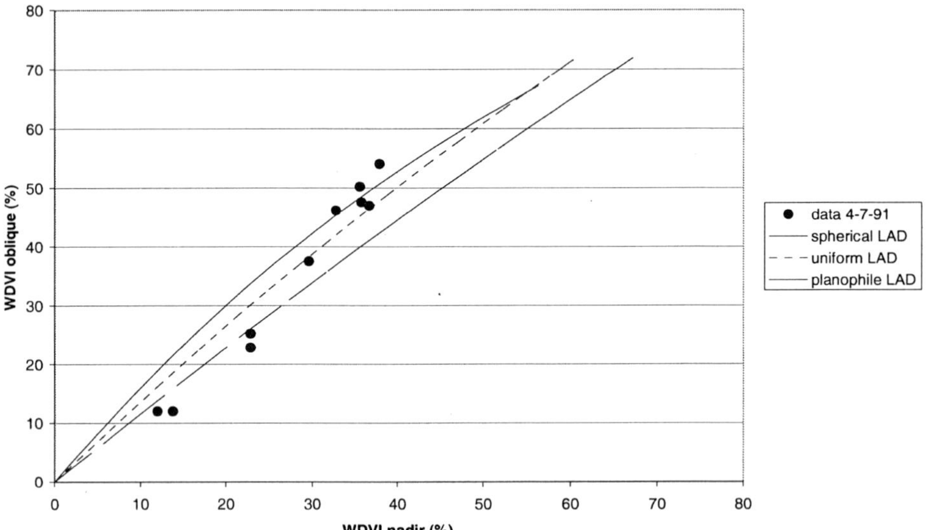


*Figure 18.* Relationship between the WDVI from nadir and the WDVI at an oblique viewing angle of 52°. Simulations for a spherical, uniform and planophile LAD (SAIL model with a hot spot size-parameter of 0.5 for sugar beet) and measurements obtained with CAESAR, July 4th - calendar day 185 -, 1991 (13.30 GMT). Solar zenith angle of 36° and azimuth angle between plane of observation and sun of 7°.

### 5.4.2    Estimating LAD

Information on LAD was obtained by means of the CAESAR scanner in dual look mode. Measured WDVI values at an oblique and nadir viewing angle are plotted into a nomogram, based on the actual recording geometry and the leaf optical properties from the previous section. This can yield estimates of both LAI and LAD. Figure 18 gives the results of July 4th for the CAESAR scanner together with simulated curves for a spherical, uniform and planophile LAD (LADs as defined by Verhoef & Bunnik, 1981). Clevers et al. (1994) present results obtained with the CAESAR scanner on July 23rd and August 29th, respectively. Results for all three dates showed that sugar beet mostly matched the curve for a spherical LAD rather well, except for the beginning of the growing season (LAI<1.5) when the LAD was more planophile. This information is important for determining the relationship between WDVI and LAI. It can also be used to derive extinction coefficients that are input for crop growth models. Currently, SUCROS uses an extinction coefficient for diffuse PAR, which is calculated to be 0.69 for a spherical LAD.

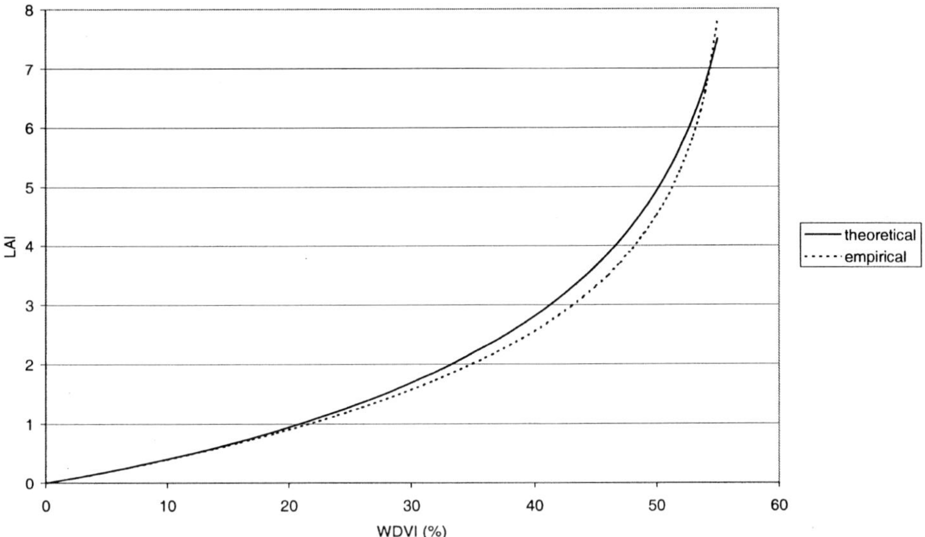

*Figure 19.* Theoretical and empirical relationship between WDVI and LAI for sugar beet.

### 5.4.3    Estimating LAI

Using the optical leaf properties measured in the laboratory and the spherical LAD found in the previous section, the relationship between WDVI and LAI was simulated using the SAIL model. The regression of LAI on WDVI (Eq. 4) yielded for $\alpha$ an estimate of 0.418 and for $WDVI_\infty$ an estimate of 57.5 (spherical LAD). Bouman et al. (1992) found for sugar beet empirically for $\alpha$ an estimate of 0.485 and for $WDVI_\infty$ an estimate of 48.4, whereby the WDVI was based on green reflectance instead of red reflectance. CropScan[TM] measurements and SAIL simulations yielded a ratio between WDVIs based on green and red reflectances, respectively, for sugar beet of 1.16. As a result, a value of 48.4 for $WDVI_\infty$ (based on green reflectances) corresponds with a value of 56.3 for a $WDVI_\infty$ based on red reflectances. Figure 19 illustrates that the simulated relationship (SAIL model) and the empirical relationship correspond well (WDVIs based on NIR and red reflectances).

### 5.4.4    Estimating leaf optical properties

By plotting both the measured WDVI and the red-edge values (both acquired with AVIRIS) into a nomogram for actual recording conditions, an estimate of both LAI and leaf chlorophyll content is obtained (see Figure 20). Results for sugar beet yielded an estimated chlorophyll content of about 30 $\mu$g.cm$^{-2}$, except for the beginning of the growing season (LAI<1.5) when the chlorophyll content was somewhat higher. By using the PROSPECT model this leaf chlorophyll content yielded an average leaf scattering coefficient in the PAR region of 0.158. This value is comparable to the one measured in the laboratory (0.144).

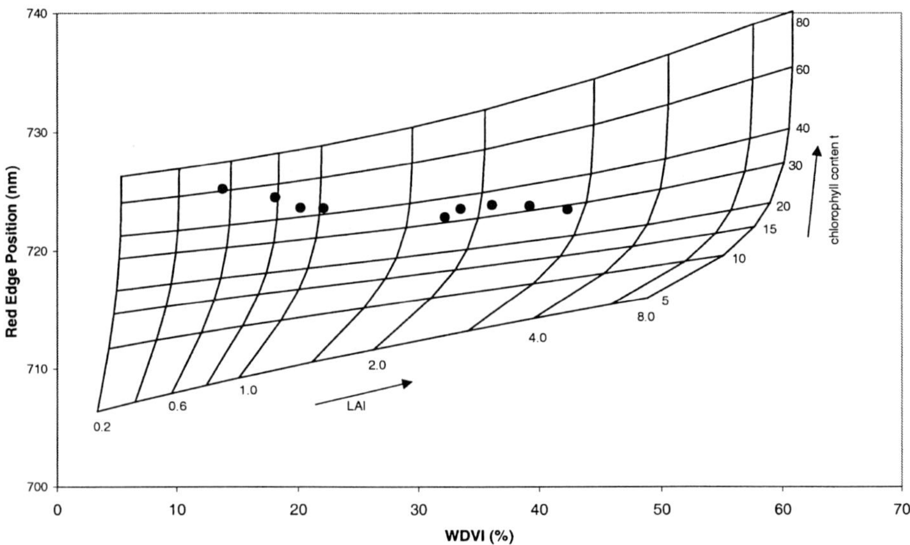



*Figure 20.* Relationship between the WDVI (from nadir) and the position of the red-edge as simulated with a combined PROSPECT-SAIL model and measurements obtained with AVIRIS, July 5th -calendar day 186 -, 1991 (13.08 GMT).

### 5.4.5    Results calibration SUCROS

The crop growth model SUCROS was run to estimate the final beet yield for ten selected farmers in the test area. First, the location parameters, weather data for the 1991 growing season and crop-specific model parameters were used as input for the model. The latter parameters were determined by Spitters et al. (1989) on many field experiments and from literature (resulting into 'standard' SUCROS). Since in most practical situations of (regional) yield prediction no actual information on sowing and harvest dates is available, the simulation was performed for a hypothetical, average sugar beet crop with a sowing date of April 15th and a harvest date of October 22nd. This resulted into an estimated beet yield of 60.0 tons/ha (fresh weight) and an average underestimation of the actual yield by 13.4 tons/ha (17.5 %). Next, the 'standard' SUCROS model was run with actual sowing and harvest dates for all 10 farmers individually. On the average, the simulation error of (fresh) beet yield now was 6.6 tons/ha (8.6 %). Figure 21 illustrates the differences between this latter estimated yield and actual beet yield. Finally, SUCROS was calibrated so that simulated LAI during the growing season matched LAI values estimated from CropScan$^{TM}$ WDVI measurements (at about 10 dates during the growing season) as close as possible for all ten fields individually. Within SUCROS a spherical LAD and a scattering coefficient for visible light of 0.15 was used for the whole growing season (these values were found in previous studies). Since in operational yield estimations information on exact sowing and harvest dates are not readily available, SUCROS was also initialized on the WDVI measurements (i.e. the sowing date was one of the calibration parameters). The harvest date was taken as October 22nd. Results for the calibrated SUCROS model using remote sensing information are also illustrated in Figure 21. For all (except one) fields, the simulated yield after calibration was closer to actually obtained yields than before calibration. On the average, the simulation error of (fresh) beet yield decreased from 6.6 t/ha (8.6%) using 'standard' SUCROS, to 3.7 t/ha (5.1%) with SUCROS calibrated to time-series of WDVI.

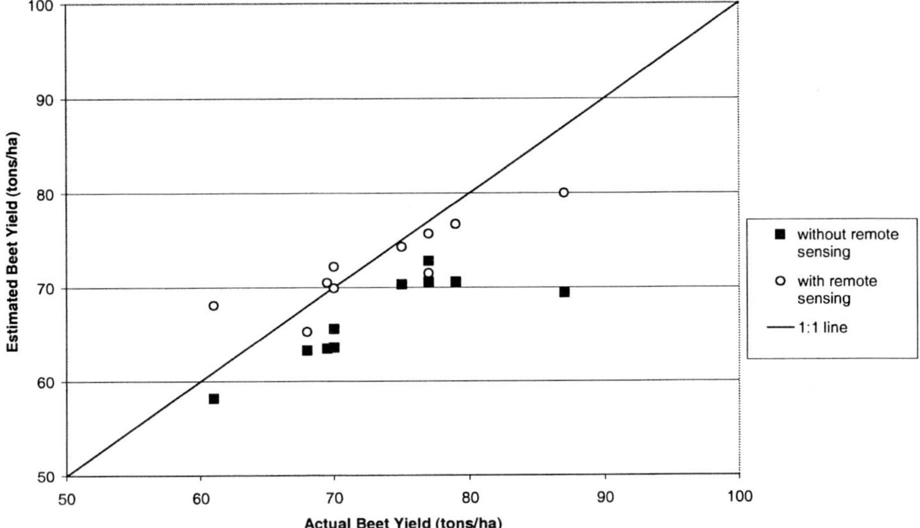


*Figure 21.* Estimated beet yield using SUCROS, with and without calibration to estimated LAI from measured WDVI, versus actually obtained beet yields.


## 5.5    CONCLUSIONS

A framework was presented to integrate crop canopy information derived from optical remote sensing with crop growth models for the purpose of growth monitoring and yield estimation. Basic for this framework is that (bi-) directional optical measurements by performing observations at two viewing angles are combined with high spectral resolution observations performed with an imaging spectrometer. The former is used for estimating LAI and LAD simultaneously; the latter are used for estimating the leaf scattering coefficient. These are the crop variables that play an important role in both the processes of crop growth and canopy reflectance. Subsequently, the estimated crop variable values were used as input into crop growth models and for calibrating crop growth models.

The framework was applied to data gathered during the MAC Europe 1991 campaign over the Dutch test site Flevoland. Results for sugar beet indicated the feasibility of estimating LAI, LAD and leaf optical properties from reflectance measurements. A critical point to consider is the precision and additional value of the variable values derived from remote sensing compared to the standard values already used in the growth model. For instance, much relative benefit might be obtained from the estimation of leaf optical properties expressed in leaf nitrogen or chlorophyll content. Especially the modelling of leaf nitrogen status in canopies is extremely complicated (but equally important through its effect on maximum leaf photosynthesis rate) and actual information derived from optical reflectance would be valuable. However, it will take more research and dedicated experiments together with crop physiologists to investigate the potentials of optical remote sensing for the assessment of leaf (or canopy) nitrogen status.

The method of model calibration was tested on sugar beet. For nine out of ten fields, the simulated yield was better in agreement with actually obtained yields after model calibration than without model calibration. Since the calibration procedure mainly concerned the calibration of the simulated LAI, these results indicate the importance of LAI for accurate growth simulation.

The framework can be extended to include other variables that are relevant to both crop growth and optical reflectance. Moreover, the framework can incorporate other remote sensing techniques as well, such as radar, passive microwave or thermal remote sensing (Bouman, 1991; Clevers & van Leeuwen, 1996).

## 6    Case study III – Using MERIS for deriving the red-edge index

### 6.1    INTRODUCTION

Within ESA's Earth Observation programme, MERIS (MEdium Resolution Imaging Spectrometer) will be one of the main payload components of the European polar platform ENVISAT-1 to be launched in 2000 (ESA, 1997). MERIS is a 15 band programmable imaging spectrometer, which includes the capability to change band position and bandwidths throughout its lifetime. It is designed to acquire data at variable bandwidth between 1.25 and 30 nm over the spectral range 390 – 1040 nm. Band choice and bandwidth depend on the width of the spectral feature observed and the amount of energy needed in a band for adequate observation (Rast & Bezy, 1990). Data will be acquired at 300-m spatial resolution over land, so vegetation monitoring can be performed at continental and global scales. Basically MERIS will be operated with a standard band setting (as shown in table 3). However, it has the capability of in-flight selection of bands for specific applications or experiments, for repositioning of bands because of new findings or to cope with partial instrument failures or deficiencies and for exploration of new issues. However, operational constraints will limit the number and frequency of band changes.

The best MERIS resolution of 300 m should be sufficient to monitor the highly heterogeneous terrestrial surfaces at scales required for global change studies (ESA, 1995; Verstraete et al., 1999). Vegetation indices using red and NIR reflectance can be used for estimating the LAI and the fraction of absorbed photosynthetically active radiation (Govaerts et al., 1999). The red-edge feature may provide useful information on the physiological status of the vegetation under observation. Verstraete al. (1999) note that the spectral and spatial resolution of MERIS will be particularly appropriate to derive this red-edge variable. However, they also state that much remains to be done in this respect to design algorithms based on MERIS data. The objective of this section is to study whether the red-edge index can be derived from the MERIS standard band setting, as this will be the setting that will be applied under normal operation. Thus, the standard band setting provides the bands that are most suitable for monitoring purposes. In this study we will pay only little attention to the interpretation and application of a red-edge index at the spatial scale MERIS is operating at.

TABLE 3: The 15 spectral bands of the standard band setting for MERIS (Rast et al., 1999).

| Band nr. | Band centre [nm] | Bandwidth [nm] | Potential Applications |
|---|---|---|---|
| 1 | 412.5 | 10 | Yellow substance, turbidity |
| 2 | 442.5 | 10 | Chlorophyll absorption maximum |
| 3 | 490 | 10 | Chlorophyll, other pigments |
| 4 | 510 | 10 | Turbidity, susp. sediment, red tides |
| 5 | 560 | 10 | Chlorophyll ref., susp. sediment |
| 6 | 620 | 10 | Suspended sediment |
| 7 | 665 | 10 | Chlorophyll absorption |
| 8 | 681.25 | 7.5 | Chlorophyll fluorescence |
| 9 | 705 | 10 | Atmospheric correction, red-edge |
| 10 | 753.75 | 7.5 | Oxygen absorption reference |
| 11 | 760 | 2.5 | Oxygen absorption R-branch |
| 12 | 775 | 15 | Aerosols, vegetation |
| 13 | 865 | 20 | Aerosols corrections over ocean |
| 14 | 890 | 10 | Water vapour absorption ref. |
| 15 | 900 | 10 | Water vapour absorption, vegetation |

## 6.2    DATA SETS

Two different data sets will be used for this study. Simulated data with a combination of the PROSPECT leaf reflectance model and the SAIL canopy reflectance model (data set I) and airborne reflectance spectra of agricultural crops obtained from AVIRIS data (data set II). The models already have been described in section 3.3 and the AVIRIS data have been described in section 2.3 and 5.3. The same test site as before in this chapter is used.

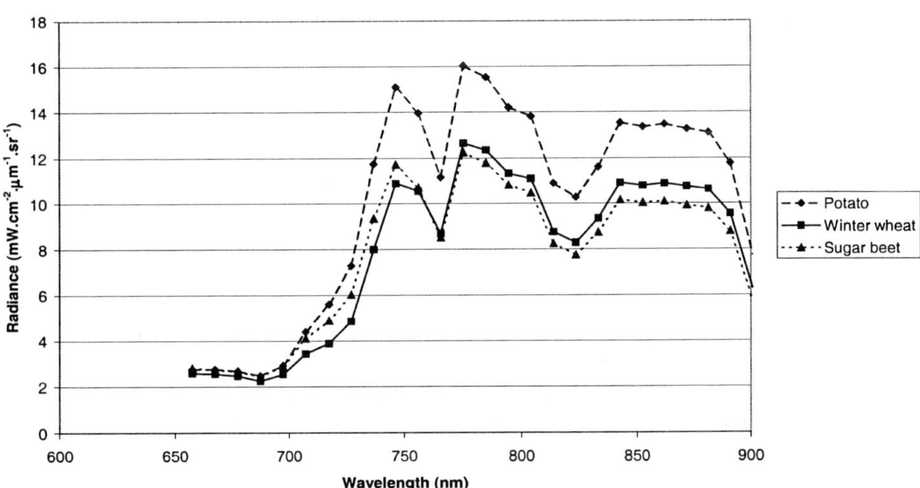


*Figure 22.* Spectral signature in radiance (mW.cm$^{-2}$.μm$^{-1}$.sr$^{-1}$) for the three major crops at the Flevoland test area, measured with the AVIRIS sensor on July 5th, 1991.

## 6.3    RESULTS AND DISCUSSION

### 6.3.1    *AVIRIS spectra*

Data set II comprises real imaging spectrometer measurements over an agricultural area. Figure 22 illustrates the radiance spectrum in the red-edge region for the main three crops in the area. For the "linear method" the following AVIRIS spectral bands can be employed: 667.4 – 697.3 – 736.7 – 785.0 nm. Clevers (1994) showed that the red-edge position determined from these radiance values was similar to the one obtained from spectral reflectance measurements performed in the field using a hand-held radiometer. Notable in Figure 22 is a dip in the radiance curve at about 760 nm. This is a well-known absorption feature caused by oxygen present in the atmosphere. Also an absorption feature at about 820 nm, caused by water vapour in the atmosphere, is obvious. Since MERIS has a spectral band at about 753 nm, biased results can be expected in applying the "linear method" of Guyot when no atmospheric correction is performed. Figure 23 illustrates the spectral signature in terms of reflectance for the main three crops in the area after an atmospheric correction has been performed. The oxygen absorption feature has been removed in this way.

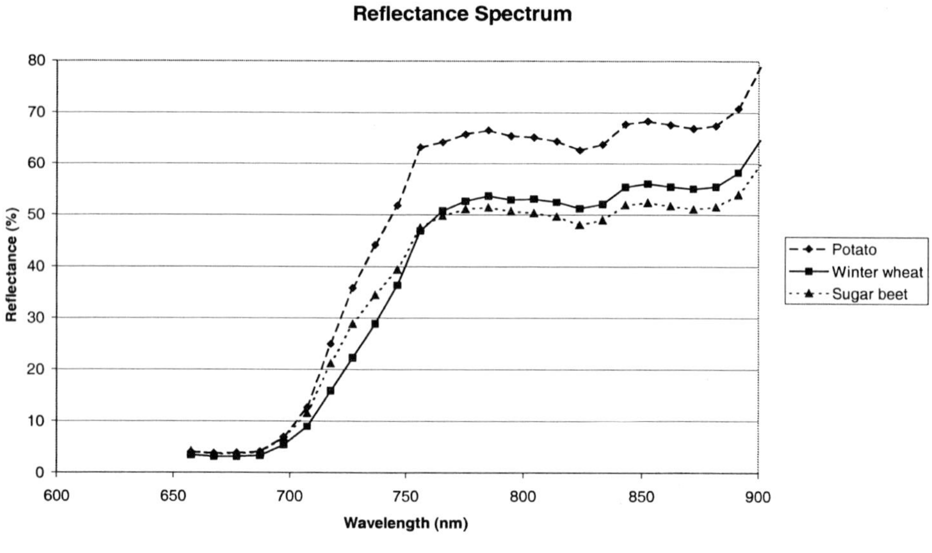

**Reflectance Spectrum**

*Figure 23.* Spectral signature in terms of reflectance for the three major crops at the Flevoland test area, measured with the AVIRIS sensor on July 5th, 1991.

### 6.3.2    *Red-edge index simulation with MERIS*

As seen in table 3, MERIS has also a spectral band in the chlorophyll absorption feature at 665 nm and at the NIR plateau at 775 nm. These are very close to the wavelengths used in the "linear method" as described by Guyot & Baret (1988). These two can be used for estimating the reflectance value at the inflexion point. The next step in the "linear method" is that this reflectance value at the inflexion point is translated to the corresponding wavelength by a linear interpolation between the reflectances measured at two wavelength positions on the red-edge slope. From the MERIS standard band setting the spectral bands centred at 705 and 753.75 nm are available. In this section it will be tested whether these bands yield useful estimates of the red-edge position. When the MERIS bands are simulated we will call the linear method the "MERIS method".

**PROSPECT-SAIL Simulation**

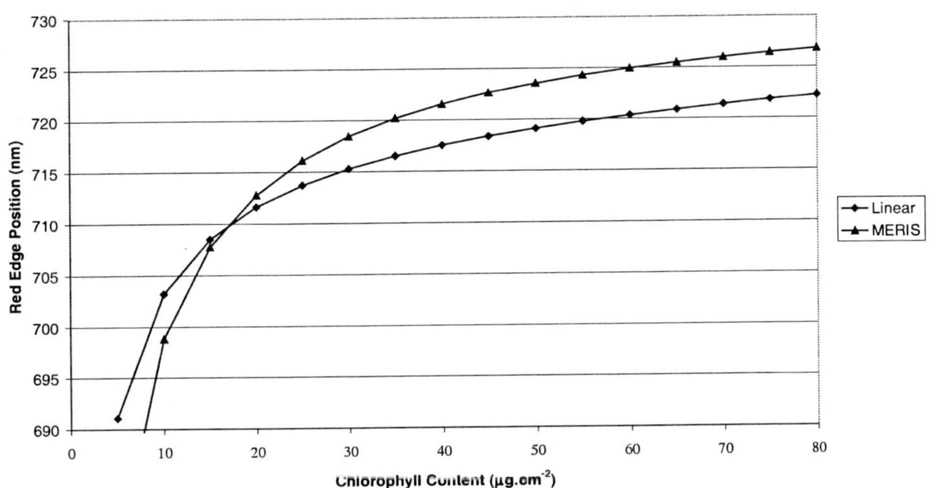


*Figure 24.* Red-edge position as a function of the leaf chlorophyll content for the "linear method" using the 700 and 740 spectral bands and for the linear method using the MERIS band setting ("MERIS method") for a canopy with an LAI of 2.0 as simulated with a combined PROSPECT-SAIL model.

Using data set I, Figure 24 illustrates the red-edge position as a function of the leaf chlorophyll content for the "linear method" using the interpolation between 700 and 740 nm and for the "linear method" using the MERIS band setting and interpolating between 705 and 754 nm. Both curves exhibit the same pattern, whereby the MERIS-based "linear method" shows somewhat larger red-edge index values, except for index values below 705 nm. For the latter range considerable deviations between the two methods may occur, which primarily can be attributed to the fact that using the MERIS bands we are performing an extrapolation instead of an interpolation. The relationship between the two methods is depicted in Figure 25, showing a good linear relationship in this case albeit that it not matches the 1:1 line. In this Figure 25 the relationship is simulated for a canopy with an LAI of 2.0. Figure 26 shows the same relationship for smaller steps in chlorophyll content and for canopies with various LAI values. This result shows that one may derive one general relationship between the two methods and that a linear relationship will yield a good fit to the data. Using all the data of Figure 26, the regression line yields

$$re_{meris} = -409.0 + 1.574 \cdot re_{700-740} \qquad (9)$$
$$r = 0.996.$$

Using data set II, MERIS spectral bands were simulated with the AVIRIS bands. The bands at 667 and 775 nm were used for estimating the reflectance at the inflexion point and the bands at 707 and the average of 746 and 756 nm were used for the linear interpolation. Figure 27 illustrates the relationship between the red-edge position calculated from the interpolation between 697 and 737 nm and that calculated using the simulated MERIS bands for the main agricultural crops at the Flevoland test site. When we compare this relationship with the regression line found for data set I, again a good match is observed. Like for the other data sets larger deviations occur at high red-edge values.

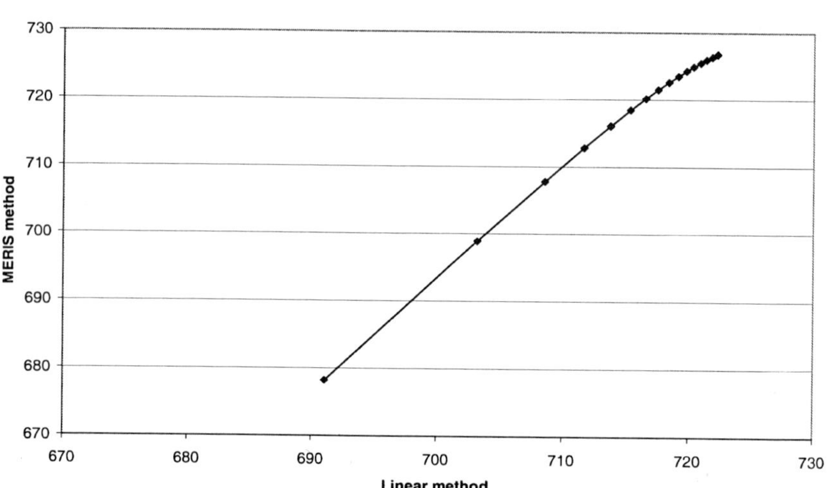

*Figure 25.* Comparison of the red-edge position calculated using the "linear method" interpolating between 700 and 740 nm and the "MERIS method", thus interpolating between 705 and 754 nm, PROSPECT-SAIL simulation for a canopy with varying leaf chlorophyll content and an LAI of 2.0.

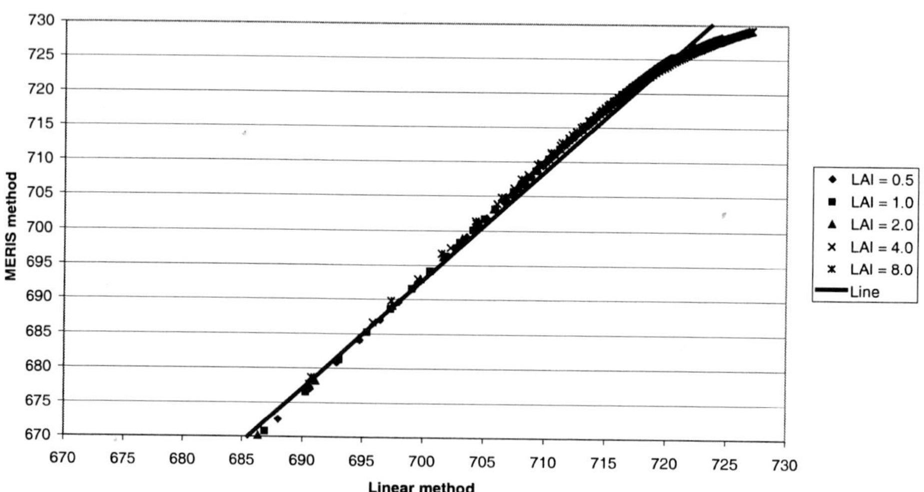

*Figure 26.* Comparison of the red-edge position calculated using the "linear method" interpolating between 700 and 740 nm and the "MERIS method", thus interpolating between 705 and 754 nm, PROSPECT-SAIL simulation for a canopy with varying leaf chlorophyll content and various LAI values. The regression line $re_{MERIS} = -409.0 + 1.574 * re_{700-740}$   is also plotted.

### 6.3.3    Upscaling to the MERIS resolution

Figure 28 illustrates the relationship between the red-edge position calculated from the interpolation between 697 and 737 nm and that calculated using the simulated MERIS bands for a number of degraded pixels at the Flevoland test site (cf. previous section). When we compare this relationship with the relationship found at 20 m spatial resolution (Figure 27), we see that the data coincide very well and that the relationships match. The range in red-edge values at 300 m is less because there are no points at the lower end in comparison to the range at 20 m. This is caused by the fact that no pure bare soil points are present in the degraded image.

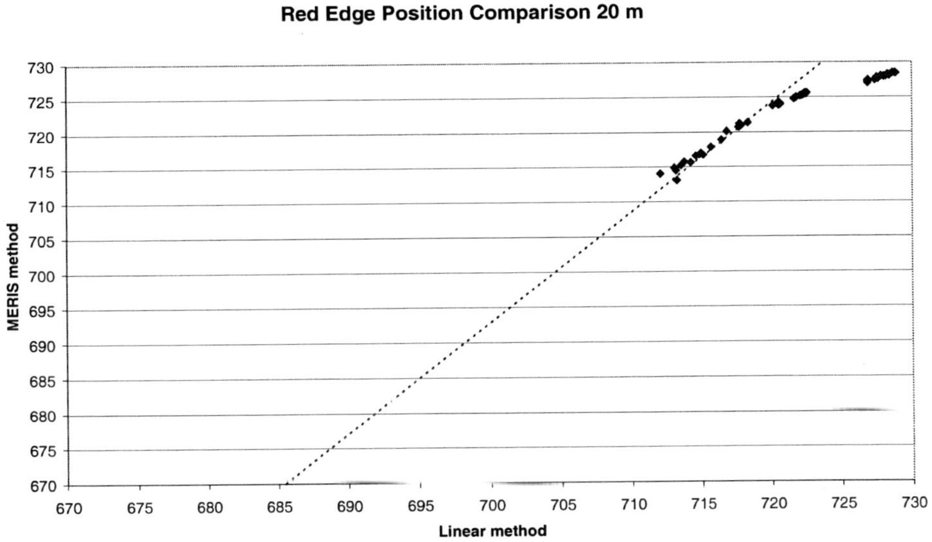

**Red Edge Position Comparison 20 m**



*Figure 27.* Comparison of the red-edge position calculated using the "linear method" interpolating between 697 and 737 nm and the "MERIS method" for the main crops and bare soils at the Flevoland site, measured with the AVIRIS sensor on July 5th, 1991. The regression line from Figure 26 is also plotted.

From this we may conclude that the red-edge index derived for the MERIS spectral bands at the MERIS spatial resolution of 300 m still provides useful information. If the objects at the earth surface are large enough, we may obtain a significant range in red-edge values. At the Flevoland agricultural test site individual fields could not be recognized, but MERIS will provide informative red-edge values on large units (groups of fields).


## 6.4    CONCLUSIONS

The "linear method" (Guyot & Baret, 1988) assumes a straight slope of the reflectance spectrum around the midpoint between the reflectance at the NIR plateau and the reflectance minimum at the chlorophyll absorption feature in the red. This midpoint is then defined as the red-edge index. This point may not coincide with the maximum of the first derivative, but it appears to be a very robust definition and it needs only a very limited number of spectral bands. Thus, this method is very useful for practical applications.

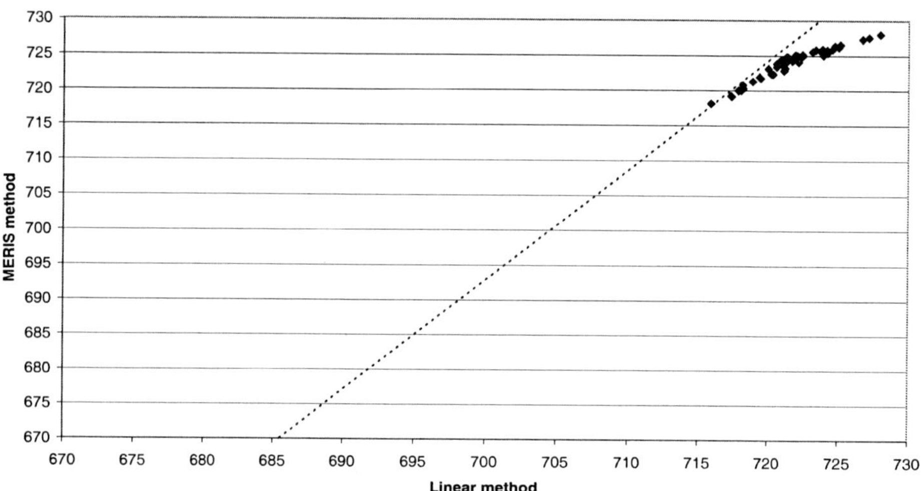


*Figure 28.* Comparison of the red-edge position calculated using the "linear method" interpolating between 697 and 737 nm and the "MERIS method" for AVIRIS pixels degraded to 300 m at the Flevoland site. The regression line from Figure 26 is also plotted.

Although the spectral bands of the MERIS standard band setting at the red-edge slope are not optimally located, they can be used for applying the "linear method" for red-edge index estimation. The bands to be used are located at 705 and 753.75 nm. However, since the latter band is located very close to the oxygen absorption feature of the atmosphere, an atmospheric correction must be applied previous to calculating the position of the red-edge using the MERIS bands. Previous studies (e.g., Clevers, 1994) have shown that for bands located at about 700 and 740 nm, an atmospheric correction is not necessary since the red-edge calculated in terms of reflectances equals the one calculated in terms of radiances.

Results for both data sets have shown that there exists a nearly linear relationship for the "linear method" using either bands at about 700 and 740 nm or the MERIS bands, albeit that it not coincides with the 1:1 line. Actually, this means that we have to work with instrument-specific red-edge indices. For MERIS, the above-mentioned linear relationship may be used for translating red-edge index values obtained with one method (using the MERIS bands) into values obtained with the other method (based on bands at 700 and 740 nm). Deviations occur mainly at high red-edge index values matching vegetation with a high chlorophyll content and with very high LAI values. To cope with this a non-linear function might be fitted. Another possibility is to fit a second linear relationship for the upper part of the curve. The breakpoint then seems to lay at about 720 nm for the standard "linear method" (coinciding with about 725 nm for the "MERIS method"). When fitting a linear function for this upper part of the curve, the regression line yields

$$re_{meris} = -243.7 + 0.668 \cdot re_{700-740} \tag{10}$$
$$r = 0.980.$$

For the remaining part of the curve still the relationship given in the previous section will be used. The two regression lines are illustrated in Figure 29 for data set I. When plotting these two regression lines with the results of data set II (Figure 30) reasonable results are found, although the "MERIS method" underestimates the highest red-edge values in comparison to the regression line. It appears, moreover, that the determination of the red-edge index using the MERIS band setting becomes less sensitive to variations in chlorophyll content and LAI than the standard "linear method" using interpolation between 700 and 740 nm.

**Red Edge Position Comparison**

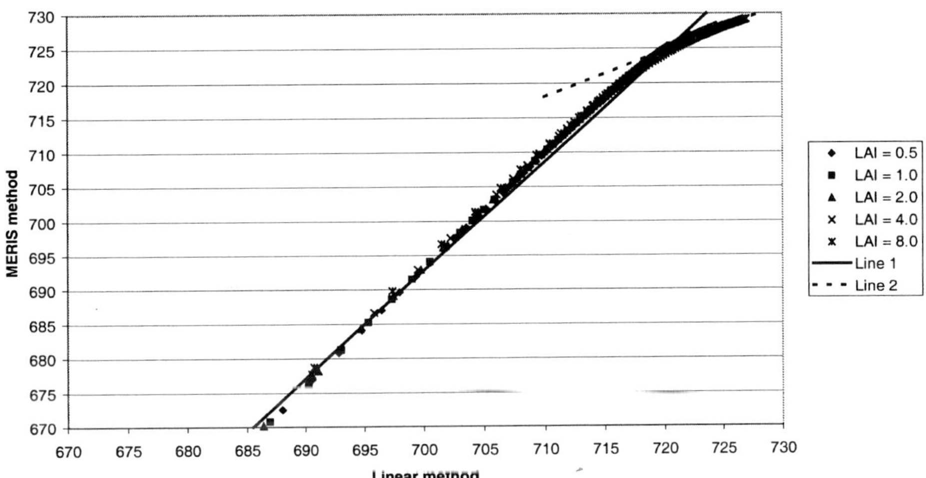


*Figure 29.* Comparison of the red-edge position calculated using the "linear method" interpolating between 700 and 740 nm and the "MERIS method", thus interpolating between 705 and 754 nm, PROSPECT-SAIL simulation for a canopy with varying leaf chlorophyll content and various LAI values. The two regression are also plotted:

Line 1: $re_{MERIS} = -409.0 + 1.574 * re_{700\text{-}740}$  Line 2: $re_{MERIS} = -243.7 + 0.668 * re_{700\text{-}740}$

This study has shown that the MERIS standard band setting can be used for deriving red-edge index values. It has paid no attention to the physical interpretation of red-edge index values at the scale level of MERIS. The spatial resolution of MERIS is 300 m, so one hardly will observe single fields or objects of just one vegetation type. MERIS application concerns more the application at, for instance, the ecosystem level. Highest sensitivity to variations in vegetation variables, like chlorophyll content and LAI, occurs at the lower values of these variables. One may expect these ranges in variables more at the ecosystem level, particularly when looking at natural ecosystems.

## 7 Conclusions

- The merit of imaging spectroscopy for agricultural applications lies in yielding information provided by some spectral bands at the red-edge region, not covered by the information provided by a combination of an NIR and a VIS broad spectral band.

- The "linear method", assuming a straight slope of the reflectance spectrum around the midpoint between the reflectance at the NIR plateau and the reflectance minimum at the chlorophyll absorption feature in the red, yields a robust method for determining the red-edge index. Four narrow (5-10 nm width) spectral bands at about 670, 700, 740 and 780 nm can be used for determining this index.
- The red-edge index may be considered as a particular "vegetation index" of interest because of its low sensitivity to disturbing factors such as atmospheric conditions or soil brightness and its high sensitivity to canopy characteristics such as chlorophyll content or LAI.

**Red Edge Position Comparison 20 m**

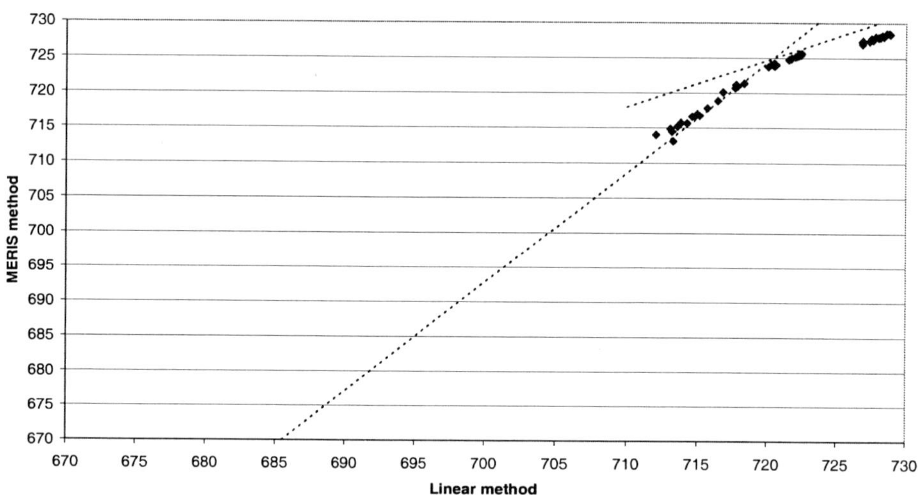


*Figure 30.* Comparison of the red-edge position calculated using the "linear method" interpolating between 697 and 737 nm and the "MERIS method" for the main crops at the Flevoland site, measured with the AVIRIS sensor on July 5th, 1991.
The regression lines from Figure 29 are also plotted.

- The red-edge index can be applied to estimate the nitrogen status of a crop canopy. In a potato crop, for instance, the red-edge index alone cannot be used as an indicator for nitrogen limited growth conditions. Additional information is needed on dry matter partitioning between above-ground and below-ground crop organs and nitrogen partitioning between above-ground and below-ground dry matter.
- By combining the red-edge information from imaging spectroscopy data with a traditional (broad-band based) vegetation index and with bidirectional reflectance measurements (at least two viewing angles), LAI, LAD and leaf chlorophyll content can be estimated simultaneously. By assimilating this information into a crop growth model, a practical framework for yield prediction is obtained.
- In potato, nitrogen fertilisation practices can be quantified by the use of spectral information. The crop nitrogen status can be determined by the use of vegetation indices derived from spectral information (red-edge index and WDVI). However, additional information is needed to determine the degree of growth reduction and the quantity of fertiliser that is needed to overcome the growth reduction. By

taking the within-field variability into account, imaging spectroscopy might contribute to the application in fields like precision farming.

- Although the spectral bands of the MERIS standard band setting at the red-edge slope are not optimally located, they can be used for applying the "linear method" for red-edge index estimation. The bands to be used are located at 705 and 753.75 nm. Since the latter band is located very close to the oxygen absorption feature of the atmosphere, an atmospheric correction must be applied previous to calculating the position of the red-edge using the MERIS bands.

- The red-edge index derived for the MERIS spectral bands at the MERIS spatial resolution of 300 m still provides useful information. If the objects at the earth surface are large enough, we may obtain specific information concerning these objects.

# CHAPTER 7

# IMAGING SPECTROMETRY AND GEOLOGICAL APPLICATIONS

Freek VAN DER MEER [α,β], Hong YANG [χ] & Harold LANG [δ]

[α] International Institute for Aerospace Survey and Earth Sciences (ITC), Division of Geological Survey, Enschede, The Netherlands.

[β] Delft University of Technology, Department of Applied Earth Sciences, Delft, The Netherlands

[χ] Shell Exploration and Production B.V., Rijswijk, The Netherlands
[δ] NASA Jet Propulsion Laboratory, Pasadena, CA, USA

## 1    Introduction

In this chapter, geological applications of imaging spectroscopy are described. We refrain from detailed discussions on each type of application, but instead provide a short description and many references to journal papers. The last part of the chapter shows two case-studies to illustrate some aspects of geological imaging spectroscopy directed to petroleum and mining applications. The first part of this chapter is partly based on a previous review by the first author (Van der Meer, 1999).

## 2    Mineral mapping; surface mineralogy

Goetz *et al.* (1982) demonstrated that spectroscopy may aid in surface mineralogy mapping. This early work was the start of a large number of similar type of works. Murphy (1995) for the first time succeeded in mapping jasperoid in the Cedar mountains. An expert system approach without *a priori* knowledge of the area was successfully implemented by Dwyer *et al.* (1995) where they applied AVIRIS data to surface mineralogy mapping in the Drum mountains of Utah (U.S.). Similarly, Gaddis *et al.* (1996) presented techniques to use only information contained within a raw AVIRIS data set to estimate and remove additive components (atmospheric scattering and instrument dark current), normalize multiplicative components (instrument gain, atmospheric transmission) and enhance, extract and map surface composition and mineralogy. Kruse *et al.* (1993) presented mineral mapping using an expert-systems approach. Ben-Dor *et al.* (1994a) and Ben-Dor & Kruse  (1994b) demonstrate that mineral maps (derived using spectral unmixing techniques) from GER 63 channel imaging spectrometer data over the Makhtesh Ramon area in Israel provide additional



*F.D. van der Meer and S.M. de Jong (eds.), Imaging Spectrometry,* 201–218.
© 2006 *Springer. Printed in the Netherlands.*

geologic details not yet revealed on existing geologic maps. This works builds on earlier studies by Kaufman *et al.* (1991). Crowley *et al.* (1993) use airborne imaging spectrometer data (AVIRIS) of the Ruby mountains, Montana for mineral discrimination using relative absorption band-depth images. This author also presented a study on mapping of evaporite minerals with AVIRIS in Death Valley (Crowley 1993). A study by Farrand & Seelos (1996) showed the possibility of mapping faults by evaluating mineral maps for linear or curvilinear features. These authors were able to delineate some previously unmapped faults in the Summitville area from mineral endmember images obtained using the constrained energy minimization technique. The use of classification methods including neural networks for mapping surface mineralogy from AVIRIS data of a volcanic terrain on Iceland is described in Benediktsson (1995). Recently, Baugh *et al.* (1998) showed the potential of AVIRIS for mapping ammonium type minerals.


## 3    Mineral mapping; exploration

Numerous studies on remote sensing to aid in mineral exploration have been conducted at Cuprite and Goldfield, Nevada (U.S.); two test sites of NASA for instrument calibration and geologic mapping. Most of these studies center on mapping of hydrothermal alteration to aid in mineral prospecting (particularly for gold). Calibration and testing of new sensors as well as validating new processing strategies is typically conducted at Cuprite. Cuprite and Goldfield are areas of extensive hydrothermal alteration within a sequence of rhyolitic welded ash flow and air fall tuffs (Albers & Stewart 1972; Ashley 1971) which can be subdivided into three mappable units: silicified rocks, opalized rocks, and argillized rocks. Silicified rocks form a large irregular patch extending from the middle to the south end of the area. The silicified core represents the most intensely altered rocks at Cuprite containing quartz, calcite and minor alunite and kaolinite. Opalized rocks contain abundant opal and as much as 30% alunite and kaolinite. Locally, an interval of soft, poorly exposed material mapped as argillized rock separates fresh rock from opalized rock. In the argillized rocks, plagioclase is altered to kaolinite, and glass is altered to opal and varying amounts of montmorillonite and kaolinite. The distribution of these alteration assemblages is characteristic for a fossilized hot-spring deposit which often contains gold. Volcanism at Cuprite began in Oligocene with eruption of rhyolite and quartz latite flows. Hydrothermal alteration is related to a second, early Miocene, period during which dacite and andesite flows were extruded and hot, acidic brines began circulating through the volcanics. The general geology of the Cuprite mining district is treated in more detail in Abrams *et al.* (1977). Mineral zones that can be mapped for rings around a central vent formed by altered silica. The alteration zones are composed of kaolinite, alunite, buddingtonite and various ferrigenous minerals characterising the host rocks. Some of the sedimentary units show mappable amounts of calcite and dolomite. The mineral zonation is indicative of a sulfuric acid-charged system with hydrothermal fluids emitted in a hot-spring type of environment. The circular distribution of the mineral zones demonstrates that alteration occurred along a central vent with the lateral mineral zoning controlled by a decrease of acidity and temperature. It should be noted that alunite is a pathfinder mineral for gold, in fact for many years gold has been explored at mines near the town of Goldfield. Some examples of recent studies at Cuprite and Goldfield include those of Resmini *et al.*(1997), HYDICE, constrained energy minimization), Okada & Iwashita (1992; waveform characterisation), Abrams & Hook (1995; Simulated ASTER data, decorrelation stretching), Nedeljkovic & Pendock (1996; AVIRIS, probability density functions), Carrere & Abrams (1988; AVIRIS), Kruse *et al.* (1990; GER 63-channel imaging spectrometer), Hook *et al.* (1991; AVIRIS, GEOSCAN), Van der Meer & Bakker (1998; AVIRIS, CCSM). In brackets, the data and (in some cases) the analytical technique are listed.

Other examples of gold mineral exploration through mapping of hydrothermal alteration zones are described by Crosta *et al.* 1996) who use GEOSCAN AMSS data in an area in the Rio Itapicuru greenstone belt in the northeastern part of Brazil, Ferrier & Wadge 1996) who use AVIRIS data from southern Spain, Hutsinpiller (1988) who used AIS-I data from Virginia City, Nevada, Feldman & Taranik (1988) who used AIS-I data to map alteration minerals in the Tybo mining district of Nevada, and Crosta *et al.* (1998) who use AVIRIS data for alteration mapping at Bodie (California). Other alteration processes that have been studied in imaging spectrometer data are the serpentinization of ultramafic rocks which relates to major asbestos deposits. Based on laboratory studies of Hunt & Evarts (1981) it has been show that the degree of serpentinization of ultramafic rocks can be estimated through spectral analysis of imaging spectrometer data (Van der Meer 1995;Beratan *et al.*1997) used AVIRIS to map potassium metasomatism, a type of alteration marked by addition of large amounts of potassium at the expense of sodium. These authors mapped hematite as a proxy for potassium since potassium itself has no distinct absorption features to be evaluated. Van der Meer *et al.* (1997) used data from the Chinese imaging spectrometer MAIS to map surface mineralogy in the Jinchuan ultramafic intrusion. This work contributed to prospecting for nickel-copper sulphide mineralizations in the Gansu province of China. Another case of hydrothermal alteration mapping in rhyolites and acid volcanics is presented in Loercher *et al.* (1994) who studied the Torfajokull volcanic complex of Iceland using AVIRIS and map zonation patterns related to active geothermal fields. Mapping of mineral-bound ammonium (buddingtonite) in hydrothermally altered rocks to permit the determination of fluid chemistry in fossil hot springs was demonstrated by Baugh & Kruse (1994) for the southern Cedar mountains using AVIRIS data.


## 4    Mineral mapping; lithology

An example of remote mineralogic and lithologic mapping using AVIRIS is presented for the Ice River Alkaline complex of British Columbia (Canada) by Bowers & Rowan (1996). These authors use spectral unmixing techniques to map mineralogy related to shales, slates and limestones incorporating vegetation and weathering products into the analysis. Integration of optical, thermal and RADAR data from AVIRIS, TIMS and AIRSAR by Keirein-Young (1997) shows the synergy of such approach while mapping surface geology and morphology in alluvial fans in the Death Valley area (California, U.S.). Alternatively, geologic imaging spectrometry may be used to assist metamorphic facies mapping for regional geologic investigations (e.g., Rowan *et al.* (1987); Van der Meer 1996). Through the analysis of the position of the absorption band in the 2.3 $\mu$m wavelength range it has been shown that mapping calcite versus dolomite is possible from imaging spectrometer data thus allowing to map dolomitization patterns from space (Windeler & Lyon 1991; Van der Meer 1998). This has potential for petroleum exploration since porosity increase due to dolomitization makes these rocks potential reservoir rocks. Mustard (1993) used the relationship between soil, grass and bedrock to map in a serpentinite melange using AVIRIS data. A last example comes from Lang & Cabral-Canoz (1998) who use AVIRIS data for tectonostratigraphic mapping in the Northern Guerrero state of southern Mexico.


## 5    Vegetation stress and geobotany

Indirect detection of mineral deposits using imaging spectrometer data has been attempted through analysis of spectra of "stressed vegetation". Collins *et al.* (1983) were the first to report the shift of the red edge (e.g. the point of maximum slope on the reflectance spectrum of vegetation between red and near-infrared wavelengths) toward the blue end of the spectrum as a result of stress due to copper in the subsurface. These authors claimed that the position and shape of the red edge could be used to guide in mineral prospecting. Their work has led to a still ongoing debate on the potential use of the red edge shift. Although the amount and direction of shift as a result of geochemical stress on vegetation is uncertain, most workers agree that the red edge can be used to asses the vitality of plant communities (see, for example, Boochs *et al.* (1990) for a recent discussion on this topic). Recent publications on the use of vegetation stress indicators derived from AVIRIS data can be found in Clark *et al.* (1997) and Lelong *et al.* (1998) [171]. An overview of techniques for delineating the red-edge can be found in Dawson & Curran (1998).


## 6    Environmental geology

High spectral resolution data of the GEOSCAN Mk II were used to delineate mercury-contaminated mill tailings within an EPA Superfund site in north central Nevada where surface geochemical data could be used for validation of the results (Fenstermaker & Miller 1994). Similarly, Lehmann *et al.* (1990) investigated GER spectra of a vegetation covered mine waste deposit. In a recent paper, Farrand & Harsanyi (1997) map the distribution of mine tailings in the Coeur d'Alene River valley, Idaho, through their Constrained Energy Minimization Technique. Farrand (1997) used AVIRIS data to map pernicious trace metals, ferric oxides and oxyhydroxide minerals in an acid water environment. The trace metals are released into the environment as a result of mining operations. Traditionally, such phases are mapped in the field using geochemical sampling and laboratory analysis. Bianchi *et al.* (1996) used data from the MIVIS imaging spectrometer to map oil spills over land by relating the ppm of oil with the oil hyperspectral information gathered by the imaging sensor. Their study focuses on the oil spilled during the blow-out from an AGIP rig located within the Ticino Regional park in Italy. Mapping of abundance images of goethite, jarosite, hematite, kaolinite, muscovite, montmorillonite along with vegetation and soil from AVIRIS over the Leadville area, Colorado aided in an environmental hazard assessment while muscovite/illite mostly corresponded to mine tailings and iron minerals often were associated with dark smelter slag materials (Kruse 1996). The combination of airborne Daedalus-ATM scanner data and imaging spectrometer (reflective and thermal) data from DAIS was used to map the area of a former uranium mining site in eastern Germany (Mueller *et al.* 1996). Mapping of temperature anomalies and differentiation of surface materials allowed assessment of the hydrologic conditions at the site and providing information relevant for recultivation. In a paper by Clark *et al.* 1996) a USGS initiative is outlined for monitoring of surficial geology, vegetation communities, and environmental materials the U.S. National Parks to aid in ecosystems environmental mapping.


## 7    Petroleum related studies

Bammel & Birnie (1994) studied spectral reflectance response of big sagebrush to hydrocarbon induced vegetation stress in the Bighorn basin of Wyoming and found a significant blue shift (to shorter wavelengths) of the green reflectance peak at 0.56 $\mu$m and red trough (0.67 $\mu$m) as stress indicators. However these authors concluded that the correlation found between geobotanical anomalies in relation to surface or subsurface hydrocarbons could not be demonstrated with statistical significance. In a recent study, Yang *et al.* (1999) surveyed two profiles in an oil rich basin in the delta of the Yellow river in China. Field spectral measurements of mono-cultural winter wheat over areas know to have light hydrocarbon gas seepages constraint from subsurface gas measurements showed an anomalous red shift (contrary to the result of Bammel & Birnie 1994) of the red-edge of the vegetation spectrum. Although a visual relationship between gas measurements and spectral anomalies could be seen, no statistical significant correlation could be demonstrated. Mapping of dolomitization patterns through mapping dolomite versus calcite concentration is a goal of remote sensing since dolomites have a 12 % greater porosity as opposed to limestones containing merely calcite thus making these interesting host-rocks for oil accumulation. Windeler & Lyon (1991) demonstrated the possibility of separating calcite from dolomite in GEOSCAN airborne data although their study centered on mineral prospecting in skarn-type deposits. In a series of publications, Van der Meer (1998) demonstrated that imaging the shift of the carbonate absorption would allow mapping compositional variations in dolomite and calcite percentages. Hydrocarbon soil geochemistry and airborne spectrometer data were integrated with Landsat TM imagery for a portion of the Sao Francisco basin of Central Brazil by De Oliveira & Crosta (1996). This approach allowed these authors to map both soil and vegetation anomalies related to hydrocarbon microseepages. Interested readers are referred to the Chapter 8 on petroleum applications in this book for further reading.


## 8    Atmospheric effects resulting from geologic processes

Most studies use imaging spectrometer data calibrated to reflectance. As discussed previously, raw imaging spectrometer radiance data shows absorption features that can be attributed to atmospheric gasses. Vice versa, these absorption features can be used to map quantities and differences in these gasses in the atmospheric column. Using ratios of absorption band depth estimates of water bands at 0.95 $\mu$m and 1.15 $\mu$m, it has been demonstrated that water vapour total column abundance can be mapped from imaging spectrometer data (Carrere & Conel 1993; Gao & Goetz 1990; 1995; Frouin *et al.* 1990). Similarly, De Jong & Chrien (1996) and De Jong (1998) detected abnormally high abundances carbon dioxide and methane in AVIRIS data from the Mammoth Mountain area which they attributed to renewed volcanic activity. A general discussion on the effect of atmospheric attenuation on signal is provided in the Chapter 2 on analytical techniques in this book.

Hydrothermally-altered rocks on stratovolcanoes are closely linked to edifice failures and the generation of destructive debris flows while altered rocks form zones of weakness along fractures and contain hydrous clay minerals that modify the physical properties of derbis flows (Crowley & Zimbelman 1996). Mapping alteration minerals on the slopes of Mount Rainier using AVIRIS data allowed Crowley & Zimbelman (1997) to delineate hazardous sectors and develop a method also applicable to other areas.

## 9    Thermal infrared studies

Geologic studies using thermal infrared multispectral data date back to the early 1980's when Kahle & Goetz (1983) and Kahle (1987) derived emissivity information and thermal inertia from the first TIMS data and used it to map mineralogy and some lithologies. Following onto these studies were the first examples that showed that the age of lava flows could be constrained from thermal data (Abrams *et al.* 1991). In 1992, Hook *et al.* (1992) presented a comparison of techniques for extracting emissivity information from thermal infrared data for geologic studies. Abrams & Hook (1995) use the TIMS data to simulate ASTER data and evaluate the use of this data for geologic mapping purposes. A comparison of thermal infrared emissivity spectra measured *in situ,* in the laboratory and similar spectra derived from NASA's airborne thermal infrared multispectral scanner (TIMS) in Cuprite (Nevada) using spectral matching techniques demonstrated the complementary nature of the thermal region and the reflective region of the EM spectrum (Ninomiya *et al* 1997).

Many silicate minerals that have no diagnostic absorption features in the reflective part of the spectrum and which can therefore be difficult mapped using VIS-SWIR spectroscopy show strong features in the thermal part of the spectrum thus allowing more accurate mapping with thermal spectrometers such as TIMS. Thermal infrared spectra of TIMS acquired over sedimentary rocks in the Tarim basin allowed Bihong & Xiaowei (1998) to accurately map sandstones, siltstones, argillites and carbonates using decorrelation stretching. In a decorrelation stretched color-composite TIMS image, temperature variations show as intensity differences while emissivity differences show as color variations in the image (Gillespie 1992). For more details please refer to the chapter 10 on thermal infrared applications in this book.

## 10    Case-study I: Petroleum case study: seepage detection at Bluff using mineral alteration

### 10.1   MINERAL ALTERATION

Many oil and gas reservoirs in the past were found due to the presence of tar sands and oil pools at the surface. These indicate that the reservoirs are leaking. Visible macroseeps occur onshore as well as invisible microseeps. Long-term leakage of hydrocarbons can establish locally anomalous redox zones that favour the development of a diverse array of chemical and mineralogical changes (Saunders et al. 1999). The bacterial oxidation of light hydrocarbons can directly or indirectly bring about significant changes in the pH and Eh of the surrounding environment, thereby influencing mineral stability and chemical reactivity. Such oxidation in the chimney above a leaking petroleum accumulation leads to dissolution or precipitation of minerals and the mobilisation or immobilisation of certain elements in the chimney, which thereby becomes mineralogically and chemically different from laterally-equivalent. The resulting alterations include: the formation of calcite, pyrite, uraninite, elemental sulfur, and certain magnetic iron oxides and iron sulfides; bleaching of red beds; clay mineral alteration; electrochemical changes; radiation anomalies; geomorphic anomaly; the edge anomaly of adsorbed or occluded hydrocarbon in soils and Delta C (ferrous carbonate); and biogeochemical and geobotanical anomalies. The presence at surface of bleached and discolored red sandstones above petroleum accumulation has been widely noted. Bleaching occurs whenever acidic, reducing fluids dissolve the ferric oxide (hematite) that gives the red bed its characteristic colour. The acidic conditions resulting from the oxidation of hydrocarbons in near-surface soils and sediments promotes the diagenetic weathering of feldspar to clay and the conversion of smectite clay to kaolinite. The kaolinite thus formed remains chemically stable unless the environment is changed. Ferrous carbonate, also called "Delta C" show highs above the edges of hydrocarbon accumulations. The bicarbonate and carbonate chemistry of calcium and iron may provide a viable explanation for edge-leakage of Delta C iron carbonate anomalies. Many of these alteration minerals can be mapped using spectroscopic data. This case history is an example of mapping such alteration zones.


## 10.2   THE BLUFF AREA AND PETROLEUM GEOLOGY

The Bluff area (Figure 1) is located on the Utah side near the borders between Utah-Arizona and Utah-Colorado. The area is bounded by 109.34°-109.63°E, 37.21-37.41°N, a total area of about 300 km$^2$. The Bluff area is reached by paved roads from Monticello (U.S. Highway 160). Numerous unpaved dirt roads and unimproved dirt roads provide access to all parts of the study area. The San Juan River runs through from the east to the west. The altitude throughout the test site ranges from about 1200 to 1600 m. The area belongs to the southwestern part of Colorado Plateau featured by sparse desert-type vegetation, multicolored generally flat-lying sedimentary rocks laid bare by erosion, and imposing escarpments, gorges, domes and vertical cliffs. The climate is semiarid, with cool winter and relatively hot summer.

From December through March, snow may cover most of the surface. The average temperature of the hottest month in July is 30°C and the average temperature of the coldest month in January is 0°C. The sparse vegetation is chiefly Blackbrush (*Coleogyne*) growing on shallow Torriorthens. There have been many exploration activities in the area resulting in the findings of the Greater Aneth, Gothic Mesa, Recapture Creek and Turner Bluff oil fields. In neighboring areas such as Lisbon Valley (100 km north of the Bluff area) in the same Paradox Basin, it has been observed that red sandstones have been bleached to gray color due to hydrocarbon microseepage (Segal *et al.*, 1984 & 1986; Conel and Alley, 1985).

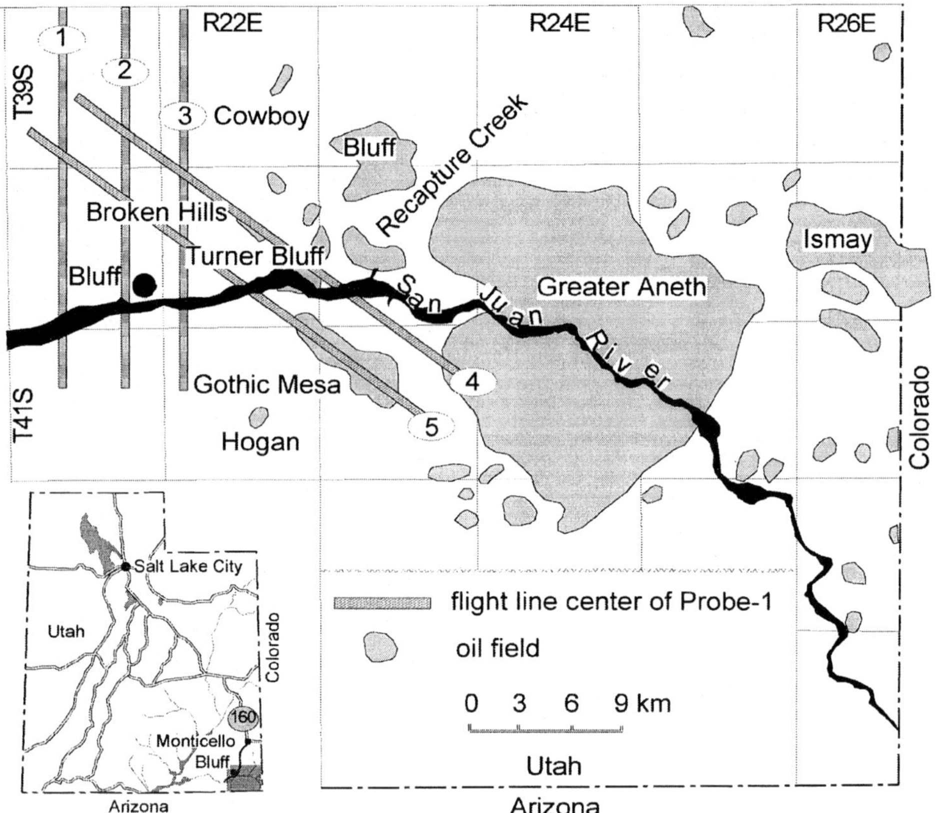


*Figure 1.* Location map of the probe data acquired at the Bluff test site. Line 5 was used in this study.

The Desert Creek zone of the Paradox Formation in southeastern Utah includes three generalized facies belts: open marine, shallow shelf and shelf margin, and intra-shelf, salinity restricted (Chidsey *et al.*, 1996). The open-marine facies belt includes open-marine buildups (typically crinoid rich), crinoid- and brachiopod-bearing carbonate muds, euxinic black shales, and detrital carbonate fans. The shallow-shelf and shelf-margin facies belt includes shallow-shelf buildups (composed of phylloid algal, coralline algal, and bryozoan buildups), platform-margin calcarenites (in beach, dune, and stabilized grain flats), and platform-interior carbonate muds and sands.

The stratigraphic column for sediments found in the Bluff area is shown in Figure 2. The intra-shelf, salinity-restricted facies belt includes platform-interior evaporites, dolomitized tidal-flat muds, bioclastic lagoonal muds, tidal-channel carbonate sands, stromatolites, and euxinic dolomites. Carbonate buildups, tidal-channel carbonate sands and other features can yield seismic anomalies. However, if these carbonate buildups are located within either the open-marine or intra-shelf, salinity-restricted facies belts, the reservoir quality is typically poor. Porosity and permeability development is limited or, if present, plugged with anhydrite in these facies belts.

Carbonate buildups and calcarenites in the shallow-shelf and shelf-margin facies belt can have excellent reservoir properties. Hydrocarbons are stratigraphically trapped in porous and permeable lithotypes within the mound-core and supra-mound intervals of the Desert Creek carbonate buildups. These intervals are effectively sealed by impermeable platform intervals at the base, marine muds on the flanks, and a 6 m thick layer of anhydrite, usually at the top of the Desert Creek zone.

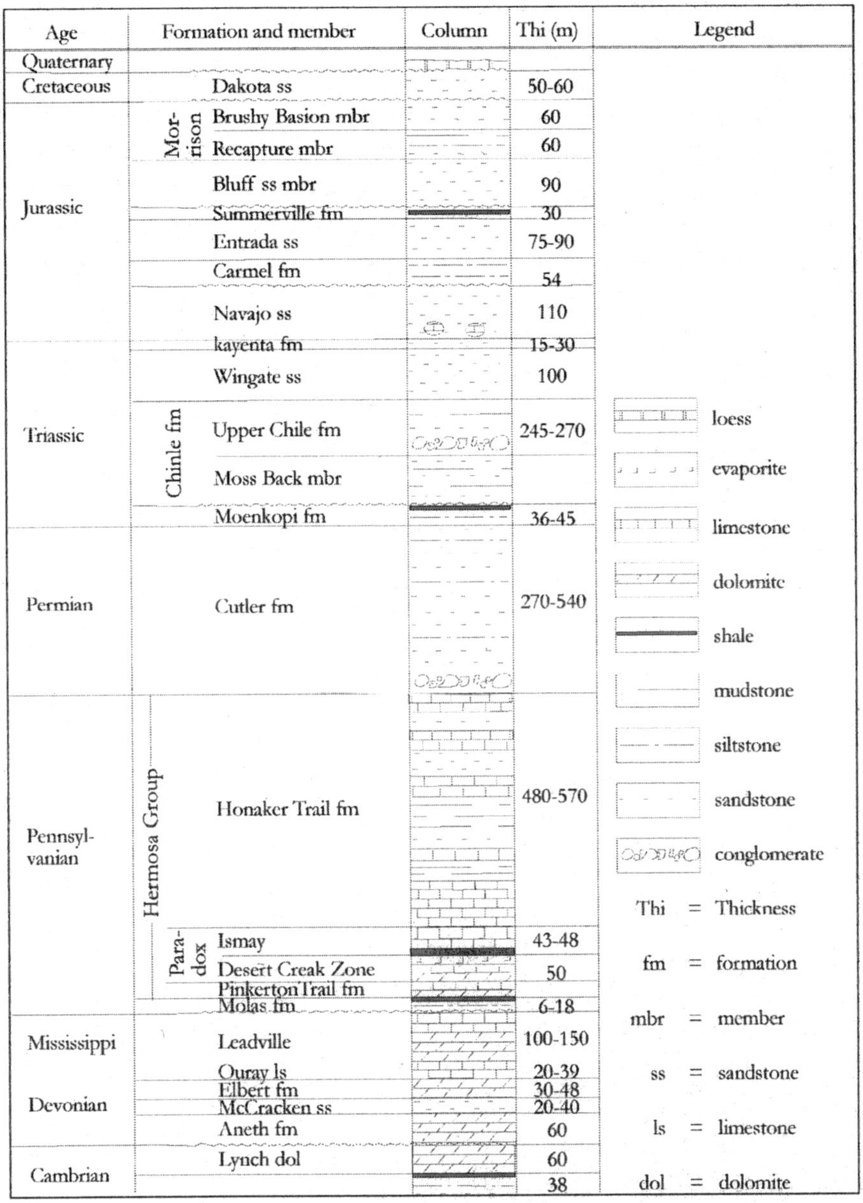

| Age | Formation and member | | | Column | Thi (m) | Legend | |
|---|---|---|---|---|---|---|---|
| Quaternary | | | | | | | |
| Cretaceous | | | Dakota ss | | 50-60 | | |
| Jurassic | Mor-rison | | Brushy Basion mbr | | 60 | | |
| | | | Recapture mbr | | 60 | | |
| | | | Bluff ss mbr | | 90 | | |
| | | | Summerville fm | | 30 | | |
| | | | Entrada ss | | 75-90 | | |
| | | | Carmel fm | | 54 | | |
| | | | Navajo ss | | 110 | | |
| | | | kayenta fm | | 15-30 | | |
| | | | Wingate ss | | 100 | | loess |
| Triassic | Chinle fm | | Upper Chile fm | | 245-270 | | evaporite |
| | | | Moss Back mbr | | | | limestone |
| | | | Moenkopi fm | | 36-45 | | |
| Permian | | | Cutler fm | | 270-540 | | dolomite |
| | | | | | | | shale |
| | | | | | | | mudstone |
| Pennsyl-vanian | Hermosa Group | | Honaker Trail fm | | 480-570 | | siltstone |
| | | | | | | | sandstone |
| | | | | | | | conglomerate |
| | | | | | | | Thi = Thickness |
| | | Para-dox | Ismay | | 43-48 | | fm = formation |
| | | | Desert Creak Zone | | 50 | | |
| | | | Pinkerton Trail fm | | | | |
| | | | Molas fm | | 6-18 | | mbr = member |
| Mississippi | | | Leadville | | 100-150 | | |
| | | | Ouray ls | | 20-39 | | ss = sandstone |
| Devonian | | | Elbert fm | | 30-48 | | |
| | | | McCracken ss | | 20-40 | | ls = limestone |
| | | | Aneth fm | | 60 | | |
| Cambrian | | | Lynch dol | | 60 | | dol = dolomite |
| | | | | | 38 | | |

*Figure 2.* Stratigraphic column of sediments in the Bluff area.

## 10.3   PROBE-1 IMAGING SPECTROMETER DATA

Data from Probe-1 (U.S. equivalent of HyMAP) were acquired over the Bluff area. The Probe 1 is a "whiskbroom style" instrument that collects data in a cross-track direction by mechanical scanning and in an along-track direction by movement of the airborne platform. The instrument acts as an imaging spectrometer in the reflected solar region of the electromagnetic spectrum (440 - 2500 nm) regions. The instrument has four spectrometers A, B, C and D covering 440-908 nm, 896 – 1357 nm, 1390 – 1796 nm and 1977 – 2501 nm respectively. Spectral coverage in 128 bands is nearly continuous in these regions with small gaps in the middle of the 1400 and 1900 nm atmospheric water bands. The IFOV of the Probe 1 is 2.5 mrad along track and 2.094 mrad across track with a FOV of 61.4 degrees. The image covers 512 pixels across track. The radiometric calibration error is less than 10% absolute across all bands. The Probe-1 has two modes: 5 m ground resolution at 2300 m above and 10 m ground resolution at 4600 m above ground. The Signal-to-noise ratio (SNR; Figure 3) was calculated as the ratio of mean signal and the standard deviation of a homogeneous area with 31 pixels. Although the image method under-estimates the SNR, it is clear that the SNR is more than 150:1 for spectrometer A and B, about 150:1 for spectrometer C and about 100:1 for spectrometer D, which states that the data is good enough for further processing.

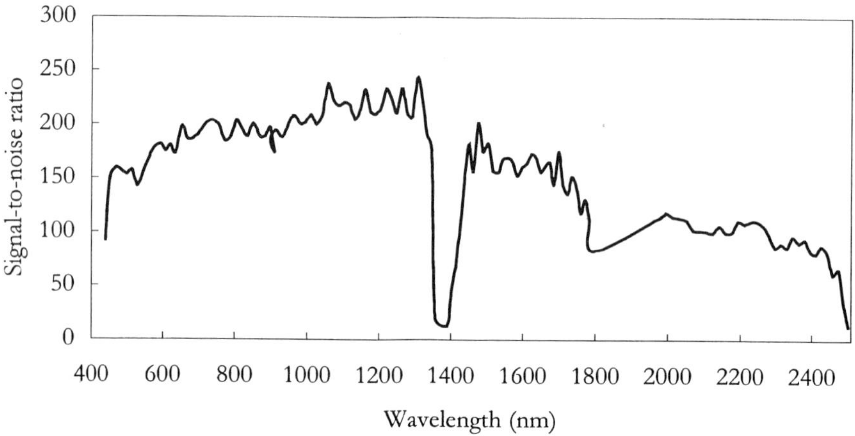

*Figure 3.* SNR values for a homogeneous 50% albedo target from the Probe data at Bluff.

To measure whether there is a shift in the position of wavelength centers, the discrete atmospheric absorption features, focusing on the $O_2$ (760 nm), $H_2O$ (1130 nm), $CO_2$ (1570/1610 nm), and $CO_2$ (2060 nm) bands, were used for spectrometers A, B, C, and D respectively. By comparing the theoretical absorption profiles and the extracted average spectra from large regions of interest from a number of the image data cubes, it is apparent that in the model results, the minimum transmission occurs at 758 nm whereas in the Probe-1 spectra, the minimum oxygen absorption occurs at 772 nm. The central wavelength position of spectrometer A has been shifted 14 nm, about one band shift to the shorter wavelength. Radiometric calibration provides a calibration file used to convert the Probe-1 digital numbers (DN) to units of radiance. The radiometric calibration of the Probe-1 data in the Bluff area was undertaken with the aid of the

Probe-1 digital raw data in Ivanpah Playa, California, and radiometrically calibrated AVIRIS data of the same area. The data was subsequently calibrated to reflectance using a simple empirical line approach.


## 10.4  MINERAL ALTERATION MAPPING FOR MICROSEEPAGE DETECTION AT BLUFF

The major units exposed are the Quaternary eolian loess deposits (Qe), the Cretaceous Dakota Sandstone (Kd), the Brushy Basin Member of the Morrison Formation (Jmb), the Recapture Member of the Morrison Formation (Jmr), and the Bluff Sandstone (Jb), among which, Brushy Basin Member and Recapture Member of the Morrison Formation are likely to be altered because of petroleum accumulation. The spectra of eolian deposits, Dakota Sandstone, Brushy Basin Member, Recapture member and Bluff Sandstone are studied and their overall reflectance, absorption features that correlate to mineralogy are described below. The dominant mineral quartz does not have absorption features in the visible through short-wave-infrared, however, the overall reflectance of rocks is high because of high percentage of quartz.

### 10.4.1  *Quaternary eolian loess deposits (Qe)*

The overall reflectance of Qe appear to be moderate among all units. It has distinctive absorption features at 0.51, 0.67, 0.91, 1.11, 1.18, 2.03, 2.08, 2.29, 2.32, 2.36, 2.44 $\mu$m. These features may infer a combination of montmorillonite, chlorite and illite.

### 10.4.2  *Dakota Sandstone (Kd)*

The Dakota Sandstone has a low reflectance comparing with other rocks. The absorption features are located at 0.64 through 0.96, 2.08 $\mu$m and strong feature at 2.205 $\mu$m indicate the presence of hematite, chlorite and montmorillonite.

### 10.4.3  *Brushy Basin Member (Jmb)*

Two rock units can be identified within this member. The 1st Jmb (solid line) has higher reflectance than the 2nd Jmb. It can be observed that the 1st Jmb may contain kaolinite and calcite and the 2nd Jmb contains hematite and montmorillonite, which infers that the 1st Jmb to be red colored rocks and the 2nd Jmb to be gray-green colored rocks that may include bleached rocks indicative to hydrocarbon microseepage.

### 10.4.4  *Recapture Member (Jmr)*

There are two units that can be distinguished within the Recapture. The 1st Jmr (solid line) has higher reflectance than the 2nd Jmr (dot-dash line). From the comparison of absorption features of the two units, it can be seen that 1st Jmr may be rich in kaolinite and the 2nd Jmr may be rich in hematite and dolomite. The magnitude of the reflectance the two rock units also suggests the 1st Jmr may be gray-green color (higher reflectance) and the 2nd Jmr to be reddish color (lower reflectance). The 1st Jmr unit may include the bleached rocks associated with hydrocarbon microseepage, one of the target rock units to be mapped.

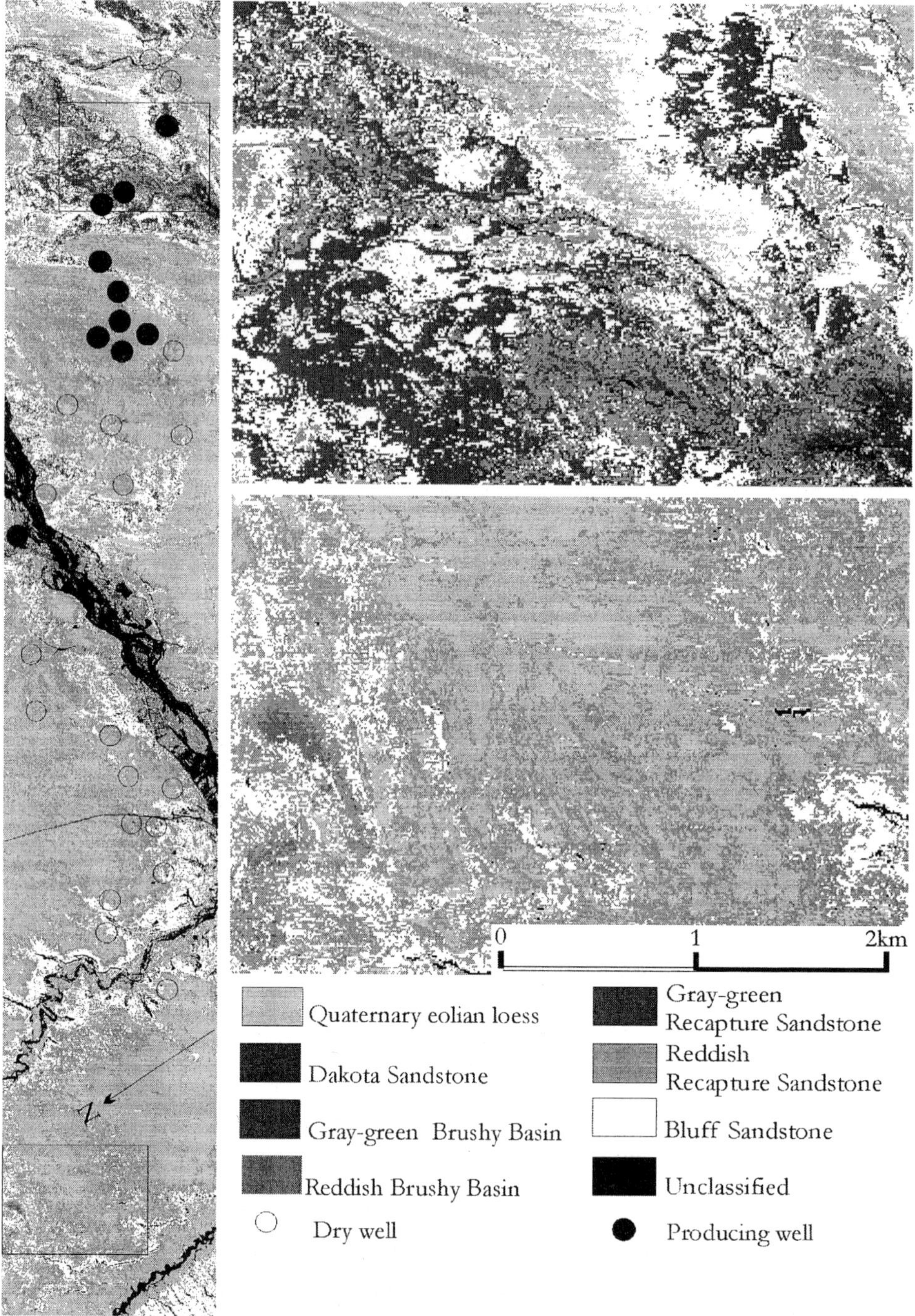



*Figure 4.* SAM classified image from probe data highlighting mineral alteration.

### 10.4.5 *Bluff Sandstone (Jb)*

The Bluff Sandstone has the high reflectance among all rock units. This may be caused by its high percentage of quartz. The absorption features are at 0.69, 1.005, 1.18, 2.09, 2.31, and 2.46 $\mu$m. There is no absorption features of clay minerals around 2.2 $\mu$m. This rock unit is not known indicative to hydrocarbon microseepage.

## 10.5 MAPPING GRAY-GREEN COLORED ROCKS THAT MAY ASSOCIATE WITH HYDROCARBON MICROSEEPAGE

According to the spectral characteristics of surface rock units, seven types of rocks are selected as inputs to Spectral Angle Mapping (SAM) to classify the image. SAM (see also the chapter 2 on analytical techniques in this book) is an automated method for comparing image spectra to a spectral library assuming that the data have been reduced to apparent reflectance (true reflectance multiplied by some unknown gain factor controlled by topography and shadows).

The algorithm determines the similarity between two spectra by calculating the "spectral angle" between them, treating them as vectors in a space with dimensionality equal to the number of bands. Seven endmembers, namely Quaternary eolian loess deposit, Dakota Sandstone, gray-green Brushy Basin member, reddish Brushy Basin member, gray-green Recapture Member, reddish Recapture Member and Bluff Sandstone were selected representing rocks to be mapped. The result is a classification image showing the best SAM match at each pixel and a "rule" image for each endmember showing the actual angular difference in radians between each spectrum in the image and the reference spectrum. Dark pixels in the rule images represent smaller spectral angles, and thus spectra that are more similar to the reference spectrum. The maximum angle thresholding of the seven classes is 0.1 radians. SAM produces a class with unclassified pixels. To assist image inspection, a color lut is used to discriminate different classes as shown in Figure 4. The results show that the gray-green colored rocks (red and brown in the classification image) are concentrated on the southeast (upper side of the image), while the reddish colored rocks (cyan and green in the classification image) are located on the northwest (lower side of the image).

These coincide with the drilling record also shown in the same image. This indicates that the Probe-1 data has the ability to distinguish the gray-green colored rocks from the reddish colored rocks of the same origin, which leads to a possibility to map areas of bleaching associated with hydrocarbon microseepage.

## 11  Case-study II: mining

### 11.1  BACKGROUND ON THE MINING PROBLEM

In this study case study, AVIRIS data acquired in 1995 from the Cuprite mining area situated some 30km. south of the town of Goldfield in western Nevada is used (Figure 5). The site is an area of extensive hydrothermal alteration within a sequence of rhyolitic welded ash flow and air fall tuffs which can be subdivided into three mappable units: silicified rocks, opalized rocks, and argillized rocks. Silicified rocks form a large irregular patch extending from the middle to the south end of the area. The silicified core represents the most intensely altered rocks at Cuprite containing quartz, calcite and minor alunite and kaolinite. Opalized rocks contain abundant opal and as much as 30% alunite and kaolinite. Locally, an interval of soft, poorly exposed material mapped as argillized rock separates fresh rock from opalized rock. In the argillized rocks, plagioclase is altered to kaolinite, and glass is altered to opal and varying amounts of montmorillonite and kaolinite. The distribution of these alteration assemblages is characteristic for a fossilized hot-spring deposit which often contains gold. Volcanism at Cuprite began in Oligocene with eruption of rhyolite and quartz latite flows. Hydrothermal alteration is related to a second, early Miocene, period during which dacite and andesite flows were extruded and hot, acidic brines began circulating through the volcanics. The raw AVIRIS data were converted to reflectance using an the empirical line approach which uses a least-squares fit between known spectra of ground targets and raw spectra of the same targets in the scene.

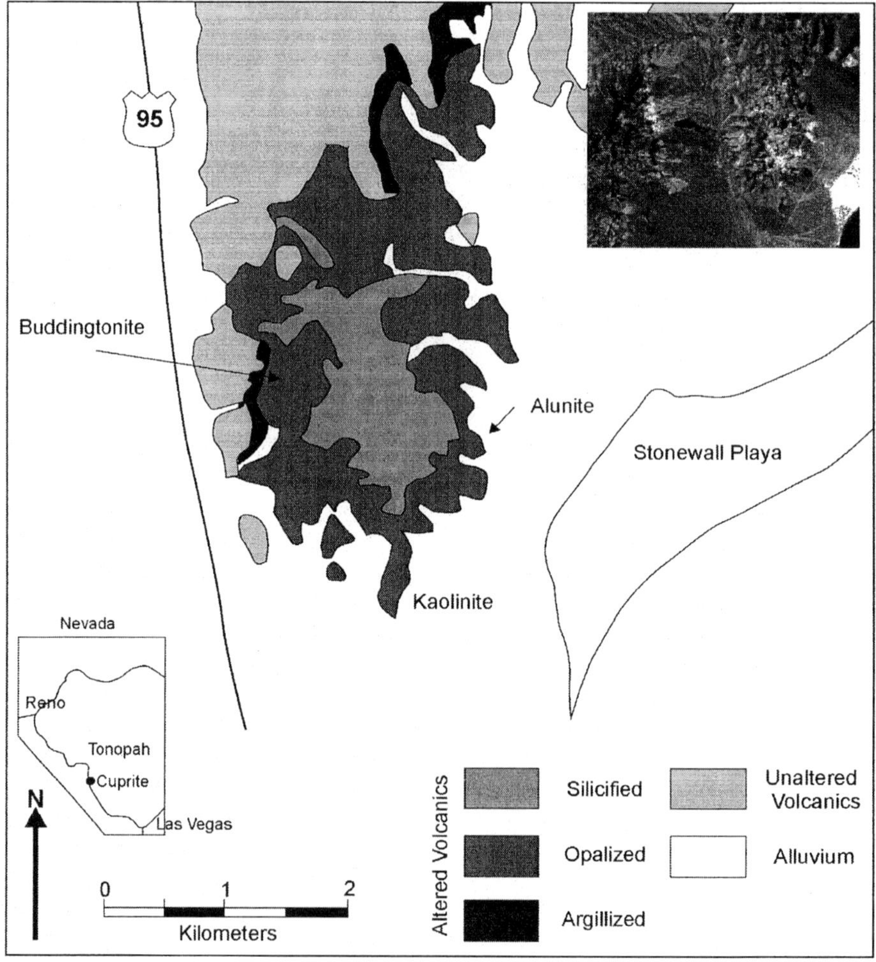


*Figure 5.* Alteration map at Cuprite, inset shows an example of the AVIRIS data.

## 11.2   DATA AND ANALYSIS

Several approaches have been published to extract information from high-spectral resolution image data, processing spectral information on a pixel by pixel basis including: binary encoding, waveform characterisation, spectral feature fitting, spectral angle mapping, spectral unmixing, constrained energy minimisation, classification, and most recently cross correlogram spectral matching (used here). These approaches are described in the Chapter 2 on Analytical Methods in this book. We use the CCSM method (Van der Meer & Bakker 1998; for details refer to the chapter 2 on analytical techniques in this book) to derive surface mineralogy maps for the minerals of interest.

*Figure 6.* AVIRIS image with subset analysed and target minerals indicated.

The method calculates the cross correlation between a spectrum to be tested (a pixel spectrum) and a known (reference) spectrum on a pixel by pixel basis. By shifting the reference spectrum subsequently over single channels and calculating the cross correlation, a correlogram is obtained (i.e., the cross correlation over channel shifts). The perfect match results in a symmetric, parabolic correlogram with a maximum correlation of 1. Skewness of the correlogram indicates that a similar absorption band exists in both reference and test spectra but with different shape and position. The statistical validity of the correlogram can be tested with a t-test. A validated surface mineralogy map is derived using several shape characteristics of the correlogram.

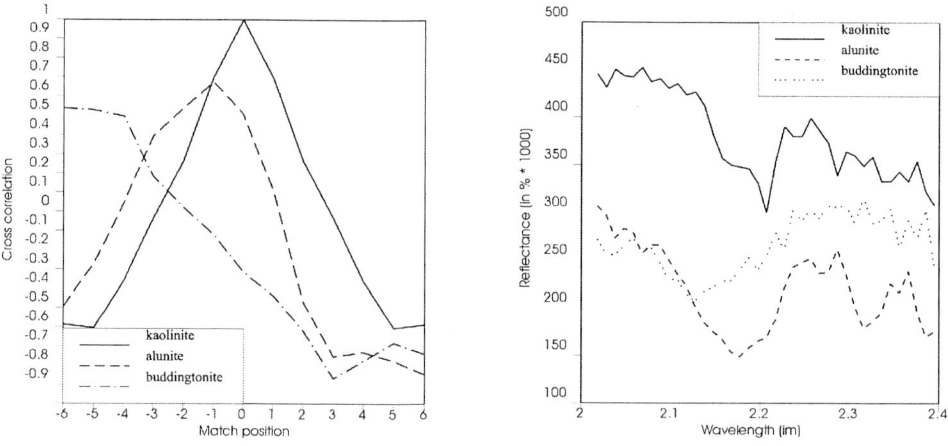

*Figure 7.* AVIRIS spectra from targets minerals (right; see Figure 6 for locations in the scene) and correlograms for a kaolinite reference (left).

The locations of known occurrences of the target minerals kaolinite, alunite and buddingtonite are shown in Figure 6, whereas Figure 7 shows the target spectra. Figure 8 shows the resulting 3D CCSM cube where the third dimension consists of correlograms.

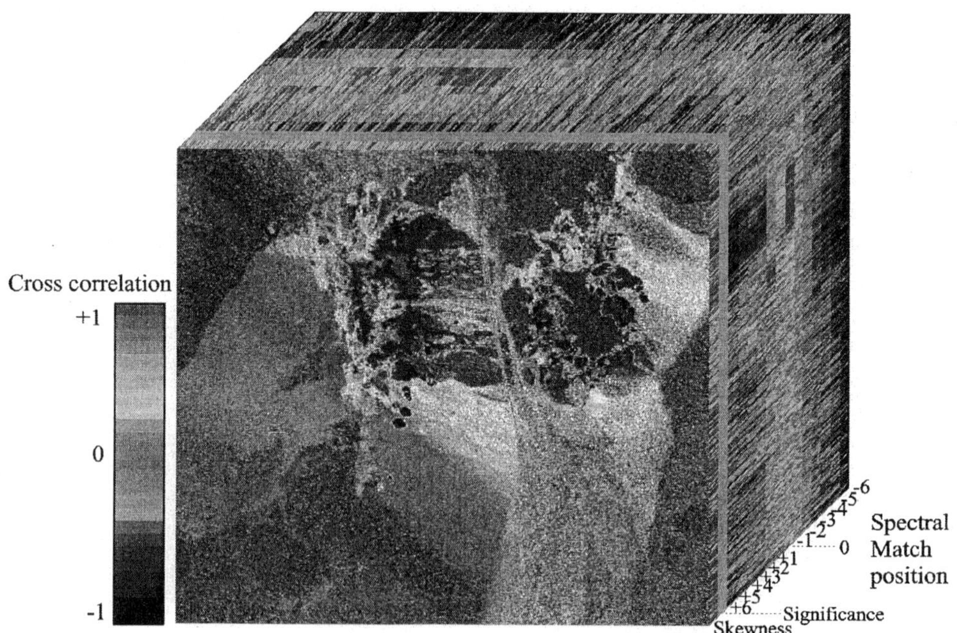


*Figure 8.* 3D CCSM cube for kaolinite. The face is a FCC with skewness, significance and match position 0.

Example correlograms are shown in Figure 7 for a kaolinite reference. From the 3D correlogram, the skewness, correlation at match position 1 and the significance of the correlation is derived (Figure 9). This is used as input to a final classification (Figure 10).

## 11.3   VALIDATION AND INTERPRETATION

The Cuprite mining site is a known area of hydrothermal alteration where in a series of rhyolitic basalt an association of the following mineral paragenesis is found:
silica $\Longrightarrow$ alunite   $\Longrightarrow$ kaolinite   $\Longrightarrow$ (buddingtonite) $\Longrightarrow$ montmorillonite.
The derived surface mineralogy maps for the minerals of interest clearly reflect this alteration pattern. It should be noted that particularly alunite is a pathfinder mineral for gold deposits. As such the resulting mineral maps guide mineral prospecting.

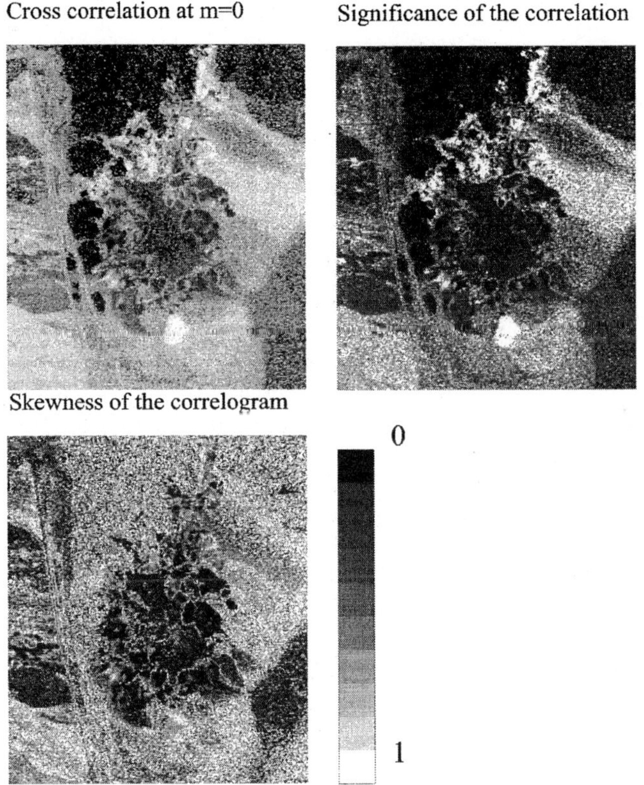


*Figure 9.* Skewness, significance and correlation at match position 0 for kaolinite as reference.

The results demonstrated in this Chapter are very promising, however it should be noted that these are ideal because (1) the sites in the case studies are relatively devoid of vegetation (with vegetation cover not often exceeding 30 % per pixel), (2) target minerals are exposed in relatively large (e.g., larger than the pixel size of the instrument), homogenous areas as opposed to other hydrothermal alteration zones, (3) only minor weathering and secondary alteration occurs, (4) target signatures have very high albedo with respect to the background (vegetation and slope materials) and (5) the imaging spectrometer data used is of very good quality.

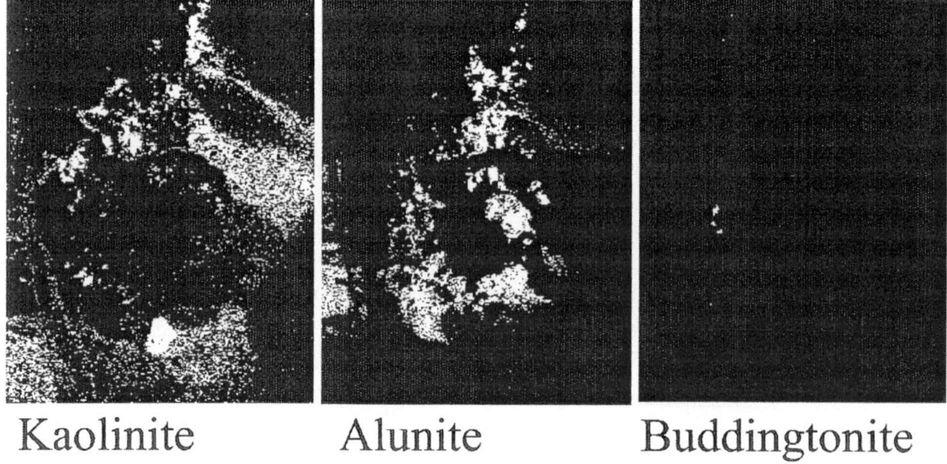



*Figure 10.* Final alteration maps at Cuprite based on CCSM.

Main problems in translating the results to areas with less favourable conditions lie in the reduction of vegetation effects and masking of the effect of weathering and cover of outcrops by lichens. Thus the success of geological exploration using airborne or spaceborne imaging sensors is dependent on the instrument characteristics as well as on the area. In areas with dense vegetation coverage, highly weathered outcrops or small (compared to the pixel size) target areas one has to rely on traditional exploration methods. However, given that imaging data of sufficient quality can be provided globally, we think that the study of spectroscopic images will contribute to mineral prospecting at a reconnaissance level in the near future. A main key to success in the future of geological imaging spectrometry will be the integration of spaceborne remote sensing data sets and complementary information. Another important aspect will be the integration of surface information obtained from remote sensing with subsurface information obtained from other geophysical and geological measuring techniques. At present surprisingly few studies make use of this possibly due to the difficulties in geocoding of the current airborne data sets.

## 12   Discussion

The high spectral resolving power combined with an acceptable spatial resolution of imaging spectrometers brings geological remote sensing from the qualitative descriptive approach to quantitative measurement of surface geochemistry and subsequent mineralogy. Airborne instruments are now capable of reproducing spectral measurements that until recently were only possible through field spectrometry. The advent of satellite-based imaging spectrometer systems opens-up a new range of possibilities for geological mapping and exploration. However the large amount of data requires techniques for surface mineralogy mapping that can be largely automated and for which the results can be evaluated in terms of accuracy.

# CHAPTER 8

# IMAGING SPECTROMETRY AND PETROLEUM GEOLOGY


Freek VAN DER MEER [α,β], Hong YANG [χ], Salle KROONENBERG [β],
Harold LANG [δ], Paul VAN DIJK [α], Klaas SCHOLTE [β] &
Harald VAN DER WERFF [α]

[α] International Institute for Aerospace Survey and Earth Sciences (ITC), Division of Geological Survey, Enschede, The Netherlands.

[β] Delft University of Technology, Department of Applied Earth Sciences, Delft, The Netherlands

[χ] Shell Exploration and Production B.V., Rijswijk, The Netherlands

[δ] NASA Jet Propulsion Laboratory, Pasadena, CA, USA


## 1    Introduction

Remote sensing has been used as a technological advance of considerable interest to earth scientists in general and exploration geologists in particular. The Multispectral Scanning System (MSS), the Thematic Mapper (TM), the SPOT satellite, Radar systems, airborne multispectral scanning systems and airborne imaging spectrometers have routinely been used to provide imagery for regional and local geological mapping (e.g., Van der Meer, 1995) identification of prospect scale structures. The method by observing the synoptic view and basin-wide assessment of favorable areas for petroleum exploration is normally called "indirect detection method" and has been carried out in many oil fields (Halbouty 1980). The key to successful application of remote sensing technology is in its integration with other exploration tools such as seismic, well, gravity and magnetic data. Many researchers (e.g. Lang *et al.*, 1985a,b) have also tried to use the spectral information to detect surface alterations which directly indicate the presence of hydrocarbon at depth because they believe that the vertical migration is the mechanism to explain hydrocarbon microseepage to the surface and the surface manifestations of hydrocarbon microseepage can be identified due to their distinct spectral characteristics.

## 2    Spectra of organics

Spectra of organics are dominated by the C-H stretch. The C-H stretch fundamental occurs near 3.4 $\mu$m, the first overtone is near 1.7 $\mu$m, and a combination band near 2.3




*F.D. van der Meer and S.M. de Jong (eds.), Imaging Spectrometry, 219–241.*


© 2006 *Springer. Printed in the Netherlands.*

$\mu$m. The combinations near 2.3 $\mu$m can sometimes be confused with OH and carbonate absorptions in minerals, especially at low spectral resolution.

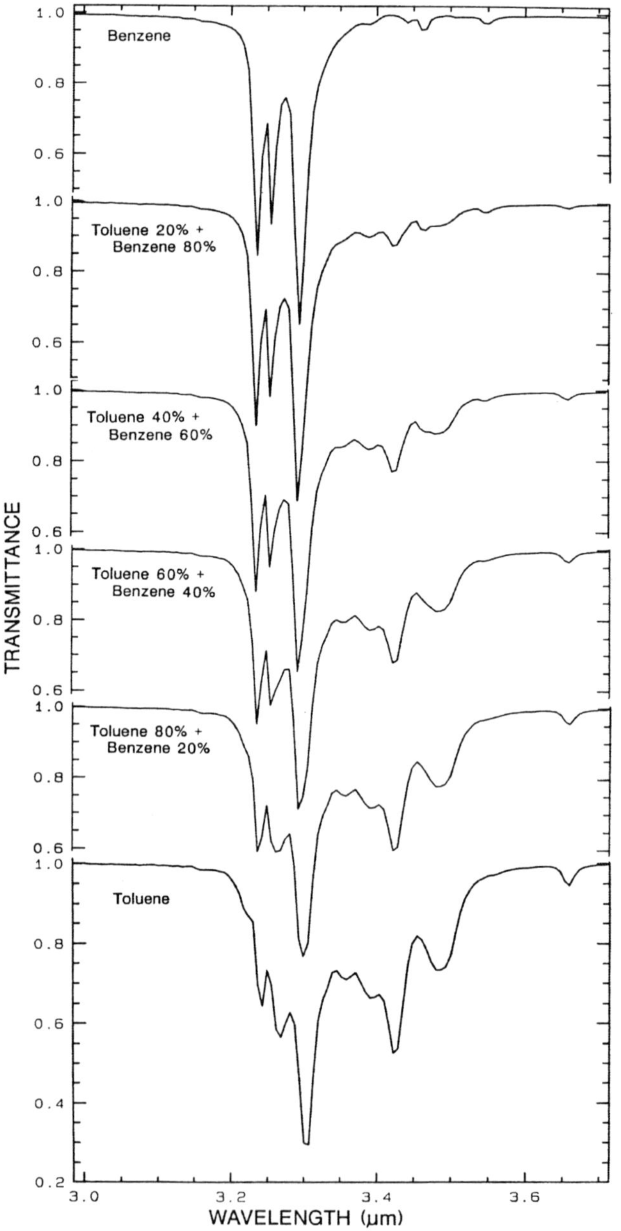



*Figure 1.* Transmittance spectra of organics and mixtures showing the complex absorptions in the CH-stretch fundamental spectral region (King and Clark, 1989).

Figure 1. shows transmittance spectra of organics and mixtures highlighting the complex nature of the absorption bands in the mid infrared where the CH-stretch fundamentals occur. Figure 2 shows reflectance spectra of montmorillonite, and montmorillonite mixed with super unleaded gasoline, benzene, toluene, and trichlorethylene. Montmorillonite has an absorption feature at 2.2 $\mu$m, whereas the organics have a CH combination band near 2.3 $\mu$m. The first overtone of the CH stretch can be seen at 1.7 microns, and the second overtone near 1.15 $\mu$m. (King and Clark, 1989).

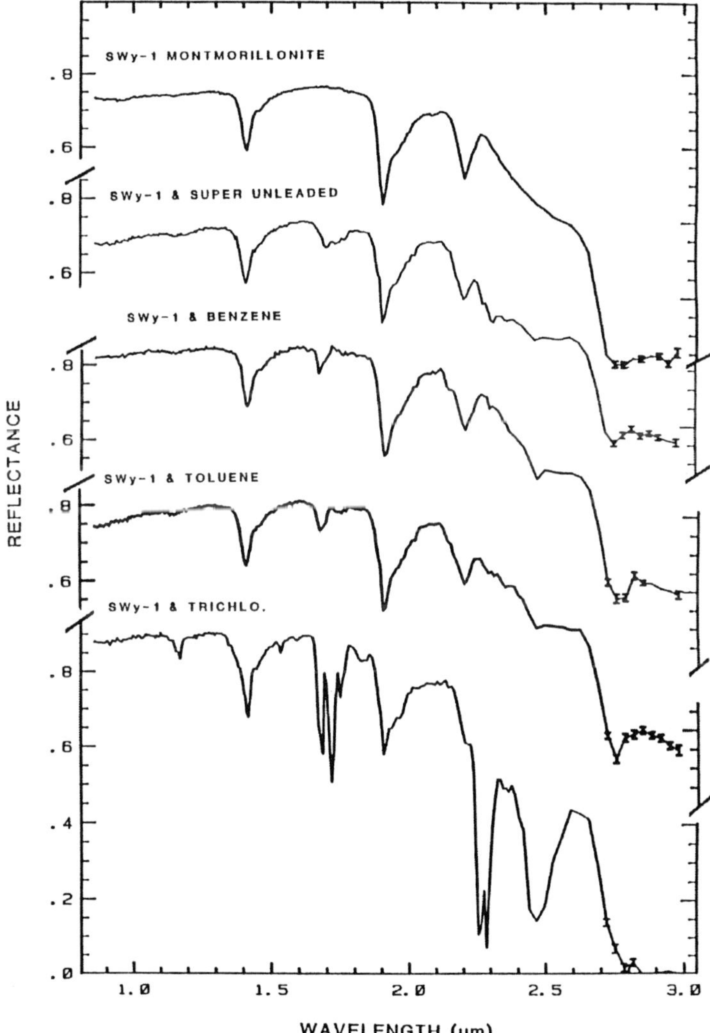


*Figure 2.* Reflectance spectra of montmorillonite, and montmorillonite mixed with super unleaded gasoline, benzene, toluene, and trichlorethylene. Montmorillonite has an absorption feature at 2.2 $\mu$m, whereas the organics have a CH combination band near 2.3 $\mu$m. The first overtone of the CH stretch can be seen at 1.7 microns, and the second overtone near 1.15 $\mu$m. (King and Clark, 1989).

Reflectance spectra of solid carbon dioxide, $CO_2$, methane, $CH_4$, and water, $H_2O$ are shown in Figure 3. These reflect the approximate fingerprints for gas absorption mapping and emission.

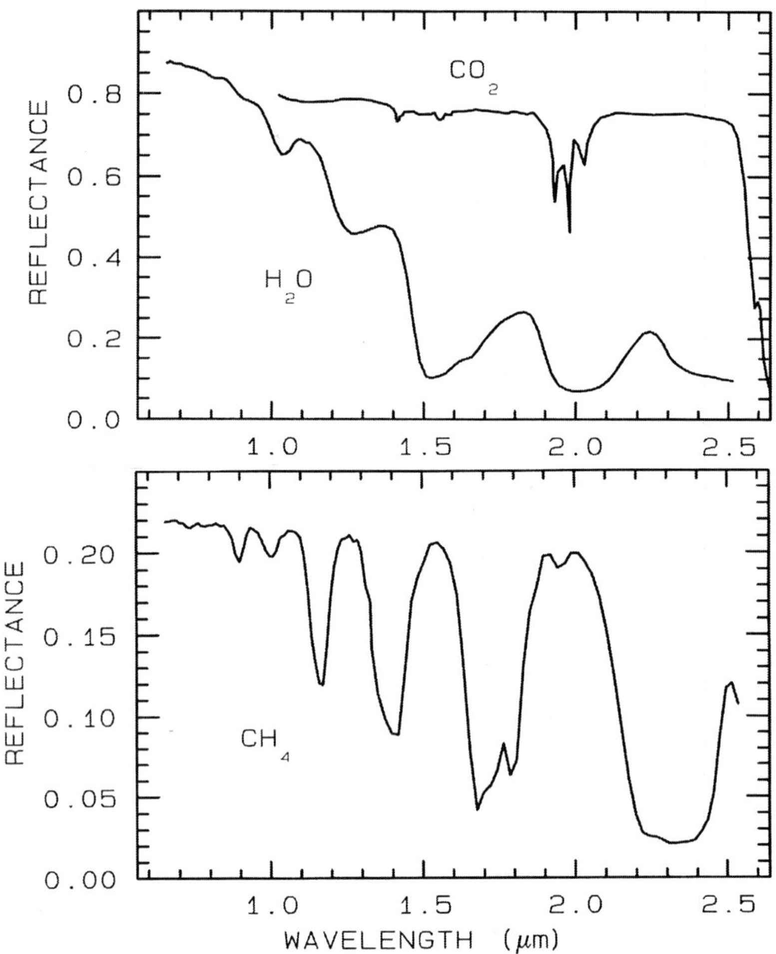



*Figure 3.* Reflectance spectra of solid carbon dioxide, $CO_2$, methane, $CH_4$, and water, $H_2O$ (King and Clark, 1989).

## 3      Hydrocarbon microseepage

### 3.1     INTRODUCTION

Macroseeps are the applied and long-accepted precursor of the microseepage concept. Macroseepage, having been documented in various parts of the world (Hunt, 1981; Tedesco, 1995), is the visible presence of oil and gas seeping to the surface.

Microseeps are hydrocarbons seeping vertically or near-vertically from the reservoir to the surface. The best evidence for near-vertical hydrocarbon microseepage is the fact that many geochemists have routinely measured statistically significant anomalous amounts of light hydrocarbons in soil gases and soils directly over petroleum deposits and successful prospects for many years. Soil gas hydrocarbons have very similar carbon isotope ratios to the gases in the underlying deposits, while ratios for biogenic hydrocarbons differ from those values. This along with the good compositional correlation of surface microseeping hydrocarbons with the type of underlying production are convincing evidence of vertical hydrocarbon migration (Saunders, *et al.*, 1991). Worldwide, there is a correlation between seeps and earthquake activity, where seeps occur predominantly in areas that are tectonically active. The amount of seepage (ppm methane/ethane) potentially is related to the pressure in reservoirs which is related to hydrostatic pressure and changes in lithospheric stress. Thus in natural seepages, a relation between the amount of seeping gas and stress could be envisaged.

Gases moving vertically through the strata are controlled by at least three seepage mechanisms: (1) effusion, as a free gas, thought to be the major factor leading to macroseeps. This is due principally to the very large pressure differential that exists across a petroleum reservoir; (2) diffusion of gases usually dissolved in vertical migrating waters that are observed passing through seemingly impenetrable barriers such as metal and glasses (Rosaire, *et al.*, 1940). This form of migration produces microseeps; (3) vertical movement of low-molecular-weight hydrocarbons dissolved in water through capping shales as a result of hydrodynamic or chemical potential drive (Duchscherer, 1980). Price concluded that the most reasonable hypothesis involves the vertical ascent of ultra-small (colloidal size) gas bubbles through a network of inter-connected , groundwater-filled "microfractures". Buoyant colloidal gas bubbles are readily displaced upward at rates up to several millimeters per second. This fast ascent explains the rapid development of soil gas light hydrocarbon anomalies over newly filled gas storage reservoirs, and rapid disappearance of such anomalies after a reservoir is depleted. Usually microseeps are invisible and so low-level that they cannot be distinguished without modern analytical methods.

Macroseeps and microseeps do not always indicate the presence of economic recoverable hydrocarbons at depth, but rather may represent leakage from a temporarily stationary source of petroleum. Therefore, even though the various methods detect hydrocarbons at the surface, the perceived anomaly is not necessarily implying an economic petroleum accumulation. The explorationist must determine if the seepage is just minor venting of petroleum. A definition of the term petroleum accumulation as used here implies an accumulation of from one to an infinite number of petroleum molecules trapped permanently or temporarily. It does not suggest either an economic or non-economic condition of the deposit (Tedesco, 1995).

The present theories on microseepage support the premise that a reservoir leaks even though it contains trapped hydrocarbons. The cross-sectional shape of the leaking pattern has conventionally been labeled a "chimney effect" (Tedesco, 1995).

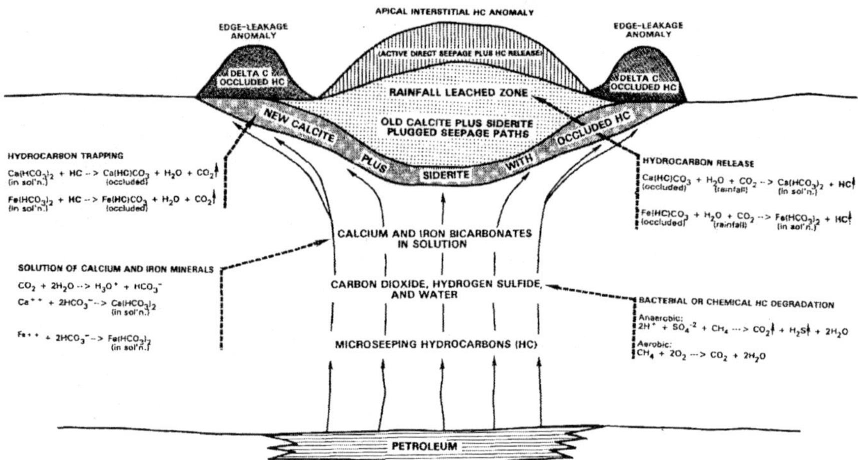


*Figure 4.* Model for edge-leakage soil hydrocarbon anomalies (Saunders et al., 1999).

## 3.2 MICROBIAL EFFECTS AND HYDROCARBON-INDUCED SURFACE MANIFESTATIONS

Long-term leakage of hydrocarbons can establish locally anomalous redox zones that favor the development of a diverse array of chemical and mineralogical changes (Saunders et al. 1999; Figure 4). Schumacher (1996) made a thorough review of the major hydrocarbon-induced changes affecting soils and sediments and their implications for surface exploration methods and applications. He believed that long-term leakage of hydrocarbons, either as macroseepage or microseepage, can set up near-surface oxidation-reduction zones that favor the development of a diverse array if chemical and mineralogic changes. The bacterial oxidation of light hydrocarbons can directly or indirectly bring about significant changes in the pH and Eh of the surrounding environment, thereby influencing mineral stability and chemical reactivity. Such oxidation in the chimney above a leaking petroleum accumulation leads to dissolution or precipitation of minerals and the mobilization or immobilization of certain elements in the chimney, which thereby becomes mineralogically and chemically different from laterally-equivalent. The resulting alterations include: the formation of calcite, pyrite, uraninite, elemental sulfur, and certain magnetic iron oxides and iron sulfides; bleaching of red beds; clay mineral alteration; electrochemical changes; radiation anomalies; geomorphic anomaly; the edge anomaly of adsorbed or occluded hydrocarbon in soils and Delta C (ferrous carbonate); and biogeochemical and geobotanical anomalies (Oehler and Sternberg, 1984). Bleaching occurs whenever acidic, reducing fluids dissolve the ferric oxide (hematite) that gives the red bed its characteristic colour. The acidic conditions resulting from the oxidation of hydrocarbons in near-surface soils and sediments promotes the diagenetic weathering of feldspar to clay and the conversion of smectite clay to kaolinite. Ferrous carbonate, also called "Delta C" show highs above the edges of hydrocarbon accumulations. The bicarbonate and carbonate chemistry of calcium and iron may provide a viable explanation for edge-leakage of Delta C iron carbonate anomalies. Many of these alteration minerals can be mapped using spectroscopic data. Problems in direct detection of microseepage via these alteration schemes is that it is difficult to separate background mineralogy from alteration mineralogy. Furthermore, the proposed alteration schemes by Saunders et al. (1999) are not clearly based on sedimentary or petrological environments, but they compose a more general classification which needs to be further refined.

Hydrocarbon-consuming bacteria and related organisms have been found in sediments, oil field waters, and crude oil at depths ranging up to 4200 m. Bacteria produce carbon dioxide and organic acids in the process of oxidizing hydrocarbons Additionally, sulfate reducing species create hydrogen sulfide. Near-surface carbonate cements and carbonate buildups on the sea floor have been attributed to carbon dioxide formed by bacterial or chemical degradation of hydrocarbons seeping upward from petroleum.

Carbon dioxide produced by anaerobic bacterial degradation (or by oxidation) of hydrocarbons produces $H2CO_3$ which in turn reacts with calcium silicates to produce secondary carbonate. Carbonate cement can also be generated directly from inorganic reduction of sulfates by

$$C_nH_m + 3CaSO_4 \rightarrow 3CaCO_3 + CO_2 + H_2O + 2H_2S \tag{1}$$

Acidic solutions created by bacterial oxidation of vertically migrating hydrocarbons can leach potassium and possibly other radioactive elements from clays (mostly illite in soils). Hence low potassium anomalies have been found associated with microseeps.

Aeromagnetic anomalies with shallow depth sources have been reported over oil and gas seeps. Hydrogen sulfide reacts with iron oxides (goethite) to produce magnetite. Sometimes goethite is altered to (magnetic forms of) hematite. In areas where red colored (hematite rich) sandstones occur at the surface on reservoirs, gas seepage results in leaching of the hematite (altered to pyrite and clay minerals) which again results in a lighter (grey-green) coloring of the sandstones (the bleaching effect). The microbial activity produces chemically reducing environments which leads to the build up of uranium concentrations. Several studies have indicated the so-called edge leakage in soils on top of reservoirs. Carbonic acid dissolves calcium and iron. Near the surface loss of carbon-dioxide causes a reversal of the oxidation reaction and yields the precipitation of secondary calcium. The secondary calcium clogs the gas escape routes and hence gasses deflect to the edges of the alteration zone.

## 4    Detection of hydrocarbon-induced surface manifestations by remote sensing

Remote sensing data most widely used in hydrocarbon exploration are aerial photography, radar, Landsat Multispectral Scanner (MSS), Landsat Thematic Mapper (TM), and airborne multispectral scanner data. Aerial photography was used to map the structure and stratigraphy to help to understand the geology since the 1930s. The use of side-looking radar images and synthetic aperture radar (Goetz and Rowan, 1981) provided the possibility to enhance subtle expression of subsurface geologic structures. Since the 1970s, with the launch of Earth Resources Satellites, for the first time data of the earth's surface observed from space were obtained in a semi-operational way. MSS, SPOT and TM images have been used for "indirect detection". The rapid development of remote sensing technology has also made it possible to carry out "direct detection" of oil resources by identifying the tonal anomaly that is the manifestation of hydrocarbon microseepage. Remote sensing of hydrocarbon-induced alteration holds great promise as a rapid, cost-effective means of detecting anomalous diagenesis in surface soils, rocks and related vegetation. The most extensive studies were about reduction of ferric iron (red bed bleaching), conversion of mixed-layer clays and feldspars to kaolinite, increase of carbonate content, and anomalous spectral reflectance of vegetation.


## 4.1    BLEACHED RED BEDS

Bleaching of red beds occurs whenever acidic or reducing fluids are present to remove ferric oxide (hematite). Such conditions also favor the formation of pyrite and siderite from the iron released during the dissolution of hematite. The possible reducing agents responsible for bleaching red beds above petroleum accumulations include hydrocarbons, $H_2S$, and $CO_2$ (Schumacher, 1996).

Donovan (1974) reported that the color of the Permian Rush Springs formation grades from reddish-brown for unaltered sandstone adjacent to the field to pink, yellow, and white along the flanks of the Cement anticline, and white along the flanks of the anticlinal axis, where maximum bleaching and iron loss occur in the Cement field area of Oklahoma. In Sheep Mountain anticline, Bighorn Basin, Wyoming, Landsat TM data isolated areas of red-bed bleaching within the Chugwater Formation which correspond spatially to known hydrocarbon deposits. Bleached areas were quantified spectrally by a decrease in the band 3/1 ratio in conjunction with an increase in total reflected radiance. The potential of this technique is greatly in areas of sparse vegetation and susceptible clays and red beds.

The geology and geochemical alteration associated with Lisbon Valley field in southeastern Utah have been described in considerable detail by Segel et al. (1986). They report that the distribution of the bleached outcrops of the Triassic Wingate formation approximates the geographic limits of the oil and gas reservoirs at depth. A ratio of TM bands 2 and 3 was used to delineate variations in ferric-iron content. Ferric-iron rich exposures exhibit very low 2/3 ratio values and alternatively, ferric-iron poor rocks have relatively high 2/3 ration values. Density slicing was used to delineate the bleached rocks (Segel and Merin, 1989). The ferric iron exhibits an absorption feature at 0.9 $\mu$m and the visible spectrum falls off sharply to the blue producing a maximum near 0.80 $\mu$m (Hunt *et al.*, 1973; Hunt, 1977). Under ideal laboratory conditions, the ferrous iron in non-transparent minerals such as pyrite and magnetite shows a low total reflectance, while in transparent mineral such as siderite, shows a broad shallow band at 1.0 - 1.1 $\mu$m (Hunt, 1970). These characteristics can be used in remote sensing data processing and separating bleached red beds from the unbleached ones.


## 4.2    CLAY MINERAL ALTERATION

The production of $CO_2$, $H_2S$, and organic acids resulting from the microbial oxidation of hydrocarbons in near-surface soils and sediments can create reducing, slightly acidic conditions that promote the diagenetic weathering of feldspars to produce clays and may lead to the conversion of normally stable illitic clays to kaolinite. Clays thus formed remain chemically stable unless their environment is changed (Schumacher, 1996).

In Utah's Lisbon Valley field, Segel et al. (1986) report that bleached portions of the Wingate Sandstone directly overlying the field contain primarily kaolinite clays, whereas the unbleached areas of the sandstone located away from the field contain fresh plagioclase and muscovite. The bleached Wingate contains three to five times more kaolinite than the unbleached rock. Segel and Merin (1989) defined the variations in clay mineral content, which have been directly correlated with differences in the relative proportion of kaolinite, using a color coded TM 5/7 ratio. Kaolinite-rich rocks associated with bleached exposures that overlie the hydrocarbon accumulation at the Lisbon Field exhibits the highest 5/7 ratio values.

Kaolinite exhibits a very strong absorption feature centered at 2.2 micrometers, as well as the presence of a diagnostic, subordinate, absorption feature (known as a doublet) centered at 2.16 micrometers. This can be indicative in clay mineral rich areas.


## 4.3    CARBONATES

Diagenetic carbonates and carbonate cements are among the most common hydrocarbon-induced alterations associated with petroleum seepage. Pore-filling calcite and replacement calcite is most common. These near-surface diagenetic carbonates are formed principally as a byproduct of petroleum oxidation, particularly of methane. Carbon dioxides evolve and react with water to produce bicarbonate. This bicarbonate bonds with calcium and magnesium in ground water and precipitates as carbonate, or carbonate cement, that has an isotopic signature matching that of the parent hydrocarbon(s) (Schumacher, 1996).

In the Palm Valley gas field, Amadeus Basin, central Australia, a distinct color anomaly in sandstones was detected by digital image processing of NASA NS-001 aircraft thematic mapper simulator (TMS) scanner and Landsat Thematic Mapper (TM) data. Laboratory spectra of brown sandstone from within the color anomaly, have only about half the brightness of the red sandstone, with a reflectance maxima of about 25% at 1850 nm. Laboratory spectra of calcrete shows a similar curve to brown sandstone but with increased brightness towards the visible wavelengths. In the calcrete, a weak to moderate absorption feature at 2322 nm is caused by the carbonate (calcite) content, and weak absorption features at 2200 nm and 1411 nm suggest the presence of clays. Based on the field spectral reflectance, a distinctive color discrimination can be obtained using calibrated TMS band-ratios.

In relatively arid regions such as West Texas, carbonates may be visible on aerial photos or satellite imagery as light tonal anomalies indicating an excessive development of caliche in the surface soils (Thompson et al., 1994).

In the Junggar Basin in Xinjiang, P.R. China, a strong delta carbon anomaly with a major increase in total carbonate in soil was identified from Airborne Short-wave Infrared Split Spectral Scanner data. Four channels at the wavelength range of 1.550-1.650 $\mu$m, 1.985-2.085 $\mu$m, 2.037-2.137 $\mu$m, and 2.039-2.193 $\mu$m were used for anomaly extraction. The most common calcite bands are from 1.8 - 2.6 $\mu$m, at 1.8 $\mu$m, 2.0 $\mu$m, 2.16 $\mu$m, 2.35 $\mu$m, 2.55 $\mu$m (Hunt, 1971) which can be used to map carbonate (calcite) concentration.

In summary, mineral alteration related to hydrocarbon microseepage can result in:

1.  creation of hydrogen sulfide due to bacterial activity
2.  production of secondary (enrichment) in carbonate (calcite, siderite) in combination with CO2 and H2S production
3.  low potassium anomalies in soils (illite)
4.  magnetic anomalies in which goethite is altered to magnetic forms of hematite
5.  build up of uranium concentrations
6.  bleaching of red (hematite rich) sandstones
7.  edge leakage (concentration of calcium around the edges of zones of microseepage) and secondary calcite enrichment

## 4.4    GEOBOTANICAL ANOMALIES

Geobotanical anomalies occur as a results of the effect of light hydrocarbons on the growth of vegetation. Reflectance properties of vegetation in the visible part of the spectrum are dominated by the absorption properties of photosynthetic pigments of which chlorophyll, having absorption at 0.66 and 0.68 $\mu m$ for chlorophyll a and b, respectively, is the most important (Elvidge, 1990). Changes in the chlorophyll concentration produce spectral shifts of the absorption edge near 0.7 $\mu m$: the red edge (also termed as inflection point, the point of maximum slope on the reflectance spectrum of vegetation between red and near-infrared wavelengths, defined as the wavelength of maximum $dR/d\lambda$. This red edge shifts toward the blue part of the spectrum with loss of chlorophyll, and shifts towards the red part of the spectrum with increase of chlorophyll. The results are (1) a decrease in the height of the infrared shoulder due to structural damage; (2) an increase in the reflectance at the chlorophyll absorption maximum due to decreased leaf chlorophyll; and (3) a shift in the position of the red edge towards shorter wavelengths (Figure 5).

Hydrocarbon microseepage creates a chemically reducing zone in the soil column at depths shallower than would be expected in the absence of seepage. Such leakage stimulates the activity of hydrocarbon-oxidizing bacteria, which decreases soil oxygen concentration while increasing the concentration of $CO_2$ and organic acids. These changes can affect pH and Eh in soils, which in turn affects the solubility of the trace elements and consequently their availability to plants (Schumacher, 1996). The environmental change will affect the root structure of vegetation, to ultimately influence those plants' vigor and hence, spectral reflectance. To seek for vegetation anomalies by remote sensing, "normal" plant variability has to be assessed by judging the plant distribution, presence of indicator plants, or morphological changes in plants induced by excesses or deficiencies in available soil nutrients. The remote sensing applications include two aspects. One is to research into the spectral characteristics of the anomalous or stress vegetation species caused by hydrocarbon microseepage; the other is to map the anomalous or stress vegetation species from imagery.

Bammel and Birnie (1994) conducted a geobotanical reflectance study at five areas in the Bighorn Basin of Wyoming to determine if the spectral response of sagebrush could be a useful tool in hydrocarbon exploration with measurements collected in the visible and near infrared regions (0.45 to 1.1 $\mu$m). The study shows that the most effective indicator of hydrocarbon-induced stress in sagebrush plants is a consistent blue shift (to shorter wavelengths) of the red-edge. This shift can only be detected where the sagebrush is actually growing in large amounts of surface-visible occurrence or perhaps in areas of profuse (but not visible) vertical seepage. Spectral reflectance intensity data were found to have no significant correlation with the presence or absence of surface or subsurface hydrocarbons.

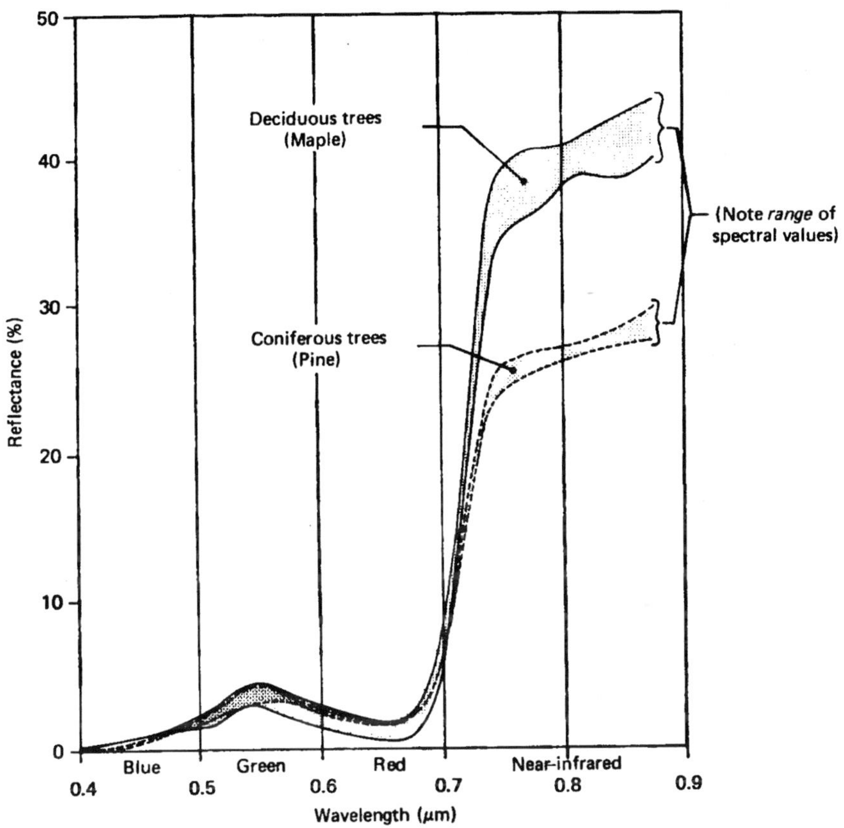


*Figure 5.* Effect of stress on the vegetation spectrum.

In the Stoney Point oil field, Michigan, USA, Fluorescence Line Imager (FLI) data were collected in both spatial mode and spectral mode that cover the spectral range of 430 to 805 nm in 288 channels. Taking advantage of high spatial resolution and spectral resolution of the two modes, the spectral changes were found occurring due to the stress (1) a decrease in the height of the infrared shoulder (Rs) due to structural damage; (2) an increase in the reflectance at the chlorophyll absorption maximum (Ro) due to decreased leaf chlorophyll; (3) a shift in the position of the red edge ($\lambda$p) towards shorter wavelengths. A stress image with low digital numbers representing anomaly was created by multiplying Rs to $\lambda$p. The encouraging relationships were found between C2-C4 soil hydrocarbons and anomalies in the Rs*$\lambda$p stress images.

High resolution reflectance spectra of Douglas Fir trees were measured at the Mist Gas Field in Oregon by using Arco's IRIS spectroradiometer built by Geophysical Environmental Research, Inc.  Results of the study indicate that Douglas Fir trees growing in areas of gas production have a different spectral signature from trees growing off-field. Trees growing off-field have higher reflectivity in the 750 nm to 1250 nm spectral region, and an absorption feature at 1200 nm is much stronger in off-field trees. Microscopic examination of needles from on field revealed the presence of stress effects at the cellular level. Chlorotic trees display higher reflectivity in the 550 nm to 650 nm region, and the sharp rise in reflectivity at 700 nm is shifted towards shorter wavelengths. The local presence of chlorosis is poorly correlated with commercial gas production. Trees characterized by both chlorosis and cell damage may be indicative of more severe stress conditions. The poor health of on-field trees may be caused by the presence of methane in the root zone of seepage-induced geochemical changes in soil and groundwater above the gas reservoirs.

In the central portion of the Sao Francisco Basin in central Brazil, where natural gas seeps occur, eucalyptus plantation specimens in the anomalous area were poorly developed, showing clear signs of nutritional deficiency. The geochemical anomalies from the soil gas survey correspond with the biogeochemical anomalies. Spectral data were collected over anomalous and non-anomalous eucalyptus stands, using an airborne non-imaging sensor, named SADA (Airborne Data Acquisition System). The analysis of the data acquired by SADA and the corresponding TM bands values show higher reflectance values in chlorophyll absorption band for the sites under hydrocarbon influence, compared to sites outside the anomalies. The presence of hydrocarbons seems also to produce a change in the internal structure of the canopy, which causes lower reflectance values at the reflectance shoulder. In gas affected areas, the spectral contribution from underlying soil energy is more evident due to a lower vegetation density.

Over the Lost River gas field in the Appalachian Mountains of West Virginia, the principal vegetation anomaly observed was the presence of maple trees over the gas field at sites where more typical oak-hickory climax vegetation would be expected (Lang et al., 1985b). The results for the Lost River site show that the maple trees occur in an area of maximum hydrocarbon seepage (methane) and minimum soil oxygen content. The anomalous maple distribution may relate to anaerobic soil conditions that directly or indirectly influence the mycorrhizal fungi living on the trees' root hairs, favoring maple trees whose fungi appear to be better able to tolerate the anaerobic soils than their counterparts living on the roots of oaks (Lang et al., 1985b). Aircraft NS-001 (The Aircraft Thematic Mapper Simulator) eight-band data were acquired during the investigation. The color-composite image of visible green, red and near infrared is the best of all tried for species and species community discrimination.

Patrick Draw field in southwest of Wyoming, sagebrush is predominant vegetation. The most pronounced anomaly observed at Patrick Draw field was an area of stunted sagebrush and an associated tonal anomaly visible on Landsat imagery. The anomaly overlies the field's gas cap and occurs in a region of strong light hydrocarbon microseepage (Lang et al., 1985a; Richers et al., 1982, 1986). The geology and production history of the field show that the sagebrush anomaly results from the upward migration of injected gases and waters used to maintain reservoir pressures in the field (Arp, 1992). These gases and waters produced anoxic, low-Eh (oxidation potential), high-pH, and high salinity soils that are toxic to the overlying sagebrush (Lang et al., 1985a; Arp, 1992).

NS-001 data were also used to map the vegetation density. The anomalous stunt sage is characterized by a decreased in vegetation cover of at least 10%, which appears differently from the surrounding region. The simple Siegel and Goetz (1977) mixing model was used to separate the vegetation and soil/rock components for spectral analysis studies (Lang et al., 1985a).

Multispectral video data, with narrow bandwidth filters were used to evaluate their sensitivity to detect environmental conditions associated with hydrocarbon microseepage at the Pollard oil field (Cwick et al., 1995). Correlation analysis reveals statistically significant relationship between Mn concentrations and vegetation growth around active producing well sites. Although single-band video data were generally not sensitive to biochemical variations, transformed video data exhibited several significant correlations with leaf Mn concentration, and the leaf Mn/Fe ratio. In particular, principal component 2 and the near-infrared/red ratio were sensitive to leaf Mn concentrations.

The spectral signatures of vegetation associated with hydrocarbon microseepage have been extensively studied. The green peak, red trough, the shift of position of red edge, the height of the infrared shoulder are the main targets that attention should be paid, after considering the factors such as bedrock geology, soil type, slope, soil moisture, and climate can have a more pronounced effect than that due to presence of hydrocarbons (Klusman et al., 1992). For applications of field and imaging spectrometry to detect and map hydrocarbon related anomalies also read Yang et al. (1999, 2000).


## 5    In Summary

To summarize the potential sources of information on hydrocarbon microseepage and oil/gas reservoirs to be implemented in a GIS based modeling system:
Surface (hyperspectral) remote sensing:
    Direct
        New formation of
            Calcite
            Pyrite
            Uraninite
            Elemental sulfur
            Magnetic iron oxides
            Iron sulfides
            Bleaching of red beds
            Clay mineral alteration
            Electrochemical changes
            Radiation anomalies
            Geomorphic anomaly
            Edge anomaly of adsorbed or occluded hydrocarbon in soils
            Delta C (ferrous carbonate);
    Indirect
        Geobotanical anomalies
        Red-edge shifts in vegetation signatures
        Biogeochemical stress indicators
Surface and sub-surface geology
    Structure
    Lithology

        Faults
Surface geochemistry
Subsurface geophysics
        Magnetic anomalies
        Anomalous temperature gradients
        Radiation anomalies (gamma-ray)


## 6 Future trends


## 6.1 TRENDS IN EXPLORATION

Exploration methods based on what are assumed to be hydrocarbon-induced soil and sediment alterations have long been popular, but the processes that produce the observed effects are not well documented. The nature and extent of the alteration can vary significantly not only laterally and vertically but also temporally. The cause of these altered soils and sediments may well be hydrocarbon related, but it is an indirect cause at best and may not be the most probable cause. We must evaluate seemingly "significant" alteration anomalies carefully to determine if they are related to hydrocarbon seepage. This requires answers to the following questions. Is the anomaly a function of geology or an artifact of culture? If geology, is the observed alteration syngenetic or authigenic? If authigenic, is the anomaly seep-related or of non-seep origin? If seep-related, does the anomaly result from an active hydrocarbon seep or a paleoseep? Finally, if the anomaly at the surface is to be related to a drilling objective at depth, does the anomaly result from mainly vertical migration or does the migration path follow a more complex route? Considerable research is needed before we understand the many factors affecting the formation of theses anomalies in the near surface (Schumacher, 1996).

Visible, near-infrared, and short-wave-infrared wavelength regions fall into the atmospheric windows mostly used for remote sensing purposes. The visible and near-infrared region is sensible for vegetation change and the spectral features of the alterations in the 2.2 $\mu$m region are particularly valuable because they are common to alteration minerals and allow discrimination from non-alteration minerals which provide features only as close as 2.4 $\mu$m (Hunt, 1979). The laboratory and field spectra of the hydrocarbon-induced altered minerals and associated vegetation have been studied in details, but the spectral resolution of MSS and TM are not high enough to be compared with the laboratory or field spectra. The broad bandwidth cannot characterize all the absorption features caused by hydrocarbon microseepage, regardless of the type of enhancements employed or the type of information extraction method applied. In short, high spectral resolution imaging data are needed for the recognition of the hydrocarbon-induced alterations and related vegetation changes. Remote sensing for direct detection of hydrocarbon microseepage holds a great promise as a "geochemistry-bio-geochemistry in the air" method that can help to recognize marginal and sub-marginal low relief structural prospects and all stratigraphic traps that are overlooked by reflection of seismograph. With better understanding of hydrocarbon microseepage and its surface manifestations, with higher spatial and spectral resolution imaging data, we have a better chance to delineate the surface anomalies associated with subsurface hydrocarbon reservoirs.


## 6.2    TRENDS IN MONITORING EMISSIONS

Oil and gas reservoirs leak. As a result large quantities of oil and gas from these reservoirs reach the surface forming seeps. When visible with the human eye we refer to macroseeps, else to microseeps. Macroseeps, particularly offshore, have been extensively studied. Microseeps are less well studied. Seeps are relevant to the oil and gas industry as a potential source of information for exploration. In addition, seeps are a source of methane emission and as such contribute a large but yet un-quantified amount of this gas to the global budget. Methane is a greenhouse gas. Hydrocarbon microseepage gives rise to expressions at the Earth surface in form of: (1) detectable trace concentrations of gasses (mainly ethane and methane), (2) mineral alteration of soils, and (3) anomalous spectral response in vegetation. Off shore, oil macroseeps can be seen forming slicks on the water surface detectable using radar imagery. Within the water column, the upward seeping gas produces bubbles that can be detected using sonar. On shore, seismics provides a means of visualising subsurface chimneys along which oil and gas migrate to the surface. At the surface, geochemical soil gas analysis provides insights into the nature and composition of the gas. High spectral resolution spectroscopic measurements in the field and from airborne or spaceborne imaging devices may potentially be used to detect and monitor hydrocarbon microseepage using mineral alteration (detection) and gas emission (monitoring using emission spectra).


## 7    A case study from Santa Barbara, southern California


## 7.1    PETROLEUM GEOLOGY OF THE SOUTHERN CALIFORNIAN BASINS

The Southern Californian Basins were formed as irregular pull-apart basins at a time of irregular rifting within the simple shear system of the San Andreas Transform Fault (Figure 6). They were initiated by rifting due to clockwise rotation of the crustal blocks about vertical axes from the Early Miocene (about 22 Ma) followed by later thermal subsidence. Paleomagnetic data has shown that 25% of the surface area of Southern California has undergone clockwise rotation and that some areas have rotated as much as 90°. The timing of this rotation is shortly before the Middle Miocene, producing 50° to 60° rotation before the beginning of the Late Miocene. Since that time an additional 30° to 40° rotation has occurred. Most of the Southern Californian Basins, including the Santa Barbara Channel-Ventura Basin, have developed as pull-apart basins due to incompatibilities within the clockwise rotating blocks. Many authors have recognized that the Southern Californian Basins, including the Ventura basin, have experienced a late stage shortening in the latest Miocene to Early Pliocene (6.5 to 5 Ma) after the initial extension in the Early Miocene. The shortening is dominated by basement involved shortening so that major anticlines were developed as fault bend folds and fault propagation folds.

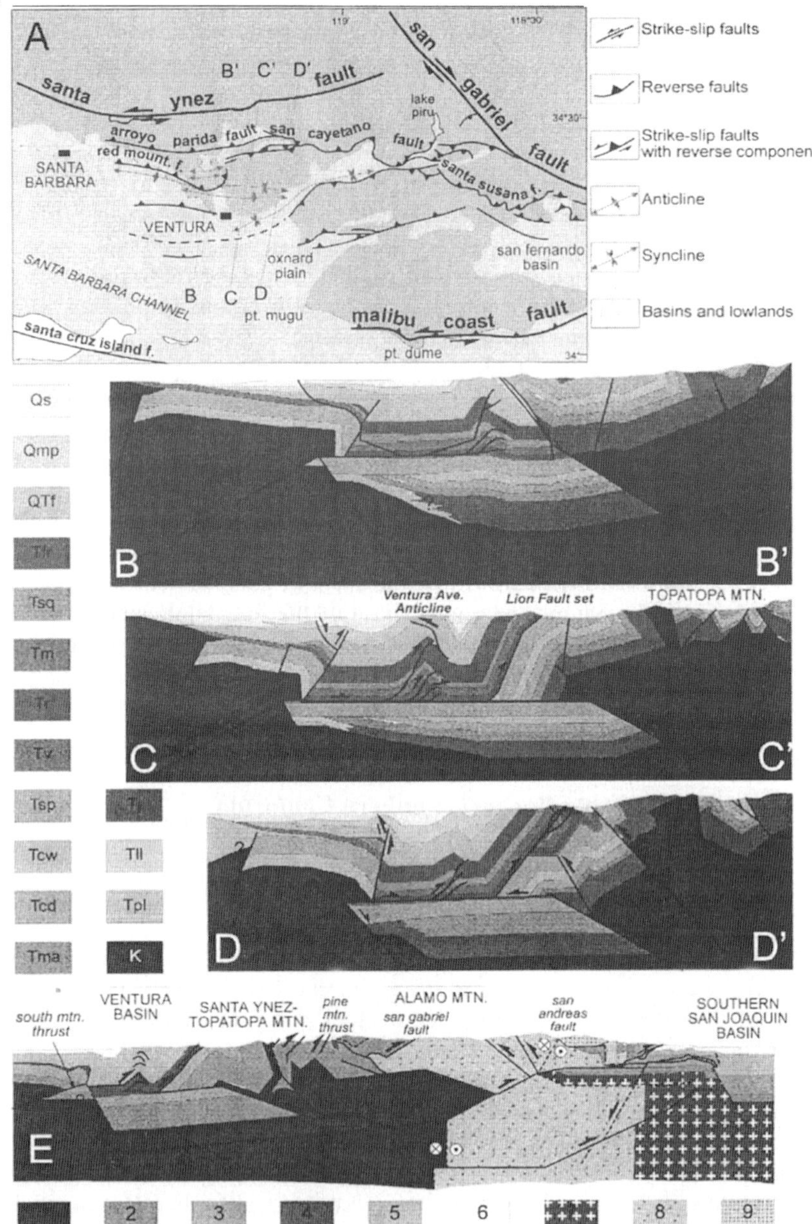


*Figure 6.* A) Simplified structural map of the Western Transverse Ranges. B-D) Simplified cross-sections across the Ventura Basin. For symbols of the units refer to the stratigraphic column of Figure 7. E) A regional simplified cross-section across the transverse ranges. Legend: 1. Cretaceous strata and Fransiscan assemblage, 2. Eocene strata, 3. Miocene and Oligocene strata (Sespe and Vaqueros Fm.). 4. Miocene strata, 5. Pliocene strata, 6. Quaternary strata, 7. Mesozoic ophiolites, 8. Mesozoic tonolite and mafic gneiss, 9. Mesozoic granite.

No major Pliocene and Quaternary strike-slip occurred in any of the faults and the major thrust faults merge with regional detachments beneath the transverse ranges and in the northern Peninsular ranges. The depth to the detachments is thought to be about 10 to 15 km. Shortening in the Ventura basin is taken up mostly by E-W trending reverse faults, blind thrusts and folds. These are the Santa Ynez, Arroyo-Parida, San Cayetano, Red Mountain and Lion Fault set in the north and the Oak Ridge Fault in the south of the basin. In the northwestern and northeastern parts the basin infill is thrusted over by the north dipping Red Mountain and San Cayetano faults. In the north central part the basin as well as its in-fill material is back-thrusted over the older units along the Lion Fault set forming a triangular zone in which the basin in-fill material is vertical to steeply dipping to the west. Shortening in the Ventura Basin accelerated to 20 ±6 to 28 ±1 mm/y since 250 ±50 ka, from 5±4 mm/y prior to 250 ka, implying that the Ventura Basin has been deforming and closing since then.

The basement to the Ventura Basin constitutes of the Cretaceous to Eocene marine strata. Paleocene units are represented by the Juncal Formation and are composed primarily of very thick arkosic sandstones inter-bedded with thin dark gray shale. The stratigraphy and petroleum geology is outlined in the stratigraphic column of Figure 7. The Eocene strata are represented by the Matilija, Cozy Dell, and Coldwater formations. These formations are exposed outside the Ventura basin and are missing in the southern part of the Oak Ridge Fault. The Matilija Formation is composed of thick bedded, tan to mottled light greenish-gray arkosic sandstone, miceaous siltstone and shale. The Cozy Dell Formation is composed of dark gray, argillaceous to silty miceaous shale alternating locally with tan arkosic sandstone. The Coldwater sandstone is composed of hard, tan, arkosic sandstone with intercalations of greenish gray siltstone and shale of marine origin.

The Sespe Formation is composed of continental red beds characterized by maroon to red and locally green silty shale, interbedded red to pinkish-gray sandstone, and pebble-cobble conglomerates. The Sespe Formation interfingers with the overlying Late Oligocene to Early Miocene Vaqueros Formation which is composed of nonmarine to near shore marine thin to thick bedded biotitic sandstone. The lower part is dark gray to black sandy siltstone and mudstone. It is the oldest unit lying on the continental Sespe Formation that indicates the onset of transgression into the region in Early Miocene. The Sespe and Vaqueros Formations constitute the lowermost competent units above which the main detachment horizons were developed. The Rincon Formation (Early Miocene) is composed of poorly bedded gray shale and siltstone. It is the least competent layer within the Ventura Basin and forms the core of the Sulphur Mountain Anticlinorium implying ductile deformation within the formation.The Monterey Formation is composed of contorted siliceous shale grading upwards into punky, diatomaceous silt, laminated diatomite and mudstone. The Sisquoc shale is composed of light gray shale, claystone, locally slightly siliceous and diatomaceous. The Repetto Formation is composed of silty sandstone, siltstone and claystone at the bottom (Repettian stage) and sandstone, pebbly sandstone and sandy conglomerate at the top (Venturian stage). It is partly correlated with the Pico Formation of Pliocene age. The Mudpit and Saugus formations are composed of non-marine weakly consolidated silt, sand, and cobble/gravel conglomerates.

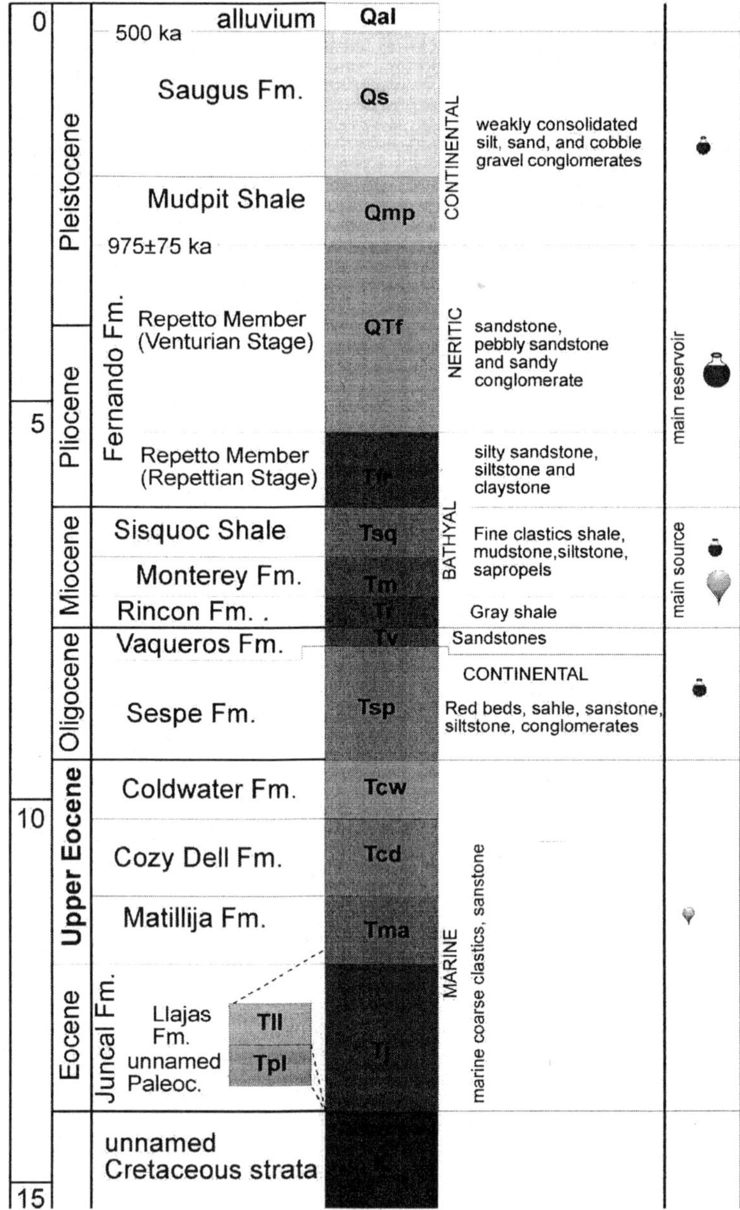



Figure 7. Generalized stratigraphical section of the units exposed in and around the Ventura Basin. The symbols in the last column indicate source and reservoir characteristics of the rocks.

It is thought that one of the primary source rocks in the region are the organic-rich rocks of the Monterey and Monterey like units (i.e. Rincon and Sisquoc formations). The intervals within these formations can be exceedingly good quality source rocks reaching TOC as high 18% dominated with sapropelic material with excellent capacity for generating liquid upon maturity. The distribution of carrier beds, potential reservoirs and seals is partly controlled by the style of basin formation and partly by broader, regional paleogeographic considerations. Migration pathways and reservoirs in the Ventura basin are dominated by the Miocene and Pliocene turbidites and associated deep marine sediments. Because of rapid syn-extensional subsidence in narrow pull-apart basins large volumes of sand and gravel have been deposited within the Southern California Basins including the Ventura basin in Miocene and Pliocene. These sands form the extensive reservoir columns of the Ventura basin. The presence of oil in the Pliocene units in the East Ventura Basin indicates that the oil migration from Miocene units crossed the stratigraphic boundaries. Seeps occur only along the vertically dipping strata where the younger units back-thrust over the older units.


## 7.2    HYPERSPECTRAL DATA ANALYSIS

Data from the Probe-1 (also known as HyMAP) imager were used in this case study. We use a vegetation index for subsetting the hyperspectral data to a vegetation subset and a bare soil/rock subset. Vegetation indices are quantitative rneasurement, based on digital values, that attempt to measure biomass or vegetative vigor. For living vegetation, a band ratioing strategy can be especially effective because of the inverse relationship between vegetation brightness in the red and infrared region. That is, absorption of red light (R) by chlorophyll, and strong reflection of infrared (IR) radiation by mesophyll tissue ensures that the red and near infrared values, will be quite different and that the ratio (IRIR) will be high. Nonvegetated surfaces, including open water, man-made features, bare soil, and dead vegetation parts, will not display the specific spectral response of vegetation, and the ratios will decrease in magnitude. Thus, the IR/R ratio can piuvide a mcasurement of the importance of vegetative reflectance within a given pixel.

For the vegetation data set we implemented two approaches for estimating the red edge position (see also the Chapters 5 and 6 in this book):
1.   the "linear method" of Guyot & Baret (1988)
2.   the "lagrangian method" of Dawson & Curran (1998)

For the soil/bare rock data set we derived spectral indicators of mineral alteration (hematite enrichment and calcite anomalies) using the SAM, spectral unmixing and cross correlogram spectral matching techniques (see the Chapter 2 in this book).

The resulting spatial patterns were integrated with known oil and gas seeps using a spatial data integration approach (SDI). The basic proposition, mathematical hypotheses which we want to test, in this study is the presence or  absence of an undiscovered oil/gas reservoir formulated as

$$F_p: \text{``p contains an \textbf{\textit{undiscovered}} oil/gas reservoir''} \qquad (2)$$

The probabilistic theory embedded in the favourability function allows to address this proposition. The favourability function method requires the two assumptions that: (i) past occurrences of a given type (i.e., a clearly identified type of process) can be characterized by sets of layers of supporting spatial data, and that (ii) new discoveries of the same type will occur in the future under similar circumstances. That is, the present should be able to predict the future. The mathematical framework of the

Bayesian implementation of the SDI is embedded in the work of Chung & Fabbri (1993). Consider a probability at any A, $prob\{F_p\}$, for the proposition $F_p$. This prior probability, a pixel p will contain a future reservoir, is obtained by:

$$prob\{F_p\} = \text{size } of\ F\ /\ size\ of\ A \qquad .(3)$$

where $F$ denotes the unknown areas which will be reservoirs within $A$ and *"size of B"* represent the size of the surface area covered by any subarea $B$ in $A$. $Prob\{F_p\}$ has the same value for all p. The purpose of the modeling is to see how the probability at p will be changed as we observe the m pixel values at p. At pixel p, the pixel value $v_l(p)$ of the $l^{th}$ layer is $c_1$ which is one of the $n_l$ classes (map units), $\{1, 2, \ldots, n_l\}$. Consider a set of all pixels whose value in the $1^{st}$, layer is $c_l$. The set is the thematic class in the $1^{st}$ layer whose pixel value is $c_l$. The set is denoted by $A_{lcl}$ and it is one of the non-overlapping $n_l$ sub-areas $\{A_{l1}, A_{l2}, \ldots, A_{lnl}\}$ in the $1^{st}$ layer. Assume that the occurrences at each pixel p can be expressed as the joint conditional probability

$$prob\{F_p \mid c_c, c_2....c_m\} \qquad (4)$$

that p will be a future undiscovered oil/gas reservoir assuming that the p contains the m values, $(c_1,c_2...c_m)$. When $prob\{F_p \mid c_c, c_2....c_m\}$ is approximately similar to $Prob\{F_p\}$ we can state that the pixel values at the p, $(c_1,c_2...c_m)$ do not add any useful information whether the pixel is a undiscovered reservoir. However, if

$$prob\{F_p \mid c_c, c_2....c_m\} >> prob\{F_p\} \qquad (5)$$

or

$$prob\{F_p \mid c_c, c_2....c_m\} << prob\{F_p\} \qquad (6)$$

then the pixel values, $(c_1,c_2...c_m)$ provide very significant information and they are highly correlated either positively or negatively with occurrences. We assume that $c_1$, will contain an undiscovered oil/gas reservoir and we assume that the $c_1,...c_m$ are conditionally independent given the condition $F_p$, (p will contain an undiscovered oil/gas reservoir). Hence, under the above conditional independence assumption, the joint conditional probability becomes

$$\frac{prob\{c_1\}...prob\{c_m\}}{prob\{c_1,...,c_m\}} prob\{F_p\} \frac{prob\{F_p \mid c_1\}}{prob\{F_p\}} ... \frac{prob\{F_p \mid c_m\}}{prob\{F_p\}} \qquad (7)$$

Under the conditional independence assumption, this joint conditional probability can be expressed in terms of three components. The first component, the ratio of $prob\{c_1\}... prob\{c_m\}$ and $prob\{c_1,...,c_m\}$ consists of the probabilities related to the input spatial data. The second component, the prior probability $prob\{F_p\}$ is the probability that a pixel p will be an undiscovered reservoir prior to having any evidence. The third component consists of m factors and each factor, the ratio of bivariate conditional probability $prob\{F_p \mid c_k\}$ and the prior probability $prob\{F_p\}$ indicates a contribution of each pixel value to undiscovered reservoirs.

The probe data used was acquired during the 1998 hyperspectral group shoot. The data was calibrated to reflectance and geocoded. Four flight lines cover an area south of Santa Barbara and north west of Ventura (Figure 8). Our initial work concentrates on line SBE4 which passes over the Sulphur Mountain anticline which is known to have several areas affected by seeps (both oil, tar and gas origins).

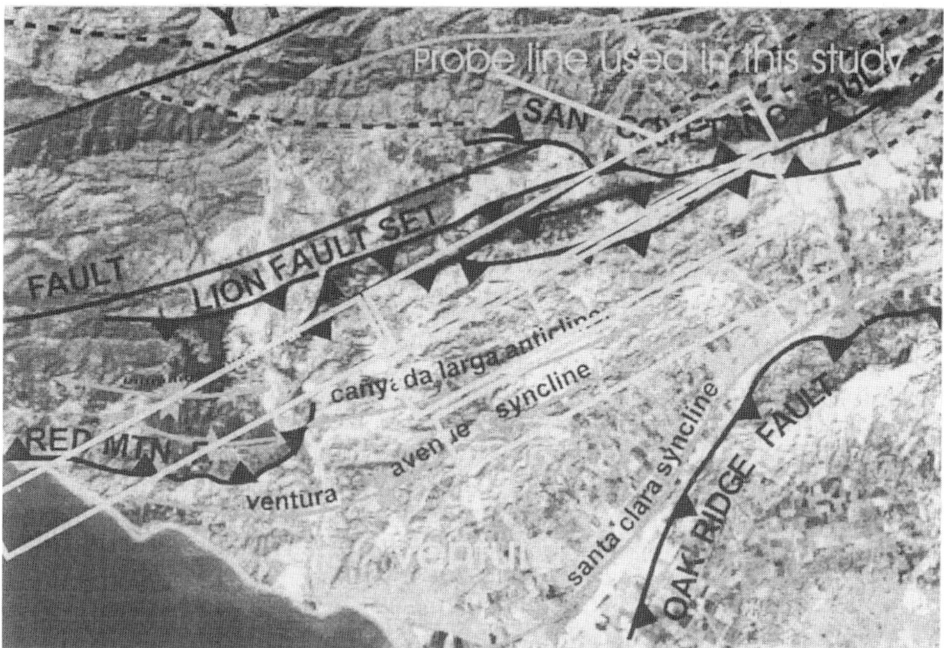


*Figure 8.* Location of the probe flightlines (backdrop image is an enhanced and interpreted Landsat 7 image).

The Sulphur mountain road sub-area (Figure 9) is located east of Oak View at around 119 degrees 17' 30'' – 34 degrees 22' 30''. Here a series of tar seeps (some ancient and some active) occur in shales of the Sisquoc formation of Late Miocene age. Local faults seem to play a role in the actual spilling sites of the oil as we inferred in the field. The area is rather flat and covered partly by light soils and grass on the actual seeps. Some agricultural fields occur here, however these seem to have not been in use during the last five years or more. Further away from the seeps, the area is covered by forest of mixed nature. From the reflectance data set of probe, a subset was extracted of 100 by 150 pixels covering the target area of interest. The results of the mineral mapping using SAM, CCSM and SMA were integrated with the locations of the known seeps in the probabilistic model described earlier (the spatial data integration approach). Because of the absence of ancillary data we were only able to integrate the results from the mineralogy mapping and the result from vegetation stress analysis with the known occurrences of seeps. The vegetation and mineral results were treated separately because they are mutually exclusive meaning that in the subsets a pixel is either assigned to the class vegetation or to the class mineral. The predictions of the probabilities were made for all remaining pixels in both cases. The results of the vegetation stress mapping are disappointing which was not surprising since the area at the Sulphur mountain has limited vegetation outcrop and is mostly covered by soil and mixed soil-vegetation. In the latter, the soil reflectance dominates the overall reflectance patterns. In the vegetation image it is particularly apparent that the areas dominated by vegetation (upper left corner of the scene for example) show very low probability values.

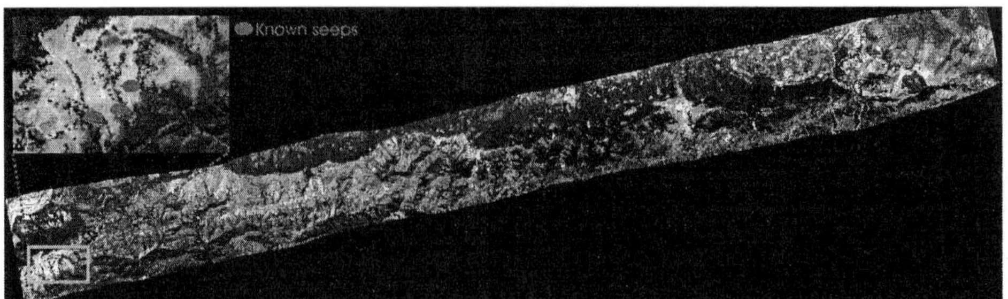


*Figure 9.* Location of the Sulphur Mountain road sub area on the geocoded Probe image.

In Figure 10 we provide and interpretation of the patterns observed. Particularly the predicted probability map for the mineral alteration provides spectacular results. The image clearly shows a elliptical pattern of high probability values around the actual seep location forming what appears to be a halo. In the center of the halo, low values give the appearance of a clogged area similar to the resulting patterns described in edge-leakage. The vegetation stress analysis in this area did reveal less information, which is due to the fact that little pure vegetation pixels are found and that large parts over the anomalous area are relatively devoid of vegetation.

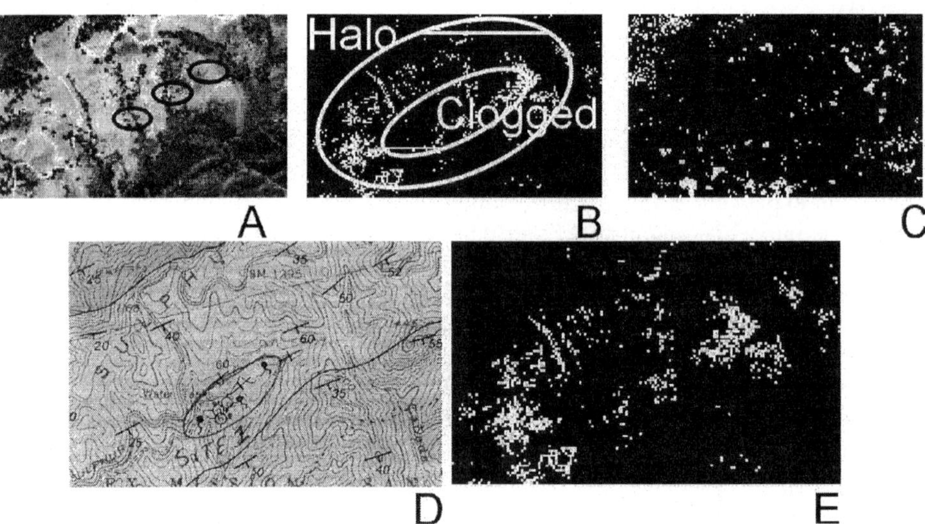


*Figure 10.* Results and interpretation. False color image of probe bands 16, 9, 4 in RGB with seeps (A), SDI probability minerals with interpretation of the anomaly (B) SDI probability vegetation (C), geological map (D) and SDI probability minerals with annotation. All probability maps display P>0.8.

The Sulphur mountain road area showed to be an ideal section for alteration mapping and mapping of vegetation anomalies in relation to microseepage. Good inferences were found from the hyperspectral data sets. The resulting probability maps can be used in an integrated exploration framework for petroleum exploration. We have presented an integrated approach toward seep mapping from hyperspectral imagery. The NDVI is used to generate a vegetated and non-vegetated subset which are analyzed separately. Mineral mapping techniques are applied to the non-vegetated areas while the red edge position is used to highlight areas of vegetation stress. The results are integrated using a spatial data integration approach which builds on Bayesian statistics. This model allows the objective combination of the results from hyperspectral data sets with known seep localities. At present, we are working on integrating other geoscience data sets into the model. These include surface geology and geochemistry and field spectroradiometer measurements.

*CHAPTER 9*

# IMAGING SPECTROMETRY FOR URBAN APPLICATIONS

Eyal BEN-DOR
Tel-Aviv University, Department of Geography, Tel-Aviv, Israel

## 1 Introduction

A city is a mixture of complex landscapes, social and economic developments and regional resources. It is therefore not surprising that the monitoring of an urban environment plays an important role in the modern era, which is governed mainly by growing urban entities. Significant rural-to-city immigration is being observed worldwide, and the center of life's gravity is now, more than ever, concentrated within the city environment. In 1800 urban residents represented only 2 percent of the total population of the world, while that number had reached 10 percent by the beginning of the $20^{th}$ century (Chen et al., 2000). Almost two hundred years later the urban population has grown to more than 3.2 billion, equaling half of the global population, and it increases by an average of 1,000,000 people per week. This dramatic growth has brought to the city environment, in addition to increased density and different effects on the atmosphere, a mixture of materials and a variety of complex structures that both directly and indirectly affect human comfort and the quality of life. At certain periods, fundamentally different processes of urbanization have emerged, resulting in an accelerated rate of urban change and the development of new, distinctly different, urban forms. Hall (1998) stated that "the only constant thing about cities is that they are always changing" and stressed the need for tracking, detecting and studying urban processes in a precise and temporal way. Obviously, this changing environment attracts the attention of many workers from different disciplines, and thus, many studies about the urban domain have been published over the past twenty years from social, economical and environmental points of view.

Urbanism has been studied from different viewpoints by many workers, giving rise to many questions regarding the changing environment and the processes it affects. In general, the rapid changes occurring in the urban environment never let this ecosystem come to a steady state, thus, making the research in this area rather important and most interesting. A few examples of this unbalanced situation are: soil bodies that are no longer exposed to the atmosphere, an ecosystem that is no longer functioning under natural conditions and anthropogenic activity that is changing diurnally and seasonally without any specific rules. Because these parameters (and others as well) differ from city to city, it could be assumed that a global view of the urban environment could become rather complicated and complex. In reality, however,

243

*F.D. van der Meer and S.M. de Jong (eds.), Imaging Spectrometry,* 243–281.
© 2006 *Springer. Printed in the Netherlands.*

the basic parameters and the main processes--such as increasing area, increasing population density and various materials used to build the city landscapes--are similar in all cities worldwide. Hence it is expected that the same tools could be valuable for analyzing the urban environment worldwide. Because the urban environment is dynamically and rapidly changing, there is a strong demand for the use of new and advanced monitoring technologies in order to allow better and more modern management of this complex environment.

## 2    Remote Sensing for Urban Applications

One of the important tools for studying the earth's landscape that has been developed during the past two decades is remote sensing. This tool enables the monitoring of varying environments from far distances, under a wide view range and on a temporal basis. Because remote sensing technology is based primarily on the interaction between sun or earth photons and atmospheric-terrestrial materials, it is obvious that it has the capability of detecting some environmental factors in the urban space, such as air pollution, vegetation status, water quality, road mapping and environmental changes (Ridd 1998). Air photos and satellite images are the basic data sources by which the urban environment is being studied via remote sensing means. In China, for example, Chen et al. (2000) reported that based on these tools an Atlas of Environmental Quality was issued for the city of Tianjin. This information is reported to be helpful to municipal authorities in tracking engineering construction activities, establishing pollution-protection activities, monitoring harbor deposits and managing coastal zone development. Another example is the city of Hong Kong, which uses an Atlas of Space View (Chen 1986) composed of high-resolution images taken from high-resolution spatial sensors such as COSMOS (2 meter) and SPOT (10 meter). These images and their corresponding interpretive maps vividly depict the achievements of Hong Kong's urban development programs, provide firsthand data for updating the urban databases and serve as a tool for city management in many respects. The potential of the remote sensing of the urban environment, using relatively large numbers of spectral bands, has been recognized by NASA, which has prompted a worldwide project, using the new ASTER sensor mounted onboard the Terra spacecraft. In this project, 1,000 cities worldwide, consisting of more than one million citizens each, have been selected as targets for monitoring by the ASTER 15-channel (30–70 m pixel size) sensors (Ramsey et al. 1999). Although data is not yet available (at the end of 2000), this project identifies a worldwide intention to focus efforts in the direction of studying and monitoring the complex urban environment from space, using advanced technologies.

In existing urban remote-sensing activities, spatial resolution seems to play a more fundamental role than does spectral resolution. This is primarily because urban areas are composed of small objects and discreet patterns that can be smeared if low spatial resolution data is used. Very fine-resolution data are useable for analysis of urban settlement areas, concerning the discrimination power of textural features with respect to the different terrain patterns present within an urban region (Jensen 1997). Nevertheless, Pesaresi (2000), by examining IRS-1C satellite data taken over the city of Athens, Greece, concluded that in spite of the importance of the spatial resolution property, a classical radiometric per-pixel approach is still required for urban mapping. Fung and Siu (2000) used SPOT HRV multispectral data to study the changing environmental qualities of Hong Kong in 1987, 1991 and from 1993 to 1995, whereas

Sultan et al., (1999) used coregistered digital mosaics of LANDSAT MSS and TM scenes over the Nile Delta from 1972, 1984 and 1990 to show that urban growth is endangering Egypt's agricultural productivity. They found that between 1972 and 1990, an increase of 58 percent in urbanization has occurred, and approximately half of this increase occurred between 1984 and 1990. This evidence can be extrapolated to predict the future, showing that, potentially, Egypt could lose 12 percent of its total agricultural area to urbanization by 2010. Remote sensing is used not only for daily management or future forecasting purposes but also for mapping and clustering wide-range areas for commercial applications. Cudlip et al. (1999) described the benefits of introducing satellite earth-observation data into the landscape assessment process by applying a special classification routine adjusted to satisfy specific customer requirements.

Sekine (1999) examined the relationship between the living environment level and land use for a residential area, using remote sensing and GIS analysis of SPOT data gathered over Morioka City. A significant relationship between the living environment level and six area classes was found, which stresses the capability of the remote-sensing approach for tracking the dynamic lived environment. Although a significant amount of information can be generated from existing satellite data despite its low spatial resolution, Konovalova (1999) suggests that the effectiveness of urban information systems can be improved by shifting emphasis from the computerized tools developed to support urban planning to true urban knowledge systems. One of the examples mentioned is solid waste management, which needs a precise technique for monitoring rapid changes in a near real-time mode. In this regard, it is obvious that new technology that can track more chemical/physical processes is strongly required.

Spectral fingerprints of objects across the solar and earth-radiation spectral range could be the source from which chemical and physical properties of sensed objects can be retrieved. However, although satellite information is based on spectral reflectance responses, current orbit sensors (including ASTER) are still lacking in their spectral capabilities in terms of what can be captured either from air-imaging sensors or field (point) spectrometers. The most significant advantages of satellite sensors (their ability to cover large areas on a routine basis and their large archives) and air photos (their ability to produce high spatial-resolution information) still cannot provide sufficient information for physical and chemical classification of the complex urban environment.


## 3    Aspects of Remote Sensing of the Urban Environment

For decades, the remote-sensing tool has enabled a better overview of the urban environment, with capabilities never before obtained. It provides a large-scale view of the expanding environment and enables a temporal spatial analysis of the changing urban areas. Ordinary remote-sensing tools enable differentiation of rural from urban areas, the definition of urban borders, the mapping of vegetation coverage, the monitoring of water bodies and the observation of urban surface temperatures.

Environmental aspects in urban areas can be grouped into two categories: short- and long-term aspects. Short-term aspects refer to those environmental changes that occur within days, such as air and water pollution, heat island effects and traffic load. The long-term aspects refer to spatial changes that emerge over months or years, such as vegetation status, built up area, surface coverage and road changes. Remote-sensing tools offer the capability of measuring both short- and long-term aspects, although short-term aspects call for a more sensitive capability for tracking small, visual, chemical/physical changes that are expected over a short time period, which can be sufficiently measured using more general probes and concepts. Ridd (1995) developed a new concept for exploring the urban ecosystem in a standard domain, using remotely sensed information and a vegetation-impervious-soil (V-I-S) model. However, even this new approach could not demonstrate fine chemical/physical recognition of objects in the urban environment. City architecture is one of the major problems preventing regular spectral  analysis in the urban environment using multispectral information. The high object-density of the urban scene causes problems such as shadowing and mixed-pixel effects (see later discussion). It is agreed that although it can be successfully applied to rural areas, conventional multispectral classification is very prone to error when applied to urbanized areas, mainly because of the high object-density of these scenes. Thus, struggling with such problems is the main task of workers who wish to remotely sense the urban environment in the most effective and precise way Hornstra, et al. (1999) and others in addition to (e.g. Casciati et al., 1996, Couloigneg 1998) proposed using very high-resolution data to map the urban environment. The idea behind this proposal is that high-resolution spatial information will shade local morphological changes, which will probably be related to short-term aspects. Another suggestion for monitoring and planning in highly urbanized areas is to use analysis of continuous streams of spatial data in a cost-effective way, which high-resolution air photos can provide. The forthcoming availability of very high-resolution data from orbit (e.g., IKONOS products) promises progress in this direction. It is, however, believed that high spatial-resolution information, either from air photos or from advanced orbital sensors, is used more as a pattern-recognition tool than as a source of information about the material's nature and composition. Therefore, in reality, this approach cannot spot information about the chemical/physical composition of the urban environment or track short- or long-term changes caused by nonvisible factors. As discussed earlier, precise and informative spectral information may be the key for short-term monitoring applications, and the new frontier from which such monitoring can be done is the use of sensors with high spectral-resolution capabilities. These sensors are characterized by a large number of channels across the active spectral region of the sun's or earth's photons, which are used to depict a spectral fingerprint of targets from short or long distances.

Because most urban changes are visible to the naked eye, it is mistakenly assumed that applying sophisticated remote-sensing technology, such as those aimed at deriving spectral information from large areas, is overloaded. Because the short-term changes in the urban environment are governed more by nonvisible chemical and physical processes that are active in the discussed spectral region, it is important to study this "overloaded" spectral approach. As discussed earlier, at present, most of the existing remote-sensing tools cannot provide precise information regarding the nature of the urban environment based on spectral recognition. In this regard, the advanced approach known as Hyperspectral Remote Sensing (HSR) technology is in a position to bridge the gap between the high spatial- and low spectral-resolution information and long- and short-term processes monitored by current remote-sensing means.


## 4    HSR and Urban Applications

Hyperspectral remote sensing (HSR, also known as image spectroscopy) has established itself as an advanced technique that further enhances the physical, biological and chemical understanding of the earth and its environment (Rast 1991). This technique is defined as the simultaneous acquisition of image data in hundreds of contiguous spectral bands, which makes it possible to produce laboratory-like reflectance spectra for each pixel in an image (Goetz 1992). The basic idea behind this concept is that a spectrum permits quantitative rather than qualitative analysis, and a combination of this feasibility with remote sensing is promising. Traditionally, urban environments could be successfully classified using simple remote-sensing techniques, such as high-resolution air photos and spatial recognition algorithms. Only a limited number of HSR-based studies have been applied to the urban environment, most likely because the highly advanced (and costly) HSR approach seems to add only minimal new information to well-known, well-mapped urban environments.

HSR sensors cover both the reflective (in the visible [VIS; 0.4–0.7 μm], near infrared [NIR; 0.7–1.1 μm] and short-wave infrared [SWIR; 1.1–2.5 μm]) and the emissive (in the thermal [TIR; 3–14 μm]) spectral regions, either together or separately. From the knowledge gathered in HSR technology for almost two decades, in areas such as geology, agriculture and geomorphology, and from recent HSR urban-based studies, it appears that the HSR approach could hold promising capabilities for urban applications. Use of sensors that are sensitive to the VIS-NIR-SWIR-TIR spectral region promises the capture of information about the composition of the urban environment, based on the fundamental chemistry of the targets. Whereas the HSR approach produces mixed spectral information regarding complex urban areas, use of sophisticated classification techniques and well-known knowledge about the spectral behavior of pure urban components are the basic keys for the successful application of HSR technology. In this regard, investigating and understanding the spectral characteristics of common materials found in urban areas is an essential step toward using the HSR technology for urban applications. In general, mixed information from the lithosphere, hydrosphere, biosphere and atmosphere can be derived from photon-matter interactions across the spectral regions in question. In many cases these interactions yield narrow absorption features, but the more finely tuned HSR technology can capture additional information, allowing for more precise classification techniques. In the terrestrial urban environment two major aspects can be remotely sensed: natural targets (e.g., soil, water, vegetation and gases) and man-made targets (e.g., buildings, pools, roads and vehicles).

The HSR sensors fall into several categories, determined mainly by the detectors and/or the data acquisition method used. From current HSR knowledge it is assumed that the SWIR and TIR spectral regions are important for urban applications, because they carry significant spectral features and information regarding minerals and thermal parameters of the city, respectively. In the VIS-NIR region, some related spectral features are weak and smeared and thus are assumed to provide only minimal information about the urban environment, although for some targets such as vegetation and water, this spectral region is ideal. From a practical perspective, the VIS-NIR region is important, because HSR-*VIS-NIR* sensors are relatively low cost as compared with HSR-*VIS-NIR-SWIR-TIR* sensors (e.g., AISA versus HyMAP). Hence, if the HSR methodology is to be introduced into a new area, this limited and narrow spectral region is generally the first to be studied. Apparently, a detailed study that examines both the narrow band spectroscopy and the VIS-NIR-*SWIR-TIR* region has not been extensively performed for urban applications. In the following section we provide spectral-based information regarding common urban targets in the refractive portion of the electromagnetic radiation. The information is divided into two categories--VIS-NIR and VIS-NIR-SWIR--using two measurement approaches: 1) artificial materials from existing libraries and 2) in situ spectral measurements within the city's environment.

TABLE 1. A brief description of the targets presented in Figure 1 ("Pure Spectra Urban Spectral Library"- PUSL) and Figure 2 (the "CASI Urban library"- CASL).

| PUSL LIBRARY | CASL LIBRARY |
|---|---|
| 1821-Clay | 1-Heavy soil |
| 2019-Leaf litter | 2-Cars |
| 2044-Basalt | 3-Red pigment, stadium |
| 2076-Granite | 4-Yellow pigment, stadium |
| 2116-New leaves | 5-Purple pigment, stadium |
| 2698-Salty water (sea) | 6-Grey pigment, stadium |
| 2699-Fresh water (lake) | 7-Running track, stadium |
| 2709-Coal | 8-Empty pool (blue pigment) |
| 2710-Concretion | 9-Dirt (dusty) road |
| 2747-Limestone | 10-Asphalt (light) |
| 2757-Mudstone | 11-Eucalyptus |
| 2803-Shale | 12-Pool |
| 2863-Reddish brown soil | 13-School roof |
| 2952-Grey paint on top of thin sheet metal | 14-School yard concrete |
| 2954-Dark green paint on top of a very thin metal | 15-Rollerblade platform |
| 2958-Black asphalt | 16-Tile roofs |
| 2984-Black asphalt core | 17-Wild vegetation with soil |
| 2995-Concrete pillar | 18-Grass |
| 3008-Grey metal | 19-Asphalt (dark) |
| 3027-Yellow paint on a thin metal | 20-Marble pavement |
| 3085-Orange epoxy | 21-Lower trees |
| 3101-Red brick | 22-Yarkon river |
| 3147-Dry grass | 23-Tennis courts |
| 2289-Quartz | 24-White roofs (treated) |
| 3399-Goethite | 25-Soil, building zone |
| org.-Organic matter | 26-Sea sand |
| | 27-Sea water |
| | 28-Yarkon lake |
| | 29-Shade |

## 5    Spectral Properties of Urban Material

As already agreed upon, the use of existing spectral information is an important step in tracking and exploring the urban environment using HSR technology. Whereas spectral libraries of natural and artificial materials have been published and are used by many workers worldwide (e.g., Grove et al., 1992 Price 1998, and so on), a spectral library of specific urban targets is still missing. This is in spite of the fact that most urban targets are made of materials that are found in nature (e.g., water, vegetation, rock, clay and sand) or are artificially manufactured (e.g., plastic, fabrics and metal). In this direction, Ben-Dor et al. (2000) built a VIS-NIR spectral library of urban objects, using a known spectral library of pure objects (Price 1995), and examined the data against real urban targets sensed by an HSR sensor (CASI sensor with 48 channels). They found good agreement between the two databases (image and synthetic) and concluded that the spectral behavior of urban objects can be predicted using existing libraries of both natural and artificial targets. In Figure 1 and Figure 2, the spectra of these urban targets are provided, and Table 1 gives a brief description of each target based on the two databases. In order to allow a better understanding and to assist in future availability of similar spectra, we have provided a full spectral-based explanation of each database for each target.

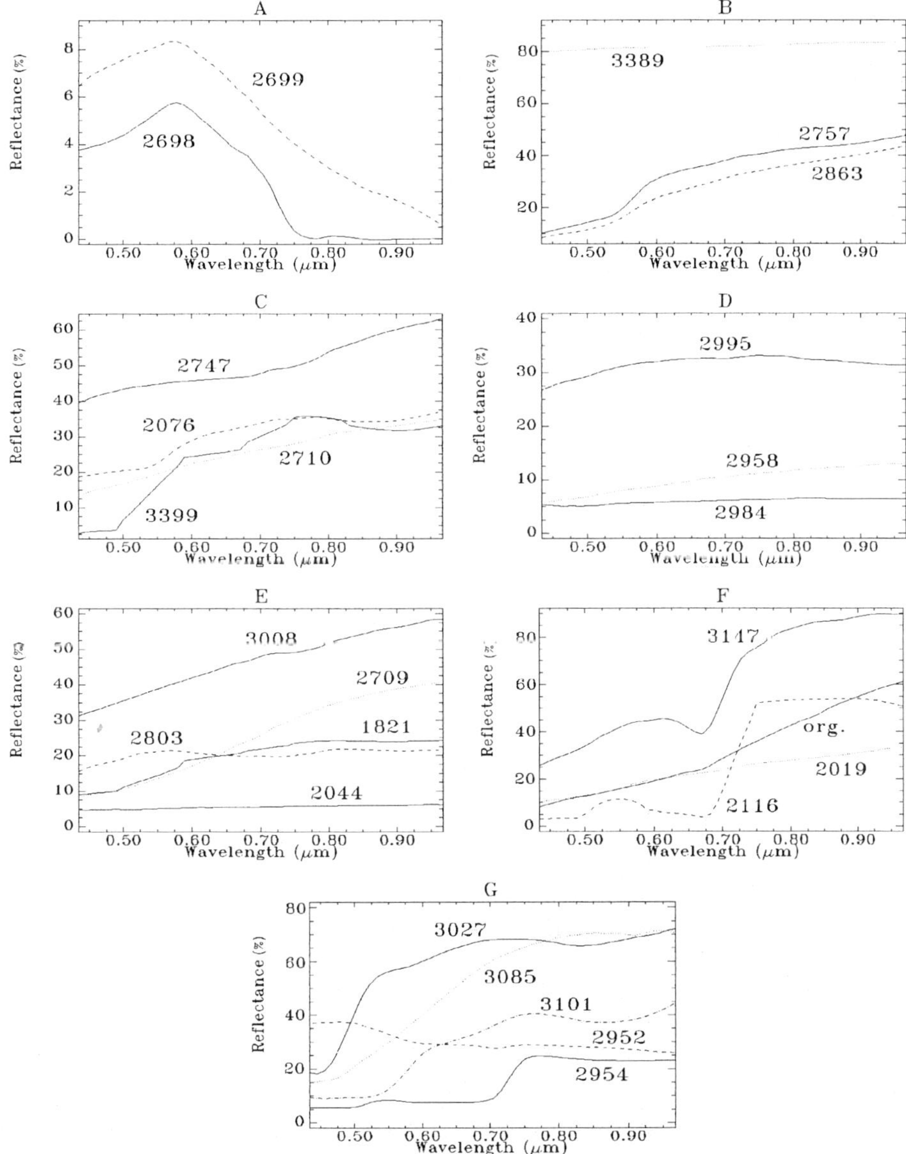

*Figure 1.* The spectral library, PUSL, in the 48 CASI channel configuration generated from the spectral library of pure objects taken from Price 1995. A brief description of each target is provided in Table 1 and in the text (Ben-Dor et al, 2000)

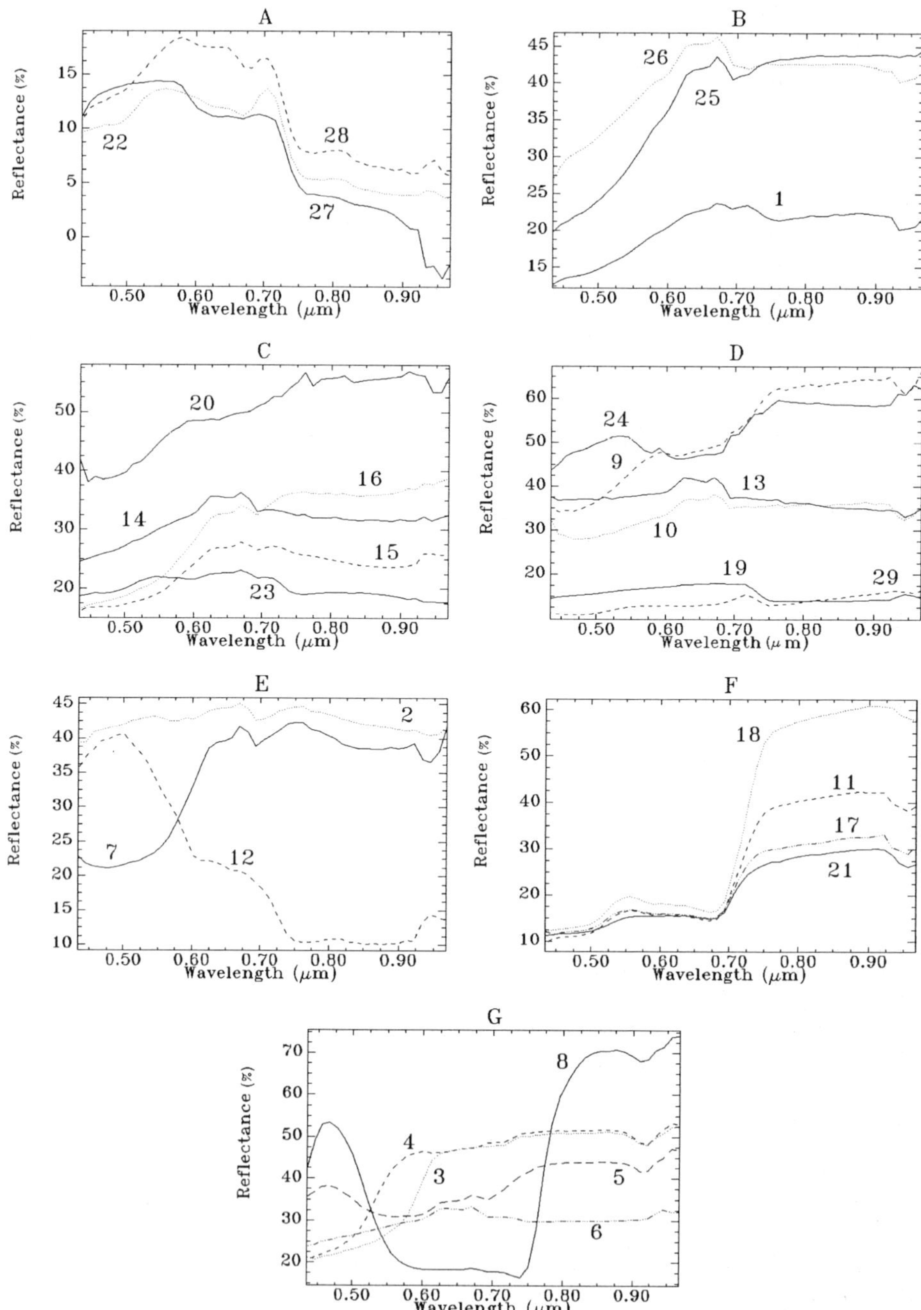

*Figure 2.* The spectral library of ground objects generated from the atmospherically corrected CASI 48 channels image (CASL). A brief description of each target is provided in Table 1 and in the text (Ben-Dor et al., 2000)

## 5.1    BUILDING A SPECTRAL LIBRARY OF URBAN OBJECTS FROM THE EXISTING DATABASE

A "synthetic" spectral library of urban-based materials that represent a typical city's coverage was generated based on Price's library (see above). The materials were: soil, dry grass, leaf litter, water, shale, plastic sheet, pigments, fiberglass, concrete, limestone, asphalt, rubber, iron cover, metal and bricks (see Table 1). The spectral library was generated from a collection of more than 3,000 different spectra of varying materials, which were resampled into 48 CASI channels and saved as a database termed "Pure Urban Spectral Library" (PUSL). The library used for comparison was generated from an atmospherically corrected CASI-image of Tel-Aviv City and was termed CASL.

## 5.2    THE PUSL SPECTRA

In Figure 1 (A–G) the PUSL spectra are presented, and in Table 1 the description of each target along with its original code in Price's library is provided. It should be noted that in some objects, the absolute spectrum baseline ("albedo") varies (e.g., gray metal [3008] and basalt [2044] in Figure 1E), whereas in others no significant variation can be observed (e.g., concrete [2710] and granite [2076]) in Figure 1C. In other materials, however, it is apparent that specific spectral fingerprints (bumps or "absorption peaks") exist (e.g., in mudstone [2757] at 0.55 µm, red brick [3101] at 0.55 µm, 0.67 µm and 0.89 µm and dry grass [3147] at 0.68 µm). In general, the total reflectance of a given object across the entire visible region (also termed albedo) is strongly related to the physical condition of the relevant targets (shadowing effects, slope and aspect, particle size distribution, refraction index, etc.), whereas the spectral peaks are more closely related to the chemical condition of the sensed target (specific absorption). Chemical and physical components that yield significant spectral features while interacting with the solar photons can be termed chromophores. In the VIS-NIR region, the spectral response of chemical chromophores is relatively weak as compared with that in the SWIR region. This is basically because the two regions are characterized by different absorption mechanisms. Whereas in the SWIR region photon absorption is governed by the combination and overtone modes of fundamental vibrations, in the VIS-NIR region, absorption is governed by an electronic process, which is relatively weak (Hunt 1977). Nevertheless, several urban-related chromophores do provide significant absorption features in the VIS-NIR, such as chlorophyll (at 0.68 µm), iron oxides (at 500nm; transition of $^5T_{2G} \rightarrow ^1A_{1g}$ in ferrous ions) 0.56 µm (transition of $^5T_{2g} \rightarrow ^3T_{1g}$ in ferrous ions) and 880 nm. (transition of $^6A_{1g} \rightarrow ^4T_{1g}$ in ferric ions) (Hunt et al. 1971) and pure color pigments (at different wavelengths across the VIS region). These are basically the most expected chromophores in the urban environment across the VIS-NIR spectra region. In addition to the specific absorption features (governed by the chemical mechanism), the overall spectral shape and nature (governed by the physical mechanism) may serve as important tools for improving the spectral recognition process. For example, water and soil, which have no specific absorption features can be distinguished based on different spectral shapes (in water: relatively high and low reflectance values in the blue and red regions, respectively, and in soil: monotonous increase of reflectance values from blue to red [Lillesand and Kiefer 1994]).

In reality, the reflectance of urban materials is a mixture of both chemical (specific absorption behavior) and physical (specific albedo behavior) chromophores. In the field, both mechanisms are active. Whereas in the laboratory the chemical effects are more pronounced (assuming that the physical effects such as illumination, particle size and sample geometry are constant); in the real urban environment physical effects can play an important role in the final spectrum behavior (e.g., shadows). A detailed discussion of the relevant pure chromophores under laboratory conditions (generated from the PUSL library) is provided below.

Vegetation (new leaves-2) holds an absorption peak at 0.68 μm because of the chlorophyll absorption. Iron oxides (red brick-3101; goethite-3399) hold a broad absorption band at 0.50 μm (transition of $^5T_{2G} \rightarrow {}^1A_{1g}$ in ferrous ions) 0.56 μm (transition of $^5T_{2g} \rightarrow {}^3T_{1g}$ in ferrous ions) and 0.88 □m (transition of $^6A_{1g} \rightarrow 4T1g$ in ferric ions [Hunt et al. 1971]). Fresh liquid water (lake-2699) is relatively featureless at the VIS-NIR spectral region, having low reflectance in the blue spectral region and almost zero reflectance in the red region. In contaminated water (with sediments or chlorophyll) more reflectance can be observed toward the red region, based on the effects of the water-matter spectra mixture (Muller et al. 1993). Dark objects (black asphalt-2958 and 2983, coal-2709, basalt-2044) are characterized by low albedo values across the entire spectral region, with no specific absorption signals. Mudstone (2757) and red brick (3101) hold an "iron-oxides"-like spectra, resulting from the relatively high content of iron in both materials. Likewise, fresh organic materials (org) and dry grass (2116) maintain relatively low reflectance in the blue region that increases toward the NIR region. Organic matter (org) consists, in general, of monotonous concave spectra in the VIS region (with chlorophyll remaining featured at 0.68 μm), whereas its slope is correlated with its stage of decomposition and conditions of aging (Ben-Dor et al. 1997). Artificial color pigments are thin layers of chemical substance that cover a given surface. This layer absorbs and reflects radiation at certain broad wavelengths across the VIS spectral region based on the chemical effects. Because in reality, reflectance is measured only from the upper 50 micrometer, painted objects cannot provide information regarding the core material (metal, plastic, etc.). The presented yellow pigment (3027) is, for example, on plastic. A significant high reflectance in both the red and green spectral regions, along with low reflectance in the blue region provides the yellow color with no plastic signals. An important conclusion that can be drawn from the PUSL stage is that although the VIS-NIR spectral region is considered in many cases to be a more "featureless" region than the SWIR spectral region (1.1–2.5 μm), this region does consist of reliable information that permits spectral recognition of targets. However, because the PUSL spectra represent chemical more than physical effects and are based on pure rather than mixed components, a proper examination of this issue must be applied in a real urban environment using IS data.


## 5.3    THE CASL SPECTRA

In order to check the accuracy of the spectral information obtained from HSR data, using a pure materials library (as was done in the previous section), the next stage was dedicated to examining specific targets under real HSR urban environment conditions. For that purpose, twenty-six targets that accurately represented the city coverage and were well distributed along a CASI 48-channel  reflectance image over Tel-Aviv City were selected. Table 1 provides a short description of each target, and the CASI reflectance spectra of each selected target are given in Figure 2 (A–G). For convenience, and to allow an easier comparison with the laboratory spectra, the CASI spectra are presented as closely as possible to the order in which the PUSL spectra are presented.

The following section provides a detailed discussion of each of the CASI spectrum. The targets were categorized into four major groups as follows: natural bodies (liquid and solid), urban surfaces, vegetation and artificial pigments.

*Figure 3*. The ground view pictures of selected targets that were used to investigate the image. Each target's number refers to its location in Figure 3 and to its detailed description in Table 2 (after Ben-Dor et al., 2000).

Natural Bodies (liquid): Figure 2(A) presents the spectra of three water bodies representing the liquid phase of the city: The Mediterranean Sea (27), the Yarkon River

(22) and the Yarkon Lake (28) (see also Figure 3 for a ground view of these targets). Each of these targets is characterized by a different water quality: Whereas the Mediterranean water is salty and relatively clean, the Yarkon River is a mixture of drainage and sea water and the Yarkon Lake is a fresh and relatively clean water body. As expected, these targets do have a typical water spectrum, that is, relatively high reflectance in the blue-green region and low reflectance in the red and near infrared regions (see spectra 2699 and 2698 of water samples in PUSL database). Significant absorption peaks at around 0.48 μm and 0.68 μm in both the Yarkon River (22) and the Yarkon Lake (28) targets are attributed to a relatively high content of chlorophyll relative to the sea water target (27), which has a different spectral fingerprint.

Natural Bodies (solid): Figure 2(B) presents the spectra of three targets representing the soils along the study area: Clayey soil (1, Typic Chromoxerets according to the USDA [Soil Survey Staff 1988]) sandy soil (26, Quartzipsamments according to the USDA [Soil Survey Staff 1988]) and red-brown sandy soil (25; Rhodoxeralfs according to the USDA [Soil Survey Staff 1988]). In general, all presented spectra hold a typical soil spectrum shape: relatively low reflectance in the green region, which monotonously increases toward the red region. The spectrum of the clayey soil shows relatively low reflectance targets ("low albedo") across the entire spectral region. This phenomenon was also observed in the field, where the selected clayey soil showed a relatively dark chroma. The relatively high content of iron oxides and organic matter in this soil in combination with the large aggregate size is responsible for this dark color. A significant peak at 0.97 μm in the clayey soil spectrum can be attributed to liquid water. At the time of the flight the field was relatively wet because of heavy rain events that took place days before the overflight. Because clayey soil is characterized by poor drainage, the relative moisture on the field was high. The spectrum of the sandy soil presents higher albedo targets than the clayey soil. Basically the sand is composed of quartz minerals (see, for example, spectrum 3389 in the PUSL database) characterized by high reflectance, featureless material across the VIS-NIR region. The spectral feature at 0.97 μm may also be related to the relatively high moisture content of this target, since it is located on the coastline). The spectral characteristics of the red-brown soil stand somewhere between the sandy (bright) and the clayey (dark) soils. In general the spectrum of the current red-brown soil matched well with the spectrum of the reddish-brown soil presented in the PUSL database (spec 2863).

Pedogenetically, the red-brown soil was formed from a sand dune during a weathering process that yielded this new soil type. This soil is composed of minerals such as quartz (main) montmorillonite (minor), iron oxides (about 0.8%) and organic matter (about 1%). Iron oxides and organic matter are basically the "coloring" agents of this soil. Whereas quartz is relatively featureless in the VIS-NIR region, the minor components (iron oxides and organic matter) are significantly spectrally active in this region, contributing to the final spectral shape. A comparison of the red-brown soil spectrum of the CASI database with the pure components of the PUSL database of organic matter (org) and iron oxides (3399, 3101) confirms this observation.

Common urban targets examined were: tile roof (16), schoolyard concrete (14), rollerblade platform (15) and marble pavement (20). Figure 2C presents the spectra of these targets, and Figure 3 provides a corresponding ground view. As can be clearly seen from the spectra, all targets provide relatively similar spectral behavior. This suggests that all of these targets be chemically related on their surfaces. In this case, a relatively high content of iron oxide in each of the targets provides the present spectral shape. Tile material (16) is a baked mixture of clay and iron oxides. Both the schoolyard concrete (14) and the marble pavement (20) showed reddish chroma on the ground, indicating iron occurrences. Similarly, the rollerblade platform is a paved, iron oxide cemented area, which shows an iron-related feature at 0.56 μm.

In Figure 2D, two asphalt targets "dark" (19) and "light" (10) are presented along with other related urban targets. Figure 3 presents the ground view of these two targets. In general, asphalt is a featureless (very) dark material (see spec 2958 and 2984 of the PUSL database). Both targets represent the asphalt condition in the city of Tel-Aviv.

The light asphalt targets present an aging (partially used) road, which was significantly contaminated by dust. In contrast, the dark asphalt target presents a recently paved and heavily used road. It is interesting to note that in another study, using another sensor (SPOT), it was possible to estimate dust contamination based on the roads' brightness values (Keller & Lamprecht 1995). This feasibility (using three SPOT bands) and the current observation (using 48 CASI bands) significantly indicates that the VIS-NIR spectral region has a promising capability for assessing dust accumulation in other areas. Further study is required, however, in order to implement this feasibility in the real world.

Shade in the urban environment is a component that plays an important role in many environmental applications (Gwinner & Schaale 1997). Shade can be found over different surface materials, which can affect the final spectrum. In the current study we examined the shade over asphalt streets only. In Figure 2D a shade spectrum (29), generated from well-defined shade-asphalt pixels, is presented. It can be clearly seen that the shade spectrum is similar to the dark asphalt but with lower reflectance values in the VIS region.

The school roof target (13) is composed of asphalt sheets that have been painted white. The asphalt sheets are intended to prevent moisture penetration during the winter, and the white color is intended to reduce heat absorption during the summer. The spectrum of the school roof target is presented in Figure 2D (13). It is composed of mixed asphalt and white color spectra, resulting from incomplete roof painting. In general, the reflectance spectrum of the current school roof is similar to the spectrum of the light asphalt (10; except for higher reflectance values in the VIS region). In both targets (light asphalt and school roof) a similar dark material (asphalt) is used for surface coverage. However, in the light asphalt the bleaching material is light dust, whereas in the school roof, it is white (calcareous) paint. Because the white paint is relatively brighter than the dust, the school roof consists of more reflection in the VIS region than the light asphalt.

In general most of Tel-Aviv's roofs are rather flat (approximately 80%), made of concrete, covered with asphalt sheets and painted white (see the previous discussion on the school roof). Treated roofs are those roofs that are homogeneously white colored (no asphalt material emerging for reflection), and untreated roofs are those roofs having mixed asphalt and white color (both materials are spectrally active, e.g., the school roof). In some cases it is also possible to find "bare" roofs (concrete or white concrete without asphalt sheets). Figure 3 (24) presents a ground view of a typical "treated roof" in Tel-Aviv, whereas, in Figure 2D, a corresponding CASI spectrum of this target is presented (24). Because this target acts as a perfect reflector, no spectral features were obtained. The significant absorption at around 0.62–0.69 μm is an artifact resulting from the detector saturation problem (see discussion in the SNR examination).

In Figures 2D and 3, a CASI spectrum and a ground view, respectively, of a dirt road target (9) are provided. The spectrum of this target (9) is quite similar to the treated roof spectrum (24) across the NIR region. Both are good reflectors (the white road is a calcareous powder road) consisting of similar detector saturation "features" at 0.62–0.69 μm. The spectral similarity between these targets is lost somewhere across the VIS region because of the iron oxide and organic matter contamination in the dirt road, which is absent in the treated road.

Figure 2E presents additional targets related to Tel-Aviv's urban environment: a running track at the city stadium (7), cars (2) and a pool (12). Figure 3  presents a ground view of these targets. The running track target (6) is composed of sand (quartz) material cemented with iron oxides. The resulting spectrum is similar to the iron oxide-rich materials, except it occurs at 0.68 □m (compare with spectra [3399-goethite and 3101-red bricks] in the PUSL database; Figure 2G), showing that even a small amount of iron in such targets can provide a significant spectral fingerprint. In the case of the targeted cars, the average of the pixels of several (well-defined) cars was used to yield the presented spectrum.

In Figure 3(2), a representative parking area in Tel-Aviv City is presented. From this figure and from long-term observation, it is postulated that Israeli drivers generally prefer a light-color car. This phenomenon is clearly observed in the cars spectrum, where a monotonous, featureless, moderate albedo spectrum is obtained.

The pool spectrum (2F) shows a higher reflection in the blue region that mimics the blue color (see Figure 3). On the other hand, the blue reflectance values are higher than expected from a pure water spectrum (compare with the fresh water spectrum [2699] in the PUSL database). At the time of the flight the selected pool was partially empty, and a spectral mixture of water and blue pigment was capture by the sensor, yielding the observed spectrum.

Figure 2F presents the spectra of four vegetation targets, which accurately represent the city's green areas, and in Figure 3, a ground view of these targets is provided. The vegetation was grass (18; in parks), lower trees (21; on a typical boulevard), wild vegetation mixed with soil (17; in open areas) and eucalyptus trees (in a wooded area). In general it can be seen that all targets are well characterized by a typical vegetation spectral shape and nature (compare to spectrum 2116 in the PUSL database). The spectral differences encountered between the selected targets are in the intensity of the chlorophyll absorption (at 0.68 μm) and in the NIR shoulder height (at 0.8–0.95 μm). These variations occurred because each target interacts differently with solar radiation. In grass (mostly dense), no soil is exposed to the sun's photons, and the cut leaves provide a closed lambertian reflection. As a result, the spectrum of grass is relatively intense across the above spectral regions. In both the boulevard tree and wild vegetation targets, the vegetation signals are mixed with asphalt or soil signals lying underneath, respectively. This obviously reduces the pure vegetation signals and, hence, appears as a weak vegetation spectrum. In the eucalyptus spectrum, on the other hand, where the leaf coverage is almost perfect (and, hence, no soil subpixels emerge), the related features are stronger than those of the wild vegetation spectrum but weaker than those of the grass.

In this case, the leaves' architecture yields a high scattering component that effectively reduces reflected photons to be captured by the sensor. The result is a spectrum that is less intense than that of the grass target. It can be summarized from the above discussion that the CASI sensor is a reliable tool for assessing vegetation status in the city of Tel-Aviv.

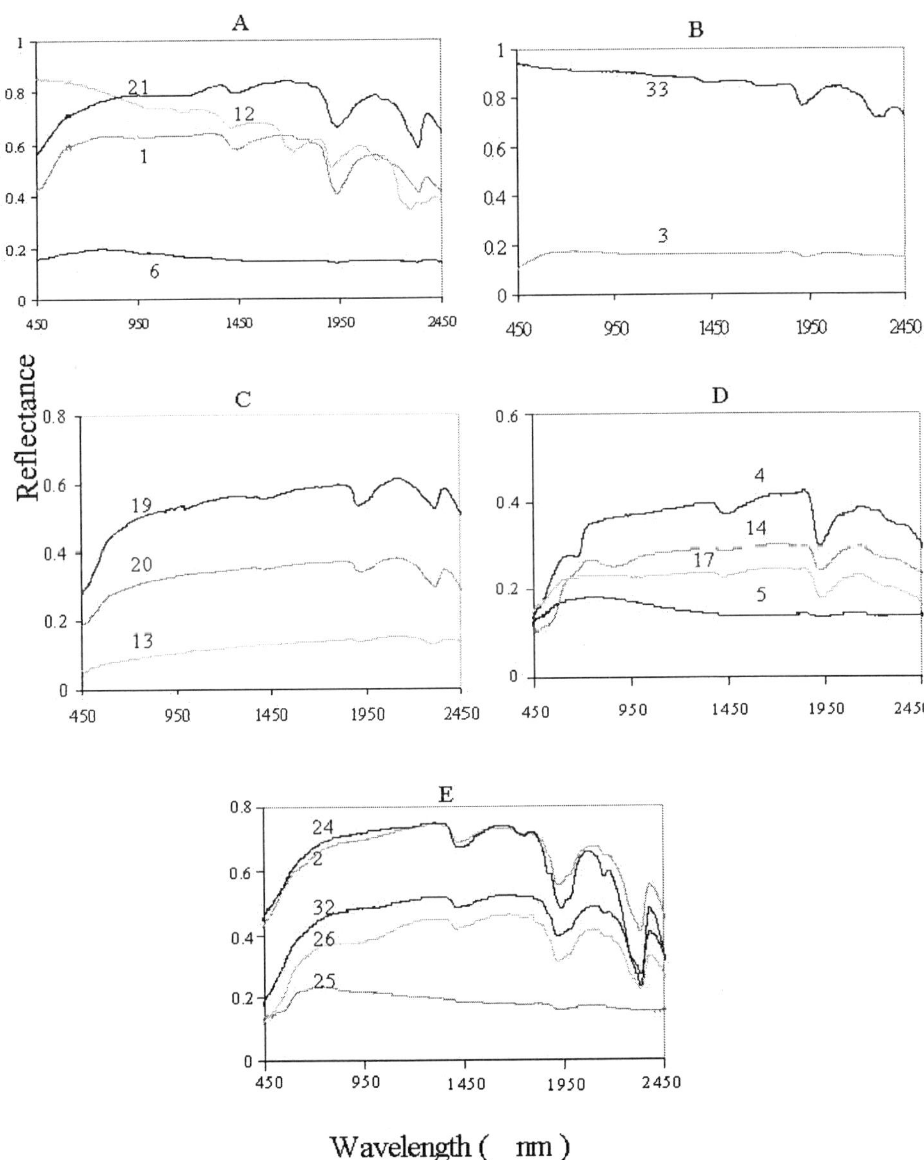


*Figure 4 (Part I).* The spectral library of ground objects generated in the field using ASD spectrometer. A brief description of each target is provided in Table 1 and in the text.

Figure 2G presents selected spectra of objects representing pure color pigments in the city. Their corresponding ground views are presented in Figure 3. The empty pool spectrum (8) presents a pure blue pigment. Because this spectrum shows a significant difference from the partially full pool spectrum (12), it indicates that the IS technique has the capability of detecting the water status in open pools during the off-season. From an environmental standpoint, this is an important finding. During the off-season, remaining water acts as a hatching ground for mosquitoes, which causes an environmental nuisance in the spring. From traditional gray or color air photos, it is impossible to locate partially full water bodies, because they all maintain a blue color. Using the IS technology as shown here, small spectral changes can be mimicked, which makes possible precise spectral recognition over a complex urban area.

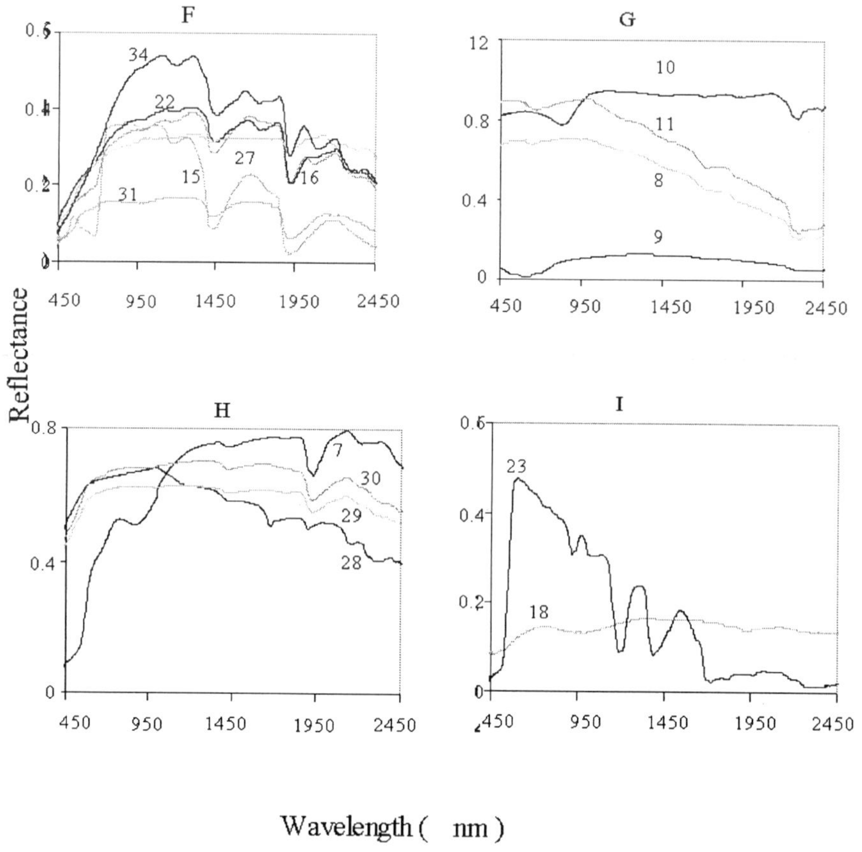


*Figure 4 (Part II).* The spectral library of ground objects generated in the field using ASD spectrometer. A brief description of each target is provided in Table 1 and in the text.

Other color pigments were taken from the plastic seats of the Olympic stadium (3, 4, 5 and 6). A good match was obtained between corresponding pigment spectra of the CASI and the PUSL databases.


## 5.4    SUMMARY FOR URBAN LIBRARY FROM EXISTING DATABASE

It can be concluded that in spite of the monotonous spectral behavior of objects in the VIS-NIR spectral region, and even though mixed pixels and materials do exist in the urban environment, this region does hold significant fingerprints of urban objects. The PUSL synthetic spectra as generated from Price's library shows good agreement with the results obtained from the HSR sensor over the urban areas. This suggests that Price's comprehensive spectral library can be used to predict, identify and validate spectral signatures over urban environments. Spectral information, as shown below, has the capability of distinguishing not only between two extreme targets and their spectra (e.g., vegetation or water) but also between similar spectra (e.g., clayey and sandy soils, grassy and wooded areas, empty and partially empty pools and shade and asphalt).

TABLE 2. A brief description of the targets presented in Figure 4.

| SAMPLE | DESCRIPTION |
|---|---|
| 1 | Cement coating of wall, cement and calcareous mixed  with light pigment |
| 2 | A floor brick (pressed cement with lime) |
| 3 | A Basalt brick |
| 4 | A cement floor ("beton") ı microphyta |
| 5 | Dark cement floor ("beton") dark |
| 6 | Cement coating of wall, dark pigment |
| 7 | Red brick, typically coated roofs |
| 8 | A dark blue car |
| 9 | A white car (with dust) |
| 10 | A silver car |
| 11 | A white car |
| 12 | Cement coating of wall, very light pigment |
| 13 | Asphalt (recently paved) |
| 14 | Red-dark brick  (used as a floor for outdoor areas) |
| 15 | Very dense grass |
| 16 | Yellow (non-healthy) grass |
| 17 | Cement brick, gray, (used as a floor for outdoor areas) |
| 18 | White -brick, used as a floor for outdoor areas |
| 19 | Iron core (used for pipe cover in streets) |
| 20 | Asphalt (light aged) |
| 21 | Calcareous stone (used as a fence) |
| 22 | Litter |
| 23 | Plastic chair (orange) |
| 24 | Marl used to pave yards |
| 25 | A large red brick paved yards |
| 26 | Red paved path  "granulate" |
| 27 | Red bare oil |
| 28 | Asbestos roof |
| 29 | White (limed) mixed asphalt paper roof |
| 30 | White (limed) roof |
| 31 | Wet soil |
| 32 | White paved path "granulate" |
| 33 | A wall paved with lime (white) |
| 34 | A wood base |
| 35 | Asphalt sheet |

## 5.5 BUILDING A SPECTRAL LIBRARY OF URBAN OBJECTS FROM IN SITU MEASUREMENTS

Following the limited spectral range of the VIS-NIR region described below, and after examining the feasibility of deriving information from the spectra of pure materials, the next stage was dedicated to building a spectral-based library of urban targets that cover the entire VIS-NIR-SWIR region, using in situ measurements. We were able to do this by conducting a controlled field campaign, using an ASD spectrometer in the open urban space. The expanded spectral region offers more information regarding a matter's chemistry composition, based on the spectral activity of functional groups governed by the overtones and combination modes of the fundamental vibrations. The in situ measurements ensure a precise recognition of urban targets in their exact environments and provide a basis against which to better compare the potential of existing artificial libraries.

A spectral library of 38 objects, carefully selected along a typical Mediterranean urban area, was generated. The following section provides a detailed description of their chromophores. Figure 4 provides the complete spectra of these objects, and Table 2 gives a description of each target. As can be seen from these figures, the spectral variation of the targets is impressive, especially in the SWIR region. This observation strongly suggests that the SWIR is a very important region in the HSR-urban domain. The spectral information of the present stage was grouped based on similarities among targets (e.g., roofs, paved materials etc.), and the following discussion is based on each selected group.

Figure 4A presents the walls spectra. The city buildings' walls are covered with cement plaster composed of different materials and pigments and are also coated with calcareous stones. It can be clearly seen that the SWIR region is, for the most part, active and informative for all samples. Sample 21 is a wall coated with calcareous stone and thus holds a significant calcite absorption feature at 2.32 $\mu$m ($3\nu_3$). The water absorption at 1.45 $\mu$m ($\nu w+2\delta w$) and 1.9 $\mu$m ($\nu w+\delta w$) indicates that this wall is relatively wet. Samples 1 and 12 are walls made of different materials mixed with the cement plaster: sample 1 is a wall plastered with a calcareous-rich cement (as seen by the typical absorption feature at 2.32 $\mu$m), whereas sample 12 is a wall plastered with a gypsum-rich cement (as seen by the typical absorption features of gypsum at 1.2 $\mu$m ($2\nu_1+\nu_2$) at 1.75 $\mu$m ($\nu_2+\nu_3$) and 2.2–2.27 $\mu$m ($\nu_3$).) Sample 6, which represents a wall plastered with cement material mixed with dark pigments, shows almost no specific absorption features. This is because the low albedo characteristic of this target inhibits the absorption potentiality of possible chromophores .

Figure 4B represents more wall variety in the city area. Sample 3 is a wall coated with basalt, showing a featureless spectrum based on the low albedo characteristic of the dark basalt material. Sample 33, which is a wall painted with a thick, white plastic paint, shows high albedo and significant absorption features across the entire region, caused by the complex material (plastic) that forms the paint. Note also that some clues are available indicating hygroscopic water at 1.9 $\mu$m ($\nu w+\delta w$), suggesting some water molecules are present in this material.

In Figures 4C–4E, the spectra of paved materials throughout the city are provided. In Figure 4E (paved material 1) it is clear that all materials are composed of calcite (absorption at 2.33 μm) and clay minerals (absorption at 1.4, 1,9 μm and 2.2 μm). Also in the visible region, the absorption of iron oxides can be seen throughout all of the samples examined (at around 630 μm) and especially in sample 26 (red granolite) and 25 (red brick), which are both composed of iron oxides added to a cement mixture. In paved material 2 (Figure 4c), there are two asphalt materials taken from recently (13) and aged (20) paved roads (dark and light material, respectively). It is evident that in the light asphalt spectrum, in addition to higher baseline values, there is a significant absorption feature at 2.34 μm (assigned to calcite [$3v_3$]) and features at 1.4 and 1.9 μm (assigned to hygroscopic water [see previous discussion]). The light tone of the aged asphalt is the result of local contamination with dust (composed of clay and calcite minerals). The features of these minerals are visible to the spectrometer, which enables observers to distinguish between the two asphalt stages. This issue has been discussed by Ben-Dor et al. (2000) who stressed the potential of the HSR technology for tracking environmental changes from spectral changes in asphalt (e.g., traffic load, air pollution etc.). Another paved material in Figure 4C is sample 18, which is composed of calcareous bricks that are used to pave yards and sidewalks in Mediterranean cities. In Figure 4D, sample 4 represents an aged cement surface that is coated with moss. The existence of moss can be seen by the absorption peak at 0.68 μm, assigned to the chlorophyll absorption. The strong absorption at 1.45 μm and 1.9 μm are assigned to water absorption in the moss tissue. In sample 14, a strong absorption in the visible region is attributed to iron oxide, which is a significant component in this particular brick, which is used to pave the roads (assigned to Fe3+5 $T2g \rightarrow 3T1g$). Samples 17 and 5 are both materials made of pure gray cement. It is evident that the brick represented by sample 17 is wetter than that represented by sample 5 based on the high intensity of the absorption peaks at 1.45 μm and 1.9 μm.

In Figure 4F, several organic-vegetation targets found in the urban environment are presented. In most of the samples (except sample 34, which represents a bulk wood target) a chlorophyll absorption signature is evident at 0.68 μm. In the high dense grass (sample 15), the chlorophyll absorption peak is very strong as is the red slope, followed by strong water absorption peaks at 1.45 μm and 1.9 μm. As vegetation becomes less green and dryer, these signatures diminish and new spectral features emerge across the SWIR region (attributed to organic compounds; Ben-Dor et al 1997). In this respect, vegetation-organic matter endmembers in the urban environment can be represented by dense grass (15) and wood (34), respectively.

One of the main targets in the urban environment is a vehicle. Figure 4G, provides the spectra of selected cars in parking areas. Four typical cars, painted silver (10), white (11), dark blue (8) and dusty white (9), are presented. It can be seen that each sample represents a different VIS-NIR reflectance response, which mimics different pigments. In addition, across the SWIR region, some signatures, which probably refer to the chemistry of the paint material, are provided. A significant difference can be seen between the white car when it is clean (11) and the same car when it is dusty (9). The dust-contaminated car is relatively darker than the non-contaminated car, suggesting that the dust is composed of dark particles (e.g., ash resulting from coal burning).

Figure 4H presents typical roof material found in Mediterranean countries. The most significant and spectrally active material in the VIS-NIR region is shown in sample 7 (red brick roof), which is composed of iron-oxide cement. This reddish cover

over the city's canopy can be very easily seen by the naked eye. Others roofs can barely be recognized from the VIS-NIR region, but enlarging the spectral region to include the SWIR region enables a finer detection of differences among roofs. For example, the asbestos roof (28) is indistinguishable from other white roofs (white painted [29, 30]) across the VIS-NIR region, whereas across the SWIR region, more spectral features of the asbestos material can be observed, which enables discreet recognition of this target from other roofs.

Figure 4I shows two nonstandard materials that can be found in the urban environment. The first is a plastic surface (23, in this case, orange colored), and the second is a metal iron cover (19). As can clearly be seen, significant fingerprints occur for these two objects. The fingerprints are assigned to electron transition in the iron (Fe3+ 5 T2g$\rightarrow$3T1g) of the metal cover, and there are several absorption features for the plastic material, which are assigned to its chemical composition.

## 5.6    SUMMARY FOR URBAN LIBRARY FROM IN-SITU MEASUREMENTS

Summarizing the ASD spectral library, it can be concluded that the spectral information across the VIS-NIR-SWIR spectral region does hold important spectral signatures for urban pattern recognition. Moreover, the additional spectral signatures relative to the common VIS-NIR sensors can in practice be used to solve, monitor or detect environmental problems or conditions, such as dust contamination, roof composition, pavement conditions, wall materials and more.

## 6    Spectral Pattern Recognition

## 6.1    EXAMINING THE SPECTRAL-BASED INFORMATION FOR URBAN MAPPING IN THE VIS-NIR REGION

Using previous VIS-NIR spectral information, the next example is provided in order to illustrate how this spectral information (major or minor) can be used to spatially recognize and map the urban environment. The ability to map the urban environment based on spectral information is essential from many aspects where and the HSR technology may practically opens this frontier. The following example shows that it is possible to perform a pixel-by-pixel spectral recognition process using the spectral information discussed previously.

Figure 5 provides a mosaic image of the MTMF results for endmember selected along the city of Tel-Aviv with a gray scale image of the study area for orientation (the MTMF procedure is clearly described in Boardman et al. 1995). The endmembers used for this procedure were the Yarkon River (EM-1), sandy soil (EM-2), grass (EM-3), asphalt (EM-4), concrete (EM-5), red-brown soil (EM-6), sea water (EM-7), total vegetation (EM-8), treated (white) roofs (EM-9) and shadow (EM-10). Figure 6 provides the spectra of the endmembers and shows that significant spectral variations exist among the endmembers. Summarizing all classified pixels in an examination subset revealed that more than 96 percent of the area could be explained by the selected endmembers. This can be seen in Figure 5, where the summarized image (SUM) represents the sum of all endmember fractions. Except for the image edges, most of the pixels revealed values close to unity (bright tones), showing that the city landscape can be clearly explained by the selected endmembers and HSR technology.

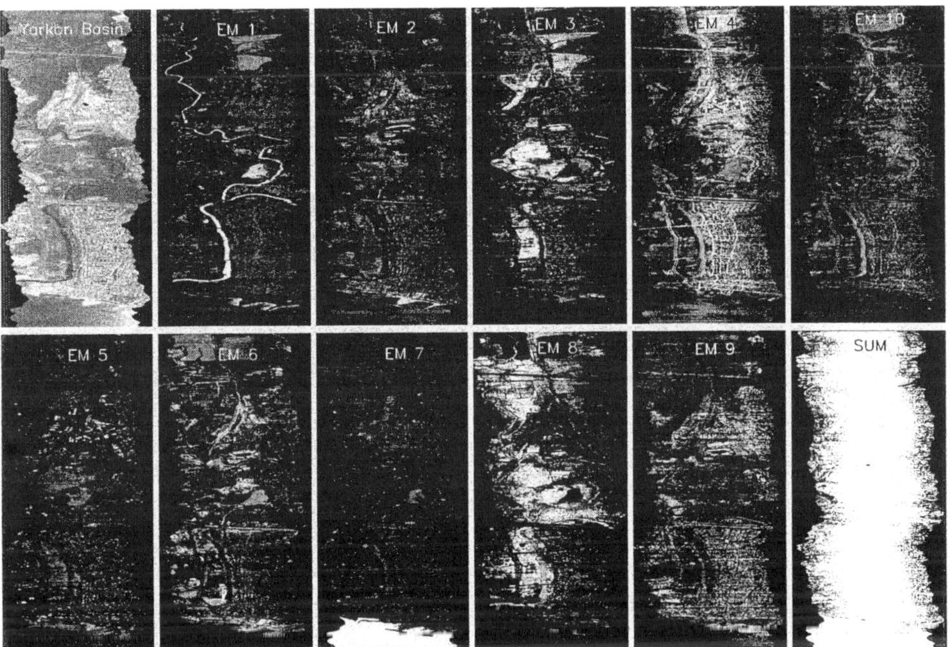


*Figure 5.* A gray scale mosaic of the MTMF analysis over Tel-Aviv. Each image stands for a particular endmember where bright tone represents more similarity to the corresponding spectrum given in Figure 6. Also provided is a gray scale image of channel 8 (0.5114 µm) for ground orientation (Ben-Dor et al., 2000).

## 6.2    EXAMINING THE SPECTRAL-BASED INFORMATION FOR URBAN MAPPING IN THE VIS-NIR-SWIR REGION

As was discussed in section 5.2, the entire spectral region (VIS-NIR-SWIR) may provide more spectral information than the VIS-NIR region. In this regard Rosner et al. (1998) examined the DAIS-7915 sensor data, using the 72 channels across the VIS-NIR-SWIR region, over Dresden City, Germany (Figure 7). They showed that a spectral-based analysis enabled the precise mapping of vegetation, bare ground, anthropogenic surfaces, buildings, water and shadows over the urban environment with a relatively high confidence level. Figure 7 provides the results of this analysis, showing an accurate and interesting classification with a relatively high significance level. Rosner et al. (1998), showed that by using a traditional classification method such as Maximum Likelihood, only 33 percent of the total area could be classified. However, applying an unmixing procedure with well-defined endmember spectra makes this classification much more effective, as about 99 percent of the area could be classified after only five iterations. Rosner et al. (1998) concluded that the spectral approach is a powerful tool for analyzing hyperspectral data of the urban environment, taking into account the great spatial and spectral variability of urban surfaces. In a follow-up study, Segel et al. (2000) improved the classification technique over a dense urban environment by developing a method for extracting endmembers from the data, taking into account the mixed pixel problem, in order to accurately identify urban surface coverage on a sub-pixel basis. The authors showed that it is possible to deal with the mixed pixel problem and encouraged future research in this direction. Fiumie and Marino (1997) also conducted a study that successfully used the HSR technology across the VIS-NIR-SWIR region for urban mapping applications. They showed that by using MIVIS sensor data it was possible to map the paved roads throughout Rome, Italy, and to generate a classification map identifying the various paving materials used, based on spectral information.

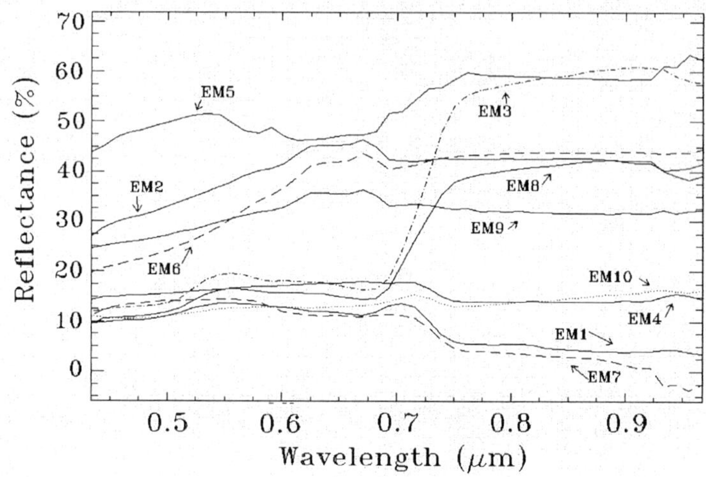


*Figure 6.* The endmember spectra of ten targets selected for the MTMF analysis (see text for more detail). Also provided is a SUM image that summarizes the endmembres calculated fractions (Ben-Dor et al., 2000).

## 6.3     EXAMINING THE SPECTRAL-BASED INFORMATION FOR URBAN MAPPING, USING THE TIR REGION

The TIR region contains thermal information of surface areas. This information, particularly the surface temperature, has been used to assess the urban heat island (UHI) phenomenon via remote-sensing means.

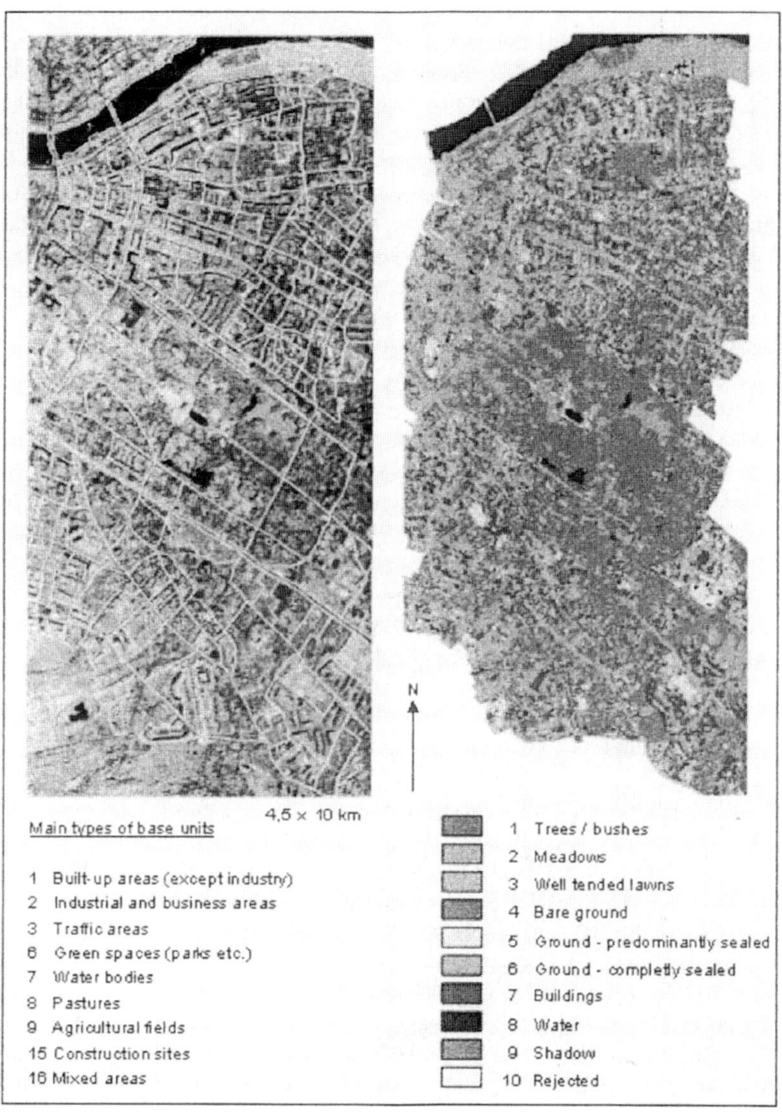


*Figure 7.* The results of spectral based classification process over the city of Dresden Germany as compare to ecological base units (Rossner et al., 1997)

The UHI is one of the most studied phenomena of a city's climate, and it has diverse environmental implications. Almost universally the modified thermal climate in cities is warmer than that in nonurban areas and leads to a set of distinct micro- and mesoscale climates, resulting from changes in the natural landscape (Roth et al. 1989). The main reasons for the UHI phenomenon are: (1) the mass of the buildings, increasing surface roughness (reducing wind speed and reducing city ventilation); (2) minimal evapotranspiration because of the loss of vegetation (evapotranspiration cools the air, therefore, reduced evapotranspiration means higher temperatures); (3) lower albedo of the city surface and (4) anthropogenic heat sources, for example, driving automobiles, industrial production, heating, air-conditioning and so on. Comprehensive reviews of typical UHIs have been provided by many UHI workers (Landsberg 1981; Atkinson 1985; Oke 1987; Oke 1995). Cities are also known to be less humid than rural areas because of the different land uses and because there is less vegetation in cities compared with rural areas. Cities are characterized by great thermal variability because of the heterogeneity of land uses. For example, the highest temperatures in cities are found in the central business districts, whereas parks and green areas are known to be "cold islands," as measured by air temperatures and surface temperatures (Balling et al. 1988; Eliasson 1996). However, Jauregui (1990) found that Chapultepec Park in Mexico City (in a tropical humid climate zone) was colder than the urban area only at night, and it had a similar temperature to the urban area during the daytime. Grimmond et al. (1996) studied the differences in energy balance between a "green" (having a lot of vegetation) neighborhood and a neighborhood with very little vegetation coverage in Los Angeles and found higher temperatures, by about 1°C, in the "green" neighborhood during the daytime. This can be explained by the lower albedo of the green neighborhood. Spatial differences in temperature and humidity can be found inside the park as well, and they are a function of conditions of location, size, vegetation types, water bodies and material coverage (Givoni 1991; Grimmond et al. 1996).

Climatic parameters such as wind speed and air and ground temperatures are the features that have been used most often in attempts to understand the characteristics, effects and generation of UHIs. In this regard, remote-sensing tools offer a significant capability for studying the UHI from a large to a local scale and can provide accurate spatial representation of the surface temperature, albedo, emissivity and surface material classifications. Air temperature, which is known to be the most significant parameter affecting the UHI from the point of view of human comfort, cannot be assessed directly by using remote-sensing tools and, hence, needs to be defined in terms of an indirect relationship or studied in a separate investigation. Other thermal parameters that are important for fully understanding the UHI are thermal heat fluxes, transpiration intensity and heat capacity of built-up areas. In practice, these parameters are difficult to assess using ordinary remote-sensing tools, and the most commonly measured parameter is surface temperature. The surface temperature does not necessarily describe the real urban heat island or the air temperature, thus, it is very important to measure heat fluxes that occur within a city. The HSR technique offers a capability for doing so, based on the informative data received across the TIR region. Richter (1996) has developed a physical-based method for determining the heat fluxes from the DAIS-7915 sensor on a pixel-by-pixel basis. In this approach, the visible and near infrared information is used to extract albedo and emissivity values, which are then incorporated with the thermal data to yield thermal fluxes on a pixel-by-pixel basis (solar flux, thermal flux, sensible heat flux, latent heat of evaporation and flux into the ground).

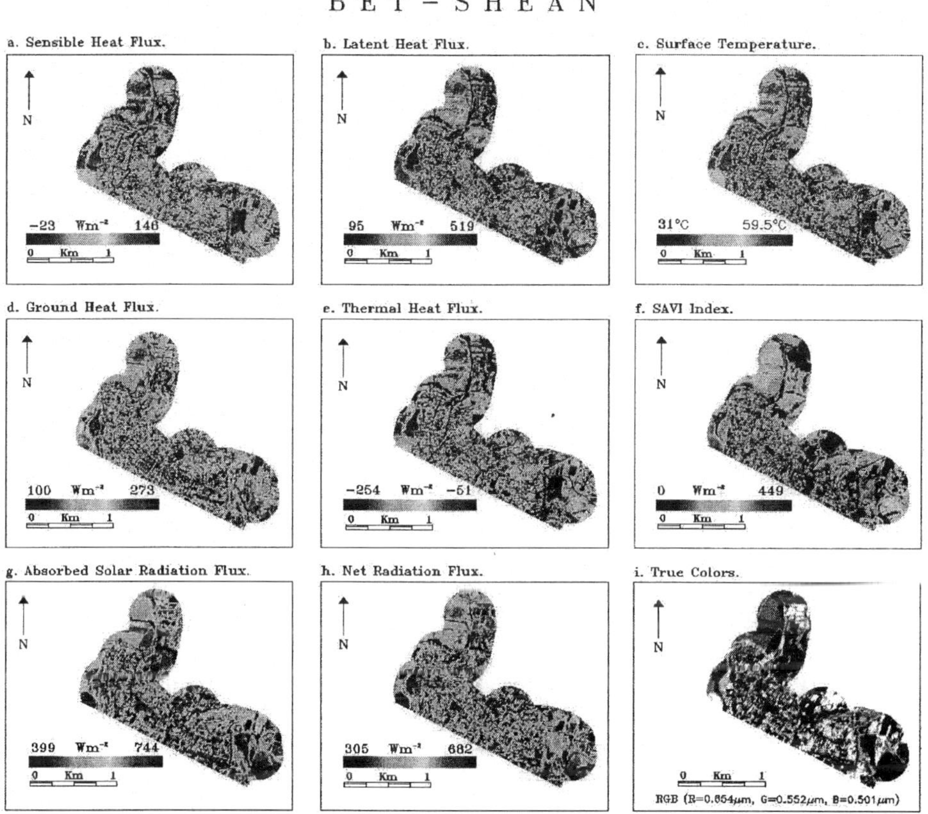



*Figure 8.* A mosaic image of the city of Beit Shean, showing surface temperature (a), ground flux, latent heat flux (b), sensible heat flux (c), temperature (d), and net radiation flux (e) as generated from the DAIS-7915 data.

Two arid cities were selected to show how Richter's HSR methodology provides additional information about the urban heat island. The cities, Afula and Beit Shean, are located in northern Israel on a flat terrain and are situated far from the cooling effect of the Mediterranean Sea. The spectral data acquired in August 1997 was accompanied by detailed ground measurements of air temperature and humidity taken during car traverses of the targeted areas. The data were processed to yield the flux parameters as described previously, using both the ground and spectral measurements. Figure 8 provides a mosaic image of the fluxes of Beit Shean as generated by applying Richter's equation pixel by pixel using the DAIS-7915 data. Whereas a traditional method for investigating the urban heat island effect from a remote-sensing perspective is to use surface temperature, the current images show that it is possible to map other important energy parameters such as solar flux, thermal flux, sensible heat flux, latent heat of evaporation and flux into the ground using an advanced airborne sensor. The thermal maps clearly show that the surface temperature phenomenon is highly correlated with the sensible heat flux. Also, a high (positive) correlation exists between vegetation coverage (SAVI) and latent sensible heat flux and a (negative) correlation exists between vegetation coverage and surface temperature, which emphasizes the necessity of vegetation coverage in urban areas as a factor in cooling. Also it can be seen that the urban heat island of both cities is significant in the center of the cities and along asphalt paved roads.

The next stage was to investigate how the urban targets specifically contribute to the UHI phenomenon. For that purpose, the city landscape was quantitatively classified, using HSR information from the VIS-NIR-SWIR region. This approach was used in order to investigate a possible correlation between thermal fluxes and landscape distribution. It is assumed that the spectrum of each pixel is a reflection of its chemistry, and, therefore, the classification, which is chemically based, should be more accurate in providing a correlation with energy fluxes.

For that purpose, the city of Afula was chosen, and six endmembers that best represent the most frequent targets along that city were selected on which to manually run the UNMIX procedure. After applying the spectral-based mapping process and generating the city landscape map, the relationship between each of the heat fluxes and the spatial distribution of the targets within the city was investigated. This was done to determine which land cover most dominantly affects the UHI and to define its magnitude. A scatter plot representing the thermal parameters of the six endmembers is given in Figure 9 which shows that there are positive correlations between vegetation and ground-absorbed latent fluxes and surface temperature. All correlations indicate a potential positive effect on the reduction of the UHI. For example, the surface temperature decreases with an increase in vegetation; ground flux decreases when vegetation increases; sensible heat flux and latent flux increase when vegetation increases and so on. In addition to these findings, a correlation can be seen between absorbed energy and all of the examined targets. An interesting phenomenon was observed in the behavior of dark soils and asphalt. The two targets are mirror-like in their scatter plot presentation, suggesting that asphalt is not a pure dark material, but is probably blanched by dust. Further examination of this issue revealed that the roads had not been paved for four years, and, thus, it is likely that dust had accumulated during that time, resulting in the asphalt being a "non-dark" target. Also it can be seen from the matrix that some roofs have similarities to dark soil, and thus we defined them as "dark roofs." The area designated "dark roofs" represents buildings that have red brick roofs. Roofs were also present in several areas of the "limestone" endmember. However, these roofs are generally flat and covered with a lime-bleaching color, and we identified them as "white roofs." From the thermal behavior of the two types of roof (dark and white), it can be concluded that from a heat (summer) standpoint, it is better to cover the city buildings with white roofs.

The ability to produce a new quantitative overview of the heat island effect using the HSR technology that was demonstrated in the previous example strongly suggests that this tool has great potential in this direction. The current results show that use of an airborne imaging spectrometer is indeed a remarkable step and one that should continue to be used to study the urban heat island effect.

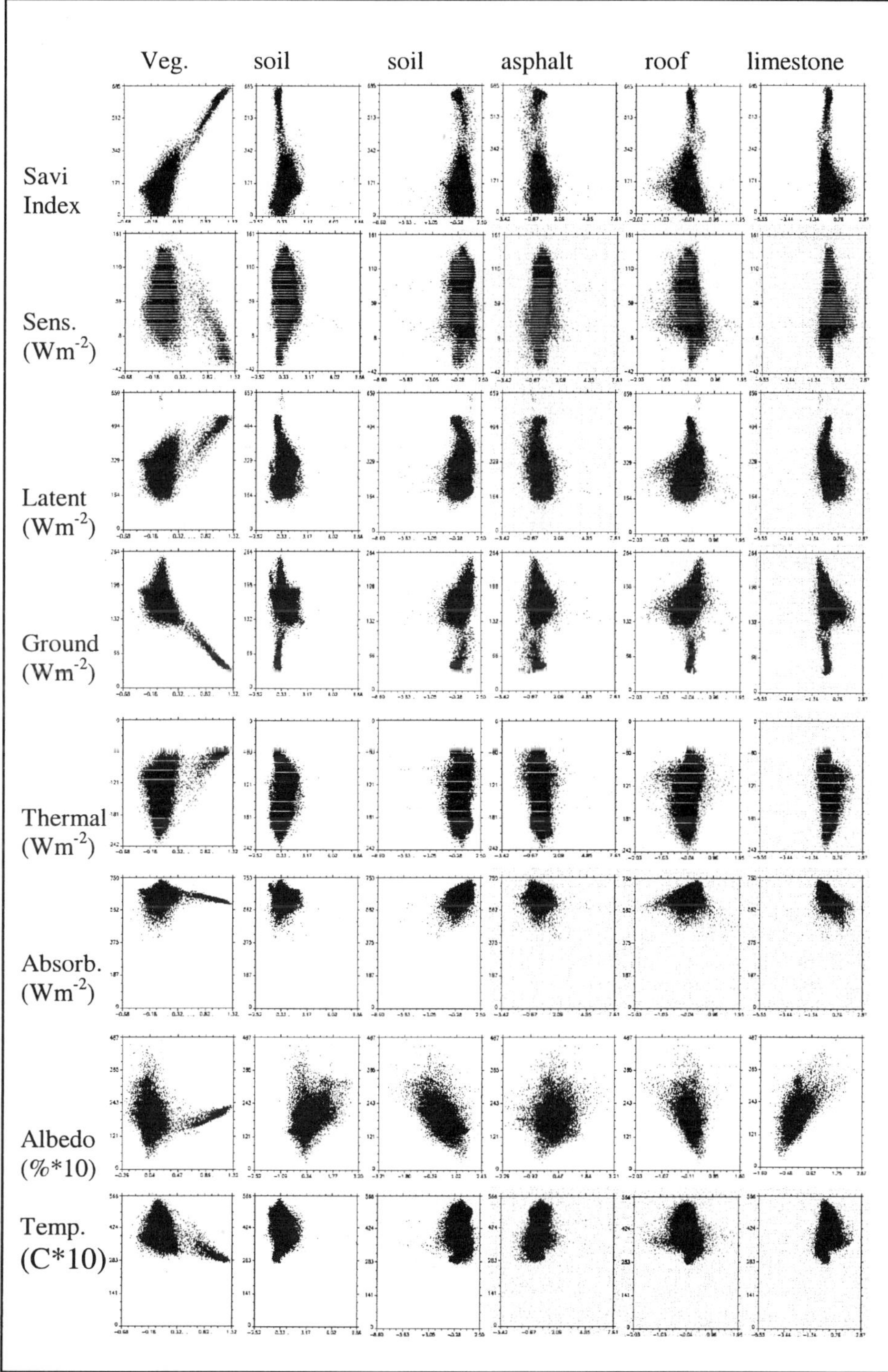


*Figure 9.* A scatter plot showing the heat fluxes (Y) over Afula city, Israel against the endmembers abundances (X) as derived using spectral unmix analysis using the atmospheric corrected DAIS-7915 data.

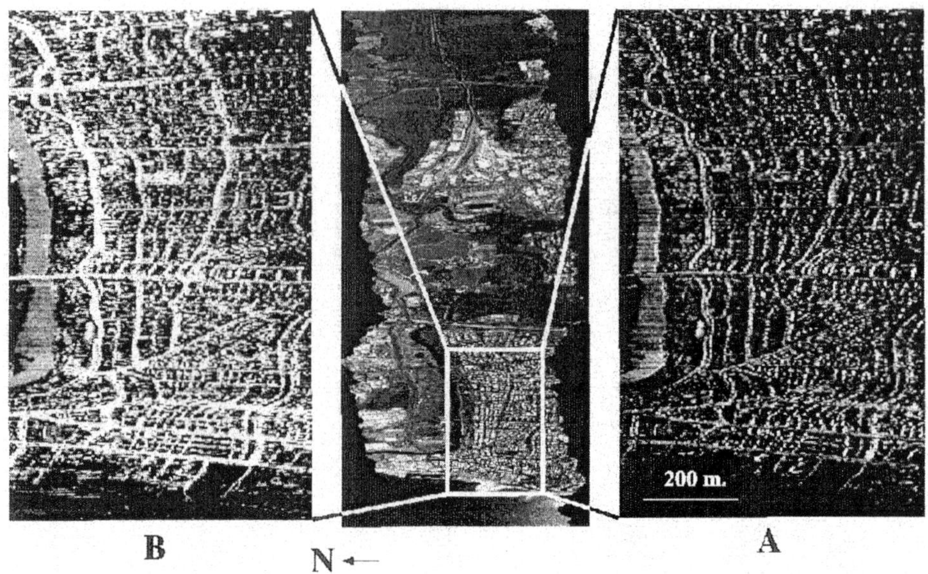



*Figure 10.* A gray scale image (middle) showing the location of the subset images for the shade–asphalt MTMF analysis and their spatial distribution (shade (A) and asphalt (B)).

## 6.4    SPECIAL BENEFIT OF THE HSR OVER URBAN AREAS: ASPHALT AND SHADE

One of the major components adding heat to a city is nonshaded asphalt roads (Ben-Dor and Saaroni 1997). In addition to changing the reflectance properties of targets, shade in the urban environment (created by both buildings and vegetation) reduces energy absorption and heat emission. Nichol (1998) pointed out that visualization of direct and shaded urban surface temperatures is important, and asphalt in this case plays an important role. It is then, very important to quantitatively assess the contribution of each component along with its interaction in the city's environment. As already pointed out, regular air photos or multispectral visible images cannot effectively discriminate between shade and asphalt. Because minor spectral variations occur between the targets, it is very important to examine whether the spectral differences allow for a reasonable and significant discrimination between the two components. Ben-Dor et al. (2000) were able to show that using the two endmembers--asphalt and shade (see Figure 2D; spectra 19 and 29, respectively)--it is possible to map the roads. Figure 10 presents a gray scale image of the selected area for both endmembers, and  Figure 11 presents a color-coded image of both endmembers that shows a possible interaction between the two components (asphalt-red, shade-yellow, partially shaded asphalt-orange). The ability to distinguish between shade and asphalt pixels and their interactions is very important in studying the urban heat island effect. This information can be used to assess heat stress caused by natural sources and may be a tool the city authorities can use to help improve the city environment (e.g., by greening highly illuminated asphalt areas). It is important to note that this finding was made possible only because of the reliable, high-quality spectral information extracted from the HSR sensor. In this respect the HSR technology shows again its great potential for monitoring environmental aspects over urban areas. The significant shadow effect over cities could be problematic in terms of extracting real reflectance data from shaded areas. In this regard, it is important to mask shaded areas from the reflectance analysis to prevent misinterpretation of data.

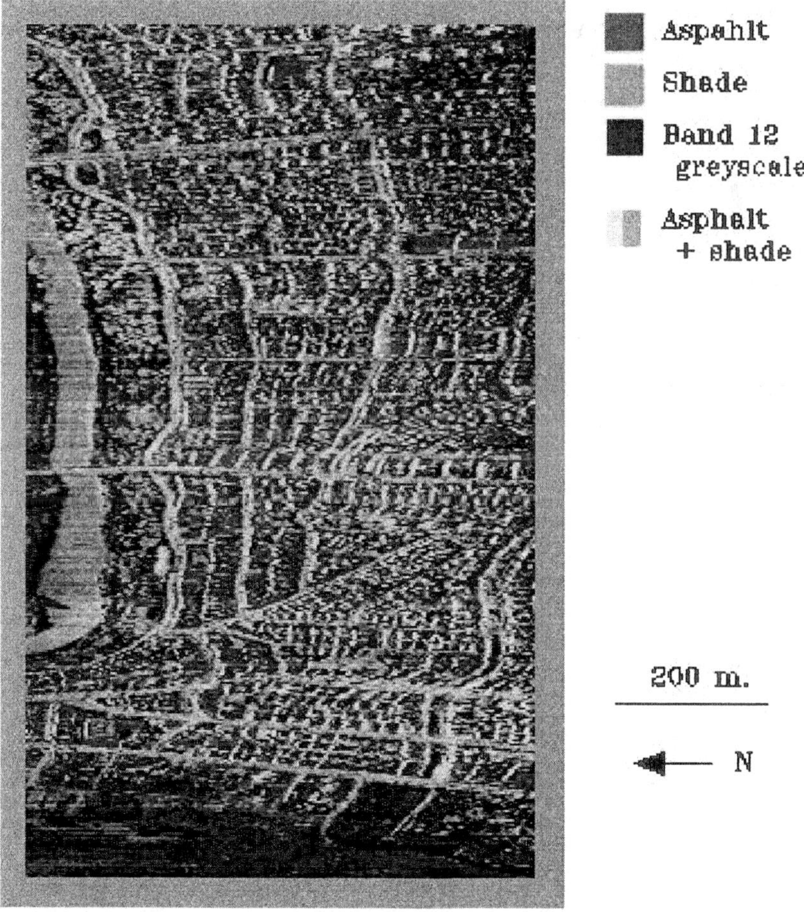


*Figure 11.* A merged RGB image of the asphalt (encoded red), the shade (encoded green) and channel 12 (0.556μm; encoded blue) as extracted from the MTMF analysis (for asphalt and shade) and from the original image).

## 6.5    SUMMARY AND CONCLUSION FOR THE SPECTRAL ANALYSIS

It can be summarized that the characteristics of an urban area can be accurately interpreted using information across the VIS-NIR-SWIR-TIR spectral region. Although the VIS-NIR spectral region is limited, it is likely to be sufficient for classifying most characteristics of an urban environment from a spectral standpoint. This capability suggests that relatively low-cost HSR sensors (e.g., CASI, AISA ) could be relevant and practical tools for assessment of the urban environment via HSR means. The use of the entire VIS-NIR-SWIR spectral region provides additional information regarding urban objects and has been found to add important information regarding the identification of objects having similar albedos (such as white and asbestos roofs), based on their spectral activity in the SWIR region. In this respect, HSR sensors that cover the entire spectral region, such as AVIRIS, HyMAP or MIVIS, could be ultimate tools for use in remote sensing of the urban environment. Another conclusion that can be drawn from the previous section is that HSR data measured throughout the urban environment contains important spectral signatures that can be used for many practical applications. A significant problem for remote sensing technology in the urban environment is the urban terrain, which prevents traditional analysis (reflectance affected by illumination). However, using the spectral information it is possible to distinguish between dark and shaded areas and to extract quantitative flux information regarding one important phenomenon of the human-urban interaction, namely, the urban heat island effect.


## 7    Remote Sensing of the Urban Atmosphere using HSR

The urban atmosphere is a giant chemical reactor in which pollutant gases such as hydrocarbons and oxides of nitrogen and sulfur react under the influence of sunlight to create a variety of products (Seinfeld 1989). Remote sensing using optical sensors demonstrated sufficient feasibility to track atmospheric pollution even with low spectral resolution capability (Wald & Baleynaud 1999). The signal received onboard the air- or space-borne sensors is a mixture of the spectral response of both atmospheric and earth surface materials. Whereas for most earth surface applications the atmospheric signals are "noise," over an urban environment these signals could be important information about air quality. Thus, if environmental aspects are to be studied using the HSR technology, automatic removal of atmospheric attenuation should be done with caution. Ozone, oxygen, water vapor, carbon dioxides and other trace gases (e.g., $N_2O$, $CH_3$) play a dominant role in the sensor's response. Also a significant effect of the aerosol scattering can be emerged across the VIS region. Mapping water vapor on a pixel-by-pixel basis is possible using Gao's method (Gao et al., 1993). The spatial distribution of water vapor can shed light on environmental processes such as pollution from factory chimneys or the possible influence of nearby water bodies on the local city climate, among other things. Another atmospheric component is the aerosol plume, which develops over an urban environment as a result of a variety of sources (e.g., smoke, dust or ice particles). Ben-Dor et al., (1994b) suggested a method for accounting for the aerosol effect on a pixel-by-pixel basis using AVIRIS data. Their method was based on the oxygen absorption peak at 0.76 $\mu$m, which is influenced by the Mie scattering effect. Assuming that oxygen is a well-mixed gas, an "oxygen image" should appear uniform when viewed. Spatial anomalies found in that uniformity could indicate dust-contaminated areas. This technique was applied in several areas (urban and rural) and was found to work well under non-visible cirrus cloud contamination because of their large particle size. Because masking the cirrus clouds was possible (by using the 1.38-$\mu$m channel method; Gao 1999), it can be assumed that it is also possible to mask dust contamination on a pixel-by-pixel basis. Although this technique still requires additional study and examination, it has been strongly demonstrated that the HSR approach has a capability for monitoring the aerosol effect of an urban environment.

Although it has not yet been studied, the high spectral resolution of some HSR sensors (e.g., AVIRIS) enables similar manipulation of $CO_2$, $CO$, $CH_3$, $O_3$ and other gases. Using MIVIS data, Boungiorno et al. (1997) were able to sense and map $SO_2$ over volcanic plumes based on the $SO_2$ absorption spectral feature in the TIR region. Although not yet used in an urban environment, this process shows that it is possible to monitor unusual gases and to map them precisely. The above thoughts are primarily based on theoretical capability as well as on the limited work that has indicated a remarkable potential in this direction. Further study on how to use the HSR technology for assessing air quality parameters over urban areas is required and strongly recommended.


## 8    Recent HSR Urban Applications: A Discussion

A dense and populated landscape is a basic characteristic of every urban environment worldwide. Although one city differs from another in its basic landscape (street pattern, building architecture, topography and geomorphology), some of the main materials that compose urban targets are similar (e.g., asphalt, cement, glass, vegetation). Therefore, reflectance or emittance from different urban environments could have common fingerprints. This fact also emerged during the previous comparison of artificial and urban spectral properties (see section 5.0). Furthermore, it is also expected that the material used in a given city will be correlated with its nearby environment. For example, if the geomorphology of a city's surrounding area is composed of limestone, more limestone products would likely appear throughout that city's landscape. A demonstration of this has been given by Bianchi et al. (1996) and later by Fiumie and Marino (1997) over the city of Rome, using MIVIS (consisting of 102 bands across the 0.43–8.18 micrometer spectral region, with 4-m spatial resolution) data. They showed that it is possible to differentiate between paving materials made with basalt and those made with marble, which are products found in close proximity of Rome. This observation stresses the importance of knowing the composition of the nearby environment of a given city--either by using an HSR tool or by studying existing information--in order to know some of the spectral endmembers to expect in the mixed urban area.

Because the HSR technology still does not offer a great spatial resolution capability, merging the high spectral resolution of the HSR technology with high spatial-resolution information from other sources is important. In this regard, the fusion of HSR and high-resolution digital camera sensors (HyMAP and HRSC-A, respectively) was studied by Lehmann et al. (1998). In their latest work they showed that merging the data obtained over Berlin from both sensors improved the urban recognition (the HyMAP consisted of 126 spectral bands across the 0.440–2.543 $\mu$m region, with 3-m spatial resolution, and the HRSC-A consisted of one-nadir spectral bands at 0.585–0.767 $\mu$m, with 10-cm spatial resolution). Recently Hoerig &Kuehn (2000) also showed that using the same technique (HyMAP data combined with HRSC image) it was possible to detect hydrocarbon materials, such as oil-contaminated soils and plastic materials, throughout the urban environment of Berlin. This work shows that using more advanced sensors in conjunction with HSR data enables better detection of the urban environment than has ever been achieved before.

The low spectral-resolution capabilities of current HSR sensors do not enable fine recognition of objects in the urban environment, causing the problem of mixed pixels to remain unsolved. In this regard, Segel et al. (2000) developed a methodology for reducing this effect. They used DAIS data taken over Dresden City and showed that in spite of generally good classification results, identification problems remained in buildings and sealed open spaces. Their improvement was in the direction of developing an approach for shape-pixel-oriented endmember selection. This was done by using both the refractive and the thermal spectral ranges of the DAIS sensor, which demonstrated that new algorithms and computations are still important for urban HSR analysis.

For the most part, airborne HSR sensors detect the canopy layer of the urban environment and thus are able to recognize the composition of roofs but not of walls. Because walls can play a major role in a city's environment, oblique HSR sensing could be necessary to provide a complete nadir view. In this regard, use of ground HSR sensors (e.g., http://www.spectral-imaging.com) that operate either from high buildings or from the ground should be seriously considered by workers. To the best of our knowledge, this idea has not yet been studied nor have ground sensors been used for this purpose within an urban environment. Mapping the composition of roofs is also an important mission for HSR technology. Asbestos, for example, is a carcinogenic material, which should be removed from structures in the populated environment. Neither regular air photos nor even HSR information across the VIS-NIR region can provide sufficient information for detecting asbestos roofs, because the spectral characteristics of this material across the VIS-NIR region appears to be similar to those of other common roof materials. Only the SWIR spectral region has a unique spectral fingerprint assigned to asbestos, and, hence, only HSR sensors having capabilities in the SWIR region are adequate tools for mapping asbestos roofs over an urban environment. Based on this evidence, one mission to map asbestos roofs over an urban area has shown remarkable results using the MIVIS sensor data integrated with GIS data (Marino et al. 2000). In this mission they were able to distinguish between aluminum, tiles, bitumen and plain concrete using the spectral information provided by the MIVIS. A mission to map asbestos roofs over the city of Milan, Italy, has been launched on the basis of the above-mentioned proven capability (Cavacini, personal communication). Improving the spectral analysis is a key area for which HSR data can be used to extract more information than other common sensors. Ridd et al. (1997) used AVIRIS data (consisting of 224 spectral bands across 0.4–2.5 μm and 20-m spatial resolution) acquired over Pasadena, California, and showed that a combination of the hyperspectral data with neural network analysis improved the classification results as compared with the SPOT (Chung 1989) and TM data (Card 1993) techniques. The HSR technology is also very capable of detecting water status using the VIS-NIR spectral region, and it is important to use this tool for this purpose. Figure 12 gives an example of the spectra of four water bodies extracted form the CASI sensor (48 channels across the VIS-NIR region) acquired over the city of Ashdod, Israel. This city is Israel's biggest harbor. The four bodies of water represent harbor water, close seawater, water from a river that crosses the city and water in an open fresh reservoir.

In all figures it can be seen that different absorption peaks exist, which are assigned for the most part to chlorophyll (0.68 μm) and sediment (0.550μm; 0.800μm). Many more absorption features emerge in the river water, which can possibly be assigned to the chemicals associated with heavy waste contamination. This example stresses the capability of HSR to track the water quality of urban water bodies.

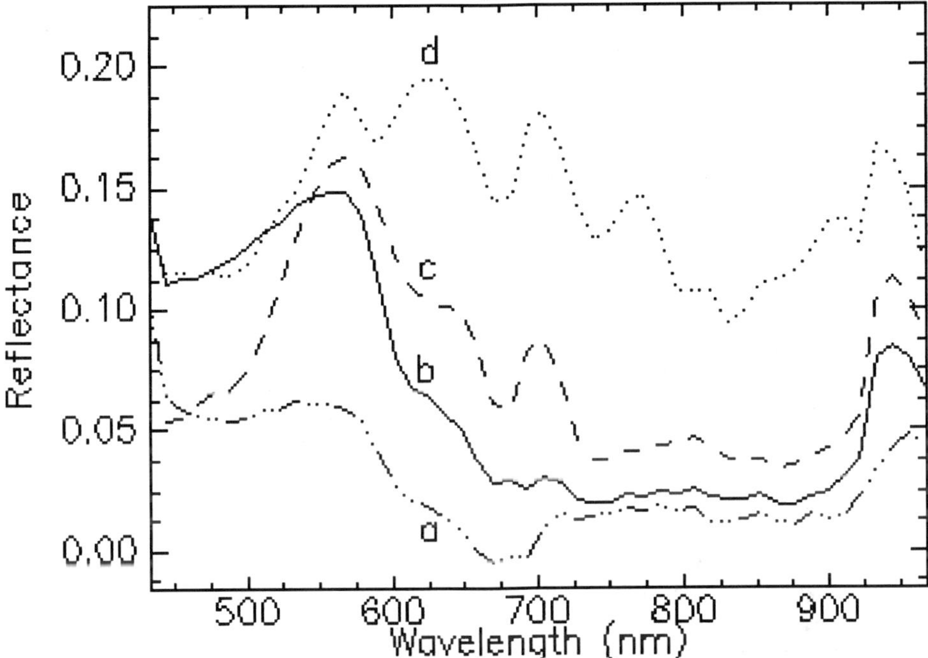


*Figure 12.* Spectra of water bodies over Ashdod city Israel: *a*-open sea water, *b*- harbor sea water, *c*-fresh water (ground water source) *d*- contaminated river.

The 3-dimensional information of a city's environment is also an essential aspect that should be considered by HSR urban workers. If 3-D views of a city were available, it would be possible to correct for the sun's illumination angle in order to derive the natural reflectance response and to determine the chemical composition of vertical targets (e.g., walls), which would serve to greatly improve HSR information. Merging HSR and radar or laser scanner data (able to generate a 3-D model) of an urban environment is an important step toward optimal sensing of this complex environment. In this regard, recent work by Hepner, et al. (1998), who merged AVIRIS with IF-SAR data over the Westwood neighborhood in the city of Los Angeles, demonstrates this capability. Figure 13 shows a 3-D view of the urban area studied by the authors, as generated by using the DTM information of the city (extracted from the IF-SAR interferometer data) and the spectral information (extracted from AVIRIS data).

As has been extensively discussed in this chapter, gathering spectral information is key to the precise study of an urban environment. However, it is interesting to note that based on significant spectral variation, Hepner et al. (1998) were able to extract coarse "footprints" using AVIRIS data collected over Los Angeles. In this work, researchers showed that a study area can be classified using crude radiance information (before atmospheric and sun illumination angle correction processes have been applied). It is, however, obvious that rectifying the data into reflectance information provides new capabilities for classification, as has been demonstrated by of Roessner et al. (1998) over Dresden (using the DAIS sensor) and Ben-Dor et al. (2000) over Tel-Aviv (using the CASI sensor). Furthermore, based on the success of these studies, a recent campaign using a better sensor (HyMap) combined with SAR data was conducted over Dresden in 2000 (Kaufmann H. personal communication). In another study, Zhang et al. (2000), developed a hyperspectral processing and analysis system (HIPAS) for analyzing HSR data in China, where one of the main goals was to map the complex urban areas. They showed that by selecting spectral-based endmembers from the data, it was possible to geographically map the urban environment and to produce a larger and better map of the city than air photos could provide.

Hence, it is widely accepted that the HSR technology has a great capability for providing new and useful information regarding the urban environment. Nevertheless, if HSR data is combined with data acquired simultaneously by other advanced sensors, the HSR approach will be able to attain its maximum potential.

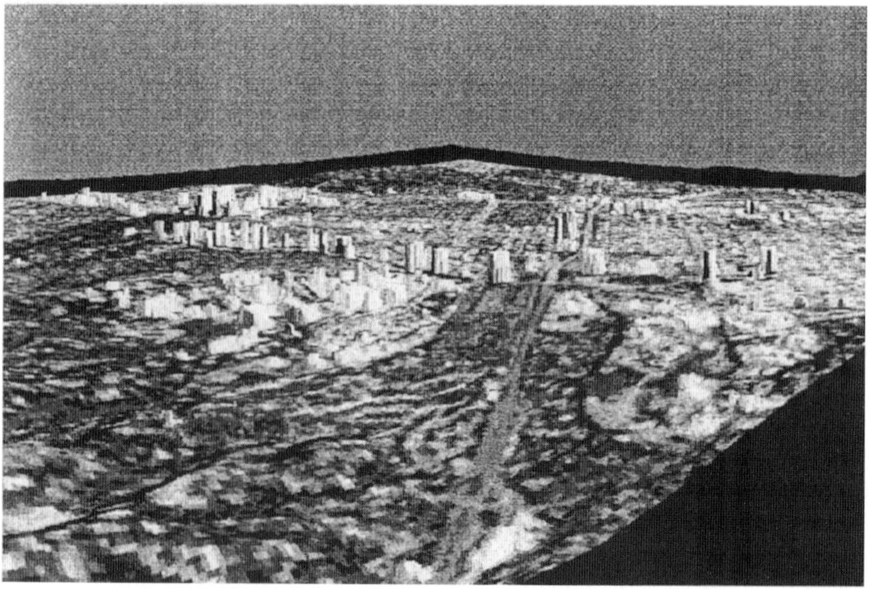

*Figure 13.* Oblique perspective created using interferometric SAR measured structure superimposed on AVIRIS image of an area in Westwood , Los Angeles. The image shows the general natural topographic surface with over 400 larger measured structures (Hepner et al., 1998).

## 9 Recommendation for HSR Utilization over Urban Areas

From the limited experience gained with regard to HSR utilization over the urban environment and with the knowledge accumulated in HSR technology over the past two decades, it is possible to provide herein basic recommendations for potential users on how to use the HSR technology for urban applications:

## 9.1    DATA EVALUATION AND PROCESSING

Obtaining adequate reflectance information is a key factor for HSR technology in general and for urban applications in particular. For that purpose, a preprocessing stage during which the raw data can be evaluated is essential, because if raw data is not properly considered, the spectral information might no longer represent the actual situation on the ground. Artifacts or nonrelated spectral features could be interpreted as terrestrial signals, greatly biasing the analysis. In the complex urban environment, where obtaining finely detailed information could be a major factor in good classification, this preprocess stage is extremely important, and both spectral and spatial preprocessing steps must be carefully considered. These steps include spatial rectification, radiance conversion, band centering and signal-to-noise examinations. The final step in this direction is the atmospheric correction stage, in which reflectance information is obtained. Each of these steps could affect the final spectral signature (and, hence, bias the conclusion), and it is extremely important that each step be carefully taken.

## 9.2    SPECTRAL PREPROCESSING STEPS:

*a) Radiance Conversion:* The raw data, recorded as digital numbers (DNs), should be converted into radiance values using gain and offset parameters that have been obtained in the laboratory prior to the flight. We recommend that users get the raw data (in DNs) and the calibration parameters separately and convert them into radiance manually. This is because in some cases the sensor calibration procedure is not performed adequately and applying the DNs to radiance conversion can in some cases result in problematic calibrated bands. The radiometric calibration should be checked after the DNs have been converted to radiance by simulating the at-sensor response using the MODTRAN code (Berk et al. 1989). Afterward, resampling the model results in the sensor band configuration.

*b) Band-Centering Examination:* A wavelength shift can occur before or during the data acquisition process. Based on well-known spectral signatures of atmospheric gases it is possible to investigate and correct linear shifts. Water vapor, oxygen and carbon dioxides are the gases that are significant in the VIS-NIR-SWIR region for ordinary data sets (water vapor at 0.658 $\mu$m, 0.727 $\mu$m, 0.819 $\mu$m, 0.910 $\mu$m, 0.934, 1.34 $\mu$m] and 1.89 $\mu$m for oxygen at 0.687 $\mu$m and 0.692 $\mu$m, 0.761 $\mu$m and 1.34 $\mu$m; for carbon d-oxide at 1.11 $\mu$m], 2.015 $\mu$m and 2.005 $\mu$m). In areas of highly contaminated gases, other gases can also be taken into consideration. Because different detectors are used for the VIS-NIR and SWIR regions, it is important to examine all the above features and adjust the spectral range accordingly.

*c) Signal-to-Noise Examination:* A signal-to-noise ratio (SNR) examination of each sensor is essential for judging the information obtained by the sensor. It is, however, important to mention that it impossible to inspect the real SNR values of the sensor and hence an empirical method, widely used in other HSR application is suggested. This examination is based on Kaufmann et al.'s (1991) suggestion, which can practically be run on most HSR data. Usually it is recommended that researchers use two uniformly dark and bright targets and examine the results. Figure 14 presents the SNR spectra of two extreme albedo targets taken from CASI data obtained over Tel-Aviv City (sand: bright; asphalt: dark) and Ashdod City, respectively, during two different campaigns, two years apart. In general the SNR values range from 10 to 50 in the bright target and from 10 to 15 in the dark target. These values are in good agreement with typical SNR values obtained by other IS sensors using a similar methodology (e.g., Ben-Dor 1994a). Note, however, that between 0.62 and 0.69 $\mu$ m, in the bright target taken from Tel-Aviv data, the SNR values are rather high. Ben-Dor et al. (2000) determined this jump to be a result of selecting an incorrect aperture for this mission and added precautionary notes regarding processing the spectral data across this particular spectral region and selecting apertures for future HSR missions over urban areas. Low SNR values in the Ashdod data set across the blue spectral region suggest that the first channels were not sensitive enough for the HSR mission and should be omitted from the thematic analysis.

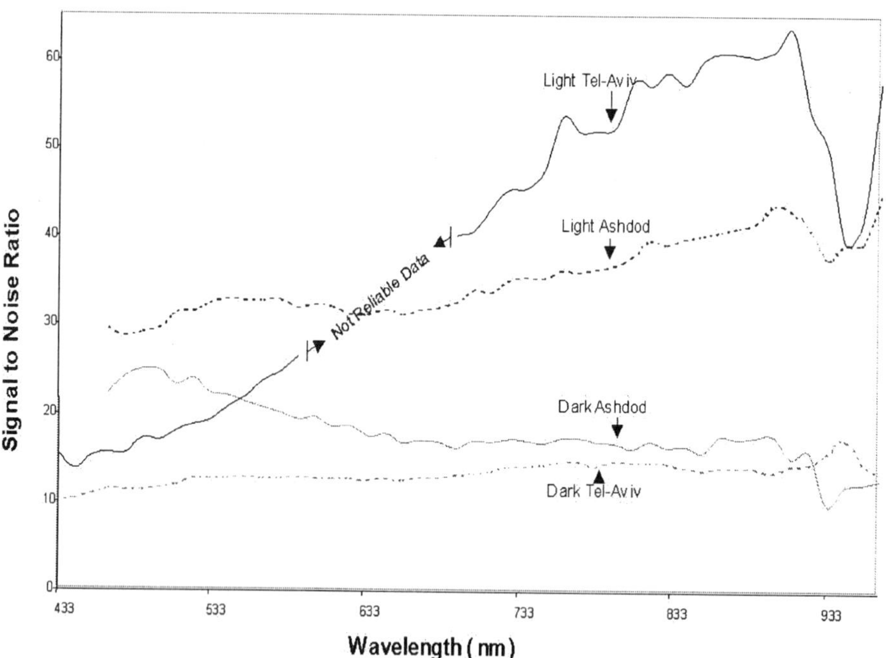


*Figure 14.* The spectra of the signal to noise ratio of two targets over the city of Tel-Aviv and Ashdod using the CASI data as generated using Kaufmann (1991).

## 9.3    SPATIAL PRE-PROCESSING

Unlike in the HSR of water bodies, vegetation, rocks or soils, in the urban environment, spatial resolution is important. We assume that a 2- to 8-meter pixel size is sufficient where lower resolution is being considered, depending on the question at issue. It should be remembered that use of high spectral and spatial resolution techniques can result in huge data sets that can inhibit smooth processing. Image deformation resulting from either systematic or nonsystematic geometric interference is also a significant problem in urban HSR applications, where well-defined patterns can be lost, preventing effective spatial recognition of the area examined. Several methods are known to be able to correct such interferences, the most effective being the use of combined GPS, INS information that is acquired onboard an aircraft simultaneously with on-the-ground data acquisition. However, in many cases, this capability is absent and, as a result, HSR data might be considered useless. Under such circumstances, registration methods should be applied. Ben-Dor et al. (2000) described a "cross-correlogram" technique (Press et al. 1988) for correcting significant role interferences that disable the thematic analysis of CASI data acquired over urban environment. This technique employed a linear regression analysis between pixel values in close rows. The procedure is repeated after shifting given rows toward each other in pixel-by-pixel increments. The shifting position that yields the highest regression scores between close rows expresses the role displacement and, accordingly, corrects its distortion without affecting the original DN values. Another technique is to use ground GPS data from several well-defined areas and applying available warping algorithm techniques to the data. As in urban areas, where many well-defined GCPs can be extracted, this capability is highly recommended. In this regard it should be remembered that the spectral information that results after such corrections have taken place might be changed because of the resampling process or the rectification method used, and therefore, whether or not to use this technique should be considered on a case-by-case basis. In summary, it can be said that when applying the HSR technique for urban applications, significant attention should be given to the spatial as well as the spectral domain.


## 9.4    ATMOSPHERIC CORRECTION

Accurate removal of the atmospheric attenuation is a critical step for precise understanding of the spectral features of terrestrial objects (e.g., Gao et al. 1993). A priori selection of a particular method for a particular data set is, in most cases, difficult (Farrand et al. 1994, Ben-Dor and Levin 1998). In this regard, and because the urban environment is recognized as a complex environment with great spectral variation, the atmospheric correction technique is very important. Judging from accumulated experience (Ben-Dor et al. 1999, Ben-Dor et al, 2000), it is essential to examine several techniques in order to select the best method for correcting data. The available methods are scene-based methods (the Internal Apparent Reflectance Ratio [IARR], Kruse 1988, and the Empirical Line [EL] Roberts et al. 1985) and model-based methods (MODTRAN [Berk et al. 1989], ATREM [Gao et al. 1993] or ATCOR [Richter 1998]). Also hybrid methods that combine scene- and model-based methods are sufficient. The judging parameters should be a comparison of the field spectra of several ground targets with their corrected images. It should be pointed out that the geometry of an urban environment might cause problems in the precise estimation of the direct reflectance information and, therefore, atmospheric rectification should be done before geometric correction. On the other hand, shadowing and shadow effects are significant throughout urban environments, and it is necessary to account for them. In ATCOR-4 the topographical effects of a varying terrain can be taken into account using DTM information. Determining the urban terrain is possible using ordinary methods, such as stereo air photos, or advanced techniques, such as laser altimeter or radar interferometers. Apparently, and to the best of our knowledge, this issue has not yet received the attention of workers, and we strongly recommend further study in this direction for precise rectification of atmospheric interference. In this step, it is mostly important to validate the results against field measurements or to use the spectral library provided in the previous section.


## 9.5    MIXED PIXEL PROBLEM

Because urban environments are mixed and complex areas from both a spectral and a spatial point of view, application of quantitative classification approaches that use spectral information to map a city can result in a mixed pixel problem. As used here, mixed pixel problem refers to the fact that a given pixel might be composed of several pure components and the resulting pixel information is a combination of the spectral responses of these individual pure materials. The larger the pixel, the more complex the problem. In this regard, Segl et al. (2000) developed a multi-technique approach that combines linear spectral unmixing and spectral classification for a complete inventory of main urban surface cover types. They concluded that in the urban environment the mixed pixel problem requires knowledge-based unmixing techniques that limit the number of possible endmember combinations. It is therefore recommended that better shape-based differentiation of various categories of urban targets might apply. The merging of HSR with other high spatial-resolution data, as described earlier in this chapter, is a step toward reducing the mixed pixel problem. Also learning the ideal pixel sizes of a given area and having better knowledge of the urban targets' spectral responses are other steps in this direction. New classification algorithms and further consideration of this problem by potential workers are also important factors for solving the mixed-pixel problem over urban areas.


## 10    General Summary

HSR technology has a remarkable potential for urban applications and has a strong foundation for extracting quantitative information about the surface coverage of such a complex area. In addition this approach has a potential for identifying atmospheric parameters such as water vapor or aerosol distribution. A detailed spectral library of urban targets is essential for HSR analysis. The complex urban areas, which are made from a variety of different materials, provide diverse spectral characteristics across the VIS-NIR-SWIR regions, which enable precise recognition of the urban environment using HSR sensors. It was shown that even the so-called featureless spectral region across the VIS-NIR has significant spectral information in this regard. Although it is not yet used as a routine method for remote sensing of the urban environment, the HSR technology is a promising tool for mapping the city from a chemical/physical standpoint. In recent studies it has been shown that this technology provides new information never before extracted from other remote sensing sensors. In this regard, surface targets, energy budgets and atmospheric conditions can be determined from HSR data. Although as a stand-alone tool the HSR approach can provide useful information about the urban environment, in some cases this information might be biased. Combining HSR data with data from other leading edge technologies, such as high-resolution digital cameras and radar interferometers, showed remarkable achievement in this direction. Data from ground HSR cameras, air- or space-borne HSR sensors and laser scanners can also be merged for urban applications. Low-cost HSR technologies, either air- or space-borne, that can provide data on a temporal basis, are also key factors for future urban applications. Field HSR cameras and oblique views of the urban landscape could detect additional information about an urban area. In addition to the requirement of advanced technology in the direction of HSR, new algorithms for extracting pure endmembers and minimizing the mixed-pixel problem are essential for studying the urban environment. It is expected that an orbit-based sensor will be the next HSR breakthrough for monitoring cities from far distances. Such a sensor would acquire data routinely and systematically, while creating an archive for future temporal analysis. The ASTER sensor on the Terra spacecraft and the worldwide urban monitoring projects that rely on it are a good example of this type of existing technology.

In summary it can be concluded that urban objects do consist of spectral features that can be used for detecting specific targets using a spectral-based analysis, and the chemical variety found throughout a city can be identified using the HSR methodology. Spectral libraries of artificial materials show that it is possible to create spectral libraries even without measuring the reflectance properties of the targets in situ. Recent studies showed that the HSR technique is very useful and powerful and thus, it is strongly recommended that further efforts in this direction be carried out. The HSR technology is a feasible technique for monitoring and detecting environmental changes in water quality, vegetation status, soils, pavement conditions, roofs, wall conditions, material composition in general and atmospheric conditions. Spectral information proved itself to be a reliable tool for precise mapping of chemical/physical properties in the complex urban environment, and thus the HSR approach is recommended for use in further future urban applications.

# CHAPTER 10

# IMAGING SPECTROMETRY IN THE THERMAL INFRARED

Michael J. ABRAMS[α], Simon HOOK[α] & Mark C. ABRAMS[β]

[α] NASA Jet Propulsion Laboratory, Pasadena (CA), U.S.A.

[β] ITT Industries, Ft. Wayne (IN), U.S.A.

## 1    Introduction

If we examine mineral spectra at different wavelengths, we find diagnostic features that can provide a way to discriminate or identify them remotely. The interaction of electromagnetic radiation with the atoms and molecules that make up a rock or mineral produce vibrational-rotational or electronic processes, creating spectral features. In the visible and near infrared, electronic processes dominate. Beyond 2.5 μm fundamental vibrational processes dominate. Between the two (0.5-2.5 μm) there is an overlap of features.

In the 0.4-2.5 μm wavelength region, iron-, hydroxyl-, sulfate-, water- and carbonate-bearing minerals display spectral features (Hunt, 1980). Silicate minerals, in contrast, have their spectral features in the thermal infrared (TIR). Data from the thermal infrared complement data from the visible to short wave infrared: the TIR is used to determine composition, and the VIS-SWIR is used to determine alteration products.

Historically, geological applications have made much more extensive use of the VIS-SWIR wavelength than the TIR. This is because of the ready availability of appropriate multispectral satellite data, such as Landsat Thematic Mapper data; and access to software designed for processing and analyzing these data. The broadening of geologic applications to using TIR data has been recently expanded with the launch in December 1999 of the Advanced Spaceborne Thermal Emission and Reflectance Radiometer (ASTER). This multispectral scanner is the first high spatial resolution satellite system, providing global coverage. Hyperspectral TIR satellite instruments are currently only operated by the US Defense community. However, several airborne scanners are now operating, and may well lead the way to a future operational spaceborne system. The following chapter will discuss the applications of hyperspectral TIR data for geological applications. The second section will review the theory of TIR spectroscopy. The third part will examine current TIR airborne systems. The fourth part will look at existing satellite systems. The final section will look into future prospects for hyperspectral TIR instruments.



*F.D. van der Meer and S.M. de Jong (eds.), Imaging Spectrometry, 283–306.*
© 2006 *Springer. Printed in the Netherlands.*

## 2    Theory

### 2.1    THERMAL EMISSION

The kinetic energy of random motion of particles of matter is known as heat energy. As particles move they collide with each other, causing changes in vibrational, orbital or rotational motions. This change results in the emission of electromagnetic radiation, and thus heat energy is changed to radiant energy. A blackbody is a material that transforms heat energy to radiant energy at the maximum rate. The spectral exitance of a blackbody was derived by Planck and is given by the following formula

$$M_\lambda = \frac{C_1}{\lambda^5[\exp(C_2 / \lambda T) - 1]} \tag{1}$$

where $\lambda$ is the wavelength, T the absolute temperature, and $C_1$ and $C_2$ are the radiation constants. With increasing temperature, the spectral exitance of a blackbody increases. The wavelength of maximum exitance at a given temperature is given by Wien's displacement law:

$$\lambda = C/T \tag{2}$$

where $C = 2.898 \times 10^{-3} m.K$. Finally, the relation between spectral radiance and spectral exitance is:

$$L_\lambda = M_\lambda/\pi \quad W\ m^{-3}\ Sr^{-1} \tag{3}$$

where L is radiance, and M is exitance.

### 2.2    SPECTRAL EMISSIVITY

Materials are not perfect blackbodies. Instead they emit radiation dependent on their physio-chemical makeup, absorbing some radiation through molecular interactions. The spectral emissivity of a material is the ratio of the spectral radiance to that of a blackbody at the same temperature:

$$\varepsilon_\lambda = L_\lambda\ (material)/L_\lambda\ (blackbody) \tag{4}$$

The spectral emissivity of most materials is between 0.9 and 1.0 in the 0.8 to 13 $\mu$m wavelength region. Some minerals, however, have much lower emissivities, and thus they are the easiest to detect and identify.

*Emissivity of rocks and minerals*

The first person to publish thermal spectra of rocks and minerals was Lyon (1965). This was followed by a series of papers by Hunt & Salisbury (1974, 1975, 1976), and by a book on infrared spectra by Salisbury et al. (1991). All silicates exhibit their most intense feature near 10 $\mu$m; this region is referred to as the reststrahlen band or Si-O stretching region. Figure 1 shows sample spectra of some silicate minerals in the 8 to 14 $\mu$m region. The emissivity minimum falls at shorter wavelengths for framework silicates (feldspar and quartz), and then longer wavelengths for sheet, chain and isolated $SiO_4$ tetrahedra minerals (muscovite, augite, hornblende, olivine). Near 11 $\mu$m is found the only other feature occurring in silicate minerals: an absorption due to H-O-Al bond found in aluminum bearing clay minerals.

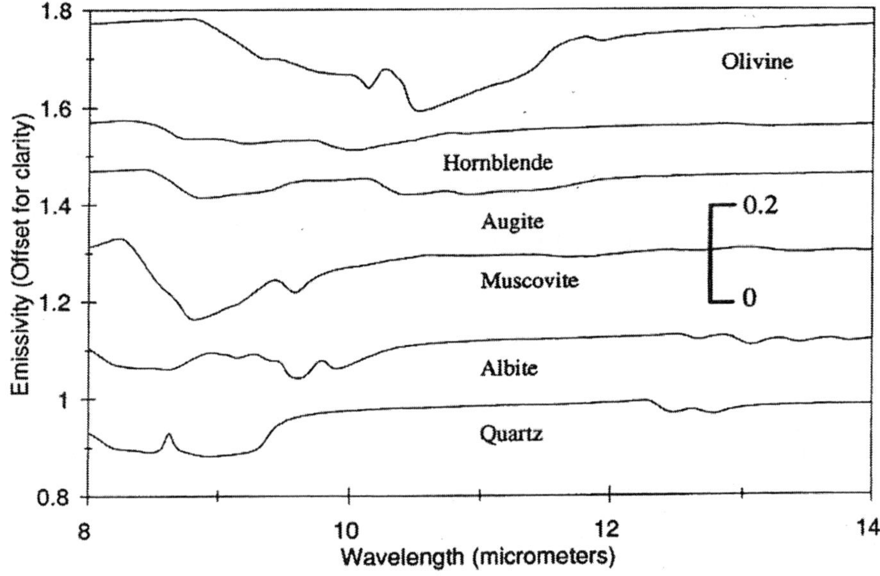

Figure 1. Sample spectra of some silicate minerals in the 8 to 14 μm region.

Nonsilicate minerals also have absorption features occurring in the 8 to 14 μm region. Spectra of selected carbonates are illustrated in Figure 2.

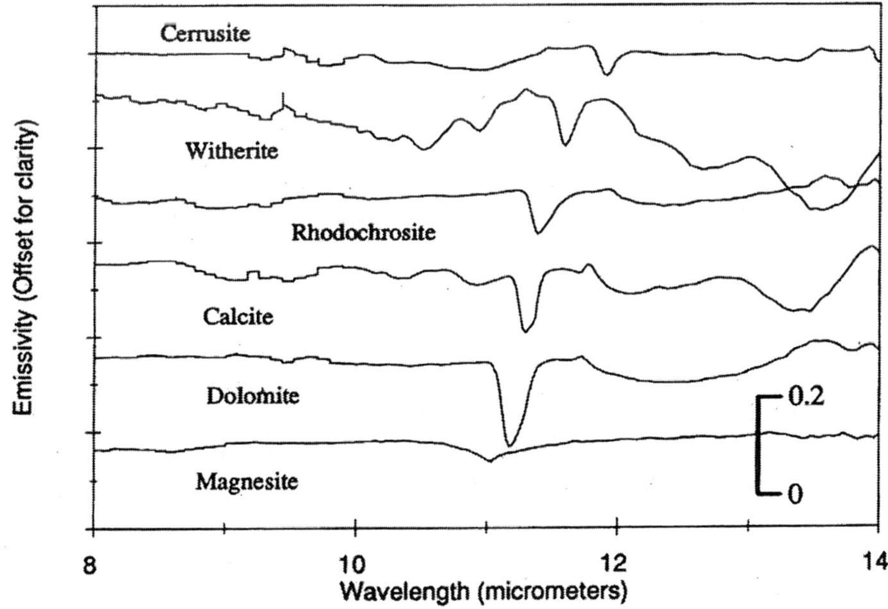

Figure 2. Spectra of selected carbonates in the 8 to 14 μm region.

Other minerals, such as sulfates, phosphates, oxides and hydroxides, also have distinct features in this region. These are minerals that typically occur in metamorphic and sedimentary rocks, providing a valuable remote sensing tool for discriminating and identifying these rock types in remote sensing TIR data.

*Figure 3.* Color composite image of the East Tintic Mountains, Utah based on principal components transformation of TIR data. A, B=quartzite; C, D, E=interbedded sandstone, limestone, quartzite, shale, dolomite and chert; F=silicified rocks; G=mine areas; H=Dragon mine; I=argillized rocks; J=quartz latite and quartz monzonite; K= latite and monzonite; L=vegetated areas; M=calcitic quartz latite; N, O=carbonate rocks; P=hydrothermal dolomite; Q=mine tailings and ponds.

# 3    Current TIR airborne systems

## 3.1    EARLY SCANNERS

The first use of multispectral TIR images for geologic mapping was by Kahle &
Rowan (1980). They obtained data from a now defunct Bendix 24-channel scanner,
that included six channels in the 8 to 13 μm wavelength region. The area they studied,
the East Tintic Mountains site in Utah, consists of complexly folded and faulted
Paleozoic sedimentary rocks that are partly covered by Tertiary volcanic rocks. Clastic
rocks are at the bottom of the section, whereas carbonates comprise the upper part of
the Paleozoic rocks. The Tertiary volcanic rocks are tuffs, flows, and agglomerates of
quartz latitic and latitic composition. Emplacement of monzonite and quartz monzonite
porphyries resulted in several types of altered rocks. The data were processed by
principal component transformations. In the resulting image (Figure 3), the following
rock types were separable: quartzites; interbedded sandstone, limestone, quartzite,
shale, dolomite; silicified rocks; mine areas; argillized rocks; quartz latite and quartz
monzonite; latite and monzonite; calcic quartz latite; carbonate rocks; and
hydrothermal dolomite. This can be compared with a geologic map of the area (Figure
4). In general, the red component of the image was related to the presence of quartz-
bearing rocks; green represented generally spectrally flat materials such as carbonate;
blue correlates with latite and monzonite without quartz as a major constituent.
Examination of image derived spectra suggested that the differences among the rock
types were in the depth of the spectral features, with the position of spectral absorption
bands being of somewhat less importance.

## 3.2    TIMS

Based on the results of this study, NASA was convinced to commission the
development and construction of a new scanner, the Thermal Infrared Multispectral
Scanner (TIMS). TIMS is a whisk-broom scanner, with a rotating mirror. Its six
channels are located between 8 and 12 μm, with a band width of about 0.5 μm. The
instantaneous field of view of 2.5 mrad and total field of view of 76 degrees produces
pixel sizes of 5 to 50 meters, depending on the altitude of the platform (Palluconi &
Meeks, 1985). This scanner has been the work horse of multispectral thermal imaging
for 15 years, until its replacement by a more advanced scanner in NASA's fleet of
operational instruments. There have been numerous studies published on the use of
TIMS for geologic mapping. Here, we will highlight a few of them to demonstrate
what can be accomplished using multispectral TIR imagery. One of the earliest studies
was by Gillespie et al. (1984) in Death Valley California. The area they studied is on
the west side of Death Valley, and included bedrock exposures of the Panamint Mtns.,
alluvial fans derived from the bedrock, and playa deposits. Dominant rock types are
sedimentary quartzite, shale, limestone and dolomite. Overlying the sediments are
Miocene volcanics of basaltic to rhyolitic composition. The daytime TIMS data were
processed using a decorrelation stretch algorithm (Soha & Schwartz, 1978), that
suppresses temperature variations and enhances emissivity features. Channels 1, 3 and

5 were displayed in red, green and blue, respectively in a color composite picture, that was then photo-interpreted (Figure 5).

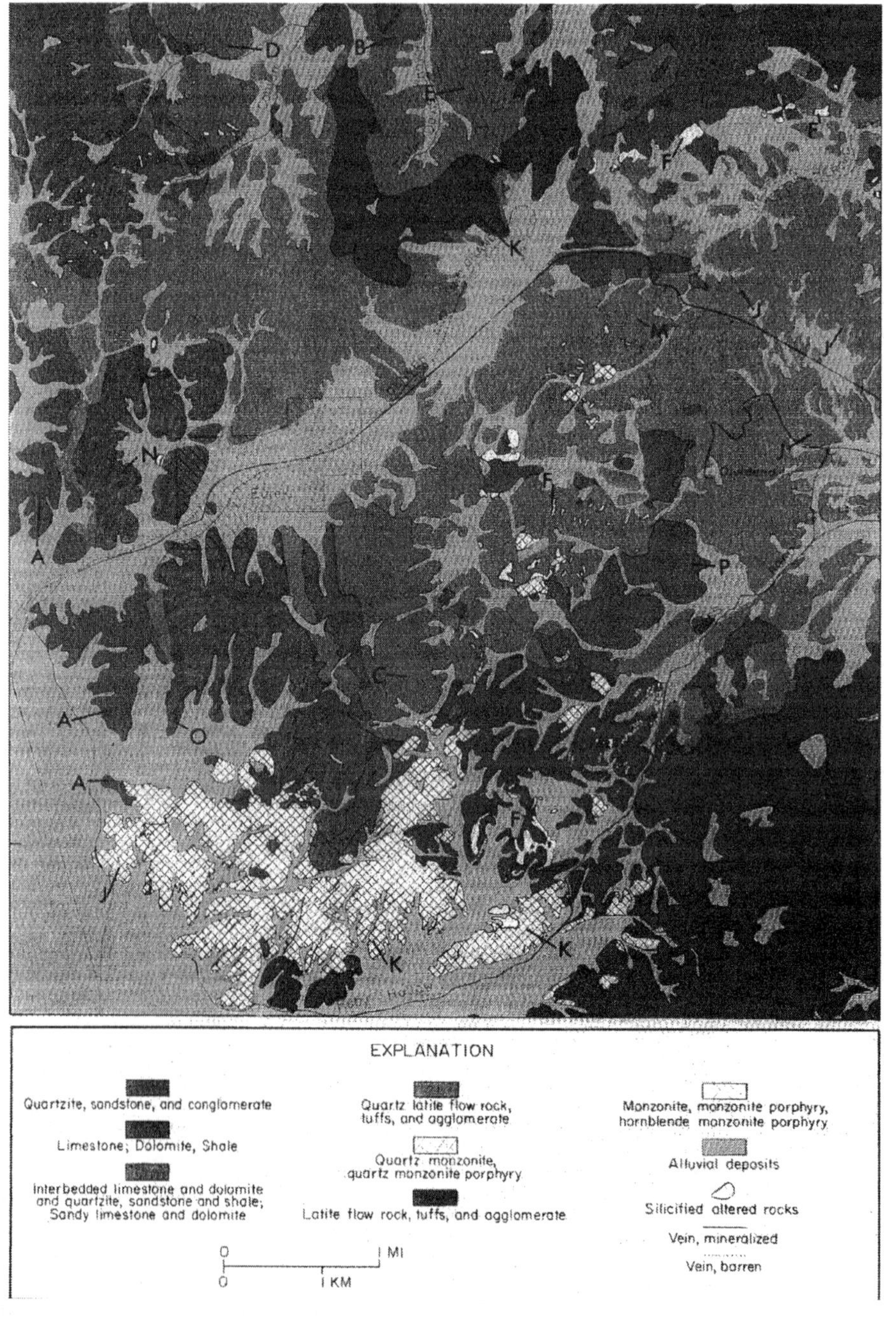


*Figure 4*. Generalized geologic map of the central part of the East Tintic Mountains, Utah (after Morris, 1964).

*Figure 5.* Enhanced TIMS decorrelation stretch image of Death Valley using bands 1, 3 and 5 displayed in red, green and blue.

Several units were easily differentiable by texture and color. The former differences related to topography separate bedrock from alluvial fans and lake deposits. Image color, on the other hand, was related to composition. An interpretation of the TIMS image is shown in Figure 6. Carbonate rocks appeared green, and quartzites are deep red. Other sedimentary rocks, such as shale, appear purple as do volcanic rocks. Basalt showed up in blue. Alluvial fans showed up in the same colors as their bedrock sources if they were from a unique source. The colors of fans of mixed lithologies was often a simple mixture of the colors of the different sources: for instance, mixing "red" quartzite and "purple" shale produces a fan magenta in color. Differential erosion of gravels within a fan contributes to compositional change over time. This formed the basis for relative age dating of fan units based on interpretation of the TIMS image. Older fans made of quartzite, for example, develop desert varnish coatings on the clasts. This has the effect of reducing the emissivity minimum (darker red colors).

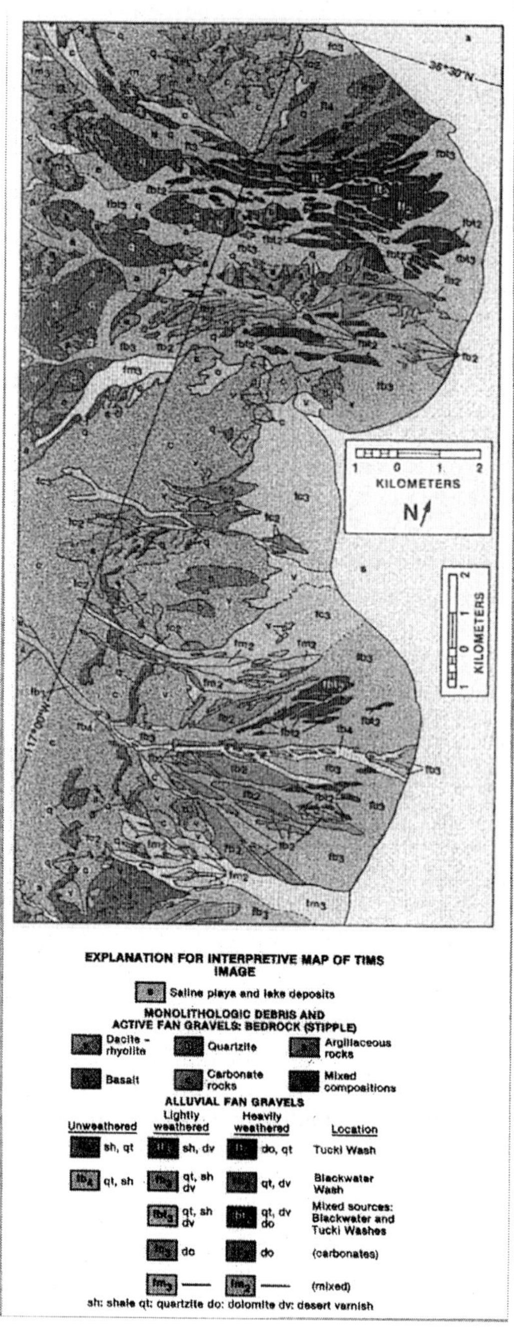



*Figure 6.* Interpretation of the TIMS image (Figure 5), showing lithologic units and fan units discriminated by composition and relative age.

In a related study published at the same time, Kahle et al. (1984) studied the same area as Gillespie et al. (1984) using an active $CO_2$ laser system operated from an aircraft. The Laser Absorption Spectrometer (LAS) was developed for remote measurement of atmospheric gases (Shumate et al., 1981). The system uses two $CO_2$ infrared lasers transmitter/receiver systems that measure the two-way transmittance between the aircraft and the ground. The aircraft was flown over the fans studied by Gillespie et al. (1984) to study the correlation between the reflectivity and the emissivity of the surface. Two bands were selected at 9.2 and 10.3 µm to achieve good geologic discrimination. The narrowness of the bands used was the first attempt to investigate the feasibility of remote sensing in the TIR with better than the broad bands of TIMS. This potentially opened up the window for hyperspectral TIR instruments.

The results of the data flight using LAS were compared to TIMS data for the same line. Kahle et al. (1984) plotted the ratio of TIMS bands 3 and 5, and the ratio of the two $CO_2$ lasers. This is shown in Figure 7. The colored bar down the middle shows the rock types interpreted and mapped by Gillespie et al. (1984). The two plots show a high negative correlation (r = -.90). According to Kirchoff's Law, the emissivity should equal one minus the reflectivity. This is in fact what was seen. Another study by Abrams et al. (1991) demonstrated the combined use of reflected visible-short wave infrared data with TIMS data to map lave flows on the flank of Mauna Loa volcano, Hawaii. The reflected data were acquired by NASA's NS001 scanner, a Thematic Mapper simulator scanner. The NS001 and TIMS data were combined using principal components analysis; emittance and reflectance properties of the rocks were correlated to some degree, as would be expected. Nevertheless, the combined data sets allowed better separation than either data set used alone. The resulting combined image, and an interpretation map are shown in Figures 8 and 9. The flows are differentiated by color changes, which are systematic with relative ages. The existence of weathering systematics implies that it is possible to estimate flow age from remotely sensed data, providing that progression of image colors has already been calibrated for a given region. Important influences on the thermal spectra appear to be surface texture, physical and chemical degradation of the glassy crusts (spalling, devitrification, alteration) accretion of silica-rich coatings and vegetation growth. Important influences on the reflectance spectra appear to be development of iron oxide minerals from weathering, vegetation growth, and surface texture. These effects can occur in characteristic sequences and at different rates, so that color pictures created from visible to thermal infrared data depict flows at different stages of their development in different colors. Hook et al. (1994) used TIMS data to map Proterozoic schists, granites and gneisses in an area of metamorphosed, structurally complex, igneous and sedimentary rocks in the Piute Mountains, California. The TIMS data were calibrated and atmospherically corrected, and emissivity variations in the form of alpha residuals were extracted (Hook et al., 1992), from which color composite images were made. There was an excellent correlation between the units revealed in the color composite image and the lithologic units mapped in the eastern side of the area. Improvements to the existing map included re-assignment of granodioritic gneiss to amphibolite; the presence of a swarm of mafic dikes was revealed. Color variations in granitoid plutons correlated with compositional variations previously unreported. In the western part of the area, the images permitted the extremely heterogeneous schists, gneisses and granites to be easily mapped. A color composite image of this area, and the interpreted geologic map, are shown in Figures 10 and 11. The Barrel Springs Pluton has been previously subdivided into areas of quartz syenite, mafic syenite, K-feldspar porphyry, and porphyry dikes (Karlstrom et al., 1993). On the TIMS image it is possible to separate the K-feldspar porphyry from the quartz syenite. In another part of the area, there is a yellow and turquoise striped unit, previously mapped as granite augen-gneiss. Analysis of the image spectra suggests that the yellow areas are muscovite-rich pelites, and the turquoise areas are quartz and biotite rich. Subsequent field work verified the image interpretation.

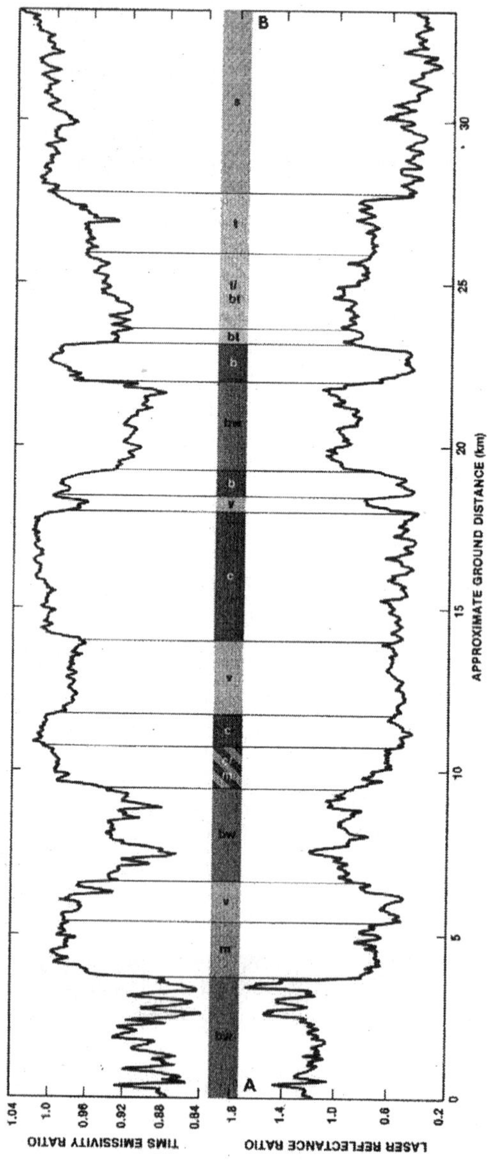



*Figure 7.* The ratio of TIMS bands 3 and 5 and the ratio of 2 LAS lasers along the Death Valley flightline.

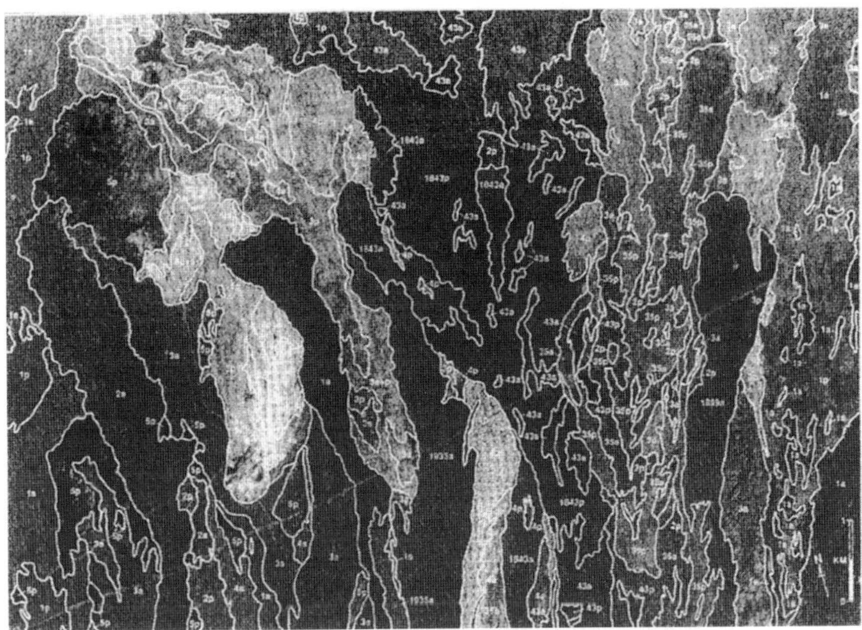

*Figure 8.* Combined NS001 and TIMS data processed using a principal components transformation for lava flows on Mauna Loa, Hawaii. The overlay shows interpreted flow contacts, with relative age assignments. p=pahoeoe; a=aa; 35a(p)=1935a(p); 43a(p)=1843a(p); I=0.2-0.5 ka; II=0.5-1.5 ka; III=1.5-4.0 ka; IV=4-8 ka; V=>8 ka.

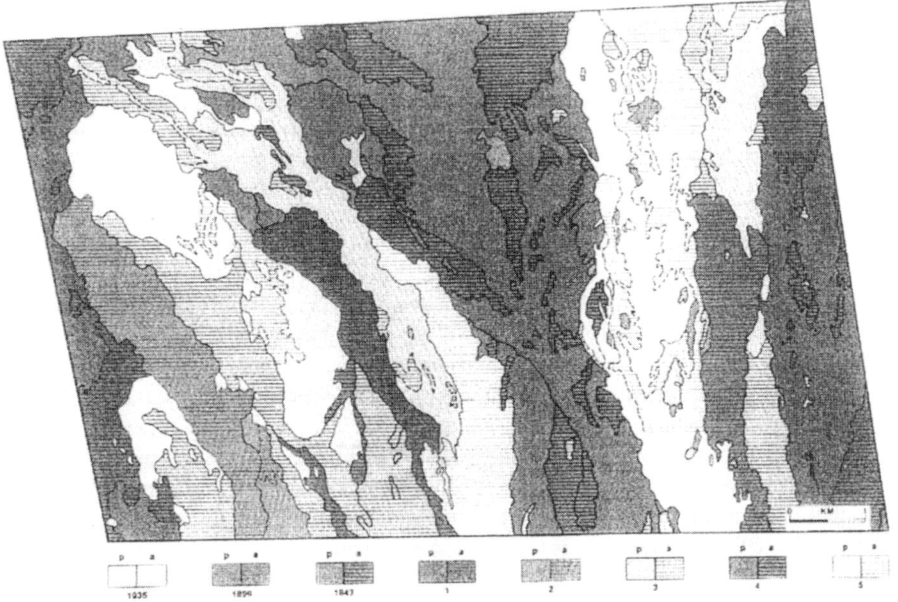

*Figure 9.* Geologic map of the Mauna Loa test site. Relative ages as indicated in the caption for Figure 8.

*Figure 10.* Color composite of the western part of the Piute Mountains, CA created by displaying TIMS alpha residual data. Labels A-E indicate areas used to extract image spectra; T is a river terrace.

## 3.3     ATLAS, AMSS, AAS, MIVIS

The Airborne Terrestrial Applications Sensor (ATLAS) was developed by NASA's Stennis Space Center. It has 15 channels, six of which are in the TIR and are similar to TIMS bands. The Airborne Multispectral Scanner (AMSS) was developed by Geoscan Pty Ltd., an Australian company. The AMSS provides 46 channels of data, of which 24 can be recorded at any time. It also has 6 bands in the TIR, similar to ATLAS.    The airborne ASTER simulator (AAS) was built by Geophysical Environmental Research for the Japan Resources Observations System Organization. The AAS has 24 channels, of which 17 are in the thermal infrared. A comparison with TIMS data is provided by Kanari et al. (1992). Since this is a one-of-a-kind scanner, very few people have had access to the data; quality, repeatability, etc. are largely unknown.

The Multispectral Infrared and Visible Imaging Spectrometer (MIVIS) is a 102 channel scanner, operating between 0.4 and 13 μm. It was built by Daedalus, and is owned and operated in Italy by the National Research Council. Early results from MIVIS are reported by Bianchi et al. (1996).

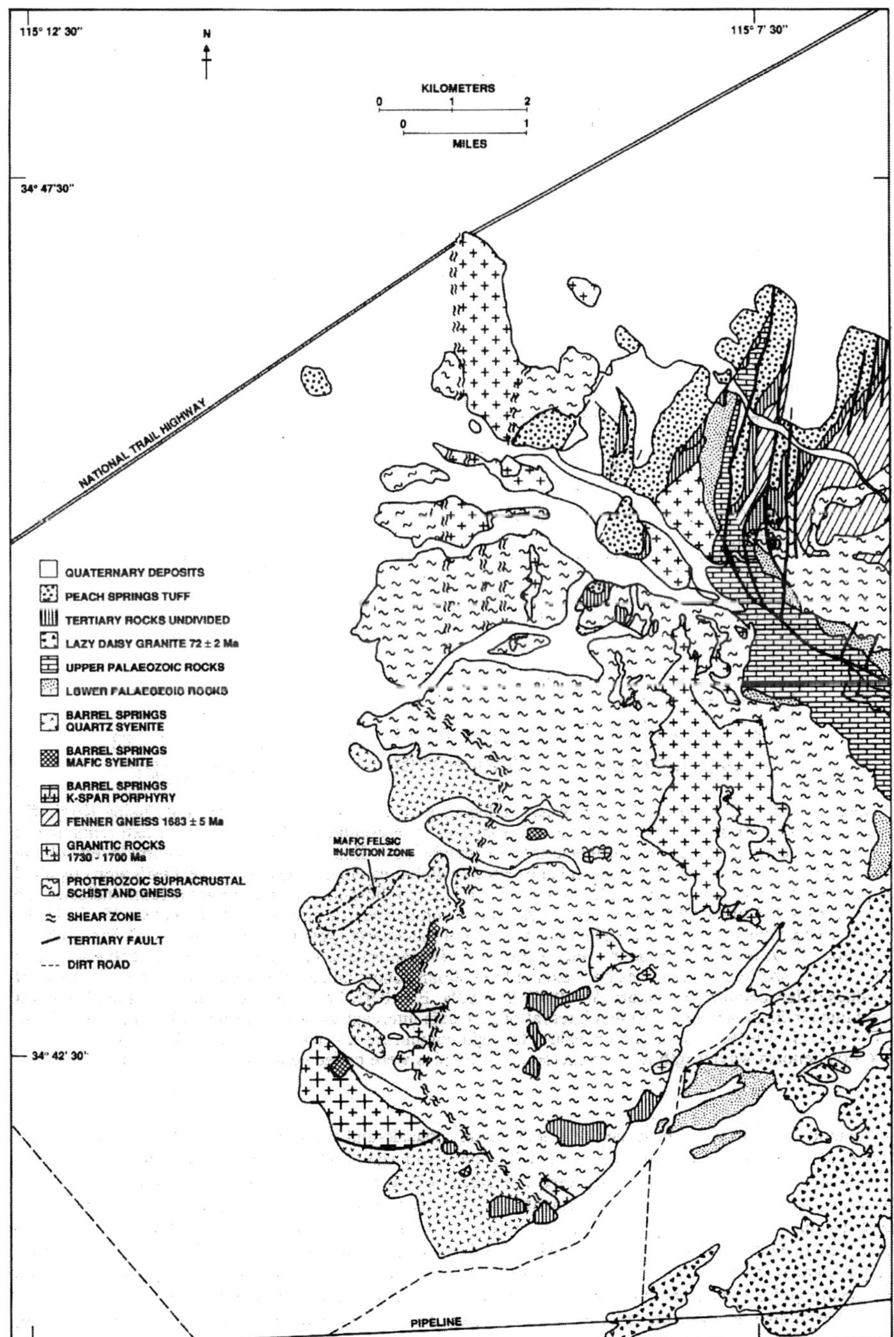



*Figure 11.* Mineralogical map of the area shown by the TIMS image in Figure 10. Some features shown as rock are alluvial-filled washes.

## 3.4 MASTER

The MODIS ASTER (MASTER) aircraft instrument is a Daedalus 50-channel scanner, owned by NASA and operated by the Ames Research Center. It has 10 bands in the TIR region, with an IFOV of 2.5 milliradians. The pixel size is variable, depending on the platform, and ranges from 5 to 50 m. This instrument was developed to simulate both ASTER data and MODIS data, so it has a number of narrow spectral bands in the VIS-SWIR, and sufficient TIR bands to simulate the 5 ASTER bands. Data are available at very low cost through the masterweb site (http://masterweb.jpl.nasa.gov). An example of the use of MASTER data for geologic mapping is given below.

The Cuprite Mining District in western Nevada has been frequently used as a test site for field instruments, aircraft scanners, and satellite instruments (Abrams et al., 1977). Basement rocks exposed at Cuprite are sedimentary rocks of Paleozoic age. Limestones, siltstones and shale are exposed on the west side. Miocene volcanic rocks include andesites and tuffs. About 7 mya, the area was subjected to intense hydrothermal alteration from the presence of an epithermal hydrothermal system. This resulted in three mappable alteration assemblages: an intense silicified package, dominated by the presence of quartz, goethite, and alunite; an opalized zone dominated by opal and alunite with abundant kaolinite; and an argillized zone dominated by hematite and kaolinite.

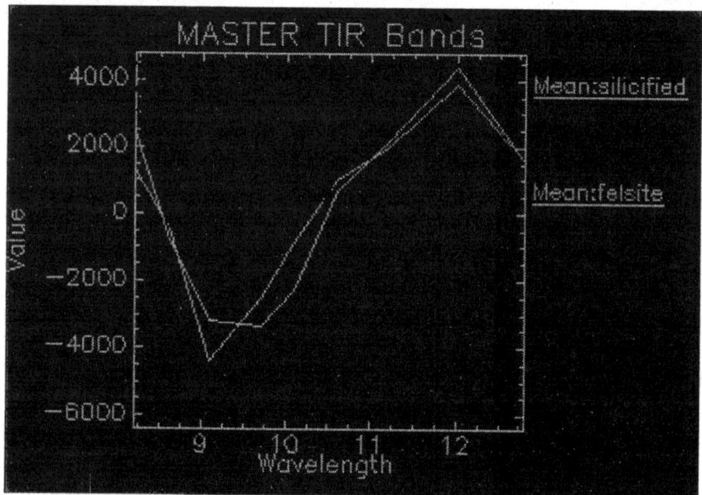


*Figure 12.* Alpha residual spectra extracted from MASTER data for Cuprite NV.

The MASTER TIR data were processed using an alpha residual technique (Kealy & Hook, 1993). This method seeks to separate temperature and emissivity by several approximations. Spectra were extracted from the alpha residual images, and plotted in Figure 12. The 10-point spectra show the diagnostic absorption features characteristic of the materials sampled. Limestone displays a minimum at 11.2 μm; silicified rocks have a minimum at 9.1 μm; tuffs have a minimum at 9.9 μm; and felsite intrusives show a double minimum at 9.1 and 9.9 μm.

A maximum likelihood classifier was used to map surface minerals for the entire scene (Figure 13). Ten different units are separated on the classified image, based on their TIR spectral characteristics. The silicified rocks are found in the opalized zone. On the west side, alunite and silicified rocks form linear zones, suggesting structural control of the alteration fluids. The playa is distinct due to its clay mineralogy. The sedimentary rocks are mappable as limestones and shales on the west side. Unaltered tuffs and unaltered alluvial fans are classifiable as distinct units. Overall, the presence of 10 TIR bands provides a clear improvement in rock type mapping compared to the earlier TIMS instrument.

*Figure 13.* Maximum likelihood classification of MASTER data for Cuprite NV. All ten channels of MASTER TIR data were used for the supervised classification.

## 3.5    SEBASS

SEBASS (Spatially Enhanced Broadband Array Spectrograph System) is an airborne push-broom imaging passive hyperspectral TIR system that comprises two helium – cooled 128x128 detector element arrays that measure in the 2.9 to 5.2 and 7.5 to 13.6 µm wavelength regions (Hackwell et al., 1996). There are 128 contiguous spectral bands for 128 pixels of each detector array. The angular resolution of the foreoptics is 1.1 mrad per pixel with a total field of view of 141 mrad. Using unpressurized aircraft, the pixel size is typically 2-3 meters. Thermal calibration of the instrument is implemented using hot and cold blackbodies, which are swung in front of the foreoptics at the beginning and end of each flight line, assuming linear drift between these blackbody measurements.

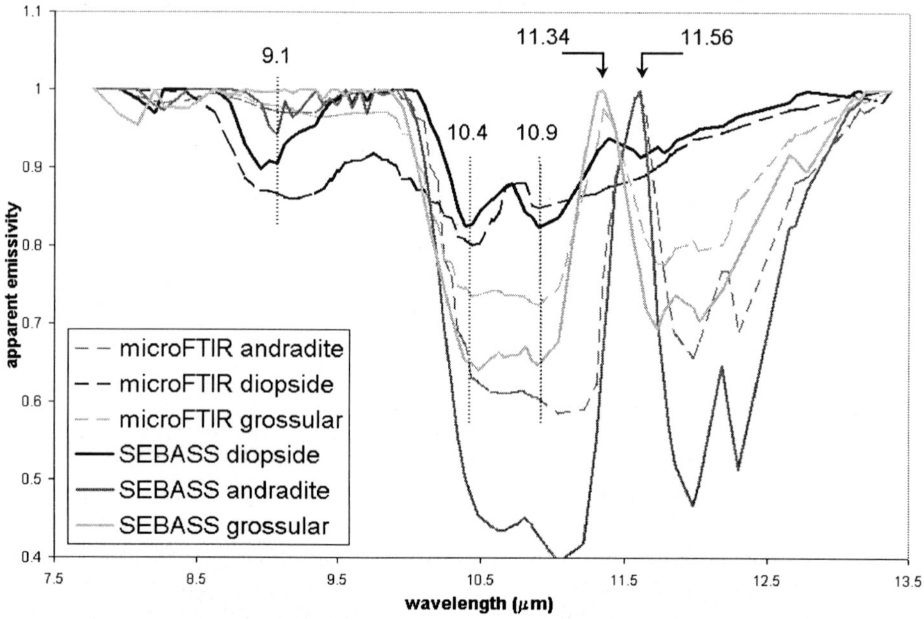


*Figure 14.* Airborne SEBASS and field Micro FTIR spectra of garnets and pyroxene samples.

The data are typically reduced using an empirical approach to atmospheric correction: a scene-based correction uses the SEBASS data themselves to derive the required input parameters for the estimation of atmospheric effects. These include an additive effect related to primarily upwelling radiation and a multiplicative effect related to attenuation by absorption of gases. The necessary assumptions include no correction for (1) heterogeneous atmosphere; (2) changes in path length; (3) reflected downwelling sky radiance; (4) atmospheric continuum absorption. This yields one gain and offset for each wavelength for all pixels.

In a series of studies, University of Nevada at Reno PhD students have been using SEBASS data to study alteration in mining districts near Reno (Calvin et al., 2000; Smailbegovich et al., 2000a, 2000b; Vaughan et al., 2000). Their approach has been to use the SEBASS data to simulate TIMS channels, then process them with a decorrelation stretch. Using these techniques, quartz, alunite, kaolinite, and carbonates could be separated. Mineral spectra extracted from the images allowed discrimination of calcite from dolomite. While interesting, these studies did not fully utilize the spectral information contained in the SEBASS data. The next study, however, shows the full potential of hyperspectral TIR data for mineral mapping.

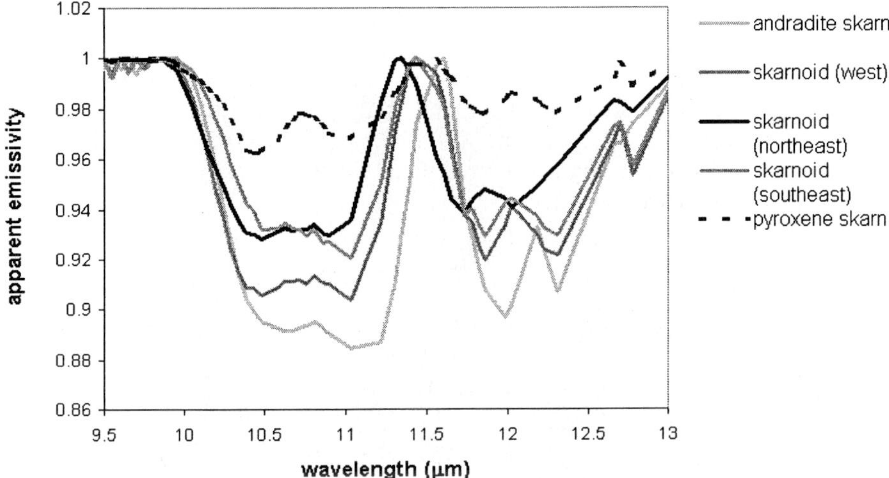


*Figure 15.* SEBASS endmember spectra of skarn-related units. The spectra have been processed to remove the continuum using a Hull technique.

Cudahy et al (2000) used SEBASS data to map mineralogy at the Yerrington test site, Nevada. Intrusion of monzonite and granite plutons formed skarn-related Cu-Zn-Pb base metal deposits in the older limestone sequence. The alteration assemblages include garnet and clinopyroxene in Ca-exoskarns; and forsterite in the Mg-exoskarns. Garnet mineralogy varies from andradite-grossular to spessartine-almadine. To compare the SEBASS mineral spectra with field emissivity measurements, the SEBASS surface radiance pixel spectra were hull-normalized to compensate for the effects of surface kinetic temperature and uncorrected atmospheric effects (Figure 14).

Spectral indices were developed involving band ratios and derivatives of polynomial fits. A ratio of SEBASS bands located at 11.4 and 11.0 µm was used to map pixels either rich in garnet or rich in carbonate. The second spectral index used was measurement of the wavelength position of the garnet and carbonate emissivity peaks near 11.4 µm. This was done by calculating the 1$^{st}$ derivative of a fitted 3$^{rd}$ order polynomial to the wavelength segment between 11.2 and 11.6 µm. These spectral indices were then used to map the garnet solid-solution chemistry.

Endmember analysis of the SEBASS data yielded diagnostic spectral signatures related to grandite, clinopyroxene, albite-oligoclase, oligoclase-andesine, quartz, dolomite, calcite, epidote, amphibole, white mica, and clay. Figure 15 shows hull-removed SEBASS endmember spectra of andradite, grossular and diopside that show the same diagnostic features apparent in the field spectra. For example, andradite has an emissivity peak at 11.56 µm, in contrast with grossular which has a peak at 11.34 µm. Diopside has two closely spaced lows at 101.4 and 10.9 µm. Figure 15 shows five endmembers that relate to skarn alteration. They all show garnet spectral signatures but with different wavelengths related to variations in solid-solution chemistry. Using these endmembers to classify the SEBASS flight line, mineral maps were generated (Figure 16). Good correlation existed around the Casting Copper and Douglas Hill Mines where andradite and pyroxene skarns are well developed. The SEBASS maps are effectively tracking a change in the Fe-Al grandite solid-solution chemistry, which becomes more Al-rich towards the contact with the intrusive.

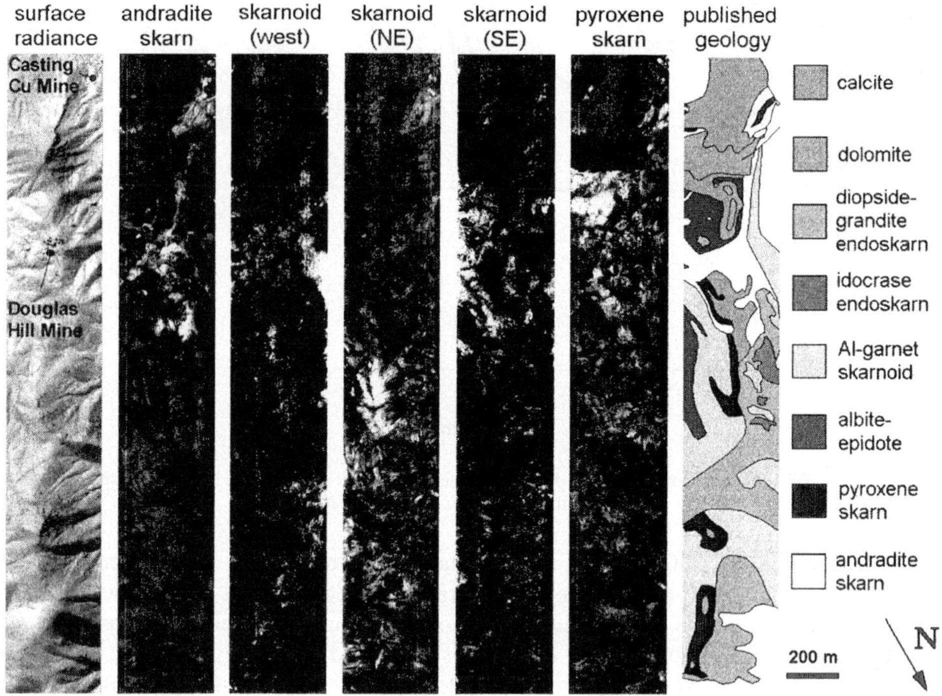


*Figure 16.* Published geology and SEBASS-derived mineral maps for the Yerrington test area.

Laboratory electron microprobe analyses of field samples show complete solid-solution variations between Ad50 and Ad100. Plotting the emissivity maximum from field spectra versus Fe-Al cation content from microprobe analyses shows the strong correlation of wavelength position versus cation substitution for andradite and grossular (Figure 17). The wavelength position of the garnet emissivity peak near 11.5 μm was calculated for the SEBASS lines to generate a map of the Fe-Al solid solution chemistry (Figure 18). The map clearly delineates those areas of exoskarn alteration and shows the progressive enrichment in Fe towards the southeast. Furthermore, all of the known skarn-related Cu mines are mapped as near-pure andradite. The map also shows that garnet can be mapped in alluvial materials that have moved at least 2 km from their source. This indicates that remote mapping of garnet chemistry by hyperspectral TIR imagery is an effective exploration tool for Ca-exoskarn alteration.

## 4    Satellite instruments

### 4.1    ASTER

The Advanced Spaceborne Thermal Emission and Reflection Radiometer (ASTER) is a facility instrument provided by the Japanese Ministry of International Trade and Industry, launched in December 1999 on NASA's Earth Observing System Terra platform. ASTER has 3 bands in the VNIR with 15m spatial resolution, 6 bands in the SWIR with 30m spatial resolution, and 5 bands in the TIR with 90m spatial resolution. An additional band in the VNIR looks backward, providing a stereo image from which DEMs can be created. Images are 60x60km in size, and any place on the earth can be revisited in 16 days or less. The TIR bands were selected to copy TIMS channels 1, 2, 3, 5, and 6. The TIMS band near 9.5 μm was eliminated because it lies directly on an ozone absorption band and was considered to be too noisy (Yamaguchi et al., 1998; Kahle et al., 1991).

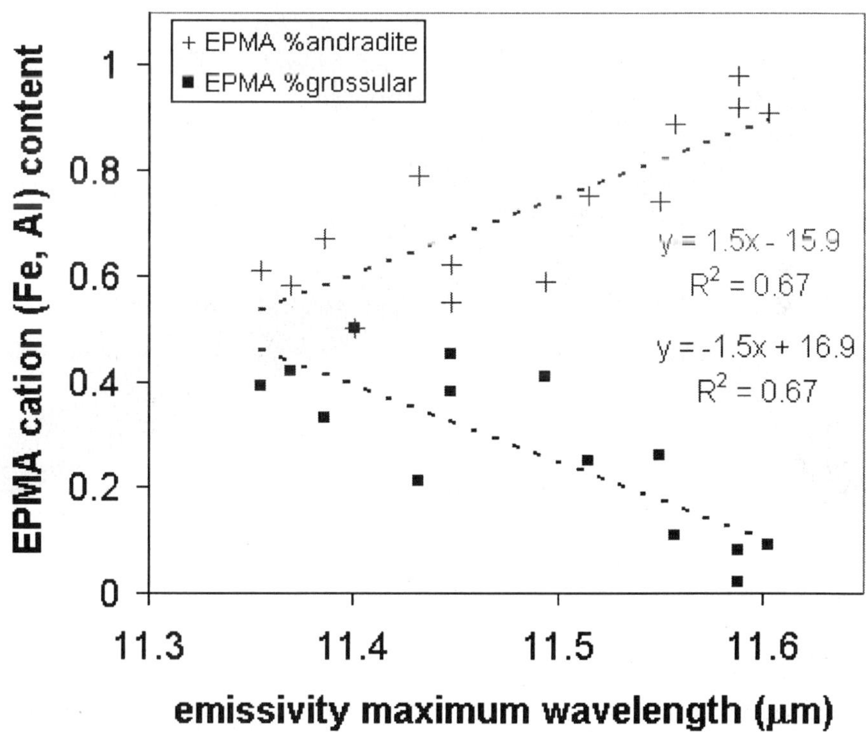


*Figure 17.* MicroFTIR versus EPMA results for grandite Fe-Al solid-solution chemistry.

ASTER is the first operational, high spatial resolution multispectral TIR instrument providing global data to anyone interested. A web-based interface allows users to search and order any ASTER scene contained in the image archive (http://edcimswww.cr.usgs.gov/pub/imswelcome/). In addition to acquiring daytime data, ASTER can be turned on at night to obtain SWIR and TIR data. This offers the advantage, in the TIR, of suppressing the effects of solar insolation and heating, and therefore focussing on emissivity differences of surface materials. An example of this capability is shown in Figure 19, obtained over Death Valley, CA. This nighttime image combines data from ASTER bands 14, 12 and 10 displayed in RGB. The data were enhanced using a decorrelation stretch algorithm. In addition, the image data were draped over a digital elevation model to create a 3-D perspective view looking north over Death Valley. Colors relate directly to compositional differences of surface materials. Red areas are outcrops and derived alluvial fans from quartzite; green areas are carbonate outcrops and fans. Purple areas are intermediate to mafic volcanics and metamorphic rocks. Variations in colors of the playa depict the presence of carbonate, sulfate and salt evaporite minerals.

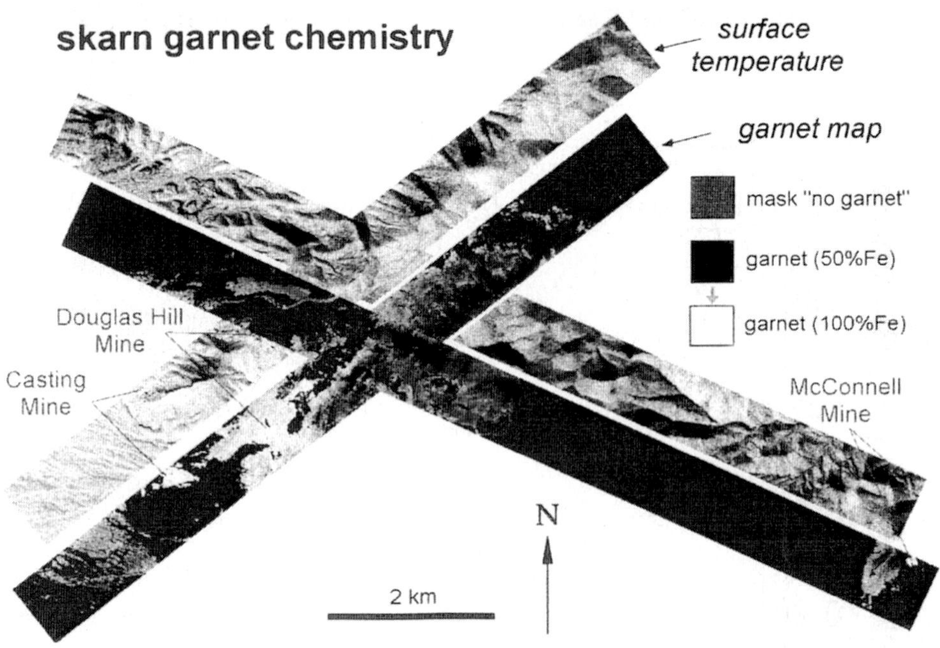


*Figure 18.* SEBASS derived garnet solid-solution chemistry map

## 4.2    MTI

The Multispectral Thermal Imager (MTI) is a space-based research and development project sponsored by the US Department of Energy (DOE) Office of Nonproliferation and National Security. MTI was successfully launched into space on March 12, 2000 on a Taurus rocket. The instrument has 15 bands in the visible through thermal infrared, with spatial resolution of 5 m in the visible, and 20 m in the other bands. MTI uses a 36-cm aperture, and a bank of three sensor chip assemblies, each carrying 15 arrays of cryogenically cooled detectors. Exact wavelength positions for each band is currently classified, as is the signal-to-noise and NE$\Delta$T of the system. Some of the bands were selected to provide simultaneous information on atmospheric water vapor, aerosol content and sub-visual cloud presence. MTI's three year mission objectives are to advance the state-of-the art in multispectral and thermal imaging, image processing and associated technologies, and to better understand the utility of these technologies. Example images acquired in the visible wavelength regions have been publicly released, and were viewable on the MTI web site. Data from the TIR remain unavailable outside the "black" community.

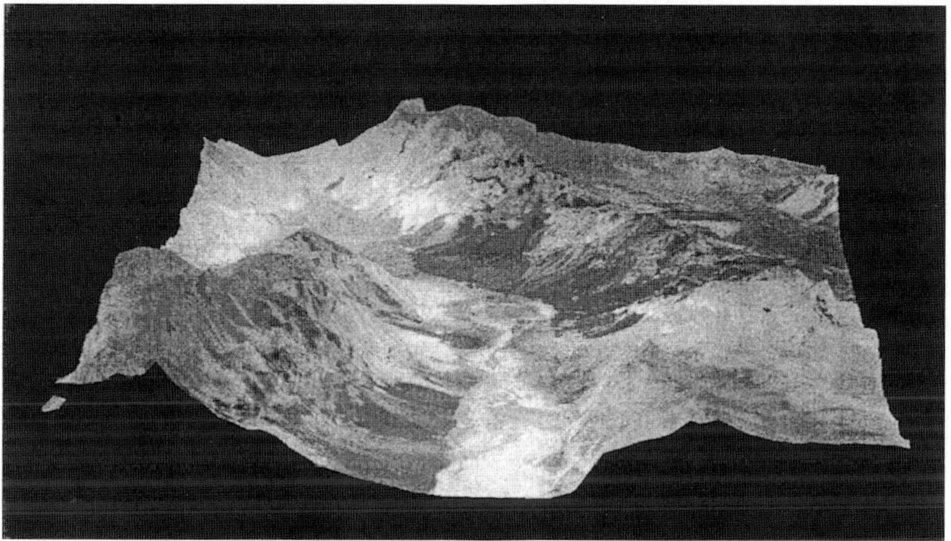

*Figure 19.* ASTER TIR composite, draped over a DEM to create a 3-D perspective view over Death Valley, CA. ASTER bands 14, 12 and 10 are displayed as RGB.

## 5 Future of Hyperspectral TIR Imaging

There are two ways of looking at the future of TIR hyperspectral remote sensing: the optimistic view that given sufficient high quality airborne demonstrations, a satellite system is inevitable; and the pessimistic view that almost nothing is currently planned. Let us examine major trends in hyperspectral remote sensing.

## 5.1 MAJOR TRENDS IN HYPERSPECTRAL REMOTE SENSING

Several major trends are driving the development and application of hyperspectral sensors in the next decade. In a practical form, the dominant trend is towards 'hyper-everything': sensors in which the user can adapt the spectral, spatial, and temporal resolution to suit the requirements of the desired data and information products. In space sensors, the current family of satellite hyperspectral sensors (EO-1/Hyperion, NEMO/COIS, Orbview-4/Warfighter, and ARIES) are striving to extend the capability of AVIRIS to space: while remarkably capable, the signal-to-noise ratios which will be achieved by these sensors is considerably below the 600-800 level which is typical of AVIRIS. In airborne demonstration systems the future is readily apparent ---

hyperspectral sensors with: (a) 1,000-10,000 channels, (b) spatial resolutions of 0.5-1.0 meters (and a goal of 0.1 m), and (c) temporal revisits of hours to days, rather than weeks to months. In practical numbers, future sensors will have ten-fold increases in spectral, spatial, and temporal capability relative to current sensors and increased signal-to-noise ratio.

Data mining, knowledge discovery, and decision-support are critical to solving the essential challenge of exploitation of hyperspectral data sources in information applications --- literally the implications of 'too much data' implicit in hyperspectral data systems. Two approaches have been articulated: adaptive information systems that tailor the data/telemetry flow to the necessary subsets of the potential data stream, and the 'Mother of All Databases'. Adaptive information systems are a necessary implication of bandwidth limitations which will always persist; in fact as we migrate to on-the-go information sourcing (dissemination to mobile users) they will only become more constrained, especially in the 'last mile' of the telemetry chain to the user. The 'Mother of All Databases' is a vision of an information system that "observes everything, never forgets, and understands it all". In this vision, the key is applications that source the right decision support information at the right time to the right customer.

Historically, the development and evolution of hyperspectral sensors have been subordinated to the progress in focal plane arrays (FPAs) and spectral separation devices. Factor of two improvements in FPA size or performance (sensitivity) typically require 5-8 years – considerably slower than the 'Moore's law' growth rate for data systems. New FPAs make 512 and 1024 channel dispersive sensors conceivable. Beyond 1024 channels remains the domain of interferometric hyperspectral sensors (the Imaging Fourier Transform Spectrometer (IFTS) and Imaging Fabry-Perot (IFP) sensors); today, 1,500-10,000 channel systems are in development. Examples of these sensors are the NPOESS Cross-track Infrared Sounder (CrIS), the Japanese Interferometric Monitor of Greenhouse Gases (IMG) and the European Interferometric Atmospheric Sounding Instrument (IASI), while these instruments have modest focal planes, the inclusion of large format focal planes is both logical and virtually inevitable.

A paradigm shift may be appropriate for hyperspectral information system: retain the redundancy at the sensor level and design reprogrammable compression and telemetry systems that are software upgradable. In this fashion, the sensor system has latent capability at launch --- it is conceivable that the data would be binned spatially, spectrally, and temporally to optimize telemetry volume. But as preliminary analysis of complete data cubes suggest improved approaches to information extraction it should be possible to alter the binning strategy to minimize the unexploitable redundancy and enable factor of two improvements in spatial/spectral/temporal coverage during the mission lifetime (nominally a factor of 8 improvement in information content should not be impossible). Similarly, onboard calibration and atmospheric correction  should allow the data to be partitioned into two parts: the atmosphere without the surface and the surface without the atmosphere. An example is an onboard hypercube that is compressed into a high spectral resolution scene average atmospheric transmission spectrum (for atmospheric correction) and high spatial resolution, modest spectral resolution, land imagery  (for surface analysis and exploitation). Complementary aerosol and water vapor maps would indicate the presence and magnitude of spatially variable atmospheric composition.

## 5.2    THE FUTURE – 1-5 YEAR FORECAST:

The major enabling factor for hyperspectral remote sensing systems in the next five years will be the availability of Commercial Off The Shelf (COTS) satellite infrastructure to support 1 meter GSD VNIR imagers.  The satellites, ground systems, sensor components, and launch vehicles are readily adaptable to hyperspectral systems --- as is the case for the Orbview-4/Warfighter system.  Given the infrastructure, the near future in space will look like the high-end airborne market today.

A true AVIRIS class sensor scaled to space is a major step beyond anything in the current pipeline: the factor of 3-4 improvement in signal-to-noise ratio significantly changes the applications basis for hyperspectral sensors.  For cardinal sensor parameters, consider a 10-30 m GSD hyperspectral landsat sensor with 256 channels in the VNIR (0.4-2.5 microns), 10 km swath, SNR > 500, calibration of 0.5% absolute, and daily temporal coverage.  Such a sensor would enable a revolution in precision agriculture and fundamentally change resource exploration, exploitation, and management strategies.

Similarly, in the thermal infrared, SEBASS scaled to a space sensor would open up new frontiers beyond the current capability of MODIS or ASTER.  A 30 m GSD thermal infrared hyperspectral Landsat with 512 channels (LWIR/MWIR) with 5 km swath and SNR >200 with multiple blackbodies for calibration would expand geomorphology and resource exploration significantly.

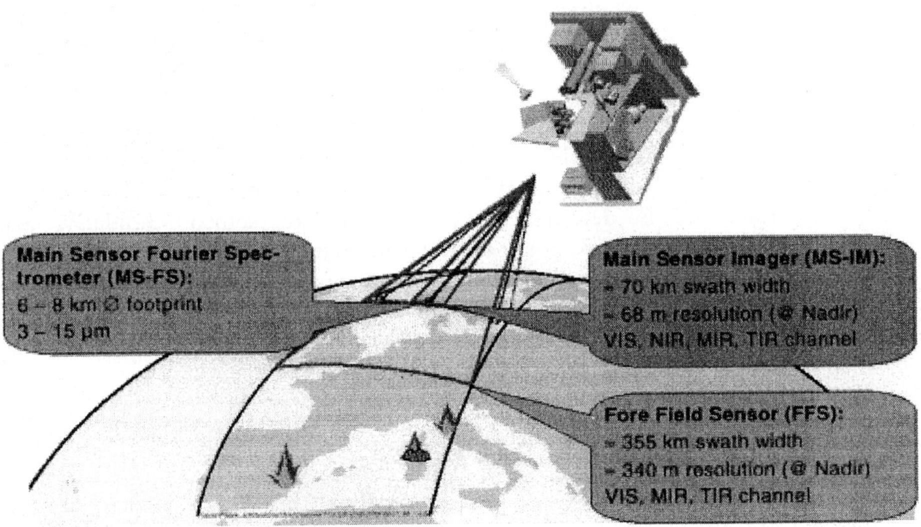


*Figure 20.* FOCUS sensor characteristics.

The only planned or proposed mission is one from ESA called FOCUS. FOCUS is an intelligent infrared sensor system for the detection of High Temperature Events (HTEs), such as vegetation fires or volcanoes. FOCUS is scheduled to be integrated on the International Space Station (ISS) during its early utilization phase. FOCUS is a scientific and technological demonstrator / precursor of an operational fire observation system and is proposed to be implemented on the externally mounted European payloads of the International Space Station (ISS).

The Main Objectives of the FOCUS Mission are: (1) Reliable autonomous on-board detection and analysis of High Temperature Events (HTE); (2) Generation of new IR data products and assessment of ecological consequences of HTE; (3) Near real-time HTE-cluster data geo-referencing and transmission to ground terminals.

The FOCUS sensor will consist of a wide-angle fore-field sensor (FFS) for autonomous hot spot detection, with 350 m spatial resolution, 350 km swath, and several bands in the VIS, MIR and TIR; and a main sensor (MS) for detailed observation and analyses of selected high temperature phenomena: higher spatial resolution, smaller swath width compared to MS; and a Fourier Spectrometer, with a 5 km footprint, and continuous spectral coverage from 3.3 to 15 microns (Figure 20).

## 5.3 THE FUTURE: 5-15 YEAR FORECAST

The technology hurdle between hyperspectral and ultraspectral sensors is defined by telemetry and computational issues --- not sensor design parameters in the traditional sense. Given an accelerated Moore's law approach to scaling computers and telemetry systems, it is reasonable to imagine a future where the data bandwidth is defined by the sensors capability (today). Hence, ultraspectral imagers – imaging Fourier transform spectrometers with large focal planes and thousands of channels are possible and desirable. In this paradigm we could imagine truly enabling the "Mother of All Databases" --- observe once, extract multiple encoded data sets tailored to specific applications and processing strategies, and design focused change detection strategies. This evolution would move towards match filter analysis, to answer specific questions: for example, the presence or absence of specific pollutants or minerals.

At the other extreme, programmable (reprogrammable) multispectral sensors may be a reasonable alternative with very high resolution imaging Fabry-Perot interferometers providing tunable spectral selection: the ability to tailor (on-orbit) the filter selection and bandwidth would represent a major step beyond current technology where the filter set is determined 10 years before launch. Similarly, active hyperspectral sensors, which combine coherent illumination with conventional passive hyperspectral sensing, and laser induced fluorescence imaging are techniques that are in laboratory development and early field trials at the present time.

# CHAPTER 11

# IMAGING SPECTROMETRY OF WATER

Arnold G. DEKKER* & Vittorio E. BRANDO* & Janet M. ANSTEE* &
Nicole PINNEL* & Tiit KUTSER** Erin J. HOOGENBOOM*** & Steef
PETERS**** & Reinold PASTERKAMP**** & Robert VOS**** & Carsten
OLBERT***** & Tim J.M. MALTHUS******

*     Environmental Remote Sensing Research Group CSIRO Land and
Water, Canberra, Australia

   **          CSIRO Marine Research, Canberra, Australia

   ***        National Institute for Coastal and Marine Management/RIKZ,
Ministry of Transport and Waterworks, The Hague, The Netherlands.

   ****      Institute for Environmental Studies, Vrije Universiteit, Amsterdam,
The Netherlands

   *****    Freie Universität Berlin, Berlin, Germany

   ****** Department of Geography, University of Edinburgh, Edinburgh,
Scotland

## 1    Introduction

Remote sensing is a suitable technique for large-scale monitoring of inland and coastal
water quality and its advantages have long been recognised. Remote sensing provides a
synoptic view of the spatial distribution of different biological, chemical and physical
variables of both the water column and if visible, the substrate. This knowledge of the
distribution is essential in environmental water studies as well as for resource management. Therefore, recent years have seen increasing interest and research in remote
sensing of water quality of inland and coastal waters ((Dekker *et al.*, 1995; Doerffer,
1992; Durand *et al.*, 1999; IOCCG, 2000; Kondratyev *et al.*, 1998; Lindell *et al.*,
1999)).

The use of water colour remote sensing for the determination of an optical water
quality variable was initially developed for the oceans, as it is virtually the only method
for assessing such vast areas. The optical properties of ocean waters are in general only
affected by phytoplankton and its breakdown products. These optically relatively
simple waters are known as Case 1 waters. Imaging spectrometry is probably overkill
for these waters as a few bands in the blue to green spectral areas are sufficient to
determine chlorophyll concentrations with sufficient precision for most oceanographic-
biological purposes. Apart from that argument, imaging spectrometers from space are
only available for civilian purposes since the launch of Hyperion in November 2000.



*F.D. van der Meer and S.M. de Jong (eds.), Imaging Spectrometry,* 307–359.
© 2006 *Springer. Printed in the Netherlands.*

Thus up till now all imaging spectrometry remote sensing work was carried out from aircraft: the scale of ocean remote sensing is not suitable for mapping from aircraft. Therefore imaging spectrometry of ocean waters are not discussed further in this paper. All other types of waters, i.e. those waters whose optical properties are influenced by more than just phytoplankton, are determined to be Case 2 waters. These other optical properties are usually a selection of dissolved organic matter from terrestrial origin, dead particulate organic matter and particulate inorganic matter. In addition if the bottom reflectance influences the water leaving radiance signal significantly, a water is also considered to be Case 2. In reality this distinction is in Case 1 and 2 waters is becoming less useful as more and more examples are being published, where this distinction doesn't hold. For instance, an algal bloom of the cyanobacterium *Trichodesmus* in ocean waters is technically a Case 1 water in this nomenclature. (Bricaud *et al.*, 1999) realise the potential of imaging spectrometry from space for deriving other algal pigments than chlorophyll *a* in oceanic waters. The optical properties of algal blooms require different remote sensing approaches requiring more spectral bands at longer wavelengths. The relatively simple band ratio algorithms for clear ocean waters do not function well any more in an algal bloom situation. A more useful approach is to describe water in terms of optically significant properties and the substances causing these properties.

Many inland and coastal waters are highly affected by anthropogenic influences. In combination with the complex hydrological situation, highly contrasted structures evolve in time and space in these aquatic environments. It is obvious that a water system with different optically active substances with temporal and spatial variations is by far more complex and requires more sophisticated models for remote sensing and separation of the water constituents than a system containing one component only like the ocean waters. Therefore airborne imaging spectrometry mainly gets applied to coastal and inland water environments and not to oceans. The vast dimension of oceans necessitates the use of ocean colour sensors on satellite platforms. Therefore this chapter focuses on imaging spectrometry as used for detection and monitoring of inland, estuarine, coastal and coral reef aquatic environments.


## 2    Light in water

### 2.1    INTRODUCTION TO THE THEORY

The colour of the water is a complex optical feature, influenced by scattering and absorption processes as well as emission by the water column and of reflectance by the substrate (Figure 1). This substrate reflectance (of seagrass, macro-algae, corals, sand, mud, benthic micro-algae etc.) is similarly a function of absorption and scattering and in a lesser degree emission of the substrate materials. Variations are essentially determined by the content of particulate and dissolved substances that absorb and scatter sky and solar radiation penetrating the water surface. The water leaving multi-spectral radiances are masked by the reflection of sun and skylight at the water surface and by extinction and scattering processes in the atmosphere. This exposes bottlenecks in the processing of remote sensing data to water quality maps. To address this bottleneck a

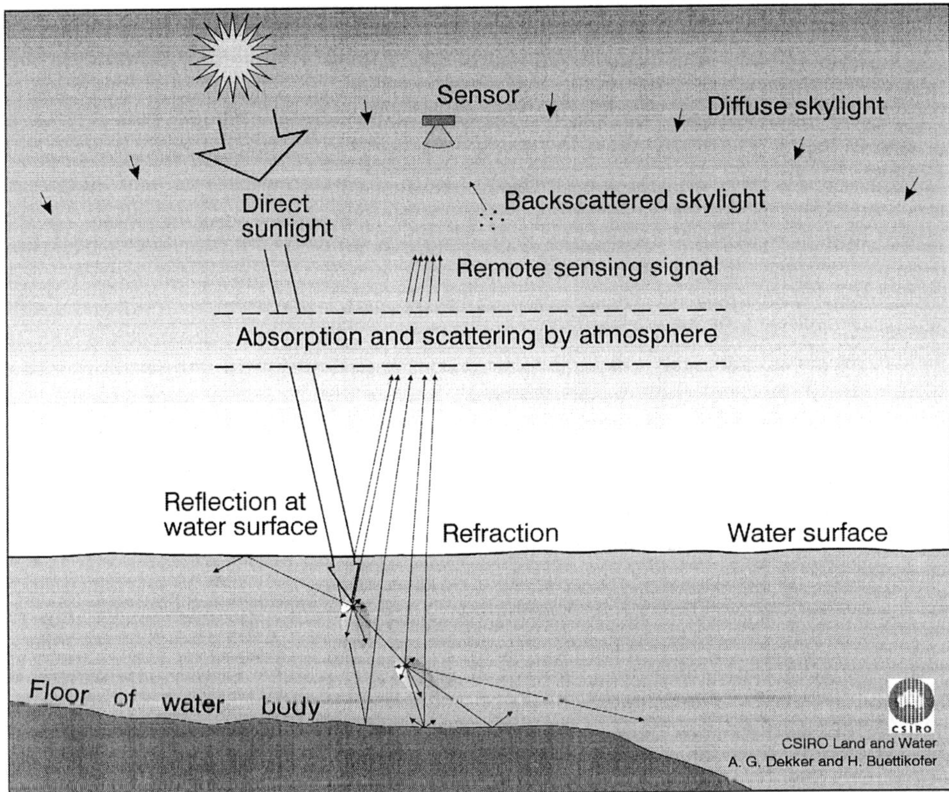


*Figure.1.* A schematic diagram of the various processes that contribute to the signal as measured by a remote sensor in an optically shallow water where the substrate has a significant effect on the water leaving radiance at the water surface.

careful and precise simulation of the radiative transfer in the water is required, at the water to air interface and in the atmosphere as a prerequisite for the improvement and development of new algorithms to retrieve the concentrations of selected water constituents. Therefore the relationship between the optical properties and the concentration units of these constituents have to be known for the water column as well as the optical properties of the substrate for substrate mapping. In regard to their optical behaviour, optically active substances can be split into distinct classes. If the inherent optical properties of these classes are sufficiently well characterised, their contribution to water column colour can be discriminated and their content quantified. For substrates there is currently insufficient information on how the optical properties influence the reflectance of substrate materials; therefore it is practice to mainly determine the reflectance of the substrate and not the concentration dependent absorption and scattering. Since the water reflected radiation depends on the quantity and specific optical properties of one or more water constituents, water colour carries spectral information about the concentration of some water quality parameters and possibly of the substrate. For the retrieval of different water constituents as well as substrate cover from a remotely sensed hyperspectral signal a suite of inversion methods are available, ranging from the often used, but less precise regression methods, through to physics based inverse modelling or inversion methods. Knowing the specific optical properties of the water constituents and of the substrate and modelling the radiative transfer through water and atmosphere as a function of these water constituents and comparing the modelled multispectral sensor signal with the measured multi-spectral sensor signal, the water colour data can be used to determine the concentrations of the water constituents and the substrate cover quantitatively. Analytical methods show better results than empirical, or semi-empirical methods which use simple correlation, or reasonable band ratios only instead of sophisticated optical models. However, the exploitation of water colour has greatly been impeded by the incapability to deal with the optical behaviour and complexity of water constituents. The range of optical water quality properties measurable in the water column that may be estimated by remote sensing has increased from suspended matter to include properties such as vertical attenuation coefficients of downwelling and upwelling light, transparency, coloured dissolved organic matter, chlorophyll *a* contents, even red tides and blue-green algal blooms. If the water column is sufficiently transparent and the substrate is within the depth where a sufficient amount of light reaches the bottom and is reflected back out of the water body it has been demonstrated that maps may be made of seagrasses, macro-algae, sand and sandbanks, coral reefs etc.

## 2.2   OPTICALLY DEEP WATERS

The large variation in the concentration of suspended sediments, phytoplankton and coloured dissolved organic matter in many inland, estuarine and coastal waters results in a highly variable light climate. Due to the optical complexity of these (often relatively turbid) waters, optical models play a key role in understanding and quantifying the effect of water composition on optical variables (obtained from either *in situ* or remote sensing measurements). Optical modelling is preferred above (semi-) empirical algorithms that have been the standard for many years in operational applications of remote sensing (and still are for ocean types of water). Many (semi-) empirical algorithms make extreme simplifications about the water composition, such as the (optical) domination of one constituent over all the others. There is a wide range of optical models available for water, from generic radiative transfer models (eg. HYDROLIGHT, (Mobley, 1994), (Mobley and Sundman, 2000)) to models based on simple analytical solutions developed for specific waters or conditions. Analytical models have the important advantage that, due to their relative simplicity, they can be solved very quickly. This is of great importance in a remote sensing application where a model must be evaluated at every pixel of an image. Thus we present an analytical optical model that describes the main light processes in both clear and turbid waters, without and with bottom visibility, taking into account highly variable optical conditions in the water column and the substrate as well as a complex geometry of the incident light field and the viewing angles of a remote sensor.

First the basic optical modelling of light in water is presented: radiative transfer theory. It explains how the radiometric properties, i.e. the radiance and irradiance, change in the water column due to the optical properties of the medium. Next we present the so-called two-flow model for irradiance, which can be solved analytically for the diffuse attenuation coefficient $K_\mathrm{d}$. With this (approximate) solution an analytical model for the subsurface irradiance reflectance $R(0-)$ is derived and compared with other analytical models from literature. R(0-) is a measure of the colour of water. It is a key parameter in the interpretation of remote sensing of water quality, because it links the measured light to the optical properties of the water. Most of the mathematics and definitions in this chapter are based on the book "Light and water" written by Mobley (Mobley, 1994). From this extensive text we have extracted those parts that are of interest within the limited scope of this study, and combined them with the modelling by (Aas, 1987). Because the scope of this chapter is imaging spectrometry applications we cannot here build a complete and consistent theoretical framework showing all the intermediate results, in stead we will state only important intermediate results and refer to others for the details.

### 2.2.1    Optical properties of the water column for optically deep waters

This paragraph introduces the optical properties and variables that are relevant for modelling the optical processes in the water column. Thus this discussion also is relevant for the optically deep water. The optical properties and variables are summarised in four tables, one table for each of the four groups that can be identified:

- The inherent optical properties (IOP) are the properties of the medium itself (i.e. water plus constituents), thus regardless the ambient light field; the IOP are measured by active (i.e. having their own light source) optical instruments (Table 1).
- The radiometric variables are the basic properties of the light that is measured by passive optical instruments (using the sun as the light source (Table 2).
- The apparent optical properties (AOP) are combinations of radiometric variables that can be used as indicators for the colour or transparency of the water, for example the reflectance (Table 3).
- The diffuse inherent optical properties are a combination of IOP and AOP and play an intermediate role in the derivation of the analytical model (Table 4).

### 2.2.2    The inherent optical properties

The inherent optical properties (IOP) depend only upon the medium. There are two main optical processes, absorption and elastic scattering, quantified by the absorption coefficient and the volume scattering function, respectively. Their definition is based on a small volume with thickness $\Delta r$, illuminated by a narrow collimated beam of monochromatic light of spectral radiant power $\Phi_i$, see Figure 2 and Table 1. Some part $\Phi_a$ of the incident power is absorbed within the volume of water. Some part $\Phi_s$ is scattered at an angle $\psi$, within a cone with solid angle $\Delta\Omega$. The remaining power $\Phi_t$ is transmitted through the volume (see(Mobley, 1995)).

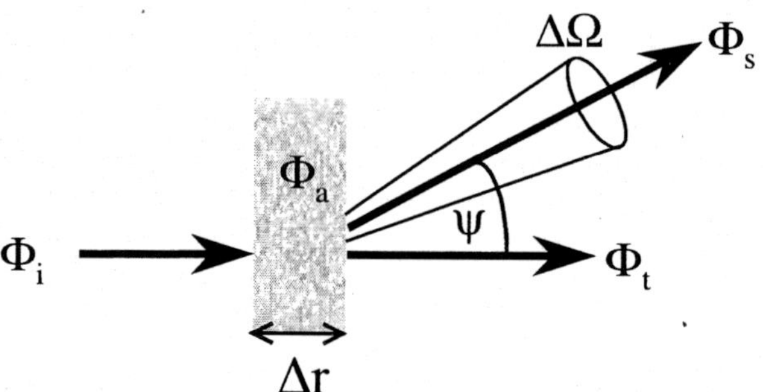


*Figure 2.* The definitions of the inherent optical properties are based on a collimated beam illuminating an infinitesimal layer  (adapted after (Mobley, 1994).)

The absorption coefficient is defined as the limit of the fraction absorbed power when $\Delta r$ goes to zero, see Table 1. Likewise the volume scattering function is defined as the limit of the fraction scattered light when both $\Delta r$ and $\Delta\Omega$ go to zero. From the absorption coefficient and the volume scattering function other IOP can be found, such as the scattering coefficient and the beam attenuation coefficient. Their definitions are summarised in Table 1. It must be kept in mind that the absorption and scattering coefficients are functions of wavelength, in other words the IOP are spectral properties. The wavelength is omitted from the definitions for brevity.

TABLE 1. Description and definition of the inherent optical properties.

| Symbol | description/definition | units/reference |
|---|---|---|
| $a$ | absorption coefficient | m$^{-1}$ |
| | $a \equiv \lim\limits_{\Delta r \to 0} \dfrac{\Phi_a}{\Phi_i \Delta r}$ | (Mobley, 1995) |
| $\beta$ | volume scattering function | sr$^{-1}$ m$^{-1}$ |
| | $\beta(\psi) \equiv \lim\limits_{\Delta r \to 0} \lim\limits_{\Delta\Omega \to 0} \dfrac{\Phi_s(\psi)}{\Phi_i \Delta r \Delta\Omega}$ | (Mobley, 1995) |
| $b$ | scattering coefficient | m$^{-1}$ |
| | $b \equiv 2\pi \displaystyle\int_{0}^{\pi} \beta(\psi)\sin\psi \, d\psi$ | (Mobley, 1995) |
| $b_b$ | backscattering coefficient | m$^{-1}$ |
| | $b_b \equiv 2\pi \displaystyle\int_{\pi/2}^{\pi} \beta(\psi)\sin\psi \, d\psi$ | (Mobley, 1995) |
| $c$ | beam attenuation coefficient | m$^{-1}$ |
| | $c \equiv a + b$ | (Mobley, 1995) |
| $\tilde{\beta}$ | normalized volume scattering function | Sr$^{-1}$ |

$$\tilde{\beta}(\psi) \equiv \frac{\beta(\psi)}{b}$$

| | | |
|---|---|---|
| $P$ | scattering phase function | Sr$^{-1}$ (Mobley, 1995) |

$$P(\psi) = 4\pi\tilde{\beta}(\psi)$$

| | | |
|---|---|---|
| $F_\alpha$ | forward scattering probability | - (Walker, 1994) |

$$F_\alpha \equiv 2\pi \int_0^\alpha \tilde{\beta}(\psi) \sin \psi \, d\psi$$

| | | |
|---|---|---|
| $B_\alpha$ | backward scattering probability | - (Walker, 1994) |

$$B_\alpha \equiv 2\pi \int_\alpha^\pi \tilde{\beta}(\psi) \sin \psi \, d\psi$$

| | | |
|---|---|---|
| $B$ | backscattering to scattering ratio | - (Dekker *et al.*, 1997) |

TABLE 1 (Cont.)

$$B \equiv 2\pi \int_{\pi/2}^\pi \tilde{\beta}(\psi) \sin \psi \, d\psi = \frac{b_b}{b}$$

| | | |
|---|---|---|
| $g$ | asymmetry parameter | - (Gordon *et al.*, 1975) |

$$g \equiv 2\pi \int_0^\pi \tilde{\beta}(\psi) \cos \psi \sin \psi \, d\psi$$

| | | |
|---|---|---|
| $\omega_0$ | single-scattering albedo | - (Mobley, 1994) |

$$\omega_0 \equiv \frac{b}{c}$$

| | | |
|---|---|---|
| $\omega_b$ | backscattering albedo | - (Mobley, 1995) |

$$\omega_b \equiv \frac{b_b}{a + b_b}$$

(Krijgsman, 1994)

## 2.3    RADIOMETRIC VARIABLES AND APPARENT OPTICAL PROPERTIES

The fundamental optical variable measured by most remote sensing instruments is radiance, $L$. From the radiance a number of other radiometric quantities can be derived, such as the downwelling and upwelling irradiance, see Table 2 (for a more detailed description see (Mobley, 1994)). Apparent optical properties (AOP) depend both on the medium and on the ambient light field, but they display enough regular features and stability to be useful descriptors of the water body. Definitions of commonly used AOP are listed in Table 3. The fact that the AOP are relatively stable and often behave well with depth, makes it easier to relate them to the water composition than (ir)radiance measurements. In particular, the reflectance just below the surface, $R(0-)$ , and the diffuse attenuation coefficient for downwelling light, $K_d$ , are very suitable, because they are sensitive to changing water compositions. In Figure 3, the geometry of the directional radiance vectors is defined.

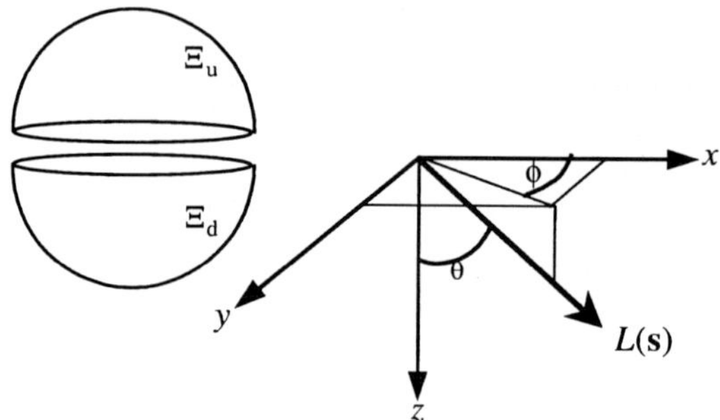


*Figure 3.* Definition of the geometry. s is a vector which gives the direction of the radiance. The vector is composed of two components: the zenith angle $\theta$ and the azimuth angle $\phi$: following the notation by (Walker, 1994)

TABLE 2 Description and definition of the spectral radiometric variables, where $\Xi$ is the unit sphere (the set of all directions **s** with solid angle $d\Omega$ ) and $\mu$ the cosine of the zenith angle of the horizontal plane. $\partial$ is the partial derivative).

| Symbol | description/definition | Units/reference |
|---|---|---|
| $L$ | (spectral) radiance | W m$^{-1}$ sr$^{-1}$ nm$^{-1}$ |
| | $$L \equiv \frac{\partial^4 Q}{\partial t \partial \Omega \partial A \partial \lambda}$$ | (Mobley, 1995) |
| $\mu$ | cosine zenith angle | - |
| | $$\mu \equiv \cos\theta$$ | (Mobley, 1995) |
| $E_d$ | downwelling irradiance | W m$^{-1}$ nm$^{-1}$ |
| | $$E_d = \int_{\Xi_d} \mu L(\mathbf{s}) d\Omega$$ | (Mobley, 1995) |
| $E_u$ | upwelling irradiance | W m$^{-1}$ nm$^{-1}$ |
| | $$E_u = \int_{\Xi_u} |\mu| L(\mathbf{s}) d\Omega$$ | (Mobley, 1995) |
| $E_0$ | scalar irradiance | W m$^{-1}$ nm$^{-1}$ |
| | $$E_0 = \int_{\Xi} L(\mathbf{s}) d\Omega$$ | (Mobley, 1995) |
| $E_{0d}$ | downward scalar irradiance | W m$^{-1}$ nm$^{-1}$ |
| | $$E_{0d} = \int_{\Xi_d} L(\mathbf{s}) d\Omega$$ | (Mobley, 1995) |
| $E_{0u}$ | upward scalar irradiance | W m$^{-1}$ nm$^{-1}$ |

$$E_{0u} = \int_{\Xi_u} L(\mathbf{s})\,d\Omega \qquad \text{(Mobley, 1995)}$$

TABLE 3. Description and definition of the apparent optical properties.

| Symbol | description/definition | units/reference |
|---|---|---|
| $R$ | irradiance reflectance<br><br>$R \equiv \dfrac{E_u}{E_d}$ | -<br><br>(Mobley, 1995) |
| $R(0-)$ | subsurface irradiance reflectance<br><br>$R \equiv \dfrac{E_u(z=0)}{E_d(z=0)}$ | -<br><br>(Gordon *et al.*, 1975) |
| $R_L(\mathbf{s})$ | radiance reflectance<br><br>$R_L(\mathbf{s}) = \dfrac{\pi L_u(\mathbf{s})}{E_d}$ | -<br><br>(Walker, 1994) |
| $R_{rs}(\mathbf{s})$ | remote sensing reflectance<br><br>$R_{rs}(\mathbf{s}) = \dfrac{L_u(\mathbf{s})}{E_d}$ | sr$^{-1}$<br><br>(Mobley, 1994) |
| $K_d$ | diffuse attenuation coefficient of downwelling light<br><br>$K_d \equiv -\dfrac{1}{E_d}\dfrac{dE_d}{dz}$ | m$^{-1}$<br><br>(Mobley, 1995) |
| $K_u$ | diffuse attenuation coefficient of upwelling light<br><br>$K_u \equiv -\dfrac{1}{E_u}\dfrac{dE_u}{dz}$ | m$^{-1}$<br><br>(Mobley, 1994) |
| $Q$ | ratio of upwelling irradiance to upwelling radiance<br><br>$Q(\mathbf{s}) \equiv \dfrac{E_u}{L_u(\mathbf{s})}$ | Sr<br><br>(Mobley, 1994) |
| $\bar{\mu}_d$ | downwelling average cosine<br><br>$\bar{\mu}_d \equiv \dfrac{E_d}{E_{0d}}$ | -<br><br>(Mobley, 1995) |
| $\bar{\mu}_u$ | upwelling average cosine<br><br>$\bar{\mu}_u \equiv \dfrac{E_u}{E_{0u}}$ | -<br><br>(Mobley, 1994) |
| $K_d^{\text{norm}}$ | normalized diffuse attenuation coefficient of downwelling light<br><br>$K_d^{\text{norm}} = \bar{\mu}_d K_d$ | m$^{-1}$<br><br>(Gordon, 1989) |

### 2.3.1    *The diffuse apparent optical properties*

In addition to the IOP and AOP there is an intermediate set of optical properties, called the diffuse apparent optical properties. They describe the absorption and scattering of down- and upwelling irradiance (Table 4). Most of these properties are only used for mathematical convenience in the derivation of the analytical model. Exceptions are the shape factors for upward and downward scattering functions (Table 4 and Figure 4), since they are not just intermediate parameters, but remain present in the final analytical model. The shape factors 'convert' the backscattering coefficient into the upward and downward scattering functions. Figure 4 illustrates that the upward scattered photons partly originate from photons that are scattered forward (shaded area). Since most particles in water scatter more light in forward directions than in backward directions, this contribution can be significant (Stavn and Weidemann, 1989).

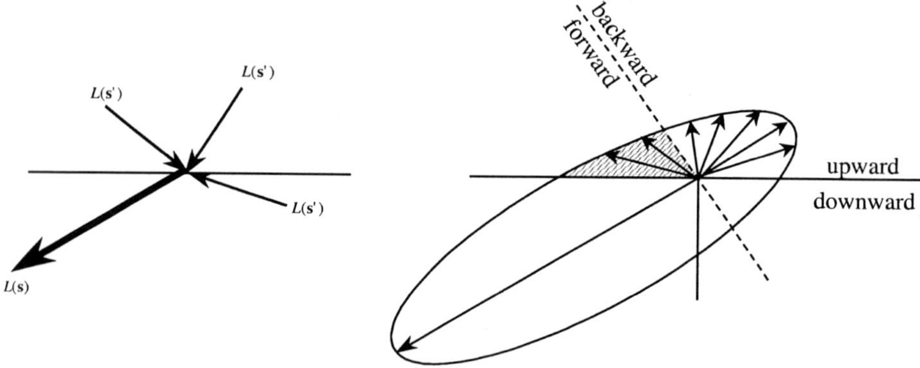


*Figure 4.* Left: The fraction of the incident radiance $L(\mathbf{s}')$ that is scattered into the directions $\mathbf{s}$ and contributes to $L(\mathbf{s})$ is given by $\tilde{\beta}(\mathbf{s}, \mathbf{s}')$. The angle between the vectors $\mathbf{s}'$ and $\mathbf{s}$ is $\Psi$. Right: The shape factor for downward scattering indicates the difference between downward and forward scattering (shaded area). Likewise, the shape factor for upward scattering indicates the difference between upward and backward scattering (see shaded area in figure).

### 2.3.2    *The two-flow model for irradiance*

### 2.3.2.1    *The radiative transfer equation*

This chapter considers the radiative transfer equation (RTE) that describes the behaviour of radiance in water. If we think of radiance as a beam of photons, six basic interactions of these photons with water can be distinguished ((Mobley, 1994)§ 5.1):
- loss of photons by conversion of radiant energy to non-radiant energy (absorption)
- loss of photons by scattering to other directions without change in wavelength (elastic scattering)
- loss of photons by scattering with change in wavelength (inelastic scattering)
- gain of photons by conversion of non-radiant energy into radiant energy (emission)

- gain of photons by scattering from other directions without change in wavelength (elastic scattering)
- gain of photons by scattering with change in wavelength (inelastic scattering)

The discussion of the radiative transfer in water will be based on absorption and elastic scattering processes only. Inelastic scattering effects, especially fluorescence of chlorophyll will only be discussed in the applications section. Now we have defined the inherent and (diffuse) apparent optical properties we can start with the radiative transfer equation (RTE). Many authors have elaborated on the RTE see for example (Aas, 1987); (Stavn and Weidemann, 1989); (Mobley, 1994) p251). Here we shall give a concise overview. The RTE shows how the radiance $L$ changes due to the optical properties of the water, hence the IOP: the beam attenuation coefficient $c$, scattering coefficient $b$ and the normalised volume scattering function $\tilde{\beta}$. As an intermediate step to the analytical model the transfer equations for upwelling and downwelling irradiance must be derived from the RTE for radiance. It is assumed that the water body is source free, i.e. inelastic scattering and true emission are neglected. In addition, it is assumed that the water body is time-independent, horizontally homogenous with a constant index of refraction (Mobley, 1994).

Finally, it is assumed that the absorbing and scattering particles are far apart with respect to $\lambda$. This latter assumption is flawed when many absorption and scattering particles are tight together, e.g. within one phytoplankton cell. In this case the scattering coefficient is not independent of absorption.

TABLE 4. Description and definition of the diffuse inherent optical properties.

| Symbol | description/definition | units/ref. |
|---|---|---|
| $a_d$ | diffuse absorption function for downwelling irradiance<br><br>$a_d \equiv \dfrac{a}{\overline{\mu}_d}$ | $m^{-1}$<br><br>(Mobley, 1994) |
| $a_u$ | diffuse absorption function for upwelling irradiance<br><br>$a_u \equiv \dfrac{a}{\overline{\mu}_u}$ | $m^{-1}$<br><br>(Mobley, 1994) |
| $b_{dd}$ | diffuse downward scattering function for downwelling irradiance<br><br>$b_{dd} \equiv \dfrac{b}{E_d} \int\limits_{\Xi_d}\int\limits_{\Xi_d} L(\mathbf{s}')\tilde{\beta}(\mathbf{s},\mathbf{s}')d\Omega'\,d\Omega$ | $m^{-1}$<br><br>(Mobley, 1994) |
| $b_{uu}$ | diffuse upward scattering function for upwelling irradiance<br><br>$b_{uu} \equiv \dfrac{b}{E_u} \int\limits_{\Xi_u}\int\limits_{\Xi_u} L(\mathbf{s}')\tilde{\beta}(\mathbf{s},\mathbf{s}')d\Omega'\,d\Omega$ | $m^{-1}$<br><br>(Mobley, 1994) |
| $b_{du}$ | diffuse downward scattering function for upwelling irradiance<br><br>$b_{du} \equiv \dfrac{b}{E_u} \int\limits_{\Xi_d}\int\limits_{\Xi_u} L(\mathbf{s}')\tilde{\beta}(\mathbf{s},\mathbf{s}')d\Omega'\,d\Omega$ | $m^{-1}$<br><br>This study |
| $b_{ud}$ | diffuse upward scattering function for downwelling irradiance | $m^{-1}$ |

$$b_{ud} \equiv \frac{b}{E_d} \int_{\Xi_u} \int_{\Xi_d} L(\mathbf{s}')\tilde{\beta}(\mathbf{s},\mathbf{s}')d\Omega' d\Omega$$

This study

$c_d$    diffuse attenuation function for downwelling irradiance    m$^{-1}$

$$c_d \equiv \frac{c}{\bar{\mu}_d} = a_d + b_{ud} + b_{dd}$$

(Mobley, 1994)

$c_u$    diffuse attenuation function for upwelling irradiance    m$^{-1}$

$$c_u \equiv \frac{c}{\bar{\mu}_u} = a_u + b_{uu} + b_{du}$$

(Mobley, 1994)

$c_{dd}$    local transmittance functions for downwelling irradiance    m$^{-1}$

$$c_{dd} = a_d + b_{ud}$$

(Mobley, 1994)

$c_{uu}$    local transmittance function for upwelling irradiance    m$^{-1}$

$$c_{uu} = a_u + b_{du}$$

(Mobley, 1994)

$r_d$    shape factor for upward scattering    -

$$r_d \equiv \frac{b_{ud}\bar{\mu}_d}{b_b}$$

(Mobley, 1994)

$r_u$    shape factor for downward scattering    -

$$r_u \equiv \frac{b_{du}\bar{\mu}_u}{b_b}$$

(Mobley, 1994)

---

Under all these assumptions the radiative transfer equation for unpolarised radiance is given by

$$\mu\frac{dL(\mathbf{s})}{dz} = -cL(\mathbf{s}) + b\int_{\Xi} L(\mathbf{s}')\tilde{\beta}(\mathbf{s},\mathbf{s}')d\Omega' \tag{1}$$

where $\Xi$ is the unit sphere (here the set of all directions $\mathbf{s}'$ with solid angle $d\Omega'$) and $\mu$ the cosine of the zenith angle. For sake of brevity the dependence on wavelength and depth is omitted. Equation 1 describes that the change in radiance over a depth interval $dz$ corrected for the zenith angle by $\mu$ is equal to that part of the radiance that is not attenuated by absorption or scattering ($c = a + b$) plus the contribution of the radiance scattered at all angles projected onto the initial direction of radiance $\mathbf{s}$.

### 2.3.2.2    *Two-flow modelling*

From eq. 1 expressions for the downwelling and upwelling irradiance can be derived by integrating over all angles in the downward and upward hemisphere respectively. With the definitions for the diffuse IOP the derivation of the radiative transfer equations for irradiance can be obtained, by integrating the RTE for radiance over all angles. Through several steps of integration and rewriting (see (Mobley, 1994)) the transfer equation for downwelling irradiance can be obtained

$$\frac{dE_d}{dz} = -\left(a_d + b_{ud}\right)E_d + b_{du}E_u \tag{2}$$

Equation 2 describes the change in downwelling irradiance with depth is equal to the downwelling irradiance that is not diffusely absorbed or diffusely scattered upwards plus the diffuse downward scattered fraction of the upwelling irradiance at that depth interval. Following the same line of reasoning, integration of the RTE over all angles in the upward hemisphere gives the irradiance transfer equation for upwelling irradiance

$$-\frac{dE_u}{dz} = -(a_u + b_{du})E_u + b_{ud}E_d \tag{3}$$

Equations 2 and 3 form the two-flow model for the source-free case, as illustrated in Figure 5. We see that the downwelling irradiance:

- decreases with depth because of absorption of $E_d$ ;

- decreases because of scattering of $E_d$ into $E_u$ ;

- increases because of scattering of $E_u$ into $E_d$ .


*2.3.2.3     An analytical solution of the irradiance RTE*


Under certain conditions the two-flow model developed in the previous chapter can be solved for the vertical attenuation coefficient. It is assumed the medium is homogeneous, i.e. it is assumed that  a set of effective IOP can be used that are constant over depth. Another assumption is that the water is optically deep so bottom effects can be neglected. Furthermore, it is assumed that the downwelling irradiance decays exponentially with depth (known as Beer's law)

$$E_d(z) = E_d(0)\exp(-K_d z) . \tag{4}$$

(Aas, 1987) derives an analytical expression for $K_d$ that goes one step beyond the single scattering approximation, since it includes a second order scattering effect in the second term. In clear waters this second term is often neglected $K_d \approx c_{dd}$

$$K_d = c_{dd} - \frac{b_{du}b_{ud}}{c_{uu} + c_{dd}} \tag{5}$$

This analytical model for $K_d$ can be rewritten in terms of the absorption and backscattering coefficients. Substituting the relevant definitions in Table 4 gives

$$K_d = \frac{a}{\overline{\mu}_d}\left[1 + r_d\frac{b_b}{a}\left(1 - \frac{r_u\overline{\mu}_d}{\overline{\mu}_u + \overline{\mu}_d}\frac{b_b}{a + kb_b}\right)\right], \qquad k = \frac{r_d\overline{\mu}_u + r_u\overline{\mu}_d}{\overline{\mu}_u + \overline{\mu}_d} . \tag{6}$$

Equation 6 can be considered as a generic model that is expected to be valid in both clear and turbid waters. In order to compare the concept of equation 6 with other models found in the literature, it is convenient to neglect the second term in equation 6

$$K_d = \frac{a}{\overline{\mu}_d}\left[1 + r_d\frac{b_b}{a}\right] \tag{7}$$

Several analytical models similar to eq. 7 can be found in literature. For instance, setting the shape factor $r_d$ to 1 gives the model of (Walker, 1994) and if, in addition, the $\overline{\mu}_d$ is approximated by $\mu_0$ (the cosine of sun zenith) we arrive at the model of Gordon *et al.* (1975).

A.G. DEKKER *ET AL.*

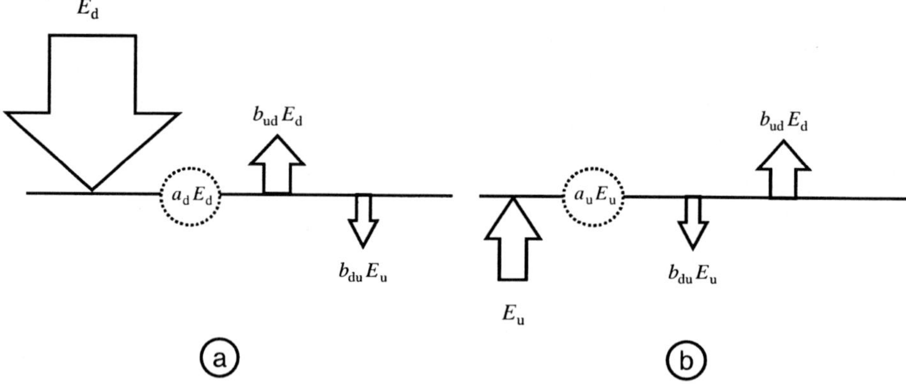


*Figure 5.* (a) The change with depth of the downwelling irradiance can be interpreted in terms of absorption and scattering functions, (b) idem for the upwelling irradiance.

TABLE 5. Several analytical models for the diffuse attenuation coefficient can be found in literature.

| Model | ref. |
|---|---|
| $$K_d = \frac{a}{\mu_0}\left[1 + \frac{b_b}{a}\right]$$ | (Gordon *et al.*, 1975) |
| $$K_d = a\left[1 + \frac{1}{6}\frac{b}{a}\right]$$ | (Wilson and Kiefer, 1979) |
| $$\overline{K}_d = \frac{a}{\mu_0}\sqrt{1 + G(\mu_0, g)\frac{b}{a}}$$ $$G(\mu_0, g) = \mu_0\left(\frac{2.236}{g} - 2.447\right) - \left(\frac{0.849}{g} - 0.739\right)$$ | (Kirk, 1991) |
| $$K_d = \frac{a}{\overline{\mu}_d}\left[1 + \frac{b_b}{a}\right]$$ | (Aas, 1987); (Walker, 1994) |
| $$K_d = \frac{a}{\overline{\mu}_d}\left[1 + r_d\frac{b_b}{a}\left(1 - \frac{r_u\overline{\mu}_d}{\overline{\mu}_u + \overline{\mu}_d}\frac{b_b}{a + kb_b}\right)\right],$$ $$k = \frac{r_d\overline{\mu}_u + r_u\overline{\mu}_d}{\overline{\mu}_u + \overline{\mu}_d}$$ | this study |
| $$K_d = \frac{a}{\overline{\mu}_d}\left[1 + r_d\frac{b_b}{a}\right]$$ | (Aas, 1987) |

An analytical model is presented for the diffuse attenuation coefficient that can be expected to be valid for turbid waters. It relates the total absorption and backscattering coefficient to the $K_d$, and to specify the model in eq.6 four parameters (AOP) are required: $\overline{\mu}_d, \overline{\mu}_u, r_d$ and $r_u$. Unfortunately relatively little is known about the values for the shape factors in turbid waters. In most analytical models the shape factors are set to 1. However, Stavn & Weidemann (1989) find that $r_d$ can vary between 1.3 and 10 and that $r_u$ can vary between 1.8 and 20, during the development of a phytoplankton bloom in Case I (ocean type) waters. These results indicate that the (variation in the) shape factors must be taken into account. As far as we know their values are not yet determined for turbid water types. Hence, research on the shape factors in these waters is highly recommended.

As will be evident from the discussion of optical models in optically shallow waters presented in paragraph 2.4 a clear understanding of the nature of attenuation with depth is essential for remote sensing of bathymetry or a substrate or a substrate cover.

### 2.3.3    An analytical model for the irradiance reflectance

We refer to (Aas, 1987) for a complete derivation of the analytical model for the irradiance reflectance. The reason for choosing this model is that it can act as a reference for understanding all other models of this kind found in literature. In terms of the backscattering and absorption coefficients the (Aas, 1987) analytical model for irradiance reflectance can be written as

$$R(0-) = \frac{r_d \overline{\mu}_u}{\overline{\mu}_u + \overline{\mu}_d} \frac{b_b}{a + kb_b}, \qquad k - \frac{r_d \overline{\mu}_u + r_u \overline{\mu}_d}{\overline{\mu}_u + \overline{\mu}_d} \tag{8}$$

This equation for $R(0-)$ states that $R(0-)$ is proportional to the backscattering divided by the sum of absorption and the second order backscattering. In more detail the equation states that the irradiance reflectance is equal to a factor times the backscattering divided by the sum of absorption and the second order backscattering (whereby the second order backscattering is multiplied by a factor that accounts for up and downwelling shape factors and the average cosines for up and downwelling irradiance). The multiplication factor takes into account the downwelling shape factor and average cosines of the up and downwelling irradiances.

Although even this model contains approximations as explained by (Aas, 1987) it may be expected to yield quite accurate results for turbid waters. Various authors have developed analytical models for the subsurface reflectance, which can be related to the remote sensing reflectance measured from (far) above the water surface. Therefore, the subsurface irradiance reflectance plays an important intermediate role in many remote sensing applications on water quality. Most of the models are developed and validated for relatively clear waters. From the comparison of the models summarised in Table 5 we see that the model in eq. 8 is generic in the sense that most of the other models can be obtained by substituting approximate values for the AOP. For instance, if we set the shape factors to unity, we get the model of (Walker, 1994). In case of a diffuse upwelling light field ($\overline{\mu}_u = 0.5$) and, $\overline{\mu}_d$ being approximated by $\mu_0$ (the cosine of sun zenith), we get the second model by (Walker, 1994). If in addition the sun zenith is 0 and the backscattering term in the denominator is neglected, we get the well-known model by (Gordon *et al.*, 1975). If we assume a totally diffuse light field ($\overline{\mu}_u = \overline{\mu}_d = 0.5$) the model of (Krijgsman, 1994) is obtained. Finally, if we assume unit shape factors and $\overline{\mu}_u = \overline{\mu}_d$ the exact solution given in (Aas, 1987)) and underlying the derivation of equation 8  simplifies to the model of (Duntley, 1963). Kirk's models (Kirk, 1984; Kirk, 1994a) were derived in a different fashion from the more analytically derived models as they are based on Monte Carlo simulations of the underwater light field. If the parameterisation of Kirk's models applies to the waters under study they are perhaps the easiest to apply as apart from the absorption and backscattering coefficients only the average cosine of all photons just under the water surface are required. In a more general applicability and because there is so little information available yet on the shape factors, we recommend using the (Walker, 1994) first model as it has the least assumptions and is easiest to use in simulation models.

TABLE 6. Analytical models for the subsurface reflectance found in literature.

| model | ref. |
|---|---|
| $$R(0-) = \dfrac{b_b}{a + b_b + \sqrt{(a + b_b)^2 - b_b^2}}$$ | (Duntley, 1963) |
| $$R(0-) = \sum_{n=0}^{3} f_n \left(\dfrac{b_b}{a + b_b}\right)^n$$ | (Gordon *et al.*, 1975) |
| $$R(0-) = 0.33\dfrac{b_b}{a}$$ | (Morel and Prieur, 1977) |
| $$R(0-) = \left(0.975 - 0.629\mu_0\right)\dfrac{b_b}{a}$$ | (Kirk, 1994b) |
| $$\dfrac{R(0-)}{Q} = 0.095\dfrac{b_b}{a + b_b}$$ | (Gordon *et al.*, 1988) |
| $$R(0-) = f\dfrac{b_b}{a}$$ $$f = 0.63 + -0.22\dfrac{b_b(w)}{b_b} - 0.05\left(\dfrac{b_b(w)}{b_b}\right)^2 - \left(0.31 - 0.25\dfrac{b_b(w)}{b_b}\right)\mu_l$$ | (Morel and Gentili, 1993) |
| $$R(0-) = f\dfrac{b_b}{a + b_b}$$ | (Dekker, 1993) |
| $$R(0-) = \left(1.018 - 0.657\mu_0\right)\dfrac{b_b}{a + 0.361b_b}$$ | [(Kirk, 1994a) |
| $$R(0-) = 0.5\dfrac{b_b}{a + b_b}$$ | (Krijgsman, 1994) |
| $$R(0-) = \dfrac{1}{1 + \dfrac{\mu_d}{\mu_u}}\dfrac{b_b}{a + b_b}$$ | (Walker, 1994) |
| $$R(0-) = \dfrac{1}{1 + 2\mu_0}\dfrac{b_b}{a + b_b}$$ | (Walker, 1994) |
| $$R = \dfrac{r_d\mu_u}{\mu_u + \mu_d}\dfrac{b_b}{a + kb_b}, \qquad k = \dfrac{r_d\mu_u + r_u\mu_d}{\mu_u + \mu_d}$$ | (Aas, 1987), this study |

Most studies neglect the variation in the shape factor and use empirical corrections based on the sun zenith angle in stead of average cosines. (Whitlock *et al.*, 1981) investigated the validity of the model by Gordon in three turbid water samples with maximum $b_b/a$ of 0.5. They were not able to fit one model due to the limitations of the Gordon model for variations in the illumination conditions. Main errors that (Whitlock *et al.*, 1981) identify are the combined effect of various solar zenith angles and skylight illumination, and the non-diffuse distribution of the upwelling light. These findings indicate that for turbid waters the values for the average cosines for downwelling and upwelling light play a significant role.

A potential problem is that it is probably impossible to use one set of typical values for $\overline{\mu}_d, \overline{\mu}_u, r_d$ and $r_u$. Near many coasts a large spatial gradient of turbidity occurs in the first 20 km. Near the coast and in intertidal area's large temporal variations in turbidity are measured, caused by the large variation in the concentration of suspended particles: from a few to more than 1000 g m$^{-3}$ within one tidal period. Since the optical conditions can be so different it may be necessary to use a two-step approach. First, the IOP (and subsequently the constituent concentrations) are calculated from the reflectance using a set of average values for $\overline{\mu}_d, \overline{\mu}_u, r_d$ and $r_u$. Second, using the calculated IOP as input these four parameters are calculated with a RTE-model such as HYDRO-LIGHT and then the IOP are calculated again with these adapted values. It is recommended to investigate the range of values for the average cosine and the shape factors that may occur in inland, estuarine and coastal waters. Most of the models in Table 5 are developed for the reflectance just below the air-water interface. The subsurface reflectance is most relevant for remote sensing applications. However, *in situ* measurements can be carried out at any depth. In many cases it is preferred to measure at some distance from the surface to minimise wave effects. From our analysis it appears that the model is valid for any depth, provided we assume that the downwelling irradiance decays exponentially with depth and that the water is optically deep.


## 2.4    OPTICAL SHALLOW WATERS

(Maritorena *et al.*, 1994) present a clear discussion of the physics of an optical shallow water body where part of the reflectance at the surface is composed of a bottom signal. They describe their analytical model for optically shallow water in the same terms used for describing the physics of the underwater light field for an optically deep system. Therefore the following text is mainly derived from their text. They use an approach derived from the two-flow equations to obtain approximate formulae based on a set of simplifying assumptions.

In optically shallow waters $E_u(0)$, is the sum of upwelling irradiance originating within the water column (where none of the photons have interacted with the substrate), $E_u(0)_C$, and the upwelling irradiance reflected from the substrate (where each of the photons have interacted with the substrate), $E_u(0)_B$

$$E_u(0) = E_u(0)_C + E_u(0)_B \tag{9}$$

To estimate the first term consider an infinitely thin layer of thickness dZ at depth Z, where the downwelling irradiance is $E_d(Z)$. At this depth the fraction of upwelling irradiance created by this layer is

$$dE_u(Z) = b_{bd}E_d(Z)dZ \tag{10}$$

$E_d(Z)$ can be expressed as in equation 4. Before it reaches the surface , d $E_u(Z)$ is attenuated along the path of Z to the surface, expressed by

$$\exp(-\kappa Z) \tag{11}$$

where $\kappa$ is the vertical diffuse attenuation coefficient for $E_u(Z)$ as defined by (Kirk, 1989). It is important to realise at this stage that $K_u$ is the vertical attenuation coefficient for diffuse upwelling light $E_u$ measuring it from the surface downwards whereas $\kappa$ is the vertical attenuation coefficient for diffuse upwelling light originating in each layer of the water column and measuring it depth upwards. The contribution of the considered layer in eq. 10 to the upwelling irradiance just below the water surface is expressed as

$$dE_u(Z \to 0) = b_{bd} E_d(0) \exp[-(K_d + \kappa)Z]dZ \tag{12}$$

If it is assumed that $b_{bd}$, $K_d$ and $\kappa$ are not depth-dependent, the contribution of all layers between Z and 0 is

$$E_d(0,Z) = b_{bd} E_d(0) \int_0^z \exp[-(K_d + \kappa)Z]dZ \tag{13}$$

Equivalent to:

$$E_u(0,Z) = (K_d + \kappa)^{-1} b_{bd} E_d(0)(1 - \exp[-(K_d + \kappa)Z]dZ) \tag{14}$$

For an infinite water depth eq. 14 reduces to :

$$E_u(0,\infty) = (K_d + \kappa)^{-1} b_{bd} E_d(0) = R(0,\infty)E_d(0) \tag{15}$$

$R(0, \infty)$ is in the case of an optical deep water equal to $R(0-)$ as given in Table 6. If we assume a totally absorbing substrate at depth H, eq. 14 becomes:

$$E_u(0,H) = R_\infty E_d(0)(1 - \exp[-(K_d + \kappa)H]) = E_u(0)_C \tag{16}$$

equivalent to the first term in eq. 9. For optically shallow water with an albedo A, the upwelling irradiance originating from reflection at the substrate at a level H (immediately above the bottom) is

$$E_u(0)_B = AE_d(0)\exp[-(K_d + \kappa)H]) \tag{17}$$

Filling in eqs 16 and 17 into eq 9 the following equation is obtained

$$E_u(0) = E_d(0)(R_\infty(1 - \exp[-(K_d + \kappa)H])) + A\exp(-(K_d + \kappa)H)) \tag{18}$$

When eq. 18 is divided with $E_d(0)$ the expression is derived for the reflectance just below the surface of a homogeneous water body with a reflecting substrate (identical to that of (Philpot and Vodacek, 1989))

$$R(0,H) = R_\infty + (A - R\infty)\exp[-(K_d + \kappa)H] \tag{19}$$

Because there are actually two upwelling light streams: one from the bottom and one from the water column $\kappa$ can be described as $\kappa_B$ and $\kappa_C$ respectively. Equation 19 then becomes

$$R(0,H) = R_\infty + \exp(-K_d H)[A\exp(-\kappa_B H) - R_\infty \exp(-\kappa_C H)] \tag{20}$$

## 2.5 LIGHT ABOVE WATER

### 2.5.1 *Water surface effects*

Previously we described what happens to downwelling irradiance once it has penetrated the water surface. Downwelling irradiance above the water surface will undergo one of two effects. It will either be reflected from the surface itself back into the atmosphere or it will pass across the air-water interface into the water, being refracted in the process (Figure 6). The surface reflected component is an unwanted signal in remotely sensed imagery used for water quality assessment. We are interested in that fraction of light which passes into the water column, interacts with it and perhaps with the substrate and then may be reflected back across the interface to be detected by a sensor.

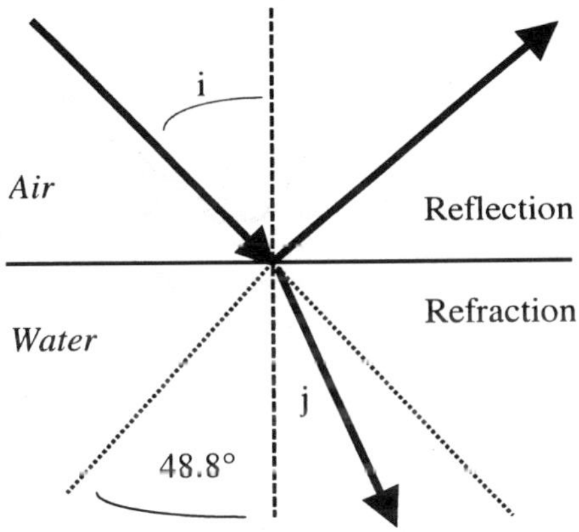


*Figure 6.* Reflectance and refraction at the water surface.

Refraction can be calculated according to Snell's law: $n_a \cdot \sin_i = n_w \cdot \sin_j$, where $n_w$ and $n_a$ are the refractive indices for both water and air, respectively. Ordinarily, $n_a$ is usually defined as equal to 1 and for most purposes the refractive index of sea water can be regarded as being 1.338 (although it is affected by both water temperature and salinity, (Plass and Kattawar, 1972)):

$$\frac{\sin \phi_a}{\sin \phi_w} = \frac{n_w}{n_a} = 1.338$$

For freshwater $n_w = 1.333$. The implications of refraction at the air water interface are that, for a flat sea surface, the whole of the hemispherical irradiance from the atmosphere which passes across the interface is compressed into a cone of underwater light with a half angle of 48.8° (Figure 6). This phenomenon also has implications for reflected radiance. Any backscattered light travelling upwards and striking the surface at angles greater than 48.8° will be totally internally reflected - they will not penetrate the surface. Similarly, the flux contained within the solid angle below the surface will be

spread out because of the refraction above the surface when it passes across the interface.

*The effects of surface roughness* - The surface of a natural water body is almost
never flat; wind driven waves will have a major effect on the ability of light to pass
across the air water interface. The effect of wind roughening is generally to widen the
solid angle through which light will penetrate, i.e. some light will penetrate the water at
angles greater than 49 degrees. The presence of slicks and/or whitecaps will further
modify the light field in different ways from that of wave action (Estep and Arnone,
1994) (Gordon and Wang, 1994)). Oil slicks, apart from having a dampening effect on
wave action will cause higher reflectance in certain regions of the spectrum.


### 2.5.2    *Atmospheric effects and atmospheric correction*


Although the physics of atmospheric correction of remote sensing data over waters is
essentially the same as for terrestrial targets, there are a few practical differences that
need to be addressed. For any water body it is the signal coming from within the water
body that is the desired signal. On land it is the surface reflected signal that is of interest. For water bodies the surface reflected signal is a signal that is considered as noise,
and is composed of the reflected component of diffuse skylight and of the direct sunlight impinging on the water surface. Water bodies in general reflect (as subsurface
irradiance reflectance) in the range of 1 to 15% of downwelling irradiance. The majority of waters reflect between 2 and 6% of downwelling irradiance. Thus to obtain e.g.
40 levels of irradiance reflectance in the range of 2 to 6% reflectance we need a minimal accuracy of atmospheric correction to 0.1% reflectance.

Water body surfaces show swell, waves and capillary waves with facets of tens of
meters to a few centimetres. Although their distribution can be predicted in a stochastic
way, the inherent chaotic nature complicates adequate removal of surface reflectance
effects. Therefore flight planning of airborne imaging spectrometry campaigns needs to
consider the solar zenith angles and azimuths that minimise (given the FOV of the
scanner) imaging of sunglint effects of the water surface. A rule of thumb is that solar
zenith angles of 30° to 60° are optimal over water targets and that flight paths should be
flown at 0° or 180° headings with respect to the solar azimuth. At low latitudes this will
mean flying in a short period around noon in summer to achieve a maximal amount of
irradiance. At mid-latitudes the flight time will be dependent on the season: as the maximum solar zenith angles increases going from summer to winter -the flight time envel-
ope decreases from approximately 6 to 8 hours surrounding noon to two hours surrounding noon. At low latitudes solar noon must be avoided to avoid sunspot effects (direct
reflectance from horizontal water surfaces into the FOV); thus a situation arises with
two periods: one in the morning and one in the afternoon.

## 3    Optically deep and shallow waters: applications and case studies

### 3.1    INTRODUCTION

The previous paragraphs described one of the theoretical approaches to describing the processes in the underwater light field. Now we will present literature reviews and case studies of spectral measurement, modelling, simulation and imaging spectrometry applications. First, inland waters and estuaries as the two most studied optically deep water systems will be discussed. Next, seagrasses and coral reefs as the two most studied optically shallow systems will be discussed. After the literature review for inland waters a case study is presented for lakes in Germany where an inverse modelling method was applied to derive images of chlorophyll and suspended matter. These two variables often confuse simpler algorithms as chlorophyll is the pigment in the algae and the algae constitute part of the biomass that is part of the suspended matter. The estuary example will be discussed as it represents an application for optically deep water where it is currently possible to parameterise most of the inherent and apparent optical properties. The estuary example is for modelling and measurement based on in situ spectra only, as there are very few publications that describe actual airborne imaging spectrometers flown over estuaries, where there is no bottom visibility.

After these optically deep waters the optically shallow waters are discussed. First a discussion on the combined effects of a water column and a substrate takes place. In this discussion bathymetry play an important role. Next a literature review and a case study on seagrass remote sensing is presented. The final subject discussed are the coral reefs. In terms of the analytical model, the seagrass and coral reef examples represent a more hybrid situation where not everything can yet be described in terms of IOPs and AOPs. Therefore, it is necessary to rely more on *in situ* measured reflectance spectra in combination with analytical modelling or numerical modelling of the effects of the water column.

### 3.2    OPTICALLY DEEP INLAND AND ESTUARINE WATERS

#### 3.2.1    *Imaging spectrometry of optically deep inland waters*

A review of satellite and airborne remote sensing of aquatic ecosystems is given in (Kirk, 1983), summarily updated in (Kirk, 1994b). (Hilton, 1984) gave a review of airborne remote sensing. (Dekker *et al.*, 1995) wrote a comprehensive review of satellite and airborne remote sensing of inland waters, including imaging spectrometry. (Bukata *et al.*, 1995) present a sound treatise on remote sensing of inland and coastal waters, where the emphasis of the applications is on the Laurentian Great lakes in the USA. (Lindell *et al.*, 1999) reviewed the literature on satellite remote sensing of lakes and (Durand *et al.*, 1999) presented a review of satellite remote sensing of inland and coastal waters.

After 1984 remote sensing of inland waters has taken place mainly using data from satellite based sensors such as Landsat Thematic Mapper and SPOT-HRV (IRS –LISS series, CZCS, NOAA-AVHRR) and airborne remote sensing using instruments varying from multispectral scanners to line spectrometers and imaging spectrometers such as the CASI, AISA, AVIRIS, HYMAP and DAIS-7915. The CASI and AISA (and in a lesser degree HYMAP and AVIRIS) systems are not each one sensor with fixed capabilities. They are a family of sensors, whereby there is a progression in sophistication of the sensor with each new model developed. E.G for the CASI there are now approximately 20 systems operational. Each one of them has slightly different capabilities, whereby each upgrade to an existing sensor (specifically the case for AVIRIS) or each new sensor outperforms the previous version. Notice must be taken that the results of a CASI or AVIRIS flown in 1990 are not the same as the results for a CASI or AVIRIS flown in 2000 because the performance of the instrument has greatly increased. Moreover for the CASI an extra complication in comparing results is that it is a programmable imaging spectrometer, meaning that each application may have a unique spectral band set applied. For the development of high spectral resolution remote sensing applications, both imaging and non-imaging (either line or point measurements) data are of interest. Ground-based surface and subsurface spectral measurements may serve as surface calibration and as the link between the remotely sensed signal and the inherent optical properties.

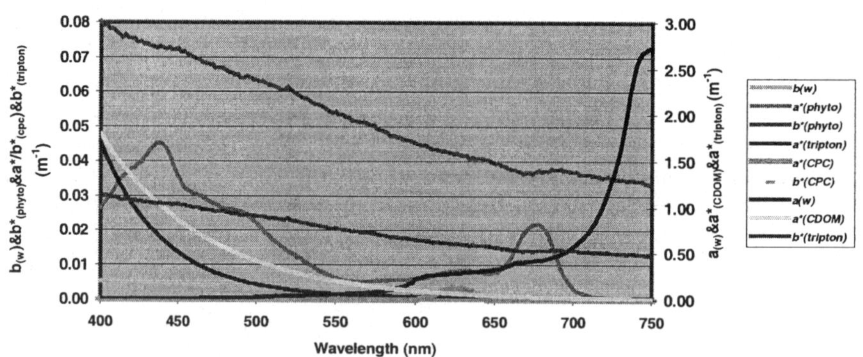


*Figure 7.* Inherent optical properties of Dutch inland waters (Dekker 1993). Left axis: b(w), a*(ph), b*(ph), a*(CPC), b*(CPC) and b*(tr); right axis: a(w), a*(CDOM)$_{norm\,440}$ and a*(tr). Units: a and b in ( m$^{-1}$) ; a* and b* of phytoplankton, CPC in  (mg  m$^{-2}$) and a*(tr) and b*(tr) in (g m$^{-2}$).

To summarise relevant information from the literature, a literature review is carried out discussing only those studies that actually made use of an airborne spectroradiometer (hand-held) or imaging spectrometer. This selection criteria is strict and excludes a significant amount of excellent work that discusses underwater and just above water measurements of inherent and apparent optical properties. Interested readers are referred to the reviews by (Dekker *et al.*, 1995), (Bukata *et al.*, 1995),(Lindell *et al.*, 1999)and (Durand *et al.*, 1999). In order to put the airborne spectrometry review into a correct perspective, simulations of reflectance using a bio-optical model are presented for algae dominated water and for total suspended matter dominated water.

### 3.2.1.1    *Simulations using a bio-optical model*

Figures 7 to 9 demonstrate the variability of inland water spectra. Figure 7 shows the inherent optical properties of inland waters as determined for Dutch lakes by (Dekker, 1993). In the case of inherent optical properties such as the absorption by the sum of chlorophyll *a* and phaeophytin (CHL) the specific inherent optical property is given (meaning the amount of absorption or scattering or backscattering per unit weight). Figure 8 shows a simulation run where CHL varied from 0-90 in 10 $\mu$ g l$^{-1}$ steps, CPC varied in 0-135 $\mu$ g l$^{-1}$ steps and a(cdom)$_{440}$ varied from 1.0 to 1.9 m$^{-1}$. The non-chlorophyllous suspended matter (tripton) was kept fixed at 1 mg l$^{-1}$. This is thus a simulation of a deep lake where a cyanobacterial dominated phytoplankton bloom is occurring.

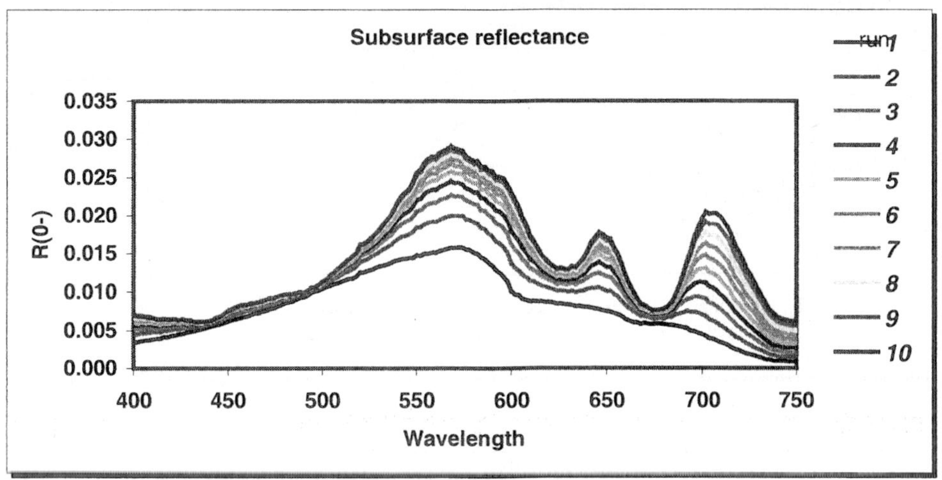


| run | 1 | 2 | 3 | 4 | 5 | 6 | 7 | 8 | 9 | 10 |
|---|---|---|---|---|---|---|---|---|---|---|
| Seston | 1.0 | 1.7 | 2.4 | 3.1 | 3.8 | 4.5 | 5.2 | 5.9 | 6.6 | 7.3 |
| Tripton | 1.0 | 1.0 | 1.0 | 1.0 | 1.0 | 1.0 | 1.0 | 1.0 | 1.0 | 1.0 |
| CHL | 0 | 10 | 20 | 30 | 40 | 50 | 60 | 70 | 80 | 90 |
| CPC | 0 | 15 | 30 | 45 | 60 | 75 | 90 | 105 | 120 | 135 |
| a(cdom)$_{440}$ | 1.0 | 1.1 | 1.2 | 1.3 | 1.4 | 1.5 | 1.6 | 1.7 | 1.8 | 1.9 |

*Figure 8.* Simulation run where CHL varied from 0-90 in 10 $\mu$ g l$^{-1}$ steps, CPC varied in 0-135 $\mu$ g l$^{-1}$ steps and a(cdom)$_{440}$ varied from 1.0 to 1.9 m$^{-1}$.

Figure 9 shows a simulation run where CHL is fixed at 1 $\mu$ g l$^{-1}$, CPC is zero and a(cdom)$_{440}$ is fixed at 1.0 m$^{-1}$. The non-chlorophyllous suspended matter (tripton) was varied from 10 mg l$^{-1}$ to 100 mg l$^{-1}$. This is thus a simulation of a lake or river where a substantial amount of suspended matter is entering the water column, either through river input or through wave-induced resuspension of bottom sediments. Figure 9 shows that if the main feature varying is TSM, reflectance increases over the entire spectrum, and this increase tends to saturate at higher concentrations of TSM. From the two figures it is clear that CDOM and pigments such as CHL and CPC as well as the tripton all contribute to lowering the reflectance at the blue wavelengths. Centred at 624 and 676 are the CPC and CHL induced reflectance troughs. These troughs are flanked by local reflectance peaks at 570-600 nm, 650 and 704-710 nm respectively. Note that no fluorescence term was required to simulate the often occurring reflectance peak at 706 nm. Many authors erroneously contribute this peak entirely to fluorescence whereas it is mostly due to a combined minimum in absorbing features at this wavelength. Indeed, the work (Gower *et al.*, 1999), who originally conceived the idea of measuring fluorescence by remote sensing in the eighties, demonstrates that above 10 to 20 $\mu$ g l$^{-1}$ CHL the fluorescence signal centred at 683-685 nm is absorbed by the broadening absorption of CHL.

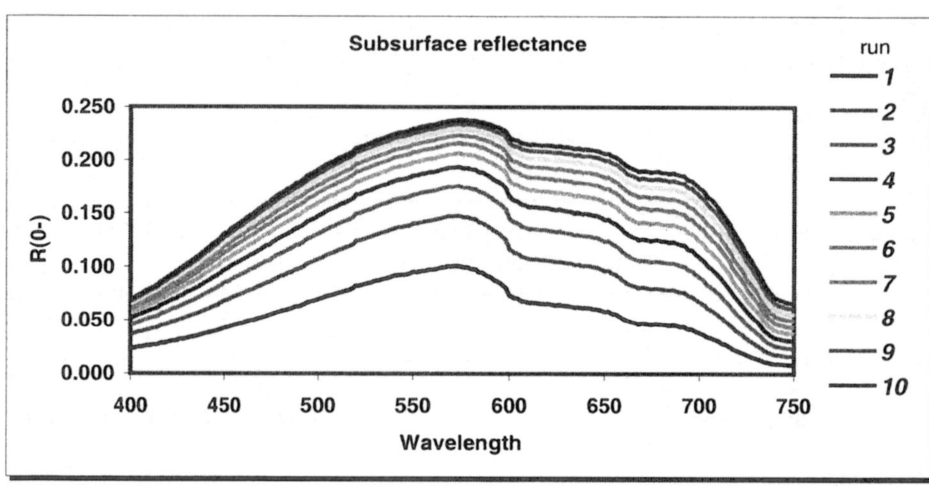



| run | ! | ; | ⊦ | ¡ | ; | ' | ; | ) | | 0 |
|---|---|---|---|---|---|---|---|---|---|---|
| Seston | 10.1 | 20.1 | 30.1 | 40.1 | 50.1 | 60.1 | 70.1 | 80.1 | 90.1 | 100.1 |
| Tripton | 10 | 20 | 30 | 40 | 50 | 60 | 70 | 80 | 90 | 100 |
| CHL | 1 | 1 | 1 | 1 | 1 | 1 | 1 | 1 | 1 | 1 |
| CPC | 0 | 0 | 0 | 0 | 0 | 0 | 0 | 0 | 0 | 0 |
| a(cdom)$_{440}$ | 1 | 1 | 1 | 1 | 1 | 1 | 1 | 1 | 1 | 1 |

*Figure 9.* Simulation run where CHL is fixed at 1 $\mu$ g l$^{-1}$, CPC is zero and a(cdom)$_{440}$ is fixed at 1.0 m$^{-1}$. Only the tripton part of seston is increased.

The simulations in Figures 8 and 9 may be used for optimal spectral band location of programmable imaging spectrometers such as the CASI or AISA.

### 3.2.1.2    *Literature review of imaging spectrometry of optically deep inland waters*

The earliest airborne imaging spectrometry results are published for the Programmable Multispectral Imager on Canadian waters in 1985; PMI and CASI campaigns over eutrophic Dutch lakes in 1988 and 1990; AVIRIS used for oligotrophic lake Mono and the saline lake Tahoe in the USA from 1990-1993; a CASI for Tennessee Valley Reservoirs mapping in the period of 1991-1992. The period 1993-1996 saw imaging spectrometers - mostly CASI's- deployed in Australia, Netherlands and German lakes. These more recent researches established a more analytical approach to hyperspectral remote sensing, and are thus to be considered as a breakthrough towards more quantitative, intercomparable, methods.

The largest data set and most consistent application was in the Netherlands where imaging spectrometry missions were flown in 1988, 1990, 1992, 1993,1995 and 1997 using PMI, CAESAR and a sequence of CASI sensors. A development in time is clear from initial more empirical and semi-empirical approaches towards more analytical approaches involving the use of parameterized bio-optical models that are subsequently inverted. Especially the work by (Jupp *et al.*, 1994) and (Hoogenboom *et al.*, 1998) and (Dekker *et al.*, 2001) illustrate the use of inversion methods, whereby optical water quality variables are inverted from the remotely sensed signal taking into account partially covering other optical water quality variables: e.g. estimating chlorophyll contents and TSM, whereby account is taken of the effect of the chlorophyll associated biomass of algae to the TSM contents.

From 1997 onwards, increased activity in Scandinavian countries and Germany is evident. Especially the work by (Olbert, 2000) and Schaale *et al.* (1998) is original and a portent of future methodologies involving the creation of massive amounts of simulated data based on (i) bio-optical modelling, (ii) air-water interface and atmospheric modelling, as well as (iii) incorporating the sensor look geometry across track, after which (iiii) either look-up tables or neural networks are created for inversion methods. See the case study of the Berlin lakes CASI images discussed later. From the literature (26 studies) it is evident that CHL is the prime variable measured in all cases (26); the blue-green (or cyanobacterial) pigment CPC in 8 cases; total suspended matter (or an equivalent thereof-the definitions and methods of measuring TSM are highly variable) in 16 cases; Secchi depth transparency in 18 cases; vertical attenuation of PAR in 8 cases and a few cases of turbidity (6) and CDOM. The CDOM determinations from the remote sensing data were not really successful as the sensors used all showed low sensitivity in the blue wavelengths, where CDOM absorption is most noticeable. The range in optical water quality variables detected during these campaigns was large:

TABLE 7. Review of measured water quality variables.

| variable | | Min | Max |
|---|---|---|---|
| CHL | ($\mu$g l$^{-1}$) | 0 | 1010 |
| CPC | ($\mu$g l$^{-1}$) | 0 | 1519 |
| TSM | (mg l$^{-1}$) | 1 | 700 |
| SD | (m) | 0.05 | 8.5 |
| Kd | (m$^{-1}$) | 0.4 | 10 |
| NTU | | 1 | 640 |

An excellent example of the use of imaging spectrometry is the determination of cyanophycocyanin, as only an instrument that can parameterise the absorption feature of CPC at 624 nm, by simultaneously measuring the local reflectance peaks at 600 and 648 nm (See Figure 8) is capable of determining the presence of cyanobacteria (Dekker, 1993) (Jupp *et al.*, 1994).


### 3.2.1.3  *Case study: inland water quality with inverse modelling*


A case study is presented of an inverse modelling approach using a radiative transfer model which is a coupled water-atmosphere model, in this case the MOMO model based on the matrix operator method (Fischer, 1983) Fischer, 1984 #382 (Fell and Fischer, in press). MOMO is a highly sophisticated model suited especially for the simulation of the radiative transfer in clear and turbid atmospheres including the water body with a rough water surface. Comparisons with a water-atmosphere Monte Carlo model as well as comparisons with airborne and underwater measurements show excellent agreement (Fell, 1997). This model allows the computation of spectral, sensor and solar zenith and azimuthal resolved radiances and reflectance. The type and concentration of atmospheric aerosols and water constituents are introduced by the extinction coefficient, single scattering albedo and the phase (volume scattering) function. The accuracy of the retrieval of water constituents depends on the choice of optical properties.

The CASI was flown on 23 April 1995 over the lake 'Tegeler See' an important fresh water reservoir for the city of Berlin, Germany. The instrument was operated in spatial mode with six bands in the visible spectral range and a spatial resolution of 2.5 x 3.5 m during a period of highest biological activity. The overall approach was to (i) convert the CASI sensor signals into radiances by a calibration and a radiometric correction (Babey and Soffer, 1992), (ii) to obtain the surface reflectance needed in the further analysis (Olbert, 1998) through atmospheric correction of the remotely sensed data, (iii) perform geometric correction, (iiii) apply the inversion to the water areas of the images.

The radiative transfer model MOMO was applied for the water constituents CHL, TSM and CDOM, using their optical properties (Olbert, 2000). The substance concentrations and the observation/illumination geometry were varied systematically in a given range (Table 8) to set up a look-up table for the further analysis whilst atmospheric parameters were fixed to realistic values.

TABLE 8. Substance concentration and observation geometry range for the MOMO run.

| Variable | Range | Step size | # of values |
| --- | --- | --- | --- |
| Chlorophyll *a* | $0 - 220 \ \mu g \ l^{-1}$ | $2 \ \mu g \ l^{-1}$ | 111 |
| Suspended matter | $0 - 100 \ mg \ l^{-1}$ | $2.5 \ mg \ l^{-1}$ | 41 |
| CDOM absorption | $0 - 35 \ m^{-1}$ @ 254 nm | $2.5 \ m^{-1}$ @ 254 nm | 15 |
| Sun zenith angle | $0° - 82.15°$ | $15.70° - 19.11°$ | 6 |
| Sun-observer azimuth distance | $0° - 180°$ | $11.25°$ | 17 |
| Observer zenith angle | $0° - 82.15°$ | $15.70° - 19.11°$ | 6 |

At the end of the model simulation runs the look-up table contained 42 million multi-spectral reflectance vectors for six bands. The inversion problem considered here may be solved by a simple look-up table approach, which is computational very time consuming, or by a faster neural network approach. The look-up table approach is straightforward and immediately applicable. The neural network approach consists of a sophisticated mathematical model and needs an intensive training of the network prior to its application.

Using the look-up table approach both the substance concentrations as well as the observation/illumination geometry has to be interpolated linearly. The neural network approach can be used as a multi-variable non-linear regression method. It was found more useful to represent look-up tables by a neural network including their non-linear interpolating and fast application properties. Each inverse modelling method needs to set up a look-up table where for a number of concentrations of water constituents based on a bio-optical model the reflectance spectra are known. Then, the concentrations, and their spatial distribution can be retrieved from remotely sensed reflectance spectra out of the database. For the inversion of the remote sensing measurements using the look-up table approach for each multi-spectral reflectance vector, or pixel, respectively, this corresponding simulated multi-spectral reflectance vector is searched for within the look-up table, where the Euclidean distance finds its minimum. The observation/illumination geometry was interpolated linearly. The concentrations for chlorophyll $a$, suspended matter, and CDOM are the result of this inversion. A significant advantage of this method is that no further pre-processing of the data is necessary. For the inversion of the remote sensing measurements using the neural network approach several manipulations were applied to the model data. Several forms of noise were added to the data.

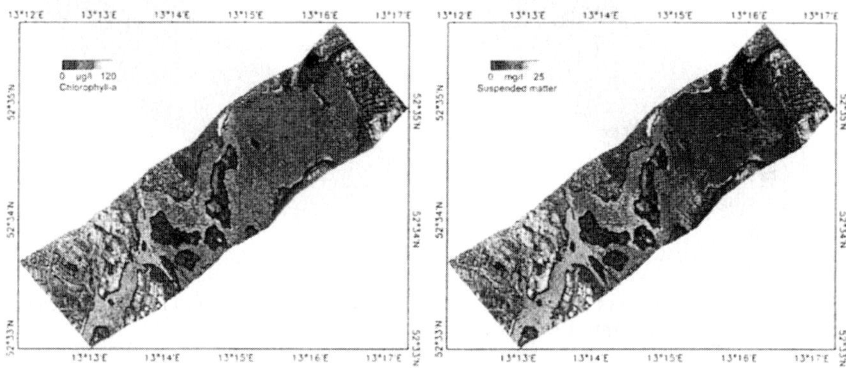

*Figure 10.* Spatial distribution of chlorophyll $a$ (left) and suspended matter (right) for lake 'Tegeler See' on 23 April 1995 retrieved by a look-up table approach.

Then, up to 1 ‰ random chosen input/output data sets were selected from the entire look-up table, and used for the training of an adapted feed-forward radial basis function (RBF) network (Bishop, 1995). 500 training steps in total were used for the training of a total of 25 neurons. The number of learning cycles should be high enough for a slow decrease of the cost function to obtain a robust solution. Three different neural networks were trained for each of the three substances: chlorophyll $a$, suspended matter, and CDOM.

The application of both inverse modelling methods to the CASI data reveals clear and well structured concentration maps for chlorophyll $a$ as well as for suspended matter (Figure 10 and 11; only chlorophyll is shown here for illustration purposes). The main (north east) part of the lake shows low chlorophyll $a$ and suspended matter concentrations while the attached waterways (south west) show higher concentrations. This is confirmed by *in situ* measurements that are in the same order of concentration ranges. Beside this main feature a variety of local features can be extracted: high gradients of concentration changes at local points (especially on the shore), traces of ships and possibly bottom effects along the shore. For CDOM no useful results were achieved. The spectral characteristics of CDOM overlap those of chlorophyll in the blue spectral region, therefore it is not simple to extract that part of absorption which is caused by CDOM.

The obtained results demonstrate the successful use of inverse modelling methods using MOMO for the retrieval of the spatial distribution of at least two optically active substances (chlorophyll $a$ and suspended matter) by multi-spectral remote sensing measurements above inland waters. Compared to the look-up table approach the neural network approach produces similar results more efficiently, by requiring less time. The higher effort for the training of the neural network in preparation of the application is the only disadvantage. Nevertheless, both methods allow the use of time consuming radiative transfer models for the simulation of more than one water constituent independently, and reasonable maps of the quantitative and spatial distribution of several inland water quality parameters can be generated.

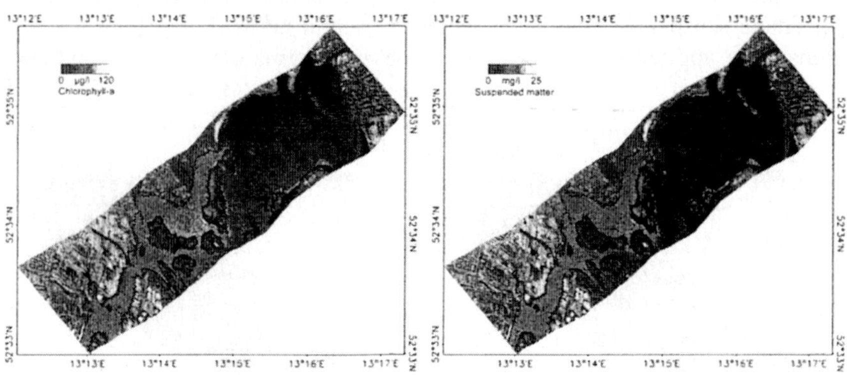

*Figure 11.* Spatial distribution of chlorophyll *a* (left) and suspended matter (right) for lake 'Tegeler See' on 23 April 1995 retrieved by a neural network approach.

### 3.2.2     *Imaging spectrometry of optically deep estuaries*

Estuaries are usually dynamic environments where freshwater meets ocean water. They are among the most productive aquatic ecosystems. Pressures from often conflicting uses, causes them to be studied extensively. (Cracknell, 1999) summarises the developments in the period of 1990-2000 in actual applied remote sensing of estuaries. However, few publications exist on imaging spectrometry applications over estuaries, based on inversion of analytical models. Some papers have been published on simple empirical methods, but these have no multitemporal or multi-site applicability. The study by (Ferrier and Anderson, 1997) is a good illustration of the frequency required for beginning to understand some of the dynamics in an estuary using remote sensing. Although they did not use imaging spectrometers, they collected in situ spectroradiometric reflectance measurements, a suite of aerial photography taken on different days as well as Airborne Thematic Mapper data (from which they mainly use the thermal band). With all this data they studied frontal systems in the Tay Estuary in Scotland. Another example is the work by (Siegel and Gerth, 2000) who studied the extent of a massive river Oder flood event using several types of satellite imagery, but basing their algorithms on previously carried out measurement and modelling of the underwater light field. They were able to produce maps of CHL, TSM and CDOM from the experimental satellite MOS sensor.

### 3.2.2.1     *Optical properties of estuarine waters*

The majority of studies that address the issues of spectral reflectance of estuarine waters, focus on the spectral optical properties of estuarine waters, with the aim of developing algorithms for processing satellite images of these estuaries. The following literature study discusses the studies that focussed on understanding the optical variability of estuarine waters. Next, the papers are discussed that proceed from these optical properties to inversion methods. After that, the literature is discussed of studies that continue this thread of analysis and actually apply it to i) non-hyperspectral satellite data, ii) airborne imaging spectrometry data and iii) studies in preparation of MERIS. MERIS will be the first dedicated imaging spectrometry system for coastal waters in space in the near future. It is arguable that MODIS and MOS sensors warrant discussion here as well. However these sensors are in essence multispectral systems designed for ocean colour analysis and less for inland and coastal waters.

(Ferrari *et al.*, 1996) studied the inherent optical properties of the Po Delta in northern Italy, focussing on the absorption coefficients of algal cultures and field samples and on the backscattering spectra of field samples. From this analysis they propose combinations of spectral bands (similar to that to be found on MERIS and MODIS) for determining CDOM absorption in deltaic waters. (Wernand *et al.*, 1997), measured 119 spectra in coastal North Sea and Dover Strait waters and were able to demonstrate that for this data set only 5 spectral bands were required to reconstruct the whole spectrum. They proved their methodology on an independent data set taken in the Baie de Somme, a turbid tidal estuary off the North France coast. They do admit that further investigation is required if this method would work for algal blooms or for chlorophyll fluorescence. (Kratzer *et al.*, 2000) describe the optical properties for the Menai Strait (Wales, UK): this strait is an excellent example of the variability in TSM and algal assemblages possible in one water body. Overall high values of TSM varies with lunar variation due to tidal action. The algal community composition starts off with a two-peaked spring bloom of first diatoms and then a mixture of diatoms and phaeocystis; after the bloom the concentrations decrease and small flagellates dominate. In summer a bloom of coccoid cyanobacteria is possible. Kratzer tried to develop algorithms based on four spectral channels only and concluded that this was an insufficient amount of bands to adequately invert such variations in concentrations.

(Frette *et al.*, 1998) present a further development of work initiated by (Jain and Miller, 1976) (Doerffer and Fischer, 1994; Doerffer and Schiller, 1994) (Tassan, 1994) which relies on an iterative matrix inversion technique for measuring three optical components in coastal waters a(CDOM), phytoplankton through CHL and TSM, often augmented by one other variable such as aerosol retrieval or a backscattering coefficient of the water. They all admit that these methods are computational intensive and probably not very useful for operational inversion of a remotely sensed signal. (Schalles *et al.*, 1998), present a study on detection of CHL, DOC and TSM in the estuarine mixing zone of Georgia coastal plain rivers. The range of concentrations for these optically active constituents was large and the source composition was highly variable ranging from terrestrial dominated, to riverine dominated to almost ocean type waters. Of particular interest in their study is the establishment of absence of covariance between most of these parameters. Their results demonstrated that many empirically established relationships would fail in remote sensing of such systems. They found poor correlation between DOC and blue wavelength reflectance (thought to be the best spectral region for CDOM/DOC measurements from remote sensing), but higher (inverse) correlation at the green wavelengths. This effect may be explained through the blue reflectance signal being confounded by multiple constituents (pigments, detritus, CDOM), whereas the green signal was dominated by the combined effect of TSM and DOC/CDOM.

(Forget *et al.*, 1999) discuss the inversion of a(cdom)$_{440}$ and the sediment refractive index for nonchlorophyllous turbid coastal waters for the Rhone river mouth in France. (Lahet *et al.*, 2000) developed a CDOM, phytoplankton and sediment bio-optical model for the Ebro River in Spain. Forget (2000) combines these two data sets (Ebro and Rhone) and discusses them in the context of two-flow radiative transfer modelling approach from (Aas, 1987). They specifically pay attention to the stratified system that may exist when a river plume flows in to relatively stable coastal water. In 66% of their inversion cases they were able to detect stratified water masses. Similar work was carried out by (Vasilkov *et al.*, 1999) who studied the spectral reflectance and transparency of river plume waters in the Black Sea and the Arctic Ocean using ship-borne and airborne spectroradiometric measurements. They measured reflectance differences of a factor 10 between river plumes and the ocean water at the other side of the front. They make a convincing case that river plumes spreading into relatively stable ocean waters have three dimensional wedge shape that needs to be considered when remotely sensing such phenomena. From airborne or satellite imagery a high concentration fresh water TSM rich layer overlays a much clearer ocean layer. (Woodruff *et al.*, 1999) carried out spectral analysis of waters of the Pamlico Sound estuary and the contributing river waters and found complex relationships relating reflectance to TSM. Their aim was to develop an algorithm for relating $K_d$ (PAR) to reflectance measured by the NOAA-AVHRR 630 nm band; this was partially successful. (Hoogenboom and Roberti, 2000) present preliminary results of a campaign intended to determine a method for optically measuring the natural variability of TSM in a heterogeneous and dynamic environment. For this purpose they use a suite of in situ optical properties measurements, ferry based spectroradiometric measurements, airborne imaging spectrometer (the EPS-A) flights and HYDROLIGHT to determine certain optical parameters. They parameterise the (Aas, 1987)/(Walker, 1994) model and thus are among the first to use the more complete model presented in the section on modelling.

(Dekker *et al.*, 1999) followed the analytical approach to study an estuary in Kalimantan (Indonesia) involving multitemporal use of SPOT and TM images. They determined the spectral IOP's of river, estuarine and ocean waters, developed a reflectance simulation model, (similar to Figs 8 and 9) to estimate the full range of TSM concentrations possible in this system and subsequently derived algorithms for analytically determining the TSM in this dynamic tidal estuarine area.

(Carder *et al.*, 1993) present the first results of AVIRIS flights carried out in 1990 over the Tampa Bay plume in the Florida coastal waters. Although at that time AVIRIS still had a relatively low S:N they were able to demonstrate the use of hyperspectral remote sensing for mapping the absorption coefficient at 415 nm and the backscattering coefficient at 671 nm. This work was followed through by (Lee *et al.*, 1994) who derived a model for modelling hyperspectral remote sensing reflectance over a variety of case 2 waters from the West Florida Shelf to the Mississippi River plume. Their model is based on the Gordon model and takes into account CDOM, TSM, CHL, Raman scattering and CDOM fluorescence, the only optical active variable missing is fluorescence by algal pigments. In (Lee *et al.*, 2000) a further development of this methodology is presented applied to AVIRIS data of 1998 (when the S:N has increased dramatically as compared to earlier versions). Because this paper discusses more the bathymetry effects of the substrate on the signal, further discussion takes place in the section on optically shallow waters. Mustard & Staid (1998) describe preliminary results of an AVIRIS flight over an estuary in New England (US). They were able to accurately model AVIRIS derived above surface reflectance. (Bagheri *et al.*, 2000) followed the same methodology as (Dekker *et al.*, 1999) and were able to closely match AVIRIS data flown over the Hudson/Raritan Estuary (New York) by simulation using a bio-optical modelling tool parameterised for the Hudson/Raritan estuary waters. (Eloheimo *et al.*, 1998) used an AISA imaging spectrometer over a coastal water area in the Archipelago of the Baltic Sea, also using satellite and in situ data. They tested algorithms proposed for MOS, SeaWiFS and MERIS for Secchi Depth transparency, turbidity, CHL and TSM. They did not undertake any spectral measurements or modelling. (Jørgensen and Edelvang, 2000) used a CASI to map the sediment plume caused by dredging for a tunnel in the Oresund, a coastal channel between Norway and Denmark. Their sediment maps were compared with a 2-D hydrodynamic model for calibration purposes. After measuring the inherent optical properties, and parameterising the simplified Gordon model it was decided to use the 544 nm spectral band as the band most suitable for determining TSM concentrations.

A special issue of the International Journal of Remote Sensing (1999, Vol 20, no 9) is dedicated to the relevant science surrounding the preparations for the MERIS sensor on board of ENVISAT: the first dedicated imaging spectrometry type sensor explicitly designed for remote sensing of optically deep coastal waters. (Buckton *et al.*, 1999) used data from the literature to develop an neural network based inversion model for MERIS. (Moore *et al.*, 1999) used the CASI airborne imaging spectrometer to develop and test algorithms for the MERIS space sensor. (Schiller and Doerffer, 1999) developed the neural network that will be implemented operationally to derive case II water properties from MERIS data on a global scale. (Gower *et al.*, 1999) presents the research leading up the CIIL fluorescence algorithm for MERIS, the first space sensor to have dedicated fluorescence band centred at 683 nm.


### 3.2.2.2    *Case study: The Western Scheldt Estuary*


Because of the complexity, an estuary is an excellent example to demonstrate the importance of the measurement of the specific inherent optical properties to parameterise the optical model. An estuary is a mix of two or more different water types, including at least river water and coastal water. Additionally, different sediment types may appear due to resuspension of bottom material from tidal flats or wind and wave induced resuspension of bed sediments. The history of wind direction and speed, tidal stages and other meteorological conditions are thus of importance for the concentrations and composition of suspended matter. The case considered deals with the analysis of the typical concentration ranges and specific inherent optical properties measured in the Western Scheldt estuary. These modelling results were applied to SPOT data for environmental baseline mapping of an intended area for construction of a (Pasterkamp *et al.*, 2000). Seasonal measurements of water quality parameters show that the primary parameter describing the optical properties in the estuary is the total suspended matter concentration (TSM). For 1998, the TSM concentration ranged between 4 and 120 mg m$^{-2}$, whereas chlorophyll *a* ranged between 2 and 20 $\mu$g m$^{-3}$ . (Maldegem, 1992) determined the mean seasonal variation of TSM. On average, mean and standard deviation increase towards the east. Dissolved organic carbon (DOC) concentration on the average increases from 1.5 near the estuary's mouth to more than 3.5 mg l$^{-1}$ towards the River Scheldt. There is only minor seasonal variability in the DOC concentration. The analytical model by (Walker, 1994) was used for the light anisotropy factor f. This model (see table 6) is given by

$$R(0-,\lambda) = f \frac{b_b}{(a+b_b)}$$

$$f = \frac{1}{\left(1+\dfrac{\mu_d}{\mu_u}\right)}$$

$$\mu_d = 0.7F + (1-F)\cos(\overline{\theta}_0) \qquad\qquad (21)$$

$$\mu_u = 0.5$$

$$\overline{\theta}_0 = a\sin(\sin(\overline{\theta}_z)/n)$$

A simulation model that calculates reflectance from the specific inherent optical properties and the concentrations is defined as follows

$$a = a_w + a_{cdom} + a^*_{TCHL}C_{TCHL} + a^*_{TSM}C_{TSM}$$

$$b_b = 0.5b_w + Bb^*_{TSM}C_{TSM} \qquad\qquad (22)$$

$$a_{cdom} = \overline{a}_{cdom}g_{440}$$

The specific inherent optical properties (SIOP) are the absorption/scattering per unit of mass. The SIOP are the fundamental properties required for optical models. It must be noted here that TSM incorporates all particles not passing through a filter pore size used to measure the CDOM concentration. Thus TSM consists of alive and dead organic particles and of inorganic particles. If possible it would be desirable to enhance the above equation of absorption to

$$a = a_w + a_{cdom} + a^*_{TCHL}C_{TCHL} + a^*_{IS}C_{IS} + a^*_dC_d \qquad\qquad (24)$$

with a$^*_{IS}$ and a$^*_d$ the specific inherent absorption of inorganic suspended sediment and detritus respectively. In the more simplified model used the main issue is to determine the backscattering ratio for the fraction of light scattered backwards towards the upper hemisphere. This ratio B is determined by fitting the measured reflectance spectra and the modelled reflectance spectra using the (Walker, 1994) optical model, the SIOP and the in-situ concentrations. In this way, an optical closure of the reflectance model is achieved. A note of caution must be issued here: an error in the inherent optical property determination or in the concentration measurements may influence the parameterization of the SIOP's and thus influence the estimation of B.

The in-situ measurements used here were taken in the scope of a project, initiated to monitor the ecological effects due to drilling for a tunnel and the resulting sediment dumping in the estuary using (among other techniques) remote sensing (Pasterkamp *et al.*, 2000). At five stations the subsurface irradiance reflectance $R(0-)$ was measured using a hand-held Photo Research (Chatsworth, CA, USA) model PR-650 spectroradiometer, and water samples were taken for subsequent laboratory analysis.

Optical properties (absorption and total scattering) were determined together with water quality parameters (TSM and total chlorophyll (Figure 12)), resulting in the specific inherent optical properties for each station (Figure 13).

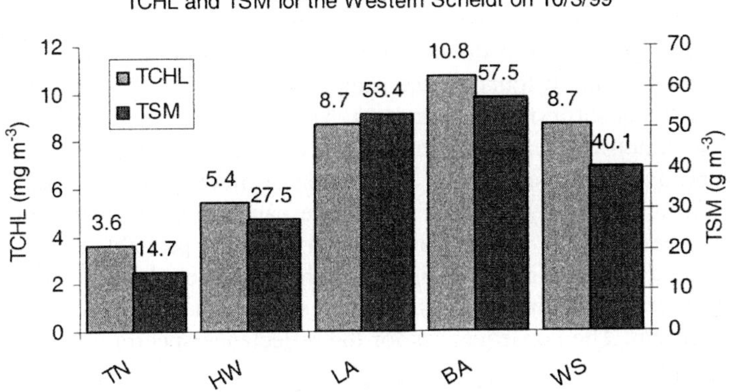



*Figure 12.* Measured water quality parameters in the Western Scheldt

For all stations absorption of coloured dissolved organic matter (CDOM) was considerable (a(cdom)$_{440}$ ranging from 1.4 to 1.8 m$^{-1}$). Apart from station 'WS' there is an apparent increase in TSM concentration in eastward direction.

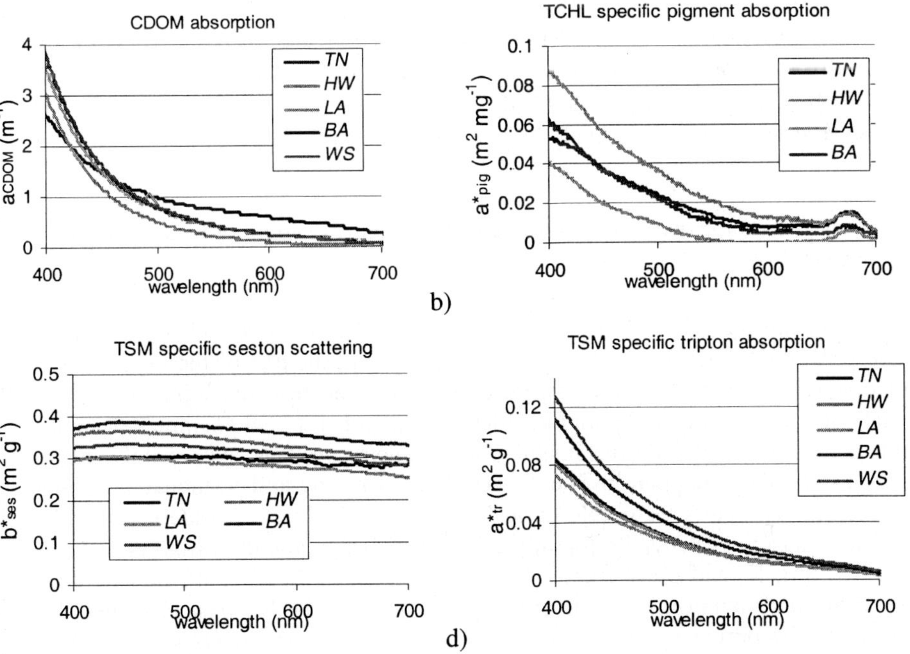



*Figure 13.* CDOM absorption (a), specific pigment absorption (b), specific seston scattering (c) and specific tripton absorption (d) measured for five stations in the Western Scheldt estuary.

For CDOM absorption (Figure 13), all exponential slopes are similar, except for the 'TN' station, which has a significant lower exponential slope than the other four. This may be attributed to the fact that 'TN' is closest to the sea, and might have a different (older) type of dissolved organic matter, more typical for seawater. For the other four stations, the CDOM absorption increases towards the river mouth. A large variation (up to a factor of 2) can be seen in the specific pigment absorption. However, because of the low CHL concentration found in the samples, relatively large measurement errors are expected in the absorption measurements. No arguments could be found to explain in detail the variation in measured specific seston scattering. A standard deviation of 10% was found over the five stations. A variation of more than 20% was found in the specific tripton absorption. A distinct difference can be noticed between the station nearest to the coast and the more inland stations. This is probably due to the higher fraction of organic matter. This analysis proves that specific inherent optical properties can differ considerably within estuaries, and that spatial gradients of water quality variables can be present. The consequences for the reflectance spectra will be discussed in the next paragraphs.

There is no unique solution for the modelling of $R(0-)$, unless the SIOP for that particular region are known. Using the specific inherent optical for stations 'TN', 'HW' and 'BA', and the mean SIOP of the five stations, the modelled spectra are compared to the measured reflectance spectrum. A standard deviation of 7 to 14% was found in the modelled R(0-) spectra, confirming that the variance in SIOP has a major impact on the modelled reflectance spectra. This analysis demonstrates that the specific inherent optical properties within an estuary can vary considerably leading to considerable differentiation in reflectance spectra. Optical modelling shows that in this estuary $R(0-)$ is optically dominated by the suspended sediment concentration, because the CHL and the CDOM concentration are relatively low.

### 3.2.3    *Conclusions for imaging spectrometry of optically deep inland and estuarine waters*

With the advent of imaging spectrometers that are suitable for water related investigations: the PMI, the AVIRIS and the CASI in particular at the end of the eighties, this field of research commenced. First missions and associated research were explorative rather than operational. From the mid-nineties onwards operational examples of multi-temporal deployments of airborne imaging spectrometry systems over mainly inland water targets started to happen in The Netherlands, Germany and Scandinavia. These studies over inland waters were able to deal with the optically deep waters quite well and produced meaningful results for up to 5 optical water quality variables such as CHL, CPC, TSM, $K_d$ and transparency. The combination of these results into ecological assessment and monitoring is gathering speed rapidly. The bio-optical models for these water are becoming more sophisticated as well as the instruments with which to measure the IOP and AOP properties. The CASI seems to be the instrument of choice, due to its flexibility in platform and its programmable band sets. The number of CASIs (20 by now) also determines the availability worldwide of course. In this field we see developments towards more complete models but also towards methods to compute an inversion of an imaging spectrometry scene using either analytical 1 to 3 band inversions, look up tables, using matrix inversion schemes or using neural networks. For turbid estuarine remote sensing, less use has been made of airborne imaging spectrometers due to the very dynamic nature and the often large size of the estuaries. Most of these studies were intended as illustrations or experiments in preparation of using satellite sensors to monitor these systems. As spaceborne imaging spectrometers become available this field of application is likely to evolve very fast.


## 3.3    OPTICALLY SHALLOW WATERS

Optically shallow waters are a special case in remote sensing of aquatic systems. It involves a measurable signal from a substrate, through the water column and through the air-water interface. In all applications where a substrate is being mapped through a water column, a bathymetry estimate is implicitly or explicitly involved. In the case of a bathymetry estimate from remote sensing a substrate reflectance and a water column optical depth is implicitly or explicitly involved. The bathymetry and the water column optical properties are thus of importance to the following remote sensing applications in aquatic environments: submerged macrophyte mapping, seagrass and macro-algae mapping, coral reef mapping, sand, coral rubble and mud floor mapping and benthic micro-algae mapping. Alternatively one may want to measure the optical properties or the concentration of substances in a water column over a visible substrate. In that case the bathymetry signal is seen as a component that must be corrected for (as with atmospheric correction) see e.g. (Gould and Arnone, 1997).

This section is composed as follows: first there will be a discussion of the literature on bathymetry and substrate mapping (usually brightly reflecting bottoms), how to simulate reflectance signals over such complex environments and what the requirements for imaging spectrometers are. The remainder of this section will discuss more in detail the state-of-the-art in seagrass, macrophyte and coral reef mapping, with a case study on seagrass, macro-algae and substrate mapping in a shallow coastal environment near Adelaide, Australia.


### 3.3.1    *Bathymetry and bright substrate mapping*

The term "measurable signal from the substrate" sums up one of the problems in performing bathymetry or measuring a substrate cover such as seagrass. It will depend on the spectral optical depth of the water column, on the brightness and spectral contrast of the substrate as well as the signal-to-noise performance of a remote sensor whether bathymetry or a substrate cover can actually be determined. If there is no measurable influence of the bottom on the remotely sensed reflectance the water is considered to be optically deep; if there is a measurable reflectance contribution from the substrate or plants in the water column the water is considered to be optically shallow.

In the theory section the model by (Maritorena *et al.*, 1994) is presented. Equation 18 expresses the $R(0-,H)$ as a function of the reflectance of an infinitely deep water column, the vertical attenuation coefficients and the irradiance reflectance of the substrate. In Figure 14 a simulation is presented of the following case. The same bio-optical model as used in the inland water simulations calculates the irradiance reflectance for an infinite water column. In this case the concentrations were: CHL = 1 µg l$^{-1}$, TSM = 2.1 mg l$^{-1}$ and the absorption by CDOM is expressed as a(cdom)$_{440}$ of 0.2 m$^{-1}$.

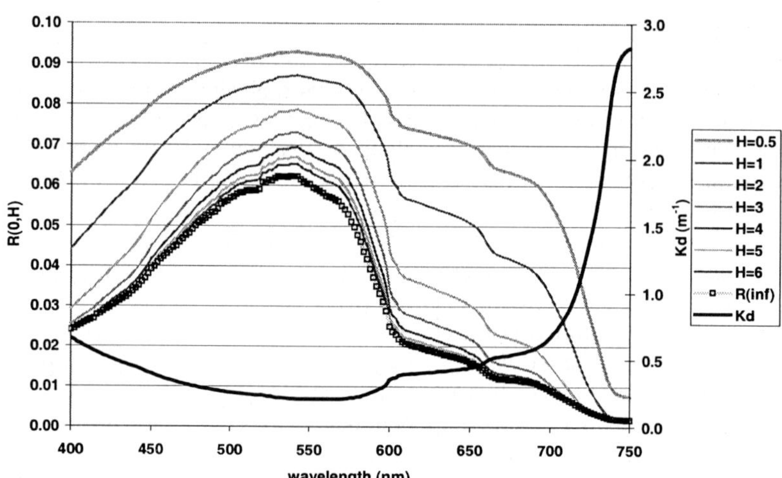



*Figure 14.* Simulation with CHL = 1 µg l$^{-1}$, TSM = 2.1 mg l$^{-1}$ and the absorption by CDOM is expressed as a(cdom)$_{440}$ of 0.2 m$^{-1}$

This same simulation software calculated the associated K$_d$, using the (Kirk, 1994a) model for relating Kd to the inherent and apparent optical properties. Next a substrate reflectance of 10% was assumed (spectral neutral- dark grey). The graphs in figure 14 show the resultant R(0-) calculated with the substrate at depths of 0.5 m, 1, m, 2m...6 m. From the results it can be deduced that for depths of 1, 2 and 3 m it would be required to measure R(0-) with an accuracy of 1 % in the green wavelengths in order to distinguish R(0-) differences due to the substrate at 1 m depth intervals. In terms of R(0-) this would equate to a discretisation S:N in terms of R(0-) of 100:1, which in turn leads to a S:N requirement of about 200:1 for R(0+) (as 48 % of upwelling irradiance just below the water surface is reflected into the water column). The required S:N in terms of R(0-) to distinguish the substrate between depths of 5 and 6 m quickly increases almost a factor 5 to 500:1 (R(0-) or 1000:1 (R(0+). These latter S:N specifications are about the maximum currently attainable by systems such as AVIRIS and CASI, flown under ideal circumstances. It is also evident from the graph that under these simulated conditions most information from the substrate reaches the water surface in the wavelengths of 500 – 600 nm. In reality the situation is more complex as we now assumed a flat water surface, a homogeneous diffuse light field and a neutral spectral behaviour of the substrate (See Table 9 for substrate reflectances).

TABLE 9. Values of high reflectance of bottom substrate from literature.

| Authors | type of substrate | $R_B$ at 400 nm | $R_B$ at 700 nm |
|---|---|---|---|
| Alberotanza *et al.* 1989 | coral sand | 28% | 60% |
| Lee *et al.* 1994 | offshore sed. | 40% | 56% |
| Maritorena *et al.* 1994 | coralline sand | 28% | 60% |
| Gould & Arnone, 1997[1] | white sand | 29% | 41% |

[1] 29 % at 412 nm and 41% at 670 nm.

For these highly reflecting bottoms in general there is an increase in reflectance towards higher wavelengths; the sharp increase of pure water absorption beyond 630 nm will effectively attenuate this highly reflected light, both on its downward as its upward journey through the water column. Some values for the clearest natural waters are those given by (Pozdnyakov and Kondratev, 1999) who give a maximal wavelength of penetration of 55 m for z90 at 475 nm in the Sargasso Sea.

In many coastal and inland water applications a bathymetry estimate from remote sensing has to deal with the following factors:

- A variable substrate coverage, in the case of macrophytes/seagrasses/macro-algae/corals with a canopy height and composition as additional variable
- A variable water column composition
- A variable air/water interface
- Variable geometry of direct sunlight, diffuse skylight, wave reflections and refraction and the sensor viewing geometry( field of view related zenith angles and azimuth angle)

For the depth of the bottom to be retrieved accurately each of these factors has to be determined through measurement or estimation. In principle a hyperspectral sensor with sufficient spectral bands has the capacity to independently determine each of these factors (Paredes & Sparo 1983), (as well as the atmosphere-which we ignore in this discussion). Further reading on the subject of the radiative transfer aspects of bathymetry may be found in: (Kohler and Philpot, 2000) (Pozdnyakov and Kondratev, 1999) (Durand *et al.*, 2000).

Lee *et al.* (1999) developed a semi-analytical model, based on above surface remote sensing reflectance $R_{rs}$, defined as the ratio of the water leaving radiance to downwelling irradiance just above the water surface, for shallow waters. They used HYDROLIGHT to parameterise the model. In the model downward and upward diffuse attenuation coefficients are explicitly described as functions of the absorption a and backscattering coefficients $b_b$, the bottom reflectance $R_b$ and the water column depth z. The only assumption made is the shape, not the magnitude, of the bottom reflectance. A set of coefficients is introduced by a computer model for a, $b_b$, $R_b$ and z. Through optimisation the modelled and the measured spectra are matched as closely as possible. This results in values for a, $b_b$, $R_b$ and z. Because the absorption coefficients are split into absorption by phytoplankton, the sum of CDOM and detritus and pure water absorption and the backscattering into particulate and pure water backscattering, this methodology also derives concentrations for algal pigments, the sum of CDOM and detritus absorption and the particulate matter concentrations (based on knowledge or estimates of the specific inherent optical properties). The model also incorporates the average cosines for up and downwelling radiance, the volume scattering function, and the sensor, solar and sky radiance geometries. The results were as followed: For computer simulated data the retrieved depth was accurate to within 5% (N=33) for a bathymetry range of 2 to 20 m. For field data in Florida bay it was accurate to within 11% (N=37) for a range of 0.8 to 25 m. For data outside of Florida Bay it was within 8% for depth (N=33). As is the case for most remote sensing analytical or semi-analytical retrieval methods the authors see most improvement possible in better measurements or estimates of the inherent optical properties of the optically active components.

Sandidge and Holyer (1998) used data from AVIRIS flown in 1993 over Tampa Bay and in 1996 over the Florida Keys in a neural network system to establish quantitative empirical relationships between depth and remotely sensed spectral radiance. No use was made of any type of physics based processing of the data to remove atmospheric or air/water interface effects. They used an extensive set of tens of thousands of depth soundings spanning a 1952 through to 1995 time range for Tampa Bay and for 1997 in the Florida Keys area. Relative to these depth data sets the RMS was 0.84 m for the Tampa Bay and 0.39 for the Florida Keys area. This RMS could have been reduced if the bathymetry *in situ* data deeper than 6m had been omitted as there was no bottom visibility beyond that depth. Also the AVIRIS was twice as sensitive in 1996 as in 1993. They also trained the neural network on both data sets which resulted in a RMS of 0.48 m. Apparently a neural network has generalisation capacity based on this research as the Tampa Bay and Florida Keys waters are very different with different substrates. These results indicate the effectiveness of neural networks for fast remote sensing data processing.

### 3.3.1.1 Bathymetry in inland waters

Hamilton *et al.* (1993) report on an AVIRIS flight in 1990 over the oligotrophic Lake Tahoe. They related the bathymetry of Lake Tahoe to upwelling spectral radiance recorded by AVIRIS using a two-channel multiple linear regression with AVIRIS data at 490 and 560 nm. They assumed a uniform bottom reflectance and were able to map bathymetry between the 3 and 10 meter contours.

George (1997) used a CASI to map bottom topography features in Lake Windermere in the UK. The data was flown in 1989 with one of the first versions of the CASI. According to George the shallow water had a higher reflectance in the nearby infrared due to increased bottom reflectance. This appears to be valid down to a depth of 20 m. If we look back at the simple simulation in Figure 14, this seems almost impossible as the $K_d$ for pure water at 700 nm is already 0.57 $m^{-1}$ and increases to 2.73 $m^{-1}$ at the wavelength interval of 746-759 used for this study. It is more likely that George found a correlation between resuspended sediment and reflectance in the nearby infrared and misinterpreted the correlation.

Tolk *et al.* (2000) performed experiments in mesocosms to investigate how brightness impacts the spectral response of a water column under varied suspended sediment conditions. They used a white aluminium panel as a bright bottom and a flat-black tank liner as the dark bottom. 16 levels of TSM were used. The major findings were that the bright bottom had the highest impact in the visible wavelengths. When suspended sediment levels reached 100 mg $l^{-1}$ there was no measurable signal from the bottom anymore. They found that substrate brightness had minimal impact at wavelengths between 740 and 900 nm, suggesting that these wavelengths are best for measuring suspended sediment for remote sensing. These findings are in direct contrast to those by George, based on the physics the conclusions by (Tolk *et al.*, 2000) seem more valid. However they should also have realised that the amount of passive solar irradiance at these longer wavelengths becomes lower and the sensitivity of Silicon based CCD arrays as used by many VIS-NIR spectrometers decreases rapidly at these wavelengths. Thus their conclusion that these wavelengths are best for remote sensing of TSM only applies to the case where the water depth is low, and the NIR wavelengths cause high attenuation due to the high water absorption.

### 3.3.1.2    Conclusions Bathymetry and substrate mapping

The (Lee *et al.*, 1999) methodology seems to be the most advanced approach if the optical properties are more or less known and there is some indication of the spectral shape of the substrate reflectance. The neural network approach by (Sandidge and Holyer, 1998) seems to be the alternative. There are two ways of training the neural network: by use of a massive amount of in situ data or by training it with a radiative transfer model such as described by (Lee *et al.*, 1999) and currently available through HYDROLIGHT. Of course the HYDROLIGHT model requires adequate parameterisation for the water bodies under consideration. The more empirical approaches loose their attractiveness once they are analysed with respect to the physics involved: often the empirical applications show ambiguous results. In many cases it is questionable whether a high signal was recorded due to substrate visibility or due to resuspension in a shallow water area.

### 3.3.2    Macrophyte/seagrass and macro-algae mapping

### 3.3.2.1    Introduction

Aquatic macrophytes provide an important habitat in both marine and freshwater environments. Species of macrophytic algae and those of higher plants can play a major role in providing a food resource for coastal and pelagic animals from zooplankton through to waterfowl. In addition, they may provide important refuges or shelters from grazing, provide nursery beds to other plant and animal communities and may help stabilise otherwise unstable or eroded shores.

Freshwater aquatic macrophytes are typically separated into three groups based upon their principal growth habit, a classification that can also be useful for marine species. Some species remain principally *submersed* throughout their life cycles. *Floating leaved* plants such as water lilies and some macrophytic marine algae have leaves or blades floating on or near the water surface. *Emergent* macrophytes penetrate through the water surface. Despite their ecological importance, human disturbance in a number of forms represents the largest threat to macrophyte-dominated communities and the life they support ((Wilcox, 1995) (Short and Wyllie-Echeverria, 1996) (Meinesz, 1999) (Pasqualini *et al.*, 2000).

The overall condition of macrophyte communities in both lacustrine and marine habitats can be directly related to the quality of the waters in which it lives Nutrients can stimulate the growth of phytoplankton and epiphytic algae that reduce light levels necessary for seagrass survival. There is evidence to suggest that it is extremely difficult, if not impossible, for seagrass species to re-colonise areas after they have died (Hick, 1997; Jernakoff and Hick, 1994). The need to map and monitor the distribution, abundance and diversity of these areas is therefore, of prime importance in assessing the status of these coastal systems.

### 3.3.2.2     *Literature review*

Traditional approaches to surveying the distribution and biomass of aquatic macrophytes have relied on potentially destructive quadrate and transect based methods similar to those typically used for ground-based survey of plant matter in terrestrial ecosystems. Such methods are also time consuming if the distribution of macrophytes over large areas are to be accurately mapped (Zhang, 1998). For mapping distribution over large areas remote sensing could offer a time-saving and potentially cost-saving non-destructive alternative (Lennon and Luck, 1990) (Mumby *et al.*, 1997) (Clark *et al.*, 1997).

A few studies have reported spectra for a range of species and growth habits (e.g. (Penuelas *et al.*, 1993) (Malthus and George, 1997) (Alberotanza *et al.*, 1999a). Figure 15 presents reflectance spectra measured over 7 types of freshwater macrophytes in Cefni Reservoir, North Wales, UK. These spectra are similar to those presented in (Penuelas *et al.*, 1993), and both studies can be used to draw general conclusions about the nature of reflectance from the different growing habitats for both freshwater and marine macrophyte species. Absolute reflectances from submersed species (e.g. *Isoetes lacustris*) are generally low, often lower than reflectance from deeper or background open water areas. Reflectance generally declines further with increasing wavelength. Red-edge increases in reflectance are generally not discernible except in the shallowest of waters as a result of high absorption by the water itself.

Remotely sensed reflectance over submersed macrophyte communities is influenced by both varying water column depth and turbidity. For example, Figure 16 illustrates a number of simulated seagrass reflectance compared to measured *in situ* reflectance of *Posidonia* around the coast of Sicily. Comparisons between the simulated and field-measured spectra indicate the relative consistency in spectral shape of the simulated output. However, much of the shape of these spectra is influenced principally by the influence of the water column rather than by seagrass reflectance, particularly the strong absorption of light in the longer wavelengths. Further simulations to investigate the influence of water column depth indicate that much of the useful signal reflected from submersed plant material is rapidly attenuated with increasing depth of the water and the bottom reflectance is diminished as it is filtered through the water column (Wittlinger and Zimmermann, 2000). All remotely sensed measurements of reflected radiance over submersed species will be similarly influenced by water column effects which will ultimately affect the accuracy with which spectral classifications of individual species can be performed (Mumby *et al.*, 1998a).

Over the last two decades a number of equations have been derived which have allowed for the either the calculation of water column depth, or in revised form the calculation of bottom reflectance after accounting for water column depth and attenuation (Lyzenga, 1981) (Moussa *et al.*, 1989) (Bierwirth *et al.*, 1993) (Jupp *et al.*, 1985) (Kirk, 1989) (Maritorena, 1996) (Maritorena and Guillocheau, 1996) (Maritorena *et al.*, 1994) (Tassan, 1996) (Pratt *et al.*, 1997). These techniques rely on the fact that the bottom reflected radiance is approximately a linear function of the bottom reflectance and an exponential function of water depth.

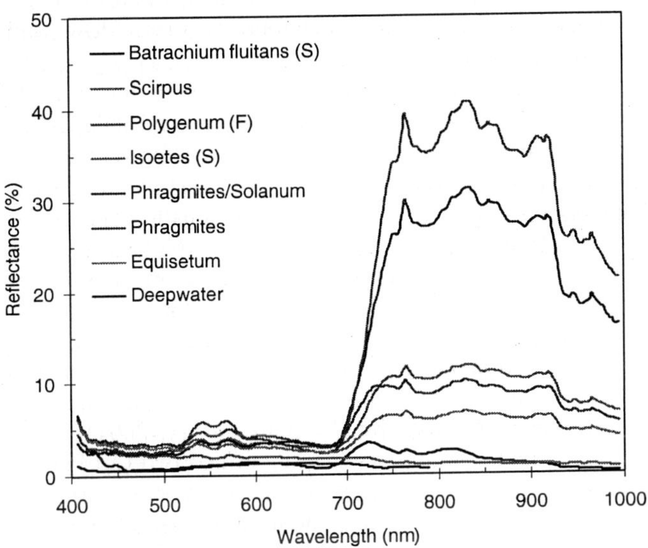


*Figure 15.* Above surface reflectance measured over different aquatic macrophyte species in Cefni Reservoir, North Wales, UK. Measurements were made using a Spectron SE590 spectroradiometer with a 15° field of view and referenced to above-surface measurements of downwelling irradiance obtained using a second sensor head fitted with a cosine correcting hemispherical receptor (from (Malthus and George, 1997).

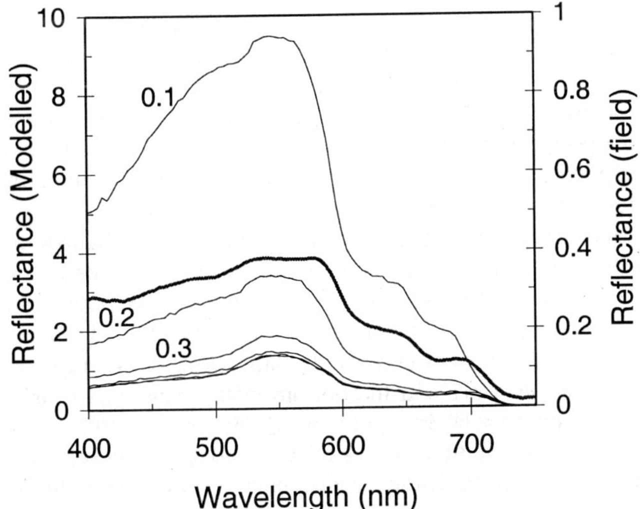


*Figure 16.* Comparison of sub-surface *Posidonia* reflectance (thick line) measured in Mediterranean coastal waters around Sicily with those simulated for increasing plant biomass using a Monte Carlo reflectance model (Malthus *et al.*, 1997). Differences between the measured and modelled spectra are attributable to the unknown parameters of the submersed vegetation in the field, the influence of epiphytic material and presence of dead leaves which were not included in the model.

Although uptake in the use of these approaches has been slow, the use of water column correction routines is now considered standard for the objective measurement of habitat change using remote sensing for routine monitoring of bottom habitats. Such a process has been shown to significantly improve the accuracy of classification of such habitats (Mumby *et al.*, 1998b) (Durand *et al.*, 2000).

In order to address the confounding influence of the water column in optically shallow waters, knowledge of the vertical attenuation for diffuse downwelling irradiance ($K_d$); the vertical attenuation for diffuse upwelling irradiance($K_u$) and for $\kappa$ the vertical attenuation coefficient for diffuse upwelling light originating in each layer of the water column, is necessary. However, few studies have used independently acquired estimates of attenuation, with most extracting $K_d$ values for relevant bands directly from their imagery in areas of uniform bottom type (e.g. sand) and known depth, assuming that these may be employed as a constant for attenuation across the image data. The work by (Lee *et al.*, 1999) presented in the section on bathymetry is an example of trying to infer IOP's and AOP's from the imagery itself making as little assumptions as possible. (Kenneth *et al.*, 2000) measured the bi-directional reflectance distribution function (BRDF) for varied benthic surfaces in Bahamas and pointed out that in specific cases a specular component does exist, while this parameter is generally assumed to be Lambertian. An investigation into the spatial variation in attenuation in a typical tropical region was undertaken by (Karpouzli *et al.*, 2001). Measurements of gross spatial variations in downwelling attenuation around two Caribbean islands indicate a four-fold variation in light attenuation in shallow littoral regions alone. Spectral attenuation measurements suggested that this variation was largely the result of scattering by particulate matter rather than varying concentrations of dissolved organic matter. This finding suggest that the results of studies where single measurements of 'average' attenuation have been used to depth-correct remotely sensed imagery should be interpreted with a high degree of caution.

Remote sensing techniques for bottom studies can include both sub-surface and above-surface techniques. Aerial photographic techniques have been commonly used for assessing the extent of macrophytes in both marine and freshwater bodies (e.g(Malthus *et al.*, 1987; Meulstee *et al.*, 1986) (Ferguson *et al.*, 1993; Marshall and Lee, 1994) (Green *et al.*, 1996) (Rutchey and Vilchek, 1999). However, if manually undertaken, photo-interpretation of such media is more often than not a subjective process and applied as an operational tool can be expensive due to the time taken for effective photo-interpretation (Jensen *et al.*, 1986).

Remote discrimination of substrate types in relatively shallow coastal waters has been limited by the strong attenuation influence of the water column (Holden and LeDrew, 1997; Holden and LeDrew, 1998b) and by the poor spatial and spectral resolution of available sensors(Dunk and Lewis, 1999). The principle component analysis (PCA), a common used tool in multispectral analysis as data reduction technique, has been successfully applied in many studies to determine the most representative spectra (Holden and LeDrew, 1998b) (Pasqualini *et al.*, 1997) (Bajjouk *et al.*, 1998). A number of researchers have investigated the application of airborne imaging spectrometers for mapping benthic plant species. (Bajjouk *et al.*, 1996) used an airborne CASI with ground based spectroradiometry, to test the ability of imaging radiometers to describe the principal seaweed and seagrass beds along the coast of

Brittany (France). Their analysis showed variation in relation to pigment characteristics, vegetation structure and environmental conditions. An algorithm was developed to discriminate the dominant species in which visible wavelengths allowed good discrimination between green, red and brown algae and infrared wavelengths allowed separation of brown species, seagrasses and floating seaweed.

The spectral reflectance characteristics of features within a submerged coastal environment are optically similar, so confusion can arise in identification. High spectral resolution sensors are required to perceive the subtle difference, which is demonstrated through analysis of in situ measurements in a study by (Holden and LeDrew, 1998a) Here the proportion of correctly identified spectra using first derivatives is 75% with the main source of error resulting from the inability to identify algae-covered surfaces. Similarly, (Malthus and George, 1997) evaluated the ability of Daedalus Airborne Thematic Mapper (ATM) imagery for mapping the distribution of freshwater aquatic macrophyte species in Cefni Reservoir on the Isle of Anglesey in the UK (see figure 15). Discriminant analysis indicated that good identification of macrophytes could be achieved by a combination of green, red and near infrared wavebands. A minimum distance supervised classifier using ATM bands 3, 7 and 8 showed separation of the species surveyed. The results indicated that airborne remotely sensed data have good potential for monitoring freshwater macrophyte species.

Hymap Imagery has been used for the mapping of seagrass distribution in a coastal environment in the upper Spencer Golf of south Australia (Dunk and Lewis, 1999). The data have been atmospherically and track illumination corrected. Three feature extraction techniques were evaluated: band ratios, PCA and spectral angle mapping (SAM). SAM was assessed to reliably discriminate features from selected endmembers (Alberotanza et al., 1999b). Another technique commonly used to extract qualitative maps of submerged aquatic vegetation is to measure reflectance spectra of different substrate in situ or in laboratory in order to create spectral libraries of seagrass meadows and substrate (Louchard et al., 2000). Higher derivative analysis has been found useful in resolving overlapping spectra, where the fourth derivative peaks occur at the same wavelength as those of the original spectrum enhancing useful information (Wittlinger and Zimmermann, 2000).

However, a radiative transfer model is an essential requirement to account for water attenuation and multiple backscatter of a reflective sea bottom(Durand et al., 2000). (Louchard et al., 2000) used a two-flow model to create an artificial spectral library of computer-simulated remote sensing reflectance and compared them to in situ spectra. In this analysis they were able to classify hyperspectral images by bottom type and bathymetry. However a two-flow model requires accurate signatures for either bottom albedo or water depths. Spectral libraries maybe a viable alternative to the two-flow model for distinguishing sediment from other bottom types, such as corals or seagrass from remote sensing images. In a study in South Australia CASI imagery has been used for mapping benthic species (see the case study at the end of this section, adapted from (Anstee et al., 2000). From the CASI imagery it was possible to discriminate different seagrass species, as well as various coverage of seagrass, *Ulva* and epiphytes within a pixel with high accuracy.

Considerable interest is being shown in hyperspectral sensor data, allowing the ability to overcome the spectral limitations of conventional spaceborne sensors such as Landsat TM and SPOT (e.g. (Jakubauskas et al., 2000)). Airborne hyperspectral sensors such as AVIRIS, HYMAP and CASI show promise and the space-borne hyperspectral instrument HYPERION is worthy of further investigation. This latter narrow-swath, 30 m spatial resolution instrument is the forerunner to several high spectral resolution sensors which will be launched in the near future (e.g. NEMO and ARIES) and with significant potential for mapping aquatic vegetation. The new generation of high spatial resolution satellites such as IKONOS, also show considerable promise for monitoring macrophytes.

In conclusion, the challenges facing digital above-surface approaches to the survey of benthic vegetation on both freshwater and marine environments, include issues associated with spatial, spectral and radiometric resolutions, surface effects and water column depth and turbidity. The ability of such techniques to monitor the extent and state of seagrass and lacustrine meadows is thus dependent on an understanding of the interaction between irradiance, the optically active constituents of the water column, the vegetation and the bottom sediment. Optical modelling research is perhaps the most effective approach in order to better understand the nature of above-surface reflectance from macrophytic stands, in particular if quantitative information is to be derived. Further research is required on the spectral signatures of both freshwater and marine macrophyte species at appropriate spatial scales for the development of appropriate indices for routine application. There is evidence that it will not be possible to accurate assess macrophytic biomass ((Malthus *et al.*, 1997), but c.f. (Armstrong, 1993) (Mumby *et al.*, 1997). New developments in high spatial and spectral resolution instruments show considerable promise for improved mapping capabilities from space. The synergy between these sensors and the information offered by subsurface acoustic instruments also warrants further investigation.


### 3.3.2.3    *Seagrass mapping case study*


A CASI was flown, in spatial mode using 12 spectral bands (see figure 17) over a portion of shallow coastal water near a sewage treatment plant near Adelaide, Australia in April 1999. The spatial resolution was 0.8 m, 2.0 m and 5.0 m. At the same time field reflectances of the benthos and estimates of the water's inherent optical properties were determined. Another field trip to validate the classification (see figure 18) was undertaken in February 2000, in similar seasonal conditions.
The overall approach was to model the atmospheric and in-water radiative transfer, the air-water interface and to use measured benthic reflectances. After atmospheric and air water interface correction, the image data was classified. A migrating means minimum distance classifier was seeded with 26 classes and then allowed to generate others to a limit of 256 classes. After final iteration 99.97% of pixels were classified. Post classification included canonical variate analyses and incremental sum of squares analyses. The 256 *basic* classes spawned from the classifier were then aggregated into 46 *super* classes using a spectral tool (*SpecTool*) and statistical summaries of the class variance. The *SpecTool* was used to model the field spectral data 'through' the optical water column model. It uses as inputs the spectral reflectance of pure specimens collected on the ground (endmembers), water depth information, and the optical water model, which uses the measured inherent water quality properties. Class signatures are compared to the end member spectra adjusted for the appropriate depth (Jupp *et al.*, 1996). In the classification there were a number of classes that were unidentifiable using the field-derived reflectance. These classes were labelled 'unknown'.

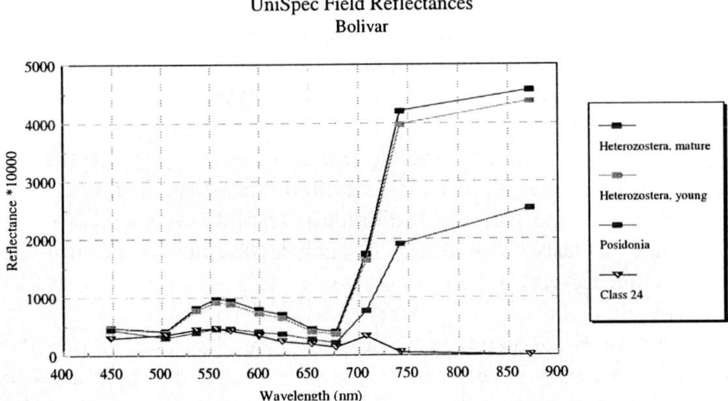


*Figure 17.* Effects of increased water absorption due to an increased water column depth. The poor fit of the *basic* Class 24, in Figure 17, to the *Posidonia* spectrum above 675 nm maybe attributed to increased water absorption due to an increased water column depth. Each of the markers is the centre wavelength location of a spectral band of the CASI.

The results of the classification are shown in Figure 18. To understand some classes that appear spectrally distinct and unmatched to library spectra a validation experiment was performed in February 2000. The region of interest was directly off the sewage treatment outlet. Random transects and point dives were carried out with differential GPS position recorded, surface and underwater photographs taken and spectral measurements taken of the benthos using a UniSpec Radiometer equipped with an underwater 10 meter cable. Particular attention was paid to the areas where the class was labeled 'unknown'. With the help of several taxonomic keys (Phillips and Menez, 1988) (Kuo and Mc Comb, 1989) it was possible to identify broad groups of seagrasses and macro-algae in the unknown areas of the 1999 image. However, it was not possible to separate *Heterozostera tasmanica* from *Zostera muelleri*, as the distinction is quite subtle. Because *H. tasmanica* is predominantly intertidal and *Z. muelleri* is predominantly subtidal, the species could be labelled correctly. The field-based separation between the *Posidonia ostenfeldii* and *sinuosa* was also quite difficult, but probably involved *sinuosa*, as it lacked the branching type structure. Quite obvious was *P. australis*, which had much broader and thicker leaves. *Amphibolis antartica* was identifiable through the alternate branching structure of the leaves.

The validation results indicate that the classification of the airborne imaging spectrometry data correctly identified 72% of the areas surveyed 11 months after the imagery was flown. The validation results indicated most of the 'unknown' classes to be a mix of the *Ulva* and *Heterozostera* (Figure 19).

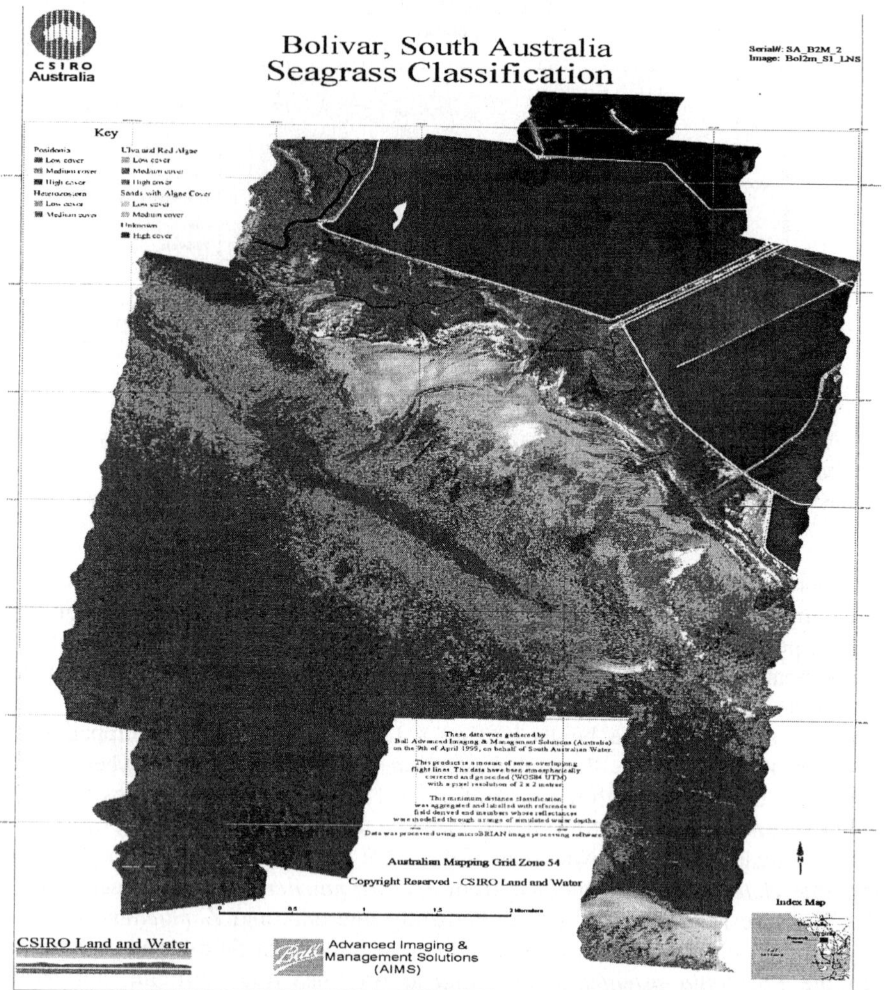



*Figure 18.* Hyperspectral Image Classification of benthic substrate species and cover for the February 1999 Bolivar Coastal site (from (Anstee *et al.*, 2000)).

### 3.3.2.4    *Conclusions imaging spectrometry for seagrass mapping*

Hyperspectral airborne data allowed discrimination of two seagrass species (*Posidonia* and *Heterozostera/Zostera spp*) and the green algae *Ulva* with 72% accuracy. It was also possible to discriminate various coverage's of seagrasses and *Ulva* and epiphytic cover within a pixel. Epiphytic infestations were found to complicate the spectral identification of the seagrasses, however linear mixing models were able to provide separation. This level of accuracy could not be obtained without modelling the atmospheric and in-water radiative transfer, air-water interface and benthic reflectance.

Aerial photography and hyperspectral imagery quality is dependent on weather. Hyperspectral imagery offers a significant improvement of aerial photos and can be temporally compared after atmospheric correction.

*Figure 19.* Intertidal zone where species are predominantly *Ulva* and *H. tasmanica* and was labeled
'unknown' in the classification

Hyperspectral imagery processed using this methodology can provide a much higher
level of spectral separation than colour aerial photography taken in the same area. This
methodology, once fully operational provides an objective classification tool, less
dependent of operator interpretation than aerial photography interpretation.


### 3.3.2.5    Coral Reefs

Coral reefs are the most spectacular and diverse marine ecosystems on the planet today
and degradation of coral reefs is a major environmental problem worldwide. Coral
reefs are often located in remote areas and may have large extent. There is a strong
management need for the mapping of coral reefs and for cost-effective assessment of
the environmental health or condition of reefs.

Remote sensing could be a valuable tool for mapping and monitoring coral reefs
and related ecosystems. A satellite image covers a large geographic area and has good
temporal coverage of most areas of interest. In the past, satellite sensors had limited
spectral and/or spatial resolution. The use of hyperspectral airborne instruments has increased the number of coral reef substrates classes that can be discriminated (Borstadt
*et al.*, 1997) (Clark *et al.*, 1997; Pratt *et al.*, 1997) (Holasek *et al.*, 1998) (Mumby *et al.*,
1998a; Mumby *et al.*, 1998b).

There are two different approaches in remote sensing analysis of coral reefs. The
first approach is image based. The image is divided into as many as possible classes
(Mumby and Harborne, 1999) and each class is named (i.e. mainly sand and rubble
with some living corals) after extensive ground truth measurements. This has been the
only method suitable for interpretation of satellite images because the 3 to 7 broad
wavelength bands (often not suitable for substrate discrimination) do not enable substrate mapping by their reflectance spectra. Hyperspectral images allow the collection
of endmembers for spectral unmixing from the image itself and this has been used to
increase number of bottom classes resolvable using remote sensing (Anstee *et al.*,

2000). The other approach is to measure reflectance spectra of different substrates *in situ* or in laboratory. This allows the creation of spectral libraries of coral reef substrates. Some research has been done to collect reflectance spectra of different coral reef benthic types (Holden and Ledrew, 1999; Maritorena *et al.*, 1994; Myers *et al.*, 1999) (Clark *et al.*, 2000) (Hochberg and Atkinson, 2000) (Louchard *et al.*, 2000) (Schalles *et al.*, 2000) (Stephens *et al.*, 2000) (Wittlinger and Zimmermann, 2000) (Zimmermann and Wittlinger, 2000). Most of the results in these papers have a limited regional scope: typically 3-4 species of living corals, a few different algae species and dead corals or sandy bottoms are measured etc. In a few cases a more systematic approach has been taken to create spectral libraries of coral reef benthic types containing reflectance spectra of hard and soft corals, sponges, seagrasses, dead coral covered with different algae, and different sand types ((Holden and LeDrew, 1998a; Holden and LeDrew, 1998b; Holden and Ledrew, 1999) (Holden and LeDrew, 2000; Kutser *et al.*, 2001; Kutser *et al.*, 2000a).

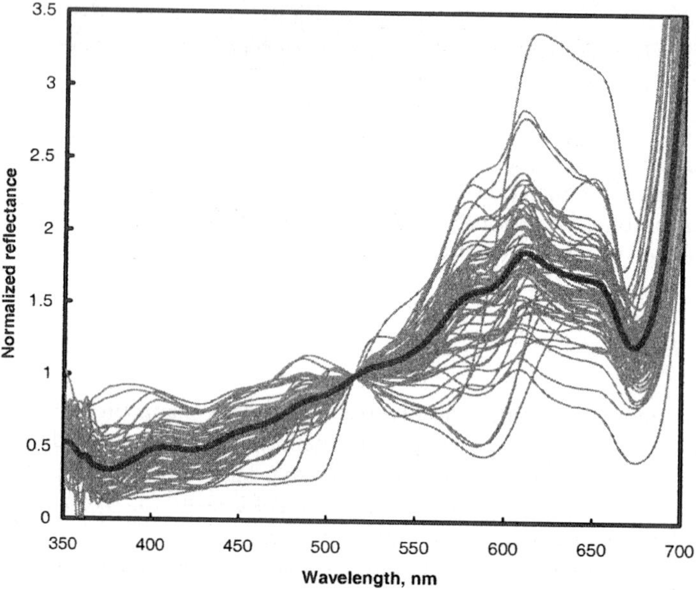

*Figure 20.* Variability in reflectance spectra of different living coral species. The "average living coral" spectrum is represented by the thick solid line. Reflectance values are normalised to a wavelength of 515 nm.

Assessing the biological state of the coral reef requires at least a capability to distinguish living coral (with symbiotic algae, *zooxanthellae*) from dead coral covered by overgrowth of benthic algae. Changes in substrate type from living corals to benthic algae, seagrasses, sponges or rubble can give valuable information to reef ecologists about the state of the reef. The spectral information required to achieve this discrimination has to be obtained through water with variable depth and optical quality. Methods have been developed to remove water column effects from remote sensing image (see section on bathymetry). Once a hyperspectral library of substrates is available and there is sufficient knowledge about the inherent optical properties of water under investigation it becomes possible to use spectral modelling or other analytical methods to remove water column effects and improve discrimination of different coral reef substrates. It may even make simultaneous detection of water depth and type of benthic substrate possible from remote sensing images (Bierwirth *et al.*, 1993) (Louchard *et al.*, 2000). Kutser *et al.* (Kutser *et al.*, 2000a) (Kutser *et al.*, 2001) have measured reflectance of about 140 different coral reef substrates in mid section of The Great Barrier Reef. About half of the studied substrates were living *Scleractinians* (reef building hard corals) of different species (Figure 20). Reflectance spectra of living hard corals are variable both in shape and value. Maxima in reflectance spectra are typically located close to 610 nm and reflectance values at this wavelength vary between 0.06 and 0.42. There are two "shoulders" in reflectance spectra of most measured living hard corals located around 580 nm and 650 nm. Reflectance is relatively lower in the shorter wavelength region of 350-500 nm. In most cases, the between species variability in the reflectance spectrum curve is relatively small in this spectral region. Almost all *Acropora* species (as well as some others) have a distinct minimum in the UV part of the spectrum (370-380 nm).

(Holden and LeDrew, 1998b) (Holden and LeDrew, 1998a) (Holden and Ledrew, 1999) (Hochberg and Atkinson, 2000) (Louchard *et al.*, 2000) (Schalles *et al.*, 2000) have measured reflectance spectra of corals living in other regions of the world ocean. In general the reflectance spectra are similar to those by Kutser *et al.* (Kutser *et al.*, 2001; Kutser *et al.*, 2000a) from the Great Barrier Reef. Exception here are reflectance spectra measured by (Schalles *et al.*, 2000) in the Red Sea. The Red Sea corals often show a peak close to 570 nm and shoulders near 510 and 630 nm.

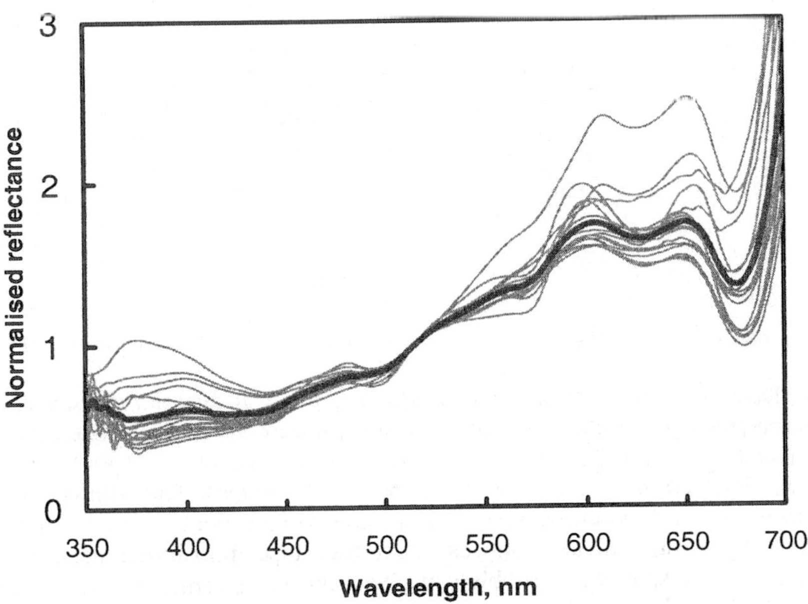


*Figure 21.* Reflectance spectra of standing dead corals and coral rubble covered with algae "turf" (normalized to 515 nm). The average reflectance spectrum is the thick solid line.

Dead coral skeletons, rocks and even sand (in places with low wave action) are often overgrown with filamentous algae (Figure 21). Most of these algae belong to red and green algae that have optical properties that differ from *zooxanthellae* (dinoflagellates) living in symbiosis with the coral polyps. The most distinctive spectral feature of algae covered substrates is a double peak in the reflectance spectra at around 600-605 and 650 nm. There is tendency that long time dead coral (covered with thick layer of algae) is darker (reflectance max 0.05-0.15) than recently dead coral (bright reflective skeleton with thin film of algae). Nevertheless, long time dead corals covered with crustose coralline algae have reflectance values (max 0.35) close to most living corals (Kutser *et al.*, 2001; Kutser *et al.*, 2000a). (Clark *et al.*, 2000) have found that long time dead corals covered with coralline algae have a higher reflectance than most living corals. This phenomenon can be explained with the low (max 0.17) reflectance of living *Porites* corals studied by them. It must be noted that algae invading the coral skeleton after the death of coral polyp and algae covering long time dead corals are not the same.

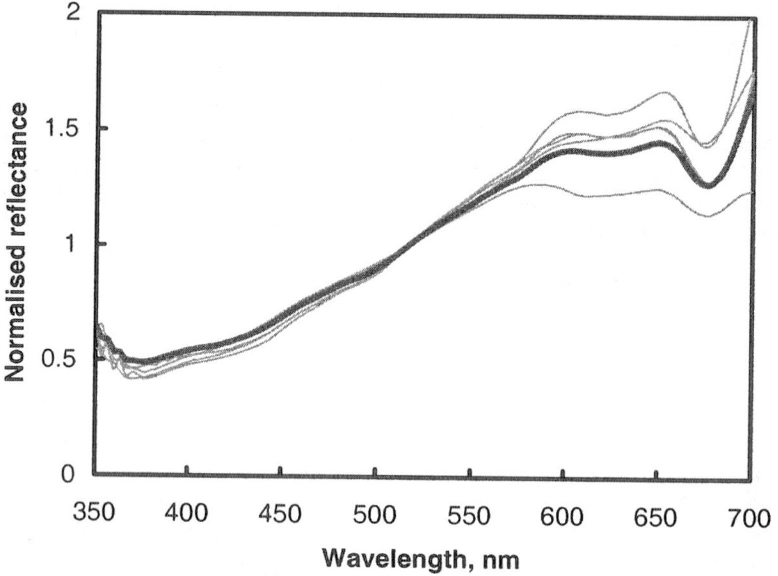

*Figure 22.* Normalized reflectance spectra of sandy bottoms in a coral reef environment.

The reflectance of bright white coral sand is much higher than the reflectance of other reef substrates (Figure 22). The shape of the spectrum is relatively flat with monotonous increase towards greater wavelengths (Maritorena *et al.*, 1994) (Holden and Ledrew, 1999) (Kutser *et al.*, 2000a; Kutser *et al.*, 2000b). This allows bright white sand to be relatively easily separated from other bottom types. Sand is often covered with coral rubble (overgrown with algae) or fine silt and other sediments (containing algae). This could explain why sandy bottom reflectances are not always higher than living or even dead coral reflectance. There is often a double peak in sandy bottom reflectance spectra, which is similar to the spectra of an algae-covered bottom, indicating the presence of algae on surface of sand particles (Kutser *et al.*, 2000a; Kutser *et al.*, 2000b) (Louchard *et al.*, 2000) (Stephens *et al.*, 2000).

(Holden and Ledrew, 1999) found the reflectance spectra of recently bleached corals to be higher than those of living corals and similar to the reflectance spectra of bright white coral sand. Kutser *et al.* (Kutser *et al.*, 2001) (Kutser *et al.*, 2000a) (Kutser *et al.*, 2000b) were able to study an *Acropora hyacintus* colony that had been dead for just one month (eaten by crown-of-thorn-starfish). This was already covered with thin film of green filamentous algae and the reflectance spectra were with the double peak typical to dead corals. The fact that the recently dead coral spectrum differs significantly from the living coral reflectance is important in monitoring of bleaching events.

Reflectance spectra of soft corals are similar to those of hard corals (Kutser *et al.*, 2001; Kutser *et al.*, 2000a) (Kutser *et al.*, 2000b). There is a minimum in the UV (370-380nm) of the reflectance spectra of most soft corals and peak close to 610 nm. The shoulder near 650 nm tends to be lower than the shoulder close to 580 nm in case of soft corals, whereas the shoulder reflectance values do not differ much in case of hard corals. Thus the difference between soft and hard coral reflectance is small. This makes differentiation between soft and hard corals difficult using optical methods. The reflectance spectra of different macro-algae species are highly variable and are amongst the lowest of all substrate types. Reflectance values do not exceed 0.09 in most cases. Some of the reflectance are similar to filamentous algae covering dead corals (red algae, blue-green algae), some to corals (bubble algae and brown algae) and some differ from all other reef substrate types corals (Kutser *et al.*, 2001) (Kutser *et al.*, 2000a) (Kutser *et al.*, 2000b). This makes reliable discrimination of macro algae from other substrate types difficult. (Wittlinger and Zimmermann, 2000) have measured reflectance of fleshy and coralline red algae and brown algae in the Caribbean. Their results are similar to Kutser *et al.* (Kutser *et al.*, 2001) (Kutser *et al.*, 2000a; Kutser *et al.*, 2000b): red algae have double peak in their reflectance spectra (near 605 and 650 nm) and brown algae reflectance spectra are similar to living corals reflectance (peak near 605 and shoulders near 580 and 650 nm). (Maritorena *et al.*, 1994) measured the reflectance of five algae species on coral reefs in French Polynesia. Their results agree well with above mentioned authors. brown algae (*Sargassum sp.* and *Turbinaria sp.*) are spectrally similar to living corals, green algae (*Boodlea sp.*) reflectance is different from all other substrate types and is similar to reflectance of *Chlorodesmis* measured by Kutser *et al.* (Kutser *et al.*, 2001) (Kutser *et al.*, 2000a) (Kutser *et al.*, 2000b) in GBR, and red encrusting algae (*Porolithon onkodes* and an unidentified algae) spectra are with the double peak (near 605 and 650 nm). (Anstee *et al.*, 2000) found that some seagrasses (*Heterozostera, Posidonia*) have reflectance spectra similar to *Boodlea* and *Chlorodesmis* (very low values below 500 nm, maximum near 550). They have also found that seagrasses covered with an epiphyte (such as *Gifforedia*) have reflectance spectra with the double peak similar to dead coral covered with filamentous algae. The number of reflectance of different sponges available in literature is too small to draw any conclusion about reflectance of sponges. Some of the spectra are close to living coral spectra, some to dead coral spectra. So, presence of sponges may decrease the level of confidence in remote sensing estimation of living and dead coral cover.


*3.3.3     Conclusions imaging spectrometry of optically shallow waters*


In the optically shallow waters developments have been somewhat different as the bio-optical or physical model describing the interaction of light in the water column and on the substrate is more complex than for optically deep waters and is less easily inverted. This inversion is required to produce meaningful maps of water variables or substrate variables. Similar to the developments in the inland waters, but increasingly complex due to the effect of the substrate, more sophisticated inversion schemes are being proposed. For bathymetry assessment over sandy bottoms and for seagrass mapping several successful examples are discussed or presented. Coral reefs are being studied increasingly as the environmental concern for coral reef health increases. Most work has been done to characterise the IOP's and especially the AOP's by establishing spectral libraries of coral reef reflectance. Very few imaging spectrometry data sets are available that were analysed for coral reef cover, health or species discrimination. Most applications were of a qualitative nature.


## 4     Conclusions

The field of imaging spectrometry is currently at a cross-roads: up till now all results for imaging spectrometry came from data gathered by airborne systems, as there were no civilian imaging spectrometry sensors in space. The successful launch of the EO-1 platform by TRW/NASA in November 2000, with the imaging spectrometer HYPERION on board marks the dawn of a new era- imaging spectrometry from space. This book and chapter is therefore marks the state-of-the-art for aquatic systems imaging spectrometry till the new space sensors results start arriving in the literature. With the advent of imaging spectrometers that were suitable for water related investigations: the PMI, the AVIRIS and the CASI in particular at the end of the eighties, this field of research commenced. First missions and associated research were explorative rather than operational. From the mid-nineties onwards, operational examples of multitemporal deployments of airborne imaging spectrometry systems over, mainly, inland water targets started to happen in The Netherlands, Germany and Scandinavia. These studies over inland waters were able to deal with the optically deep waters quite well and produced meaningful results for up to 5 optical water quality variables such as CHL, CPC, TSM, $K_d$ and transparency. The combination of these results into ecological assessment and monitoring is gathering speed. The bio-optical models for these water are becoming more sophisticated as well as the instruments with which to measure the IOP and AOP properties. The CASI seems to be the instrument of choice, due to its flexibility in platform and its programmable band sets. The number of CASI's (20 by now) also determines the availability worldwide of course.

For imaging spectrometry of aquatic ecosystems developments are towards more complete physics-based models but also towards methods to compute an inversion of an imaging spectrometry scene using either analytical 1 to 3 band inversions, look up tables, using matrix inversion schemes or using neural networks. For turbid estuarine remote sensing, less use has been made of airborne imaging spectrometers due to the very dynamic nature and the often large size of the estuaries. Most of these studies were intended as illustrations or experiments in preparation of using satellite sensors to monitor these systems. As spaceborne imaging spectrometers become available this field of application is likely to evolve very fast.

In the optically shallow waters developments have been somewhat different as the bio-optical or physical model describing the interaction of light in the water column and on the substrate is more complex than for optically deep waters and is less easily inverted. This inversion is required to produce meaningful maps of water variables or substrate variables. Similar to the developments in the inland waters, but increasingly complex due to the effect of the substrate, more sophisticated inversion schemes are being proposed. For bathymetry assessment over sandy bottoms and for seagrass mapping several successful examples are discussed or presented. Coral reefs are being studied increasingly as the environmental concern for coral reef health increases. Most work has been done here to characterise the IOP's and especially the AOP's by establish

ing spectral libraries of coral reef reflectance. Very few imaging spectrometry data sets are available that were analysed quantitatively for coral reef cover, health or species discrimination.

The last 10 years have seen a development towards more understanding of the physics of how spectral irradiance interacts with the water column and the substrate. Simultaneously the remote sensing sensors have developed towards systems with higher sensitivity and more (flexible) spectral bands available. A concurrent increase in computing power has led to a situation, (together with space imaging spectrometers being launched successfully) where we anticipate an fast expanding field of research, development, demonstration, operationalisation and commercialisation of imaging spectrometry of aquatic ecosystems.

As bio-optical and physics based models become more accurate, inversion schemes can become more sophisticated, enabling near real time processing of imaging spectrometry data. Simulations of the reflectance spectrum from waters, several examples given in this chapter, reduce the requirement for in situ measurements drastically in the long term. They also enable pre-flight determination of optimal spectral band configurations for specific tasks. As in situ detection and monitoring becomes more expensive (due to rising labour costs) and is shown to be less exact, a remote sensing based approach is beginning to make more and more economical sense. It will be necessary to have available local airborne imaging spectrometry systems or the availability of data from space sensors. The fact that the water column is an ever changing medium, both in space and time as well as in reflectance signature, indicates that imaging spectrometry will be the remote sensing instrument of choice for the future for accurate detection and monitoring of optical water quality and substrate variables.

# ACRONYMS

| | |
|---|---|
| 5S | Simulate the Satellite Signal in the Solar Spectrum |
| 6S | Second Simulation of the Satellite Signal in the Solar Spectrum |
| ADS | Absolute Diffuse Source |
| AIG | Analytical Imaging and Geophysics LLC, Boulder, U.S.A. |
| AIS | Airborne Imaging Spectrometer |
| ANPP | Aboveground net primary production |
| ANSWERS | Areal Non-point Source Watershed Environment Response Simulation |
| AOCI | Airborne Ocean Color Imager |
| AOP | Apparent optical properties |
| APAR | Absorbed photosynthetically active radiation |
| APEX | Airborne PRISM Experiment |
| ARCP | Absolute Radiometric Calibration Part |
| ARIES | Australian Resource Information and Environment Satellite |
| ARS | Average Reference Spectrum |
| ARS | Average Reference Spectrum |
| ASAS | Advanced Solid-state Array Spectroradiometer |
| ASCE. | American Society of Civil Engineers |
| ASD | Analytical Spectral Devices |
| ASTER | Advanced Spaceborne Thermal Emission and Reflectance Radiometer |
| ASTM. | American Society for Testing and Materials |
| ATCOR | Atmospheric Correction Program |
| ATLAS | Airborne Terrestrial Applications Sensor |
| ATMS | Airborne Thematic Mapper Simulator |
| AVIRIS | Airborne Visible/Infrared Imaging Spectrometer |
| BL | Beer-Lambert's  law |
| BOREAS | Boreal Ecosystem Atmosphere Study |
| BRDF | Bidirectional Reflectance Distribution Function |
| BRF | Bidirectional Reflectance Factor |
| CAESAR | CCD Airborne Experimental Scanner for Applications in Remote Sensing |
| CASI | Compact Airborne Spectrographic Imager |
| CCD | Charge Coupled Device |
| CCRS | Canada Center for Remote Sensing |
| CCSM | Cross Correlogram Spectral Matching |
| CDOM | Coloured Dissolved Organic Matter |
| CEM | Constrained energy minimization |
| CHL | Chlorophyll |

| | |
|---|---|
| CIRES. | Cooperative Institut for Research in Environmental Science |
| CIS | Chinese Imaging Spectrometer |
| CMOS | Complementary Metal Oxide Semiconductor |
| CPC | Cyanobacterial) pigment |
| CSES. | Center for the Study of Earth from Space, Boulder, U.S.A. |
| CSM. | Colorado School of Mines, Golden, U.S.A. |
| DAIS | Digital Airborne Imaging Spectrometer |
| DEM | Digital Elevation Model |
| DGPS | Differential Global Positioning System |
| DLR | Deutsche Forschungsanstalt führ Luft- und Raumfahrt e.V. |
| DN | Digital Number |
| DOC | Dissolved organic carbon |
| EL | Empirical Line |
| EM | Electromagnetic |
| EMS | Electromagnetic Spectrum |
| ENVI | Environment for Visualizing Images (RSI Inc.) |
| ENVISAT | ESA Envisat-1 satellite |
| EPA | Environmental Protection Agency |
| ESA | European Space Agency |
| $f_{APAR}$ | Fraction of absorbed photosynthetically active radiation |
| FLI | Fluorescence Line Imager |
| FOV | Field of View |
| FPA | Focal Plane Array |
| FWHM | Full Width at Half the Maximum |
| GEO | Geostationary orbit |
| GER | Geophysical Environmental Research |
| GERIS | Geophysical Environmental Research Imaging Spectrometer |
| GIFOV | Ground Instantaneous Field of View |
| GPS | Global Positioning System |
| HIPAS | Hyperspectral processing and analysis system |
| HIRIS | High Resolution Imaging Spectrometer |
| HRIS | High Resolution Imaging Spectrometer |
| HSR | Hyperspectral Remote Sensing |
| HYDICE | Hyperspectral Digital Imagery collection Experiment |
| HyMAP | Hyperspectral Mapper |
| IARR | Internal Average Relative Reflectance |
| IARR | Internal Average Relative Reflectance |
| IDL | Interactive Data Language (RSI Inc.) |
| IFC | In-flight Calibration |
| IFOV | Instantaneous Field of View |
| INS | Inertial Navigation System |
| InSb | Indium-antimonide |
| IOP | Inherent optical properties |
| IR | InfraRed (wavelength range) |
| IS | Imaging Spectrometry |
| IS | Integrating Sphere |
| ISU | Iterative spectral unmixing |
| JPL | Jet Propulsion Laboratory |
| LAD | Leaf angle distribution |

| | |
|---|---|
| LAI | Leaf Area Index |
| LEAFMOD | Leaf Experimental Absorptivity Feasibility MODel |
| LEO | Low-earth orbit |
| LSPIM | Land Surface Processes and Interactions Mission |
| LUE | Light Use Efficiency coefficient |
| LUT | Look Up Table |
| MAF | Minimum/Maximum Autocorrelation Factors |
| MAIS | Modular Airborne Imaging Spectrometer |
| MAMS | Multispectral Atmospheric Mapping Sensor |
| MAS | MODIS Airborne Simulator |
| MASTER | MODIS/ASTER Airborne Simulator |
| MDTM | Minimum Distance to Mean classifier |
| MERIS | Medium Resolution Imaging Spectrometer |
| MIR | Middle InfraRed |
| MISR | Multi-Angle Imaging Spectroradiometer |
| MIVIS | Multispectral Infrared and Visible Imaging Spectrometer |
| MLR | Multiple linear regression |
| MMF | Morgan, Morgan and Finney method |
| MNF | Maximum Noise Fraction |
| MODIS | Moderate Resolution Imaging Spectrometer |
| MODTRAN | Moderate Resolution Model of LOWTRAN7 |
| MOMO | Matrix operator method |
| MOS | Modular Optoelectronic Scanner |
| MSS | MultiSpectral Scanner |
| MTF | Modulation Transfer Function |
| MTI | Multispectral Thermal Imager |
| MTMF | Mixture Tuned Matched Filtering |
| NASA | National Aeronautics and Space Administration |
| Ne DeltaT | Noise Equivalent Temperature Difference |
| NEdL | Noise Equivalent Radiance Difference |
| NEMO | Naval EarthMap Observer |
| NER | Noise Equivalent Radiance |
| NES | Noise Equivalent Signal |
| NIR | Near InfraRed |
| NIRS | Near-Infrared Spectroscopy |
| NMP | New Millennium Programme |
| NOAA | National Oceanic and Atmospheric Administration |
| OSA | Optical Sensor Assembly |
| PAR | Photosynthetically active radiation |
| PARGE | Parametric Geocoding Application |
| PCA | Principal Component Analysis |
| PDF | Probability Density Function |
| PLI | Programmable Line Imager |
| PLS | Partial least squares |
| PPI | Pixel Purity Index |
| PRISM | Process Research by an Imaging Space Mission |
| PSF | Point Spread Function |
| PTIM | Photon transport inverse modelling approach (PTIM |
| PUSL | Pure Urban Spectral Library |

| | |
|---|---|
| RBD | Relative Absorption Band-Depth |
| RBF | Radial basis function |
| REP | Red edge position |
| ROSIS | Reflective Optics System Imaging Spectrometer |
| RRCP | Relative Radiometric Calibration Part |
| RT | Radiative Transfer |
| RTE | Radiative transfer equation |
| S/N | Signal to Noise Ratio |
| SAM | Spectral Angle Mapping |
| SCP | Spectrometric/geometric Calibration Part |
| SD | Secchi depth |
| SEBASS | Spatially Enhanced Broadband Array Spectrograph System |
| SFF | Spectral Feature Fitting |
| Si | Silicon |
| SIOP | Specific inherent optical properties |
| SIPS | Spectral Image Processing System |
| SIS | Scanning Imaging Spectroradiometer |
| SLEMSA | Soil Loss Estimator for Southern Africa |
| SMIRR | Shuttle Multispectral Infrared Radiometer |
| SNR | Signal to Noise Ratio |
| SPM | Sun Photometer |
| SPOT | Système Pour l'Observation de la Terre |
| SRF | Spectral Response Function |
| SSC | Spatial & Spectral Classification method. |
| SSTL | Surrey Satellite Technologies Limited |
| SUCROS | Simplified and Universal Crop Growth Simulator |
| SVAT | Soil-Vegetation-Atmosphere-Transfer |
| SWIR | ShortWave InfraRed |
| TACP | Thermal Absolute Calibration Part |
| TCHL | Total Chlorophyll |
| TDI | Thermal Delay integration Imagers |
| TIMS | Thermal Infrared Multispectral Scanner |
| TIR | Thermal InfraRed |
| TM | Thematic Mapper |
| TSM | Total Suspended Matter |
| UHI | Urban heat island |
| USGS | United States Geological Survey |
| UV | Ultraviolet |
| VIS | Visible part of the electromagnetic spectrum |
| VIS | Vegetation-impervious-soil |
| VNIR | Visible and Near Infrared wavelength range |
| WEPP | Water Erosion Prediction Project |
| WIS | Wedge Imaging Spectrometer |

# INDEX

## 5

5S, 40, 361

## 6

6S, 40, 361

## A

Abrams, M.J., XIII, XVII, 202, 206, 291, 296
Adams, J.B., 3, 49, 94, 145
aerosols, XXII, XXIII, 40, 332
agriculture, XXII, XXIII, 157, 158, 162, 175, 176, 247, 305
AIS, 18, 361
AISA, 18, 247, 272, 327, 337
albedo, 5, 6, 15, 16, 39, 41, 67, 68, 75, 149, 210, 217, 251, 252, 254, 256, 260, 266, 278, 313, 324, 332, 349
Allen, W.A., 116, 121, 126, 127, 128, 129, 137, 165
ANSWERS, 68, 361
AOP, 311, 313, 316, 320, 321, 340, 348, 358, 359, 361
APEX, 21, 361
ARIES, 23, 303, 350, 361
ARS, 40, 361
ASAS, 18, 361
ASD, 98, 102, 257, 258, 260, 262, 361
ASTER, 20, 21, 22, 203, 206, 244, 245, 281, 283, 294, 296, 301, 303, 305, 361, 363
Atkinson, P.M., 72, 84, 266
Atmospheric correction, 39, 184, 191
Average Reference Spectrum, 40, 361
AVIRIS, XIII, XVI, 18, 19, 24, 25, 52, 56, 58, 60, 76, 78, 93, 94, 102, 103, 104, 105, 106, 107, 108, 136, 145, 146, 148, 151, 152, 153, 154, 159, 160, 161, 162,
179, 184, 187, 188, 191, 192, 193, 195, 196, 198, 201, 203, 204, 205, 206, 211, 213, 214, 215, 216, 272, 273, 274, 275, 276, 303, 305, 327, 328, 330, 336, 340, 342, 344, 349, 358, 361

## B

Bakker, W., XIII, XVII, 56, 57, 203, 215
Bammel, B.H., 205, 228
Banninger, C., 146
Baret, F., 80, 117, 121, 128, 129, 137, 138, 139, 147, 157, 164, 165, 167, 175, 183, 192, 195, 237
Bathymetry, 341, 344, 345
Baugh, W.M., 93, 202, 203
Baumgardner, M.F., 12, 73, 74
Ben-Dor, E., XVII, 72, 73, 201, 248, 249, 250, 252, 253, 261, 263, 264, 270, 272, 276, 278, 279
Benediktsson, J.A., 56, 202
Bentonite, 89, 101
Binary encoding, 44, 45
Bleaching, 207, 224, 226, 231
Bluff, 206, 207, 208, 209, 210, 211, 213
Boardman, J.W., 49, 61, 94, 104, 262
BRDF, 4, 11, 27, 119, 120, 121, 138, 150, 154, 348, 361
BRF, 120, 361
Bunnik, N.J.J., 129, 165, 181, 186
Burrough, P.A., 70, 71, 84

## C

CAESAR, 179, 182, 184, 185, 186, 330, 361
carotenoid, 115, 116, 123, 142
CASI, 19, 146, 147, 248, 249, 250, 251, 252, 253, 254, 255, 256, 257, 259, 272, 274, 276, 278, 279, 327, 330, 331, 337, 340, 342, 344, 348, 349, 358, 361
CCRS, 31, 361

CCSM, 44, 56, 203, 215, 216, 218, 239, 361
CDOM, 329, 331, 332, 333, 334, 335, 336, 338, 339, 340, 342, 343, 361
Cellulose, 10, 130, 131
CEM, 44, 55, 361
Cement field, 226
C-H stretch, 131, 133, 219
Chabrillat, S., XIII, XVII, 97, 99, 100, 102, 105, 107, 108
CHL, 169, 170, 171, 172, 173, 174, 329, 330, 331, 332, 335, 336, 337, 340, 342, 358, 361
Chlorophyll, 114, 116, 123, 125, 130, 131, 182, 191, 361, 364
Chromoxerets, 254
Chung, J.J., 238, 274
Clark, R.N., 6, 9, 13, 14, 46, 47, 91, 92, 93, 94, 136, 204, 205, 220, 221, 222
Classification, 44, 56, 84, 352, 364
Clevers, J.G.P.W., XIII, XVII, 117, 147, 148, 152, 157, 163, 164, 168, 173, 174, 175, 176, 179, 181, 182, 184, 186, 190, 192, 196
Conel, J.E., 137, 163, 205, 207
Constrained energy minimization, 44, 55, 361
COSMOS, 244
CPC, 329, 330, 331, 340, 358, 362
Crowley, J.K., 9, 46, 47, 202, 206
Cuprite, 52, 58, 60, 202, 206, 213, 214, 217, 218, 296, 297
Curran, P.J., 36, 112, 121, 131, 132, 133, 134, 135, 137, 139, 142, 146, 153, 154, 158, 163, 204, 237

## D

DAIS, 19, 84, 204, 274, 276, 362
DART, 138
Dawson, T.P., 137, 138, 151, 163, 204, 237
De Jong, S.M., XVI, 61, 65, 70, 71, 82, 84, 205
De Roo, A.P.J., 68, 70
Dekker, A.G., XVIII, 31, 328
DOC, 335, 337, 362
Duchscherer, W., 223
Dury, S., XVIII, 135

## E

Elvidge, C.D., 10, 11, 80, 124, 130, 131, 132, 134, 146, 147, 148, 158, 228
Emissivity, 284

Environmental geology, 204
environmental hazard, 87, 94, 204
ENVISAT, XXIII, 115, 337, 362
EO, 22
Epema, G.P., XVIII, 75
Escadafel, R., 72
Evapotranspiration, 68
Expansive clays, 100

## F

Fabbri, A.G., 238
FAO, 65, 66, 71, 72, 73
Farrand, W.H., 55, 202, 204, 279
FLI, 18, 229, 362
Focal Plane Array, 362
*FOCUS*, 305, 306
Foreground-background analysis, 44, 55
FWHM, 41, 43, 362

## G

Gao, B.C., 104, 137, 139, 163, 205, 272, 279
Gates, D.M., 10, 11, 115, 118, 119, 121, 122, 123, 134
Geocoding, 35, 363
Geophysical inversion, 44, 57
GER, 19, 112, 201, 203, 204, 362
GIFOV, 19, 20, 362
Gillespie, A.R., 206, 287, 291
Goel, N.S., 80, 129, 149, 165, 182
Goetz, A.F.H., XVIII, XXI, 3, 17, 22, 44, 72, 81, 93, 102, 107, 137, 139, 142, 153, 158, 163, 201, 205, 206, 225, 230, 247
Goldfield, 202, 213
GPS, 24, 35, 279, 351, 362
Green, R.O., 45, 61, 94, 102, 115, 124, 153, 163
Guyot, G., 117, 147, 164, 172, 173, 175, 183, 192, 195, 237

## H

Hapke, B., 15
*Heterozostera*, 351, 352, 357
Hill, J., 65, 76, 77, 78, 97, 300
Hook, S., XVIII, 203, 206, 291, 296
Horler, D.N.H., 142, 145, 158, 163
Huete, A.R., 82, 137, 149, 157
Hull, 6, 45, 46, 299
Hunt, G., 3, 7, 9, 17, 73, 74, 91, 143, 203, 222, 226, 227, 232, 251, 252, 283, 284

Huntington J.F., 78, 104
Hydrocarbon, 205, 222, 228, 233
HYDROLIGHT, 323, 343, 345
Hydrothermal alteration, 202, 214
HyMAP, 19, 102, 107, 210, 237, 247, 272, 273, 362

## I

IARR, 40, 279, 362
IFOV, 19, 20, 21, 28, 31, 32, 33, 210, 296, 362
IKONOS, 246, 350
INS, 24, 279, 362
Interception, 68
Internal Average Relative Reflectance, 40, 362
*Intimate Mixture*, 16
IOP, 311, 312, 316, 317, 318, 323, 336, 340, 348, 358, 359, 362
Iqbal, M., 173
ISU, 51, 53, 362
Iterative spectral unmixing, 44, 51, 55, 362

## J

Jacquemoud, S., 121, 124, 128, 129, 134, 136, 137, 138, 139, 150, 165, 167
Jongschaap, R.E.E., XVIII, 175, 178
Junggar Basin, 227
Justice, C.O., 20

## K

Kahle, A., 20, 206, 287, 291, 301
Keirein-Young, K.S., 203
Kerekes, J.P., 25
K-M theory, 128, 129
Kroonenberg, S.B., XVIII
Krosley, L., XVIII
Kruse, F.A., 47, 93, 104, 201, 203, 204, 279
Kubelka-Munk theory, 127, 128
Kumar, L., XIX, 128

## L

LaBaw, C, 18
LAD, 151, 153, 159, 162, 165, 168, 169, 174, 179, 180, 181, 182, 183, 186, 187, 188, 189, 198, 362
LAI, 136, 138, 139, 146, 147, 148, 149, 150, 151, 152, 154, 157, 158, 159, 162, 163, 165, 168, 170, 171, 172, 174, 176, 179, 180, 181, 182, 183, 184, 186, 187, 188, 189, 190, 193, 194, 196, 197, 198, 363
Lambertian, 5, 114, 120, 126, 129, 348
Lambertian surface, 5
Lang, H.R., XIX, 204, 219, 230, 231
Leaf optical models, 127
Leaf optical properties, 115, 121, 139, 182, 185
Leaf water, 127, 143
LEAFMOD, 139, 363
Lignin, 131, 132
*Linear Mixture*, 16
Linear unmixing, 44
Lisbon Valley, 207, 226, 227
Lost River, 230
LOWTRAN, 40, 160, 184
LSPIM, 363
LUT, 363
Lyon, R.J.P., 203, 205, 284

## M

Macroseeps, 222, 223, 233
MAF, 38, 363
MAS, 20, 363
MASTER, 20, 296, 297, 363
MDTM, 86, 363
MERIS, 23, 33, 41, 43, 44, 190, 191, 192, 193, 194, 195, 196, 197, 198, 199, 335, 337, 363
Methane, 233
MLR, 133, 135, 139, 363
MMF, 68, 363
MNF, 38, 61, 104, 363
MODIS, 20, 22, 154, 296, 305, 335, 363
MODTRAN, 40, 277, 279, 363
*Molecular Mixtures*, 16
MOMO, 331, 332, 333, 363
MTI, 302, 363
Mustard, J.F., 94, 204, 337

## N

NDVI, 82, 125, 142, 145, 146, 147, 148, 149, 150, 154, 240
NEMO, 23, 303, 350, 363
NER, 31, 363
NIRS, 114, 132, 133, 134, 363
Nitrogen, 144, 175, 177, 178
NMP, 22, 363
NS001, 291, 293

**O**

Olsen, H., XIX, 97, 99

**P**

PAR, 115, 150, 180, 182, 186, 187, 331, 336, 363
Patrick Draw, 230
PCA, 60, 159, 348, 349, 363
Peñuelas, J., 112, 142, 143
Petroleum, 205, 206, 233
Photosynthesis, 124
Physiological structure, 80
Pieters, C.M., 73, 74, 94
Pigment, 141
Planck, M., 4, 5, 284
Plant stress, 144
Plate model, 129
PLS, 135, 363
Posidonia, 346, 347, 351, 352, 357
PPI, 61, 104, 363
PRISM, 21, 23, 33, 361, 363
Probe, 19, 210, 240
PROSPECT, 129, 137, 138, 165, 166, 167, 187, 191
Protein, 131
PSF, 33, 43, 44, 363
Pushbroom, 33

**R**

Radiance, 4, 31, 277, 363
Radiation, 4, 5, 39, 113, 130, 231, 232
Radiative transfer, 39, 137, 364
Rast, M., 23, 190, 191, 247
Ray tracing, 127
RBF, 333, 364
Red-edge, 163, 175, 192, 193, 231, 346
Red-edge index, 163, 175, 192
REP, 147, 364
Rhodoxeralfs, 254
Richter, R., 266, 267, 279
Rock, B.N., 140, 141, 146
ROSIS, 20, 33, 364
Rowan, L.C., 17, 203, 225, 287
RSADU, 138
RTE, 316, 317, 318, 319, 364

**S**

SAIL, 129, 137, 138, 139, 150, 165, 166, 169, 186, 187, 191

Salisbury, J., 3, 9, 74, 91, 284
SAM, 44, 47, 51, 58, 60, 104, 212, 213, 237, 239, 349, 364
Santa Barbara, 233, 238
Sao Francisco Basin, 230
Saunders, D.F., 207, 223, 224
SAVI, 82, 148, 268
Schmidt, K., XIX
Scholte, K.H., XIX
Schumacher, D., 224, 226, 227, 228, 232
SD, 331, 364
Seeps, 233, 237
SFF, 44, 47, 364
SIOP, 338, 340, 364
Skidmore, A.K., XX
SLEMSA, 68, 364
smectite, XXII, 89, 91, 93, 95, 97, 98, 100, 101, 104, 106, 107, 108, 207, 224
Soil erosion, 66, 82
Spectral Angle Mapping, 44, 47, 213, 364
Spectral Feature Fitting, 44, 47, 364
Spectral radiance, 4
Spectral unmixing, 44, 47, 53, 55
SRF, 34, 41, 42, 43, 364
SSC, 84, 85, 364
Starch, 11, 131, 132
Stefan-Boltzmann, 4
Stochastic model, 130
Stoner, E.R., 12
SUCROS, 152, 179, 180, 183, 184, 187, 188, 189, 364
SVAT, 364
Swelling soils, 87

**T**

Tasselled Cap, 82
TCHL, 338, 364
Tedesco, S.A., 222, 223
TERRA, XXIII, 20, 22
Transmittance, 137, 220
TSM, 329, 331, 332, 334, 335, 336, 337, 338, 339, 340, 342, 344, 358, 364
Tucker, C.J., 82, 130, 134, 181

**U**

Urbanism, 243

**V**

Van der Meer, F., XV, 51, 53, 56, 57, 58, 59, 61, 82, 201, 203, 205, 215, 219

Vane, G., XXI, 18, 81, 93, 158, 160
Vaughan, R.,, 298
vegetation, XXII, 10, 11, 18, 23, 65, 66, 67,
68, 69, 70, 72, 76, 77, 79, 80, 81, 82, 83,
85, 86, 100, 102, 105, 107, 108, 109,
111, 112, 114, 115, 116, 117, 118, 119,
120, 121, 124, 126, 132, 134, 135, 136,
138, 139, 140, 142, 144, 145, 146, 147,
148, 149, 150, 151, 155, 157, 158, 159,
162, 163, 173, 174, 178, 179, 181, 182,
190, 191, 196, 197, 198, 203, 204, 205,
207, 217, 218, 226, 228, 229, 230, 231,
232, 233, 237, 239, 240, 244, 245, 247,
248, 253, 256, 257, 259, 261, 262, 263,
266, 268, 270, 273, 279, 281, 291, 305,
348, 349, 350
Vegetation stress, XXIII, 204
Ventura, 233, 234, 235, 236, 237, 238
Verhoef, W., 129, 137, 139, 165, 168, 173,
181, 186

**W**

Water vapor, 277
Waveform characterization, 44, 45
Wessman, C.A., 11, 80, 81, 132, 133, 134,
135, 136, 137, 138, 149
Whiskbroom, 32
Wien, 4, 284
Windeler, D.S., 203, 205

**X**

Xanthophyll, 116
X-ray diffraction, 97

**Y**

Yang, H., XX, 142, 205, 231

# REFERENCES


A

Aas, E. 1987. Two-stream irradiance model for deep waters. *Appl.Opt.* **26**, 2095-2101.

Aber, J.D. & Martin, M.E. 1995. High spectral resolution remote sensing of canopy chemistry. *In:* R.O. Green (Ed.) *Summaries of the Fifth Annual JPL Airborne Earth Science Workshop.* January 23-26, California. JPL Publication 95-1, pp. 1-4.

Abrams, M.J., Abbott, E. & Kahle, A. 1991. Combined use of visible, reflected infrared and thermal infrared images for mapping Hawaiian lava flows. *J. Geophys. Res.,* **96,** 475-484.

Abrams, M.J., Ashley, R., Rowan, L., Goetz, L. & Kahle, A. 1977. Mapping hydrothermal alteration in the Cuprite Mining District, Nevada using aircraft scanner data for the spectral region 0.46 to 2.36 microns. *Geology* **5,** 713-718.

Abrams, M.J. & Hook, S.J. 1995. Simulated Aster data for geologic studies. *IEEE Trans. Geosci. Rem. Sens.,* **33,** 692 – 699.

Adams, J.B. 1974. Visible and near-infrared diffuse reflectance: Spectra of pyroxenes as applied to remote sensing of solid objects in the solar system. *J. Geophys. Res.* **79,** 4829-4836.

Adams, J.B. 1975. Interpretation of visible and near-infrared diffuse reflectance spectra of pyroxenes and other rock forming minerals. *In:* Karr, C., (Ed.), *Infrared and Raman Spectroscopy of Lunar and Terrestrial Materials.* Academic Press, New York, pp: 91-116.

Adams, J.B., Smith, M.O. & Gillespie, A.R. 1993 Imaging spectroscopy: Interpretation based on spectral mixture analysis. *In:* C.M. Pieters and P.A.J. Englert (Eds.), *Remote Geochemical Analyses: Elemental and Mineralogical Composition,* Cambridge University Press, Cambridge, pp. 145-166.

Adams, J.B., Smith, M.O. & Johnston, P.E. 1985. Spectral mixture modelling: a new analysis of rock and soil types at the Viking Lander 1 site. *J. Geophys. Res.,* **91,** 8098-8112.

Alberotanza, L. et al., 1999a. Eutrophication and environmental recovery of the Orbetello Lagoon. *In:* F.M. Faranda, L. Guglielmo & G. Spezie (Eds.), *Structure and processes in the Mediterranean ecosystems.* Springer Verlag, Berlin.

Alberotanza, L., Brando, V.E., Ravagnan, G. & Zandonella, A. 1999b. Hyperspectral aerial images. A valuable tool for submerged vegetation recognition in the Orbetello Lagoons, Italy. *Int. J. Remote Sensing* **20,** 523-533.

Albers, J.P. & Stewart, J.H. 1972. Geology and mineral deposits of Esmeralda County, *Nevada. Nevada Bureau of Mines and Geology Bulletin,* **78,**. 1-80.

Allen, W.A. & Richardson, A.J. 1968. Interaction of light with a plant canopy. *Journal of Optical Society of America,* **58,** 1023-1028.

Allen, W.A., Gausman H.W., Richardson A.J. & Thomas J.R. 1969. Interaction of isotropic light with a compact plant leaf. *Journal of Optical Society of America,* **59,** 1376-1379.

Allen, W.A., Gausman, H.W. & Richardson, A.J. 1970. Mean effective optical constants of cotton leaves. *J. Opt. Soc. Am.* **60,** 542-547.

Allen, W.A., Gausman, H.W. & Richardson, A.J. 1973. Willstätter-Stoll theory of leaf reflectance evaluation by ray tracing. *Applied Optics,* **12,** 2448-2453.

Allen, W.A., Gausman, H.W., Richardson, A.J. & Thomas, J.R. 1969. Interaction of isotropic light with a compact plant leaf. *J. Opt. Soc. Am.* **59,** 1376-1379.

Allen, W.A., Gayle, T.W. & Richardson, A.J. 1970. Plant Canopy Irradiance Specified by the Duntley Equations. *Journal of Optical Society of America,* **60,** 372-376.

Ammer, U., Koch, B., Schneider, T. & Wittmeier, H. 1991. High resolution spectral measurements of agricultural crops in the laboratory and in the field. *Proceedings of the 11th International Geoscience and Remote Sensing Symposium (IGARSS'91),* Helsinki (Finland), 3-6 June 1991, pp. 1937-1940.

Anstee, J.M. et al., 2000. Use of hyperspectral imaging for benthic species mapping in South Australian coastal waters. *Proceedings of the 10th Australian Remote Sensing and Photogrammetry Conference. The Remote Sensing & Photogrammetry Association of Australia,* Adelaide, Australia, pp. 1051-1061.

Armstrong, R.A. 1993. Remote sensing of submerged vegetation canopies for biomass estimation. *Int. J. Remote Sensing* **14,** 621-627.

Arp, G.K. 1992. An integrated interpretation for the origin of the Patrick Draw oil field sage anomaly. *AAPG Bulletin,* **76,** 301-306.

Ashley, R.P. 1971. Geologic map and alteration map, Cuprite mining district, Nevada (unpublished).

ASTM 1993. *Annual book of A.S.T.M. standards,* 04.08, American Society for Testing and Materials, Philadelphia, U.S.A.

Atkinson, B. J. 1985. *The urban atmosphere,* Cambridge University Press, Cambridge.

Atkinson, P.M. & Lewis, P. 2000. Geostatistical Classification for Remote Snsing: an Introduction. *Computers & Geosciences,* **26,** 361-371.


B

Babani, F. & Lichtenthaler, H.K. 1996. Light-induced and age dependent development of chloroplasts in etiolated barley leaves as visualised by determination of photosynthetic pigments, $CO_2$, assimilation rates and different kinds of chlorophyll fluorescence ratios. *Journal of Plant Physiology,* **148**, 555-566.

Babani, F., Lichtenthaler, H.K. & Richter, P. 1996. Changes of chlorophyll fluorescence signatures during greening of etiolated barley seedlings as measured with the CCD-OMA fluorometer. *Journal of Plant Physiology,* **148**, 471-477.

Bagheri, S., Rijkeboer, M., Pasterkamp, R. & Dekker, A.G. 2000. Comparison of field spectroradiometers in preparation for bio-optical modelling. *In:* R.O. Green (Ed.), *9th JPL Airborne Earth Science Workshop*, California, pp. 45-53.

Bajjouk, T., Guillaumont, B. & Populus, J. 1996. Application of airborne imaging spectrometry system data to intertidal seaweed classification and mapping. *Hydrobiologia* **327,** 463-471.

Bajjouk, T., Populus, J. & Guillaumont, B. 1998. Quantification of subpixel cover fractions using principal component analysis and a linear programming method: application to the coastal zone of Roscoff (France). *Rem. Sens. Environ.* **64,** 153-165.

Balling, R. & Brazel, S.W. 1988. High-resolution surface temperature patterns in a complex urban terrain. *Photogrammetric Engineering and Remote Sensing,* **54,** 1289–1293.

Bammel, B.H. & Birnie, R.W. 1994. Spectral reflectance response of big sagebrush to hydrocarbon-induced stress in the Bighorn basin, Wyoming. *Photogramm. Eng. and Remote Sensing,* **60,** 87-96.

Banninger, C. 1990. Fluorescence Line Imager (FLI) measured "red edge shifts" in a metal-stressed Norway spruce and their relationships to canopy biochemical and morphological changes, *In:* Vane, G. (Ed.), *Imaging Spectroscopy of the Terrestrial Environment*, SPIE 1298, Orlando (Florida), 16-17 April 1990, pp. 234-243.

Banninger, C., Johnson, L. & Peterson, D. 1994. Determination of biochemical changes in conifer canopies with airborne visible/infrared imaging spectrometer (AVIRIS) data. *In:* Chavez, P.S., Marino, C.M. & Schowengerdt, R.A. (Eds.), *Recent Advances in Remote Sensing and Hyperspectral Remote Sensing*, SPIE Vol. 2318, pp. 2-9.

Baret, F. 1994. Imaging Spectrometry in Agriculture, Comparison of Modellistic Approaches and Experimental Data. *In:* J. Hill & J. Mégier Eds., Imaging Spectrometry - a Tool for Environmental Observations. Dordrecht, Kluwer Academic, 20 pp.

Baret, F., Guyot, G. & Major, D. 1989. TSAVI: a vegetation index which minimizes soil brightness effects on LAI and APAR estimation, *Proceedings 12th Canadian Symposium on Remote Sensing*, Vancouver, Canada, July 10-14, 1989.

Barrow, C.J. 1991. Land Degradation. Development and Breakdown of Terrestrial Environments. Cambridge University Press Cambrige, 278 pp.

Bauer, M.E., Daughtry, C.S.T. &Vanderbilt, V.C. 1981. Spectral-agronomic relationships of corn, soybean and wheat canopies. SR-P1-04187. Laboratory for Applications of Remote Sensing, Purdue University, West Lafayette, Indiana, 17p.

Baugh, W.M. & Kruse, F.A. 1994. Quantitative geochemical mapping of ammonium minerals using field and airborne spectrometers, Cedar mountains, Esmeralda County, Nevada. *Proceedings Tenth Thematic Conference and Workshop on Applied Geologic Remote Sensing*, San Antonio, Texas, 9-12 May, Vol. II, pp. 304-315.

Baugh, W.M., Kruse, F.A. & Atkinson, W.W.Jr. 1998. Quantitative geochemical mapping of ammonium minerals in the southern Cedar Mountains, Nevada, using the Airborne Visible/Infrared Imaging Spectrometer (AVIRIS). *Rem. Sens. Environ.,* **65**, 292-308.

Baumgardner, M.F., Stoner, E.R., Silva, L.F., & Biehl, L.L., 1985. Reflectance properties of soils. *In:* N. Brady, N., (Ed.), *Advances of Agronomy*. Academic Press, New York, pp 1-44.

Beasley, D.B. & Huggins, L.F. 1982. ANSWERS: User's Manual. USEPA, Region V, Chicago, Illinois, Purdue University, West Lafayette, IN., 54 pp.

Belward, A.S. 1991. Spectral characteristics of vegetation, soil and water in the visible, near-infrared and middle-infrared wavelengths. *In:* Belward, A.S. & Valenzuela, C.R. (Eds.), *Remote Sensing and Geographical Information Systems for Resource Management in developing countries.* Kluwer, Netherlands.

Ben-Dor, E., Goetz, A.F.H. & Shapiro, A.T. 1994b. Estimation of cirrus cloud and aerosol scattering in hyperspectral image data, *Proceedings of the International Symposium on Spectral Sensing Research,* San Diego, California, USA, pp. 582-593

Ben-Dor, E, Inbar, Y. & Chen, Y. 1997. The Reflectance Spectra of Organic Matter in the Visible Near-Infrared and Short Wave Infrared Region (400-2500nm) during a Controlled Decomposition Process. *Rem. Sens. Environ.* **61,**1-15.

Ben-Dor, E., Irons, J.A. & Epema, G.F. 1998. Soil Spectroscopy. *In:* Rencz, A. (Ed.), *Manual of Remote Sensing, Third Edition,* Wiley & Sons Inc. New-York, Chichester, Weinheim, Brisbane, Singapore, Toronto, pp.111-189

Ben-Dor, E. & Kruse, F.A. 1994. Mineral mapping of Makhtesh Ramon, Negev, Israel using GER 63-channel scanner data and linear unmixing techniques. *Proceedings Tenth Thematic Conference and Workshop on Applied Geologic Remote Sensing,* San Antonio, Texas, 9-12 May, Vol. I, pp. 215-226.

Ben-Dor, E., Kruse, F.A., Lefkoff, A.B. & Banin, A. 1994a. Comparison of three calibration techniques for utilization of GER 63-channel aircraft scanner data of Makhtesh Ramon, Negev, Israel. *Photogramm. Eng. and Remote Sensing,* **60,** 1339-1354.

Ben-Dor, E. & Levin, N. 1999. Determination of Surface Reflectance from Raw Hyperspectral Data without Simultaneous Ground-Truth Measurements. A Case Study of the GER 63-Channel Sensor Data Acquired over Naan, Israel. *Int. J. Remote Sensing,* **21,** 2035-2074.

Ben-Dor, E., Levin, N. & Saaroni, H. 2000. A Spectral Based Recognition of the Urban Environment Using the Visible and Near Infrared Spectral Region (0.4-1.1 µm). A Case Study over Tel-Aviv, Israel. *Int. J. Remote Sensing* (in press).

Ben-Dor, E. & Saaroni, H. 1997. Airborne Video Thermal Radiometry as a Tool for Monitoring Micro scale Structures of the Urban Heat Island. *Int. J. Remote Sensing,* **18,** 3039-3053.

Benediktsson, J.A. 1995. Classification and feature extraction from AVIRIS data. *IEEE Trans. Geosci. Rem. Sens.,* **33,** 1194 - 1205.

Beratan, K.K., Peer, B., Dunbar, N.W. & Blom, R. 1997. A remote sensing approach to alteration mapping: AVIRIS data and extension-related potassium metasomatism, Socorro, New Mexico. *Int. J. Remote Sensing,* **18,** 3595-3609.

Derberoglu, S., Loyd, C.D., Atkinson, P.M. & Curran, P.J. 2000. The Integration of Spectral and Textural Information using Neural Networks for Land Cover Mapping in the Mediterranean. *Computers & Geosciences,* **26,** 385-396.

Berk, A., Bernstein, L.S. & Robertson, D.C. 1989. MODTRAN: A moderate resolution model for LOWTRAN7. Final report, GL-TR-0122, AFGL, Hanscom AFB, MA, 42 pp.

Bianchi, R., Cavalli, R., Fiumi, L., Marino, C., Pignatti, S. & Pizzaferi, G. 1996. The 1994-1995 CNR LARA Project airborne hyperspectral campaigns, *Proc. 11th Thematic Conf. Applied Geol. Rem. Sens.,* Las Vegas, February.

Bianchi, R., Cavalli, R.M., Fiumi, L., Marino, C.M., Panuzi, S. & Pignatti, S. 1996. Airborne Remote Sensing in Urban Areas: Examples and Considerations on the Applicability of Hyperspectral Surveys over Industrial, Residential, and Historical Environments. *Proceedings of the The Second International Airborne Remote Sensing Conference and Exhibition,* San Francisco, California, pp. 439-444.

Bierwirth, P.N., Lee, T.J. & Burne, R.V. 1993. Shallow sea-floor reflectance and water depth derived by unmixing multispectral imagery. *ISPRS Journal of Photogrammetry and Remote Sensing* **59,** 331-338.

Bihong, F. & Xiaowei, C. 1998. Thermal infrared spectra and TIMS imagery features of sedimentary rocks in the Kalpin uplift, Tarim basin, China. *Geocarto Int.,* **13,** 69-75.

Billings, W.D. & Morris R.J. 1951. Reflection of visible and infrared radiation from leaves of different ecological groups. *American Journal of Botany,* **38,** 327-331.

Birk, E. M. & Matson, P.A. 1986. Site fertility affects seasonal carbon reserves in loblolly pine. *Tree Physiology,* **2,** 17-27.

Birnie, R.W. & Dykstra, J.D. 1978. Application of remote sensing to reconnaissance geologic mapping and mineral exploration. *Proceedings of the 12th International Symposium on Remote sens. environ.* 20-26 April, Ann Arbor, Michigan. Vol. 2, pp. 795-804.

Bishop, C.M. 1995. *Neural Networks for Pattern Recognition.* Oxford University Press, New York.

Boardman, J.W. 1998. Leveraging the high dimensionality of AVIRIS data for improved sub-pixel target unmixing and rejection of false positives: Mixture Tuned Matched Filtering, *Summaries of the Seventh JPL Airborne Earth Science Workshop,* Jet Propulsion Laboratory, Pasadena, U.S.A., 12-16 January 1998, JPL Publication 97-21, pp. 55.

Boardman, J.W. 1989. Inversion of imaging spectrometry data using singular value decomposition. *Proceedings 12th Canadian Symposium on Remote Sensing* (IGARSS'89), Vol. 4, pp. 2069-2072.

Boardman, J.W. & Huntington, J.F. 1997. Mineralogic and geochemical mapping at Virginia City, Nevada, using 1995 AVIRIS data, *Proceedings of the 12th International Conference on Applied Geologic Remote Sensing*, Denver, U.S.A., 17-19 November 1997, pp. I, pp.191-198.

Boardman, J.W., Kruse, F.A. & Green, R.O. 1995. Mapping target signatures via partial unmixing of AVIRIS data. *Proceedings of the Fifth JPL Airborne Earth Science Workshop*, (Pasadena: JPL Publication 95-1), pp. 23-26.

Bolle, H.J. 1995. Remote Sensing in Desertification Studies. *In:* R. Fantechi, D. Peter, P. Balabanis & J.L. Rubio, Desertification in a European Context: Physical and Socio-economic Aspects. EUR Report 15415, Brussels, pp.227-246.

Bonham-Carter, G.F. 1988. Numerical procedures and computer program for fitting an inverted gaussian model to vegetation reflectance data. *Computers & Geosciences* **14 (3)**, 339-356.

Bonn, F., Megier, J. & Ait Fora, A. 1997. Remote sensing assisted Spatialization of Soil Erosion Models within a GIS for Land Degradation Quantification: Expectation, Errors and Beyond. *In:* A. Spiteri (Ed.), *Integrated Applications for Risk Assessment and Disaster Prevention for the Mediterranean*. Balkema, Rotterdam, pp.191-198.

Boochs, F., Kupfer, G., Dockter, K. & Kuhbauch, W. 1990. Shape of the red edge as vitality indicator for plants. *Int. J. Remote Sensing*, **11**, 1741-1753.

Borstadt, G.A., Brown, L., W., C., Nallee, M. & Wainwright, P. 1997. Towards a management plan for a tropical reef-lagoon system using airborne multispectral imaging and GIS. *Proceedings Fourth International Conference on Remote Sensing for Marine and Coastal Environments. ERIM*, Orlando, Florida, pp. 605-610.

Bouman, B.A.M. 1991. *Linking X-band radar backscattering and optical reflectance with crop growth models*, PhD Thesis, Wageningen Agricultural University, Wageningen, The Netherlands.

Bouman, B.A.M. 1992. SBFLEVO and WWFLEVO, *Growth models to simulate crop growth, optical reflectance and radar backscatter of sugar beet and winter wheat, calibrated for Flevoland*, CABO-DLO Report 163, Wageningen, The Netherlands, 61 pp plus appendices.

Bouman, B.A.M. 1994. Yield prediction by crop modelling and remote sensing, *Proceedings Conference on Yield Forecasting*, Villefranche-sur-Mer, France, 24-27 October 1994, 9 pp.

Bouman, B.A.M. & Goudriaan, J. 1989. Estimation of crop growth from optical and microwave soil cover. *Int. J. Rem. Sensing* **10**, 1843-1855.

Bouman, B.A.M., Van Kasteren, H.W.J. & Uenk, D. 1992. Standard relations to estimate ground cover and LAI of agricultural crops from reflection measurements. *European Journal of Agronomy* **1(4)**, 249-262.

Bowers, S.A. & Hanks, R.J. 1965. Reflection of Radiant Energy from Soils. *Soil Science*, **100**, 130-138.

Bowers, T.L. & Rowan, L.C. 1996. Remote mineralogic and lithologic mapping of the Ice River Complex, British Columbia, Canada, using AVIRIS data. *Photogramm. Eng. and Remote Sensing*, **62**, 1379 - 1385.

Bowman, W.D. 1989. The relationships between leaf water status, gas exchange, and spectral reflectance in cotton leaves. *Remote sens. environ*, **30**, 249-255.

Boyer, J.S. 1982. Plant productivity and environment. *Science*, **218**, 443-448.

Boyer, M., Miller, J., Belanger, M., Hare, E. &Wu, J. 1988. Senescence and spectral reflectance in leaves in Northern Pin Oak (Quercus palustris Muench.). *Remote sens. environ*, **25**, 71-87.

Bricaud, A., Morel, A. & Barale, V. 1999. MERIS potential for ocean colour studies in the open ocean. *Int. J. Remote Sensing* **20,** 1757-1769.

Brouwer, L.C. & de Jong, S.M. 1998. De Woestijn Rukt Op. *Natuur en Techniek, **8,** 72-83.

Buckton, D., Omongain, E. & Danaher, S. 1999. The use of Neural Networks for the estimation of oceanic constituents based on the MERIS instrument. *Int. J. Remote Sensing* **20,**1841-1851.

Bukata, R.P., Jerome, J.H., Kondratyev, K.Y. & Pozdniakov, D.V., 1995. *Optical properties and remote sensing of inland and coastal waters*. CRC Press, Boca Raton, USA.

Büker, C. & Clevers, J.G.P.W. 1992. *Imaging spectroscopy for agricultural applications*, Dept. Landsurveying and Remote Sensing, Wageningen Agricultural University, Report LUW-LMK-199206, 88 pp.

Büker, C. & Clevers, J.G.P.W. 1992. Imaging spectroscopy for agricultural applications. Report LUW-LMK-199206, Dept. Land Surveying and Remote Sensing, Wageningen Agricultural University.

Büker, C., Clevers, J.G.P.W. & Van Leeuwen, H.J.C. 1992b. *Optical component MAC Europe, Optical data report Flevoland 1991*, BCRS, Delft, report 92-28, 49 pp.

Büker, C., Clevers, J.G.P.W., Van Leeuwen, H.J.C., Bouman, B.A.M. & Uenk, D. 1992a. *Optical component MAC Europe, Ground truth report Flevoland 1991*, BCRS, Delft, report 92-27, 55 pp.

Bunnik, N.J.J. 1978. *The multispectral reflectance of shortwave radiation by agricultural crops in relation with their morphological and optical properties*, PhD Thesis, Mededelingen Landbouw-hogeschool Wageningen 78-1, 175 pp.

Bunnik, N.J.J. 1984. Review of models and measurements of multispectral reflectance by plant canopies: Recommendations for future research. *Remote Sensing, SPIE vol 475*, pp, 2-11.

Buongiorno, M.F., Bobliolo, M.P., Teggi, S. & Pugnaghi, S. 1997. Combining optical airborne image data and atmospheric measurements to estimate and map SO2 in tropospheric volcanic plumes. European Geophysical Society XXII General Assembly – Wien.

Burns, R.G. 1970. *Mineralogical Application to Crystal Field Theory*. Cambridge University Press, Cambridge, 224 pp.

Burns, R.G. 1993. Origin of Electronic Spectra of Minerals in the Visible and Near-Infrared Region. *In:* C.M. Pieters & P.A.J. Englert, Remote Geochemical Analysis: Elemental and Mineralogical Composition. Cambridge University Press, New York, pp. 3-30.

Burrough, P.A. 1986. Principles of Geographical Information Systems for Land Resources Assessment. Clarendon Press, Oxford, 186 pp.

Burrough, P.A. & MacDonnell, R.A. 1998, Principles of Geographical Information Systems. Oxford University Press, Oxford, 333 pp.

Buschmann, C. & Nagel, E. 1991. Reflection spectra of terrestrial vegetation as influenced by pigment-protein complexes and the internal optics of the leaf tissue. *Proceedings of the 11th International Geoscience and Remote Sensing Symposium (IGARSS'91)*, Helsinki (Finland), 3-6 June 1991, Vol. IV, pp. 1909-1912.

Buschmann, C. & Nagel, E. 1993. *In vivo* spectroscopy and internal optics of leaves as basis for remote sensing of vegetation. *Int. J. Remote Sensing*, **14**, 711-722.Babey, S.K. & Soffer, R.J. 1992. Radiometric calibration of the compact airborne spectrographic imager (CASI). *Can.J.Remote Sensing* **18**, 233-242.


C

Calvin, W., Vaughan, R., Taranik, J. & Smailbegovic, A. 2000. Spectral analyses of SEBASS data. Atmospheric correction, calibrated surface reflectance and mineral mapping. *Proc. ERIM 14$^{th}$ Int'l. Conf. Applied Geologic Remote Sensing*, Las Vegas, Nov., pp. 20-26.

Card, D.H. 1993. Example of a Simple Surface Composition Model for Urban Environment Using Remote Sensing. Ph.D. Dissertation, University of Utah, Salt Lake City. 136 pp.

Carder, K.L. et al., 1993. Aviris calibration and application in coastal oceanic environments. *Remote Sens.Environ.* **44**, 205-216.

Carrere, V. & Abrams, M.J. 1988. An assessment of AVIRIS data for hydrothermal alteration mapping in the Goldfield mining district. *In:* G. Vane (Ed.), *Proceedings of the Airborne Visible/Infrared Imaging Spectrometer (AVIRIS) Performance Evaluation Workshop*, Pasadena, USA, NASA-JPL Publication, 88-38, pp. 134-154.

Carrere, V. & Conel, J.E. 1993. Recovery of atmospheric water vapour total column abundance from imaging spectrometer data around 940 nm - sensitivity analysis and application to Airborne Visible/Infrared Imaging Spectrometer (AVIRIS) data. *Remote Sens. Environ.,* **44,** 179-204.

Carson, M.A. & Kirkby, M.J. 1972. Hillslope Form and Process. Cambridge, Cambridge University Press.

Carter, G.A. 1991. Primary and secondary effects of water content on the spectral reflectance of leaves. *American Journal of Botany,* **78**, 916-924.

Carter, G.A. 1994. Ratios of leaf reflectances in narrow wavebands as indicators of plant stress. *Int. J. Remote Sensing,* **15**, 697-703.

Carter, G.A., Cibula, W.G. & Miller, R.L. 1996. Narrow-band reflectance imagery compared with thermal imagery for early detection of plant stress. *Journal of Plant Physiology,* **148**, 515-522.

Casanova, D., Epema, G.F. & Goudriaan, J. 1998. Monitoring rice reflectance at field level for estimating biomass and LAI. *Field Crops Research* **55**, 83-92

Casciati, F., Gamba, P., Giorgi, F. & Mecocci, A. 1996. Planning a RADATT Extension. *Proceedings of GIS and Applications of Remote Sensing to Damage Management*, Greenbelt, Maryland, pp. 1-10.

Cetin, H. & Levandowski, D.W. 1991. Interactive classification and mapping of multi-dimensional remotely sensed data using n-dimensional probability density function (nPDF *Photogramm. Eng. and Remote Sensing,* **57**, 1579-1587.

Cetin, H., Warner, T.A. & Levandowski D.W. 1993. Data Classification, Visualization, and Enhancement Using n-Dimensional Probability Density Functions (nPDF): AVIRIS, TIMS, TM, and Geophysical Applications. *Photogramm. Eng. and Remote Sensing,* **59,** 1755-1764.

Chabrillat, S. & Goetz, A.F.H. 1999. The search for swelling clays along the Colorado Front Range: The role of AVIRIS resolution in detection, in *Summaries of the Eighth JPL Airborne Earth Science Workshop*, Jet Propulsion Laboratory, Pasadena, U.S.A., 9-11 February 1999, JPL Publication 99-17, pp. 69-78.

Chabrillat, S., Goetz, A.F.H., Olsen, H.W. & Krosley, L. 2000c. Use of hyperspectral imagery in identification and mapping of expansive clays in Colorado. *Rem. Sens. Environ.*, submitted.

Chabrillat, S., Goetz, A.F.H., Olsen, H.W. & Noe, D.C. 1997. Field spectrometry techniques for identification of expansive clay soils, *Proceedings of the 12th International Conference on Applied Geologic Remote Sensing*, Denver, U.S.A, 17-19 November 1997, pp. I:141-148.

Chabrillat, S., Goetz, A.F.H., Olsen, H.W., Krosley, L. & Noe, D.C. 1999. Use of AVIRIS hyperspectral data to identify and map expansive clay soils in the Front Range Urban Corridor in Colorado, *Proceedings of the 13th International Conference on Applied Geologic Remote Sensing,* Vancouver, Canada, 1-3 March 1999, pp. I:390-397.

Chabrillat, S., Goetz, A.F.H., Olsen, H.W., Krosley, L. & Noe, D.C. 2000a. The search for swelling clays along the Colorado Front Range: Results from field spectrometry and hyperspectral imagery, in *Summaries of the Nineth JPL Airborne Earth Science Workshop*, Jet Propulsion Laboratory, Pasadena, U.S.A., 23-25 February 2000, in press.

Chabrillat, S., Goetz, A.F.H., Olsen, H.W., Krosley, L. & Noe, D.C. 2000b. Near-infrared spectra of expansive clay soils and their relationships with mineralogy and swelling potential, *Clays and Clay Minerals*, in revision.

Chang, S.H. & Collins, W. 1983. Confirmation of the airborne biogeophysical mineral exploration technique using laboratory methods. *Economic Geology*, **78**, 723-736.

Chapin, F.S., McKendrick, J.D. & Johnson, D.A. 1986. Seasonal changes in carbon fractions in Alaskan tundra plants of differing growth form: implications for herbivory. *Journal of Ecology*, **76**, 707-732.

Chapin, F.S., Schulze, E.D. & Mooney, H.A. 1990. The ecology and economics of storage in plants. *Annual Review of Ecology and Systematics*, **21**, 423-447.

Chappelle, E.W., Kim, M.S. & McMurtrey, III J.E. 1992. Ratio analysis of reflectance spectra (RARS): an algorithm for the remote estimation of the concentrations of chlorophyll A, chlorophyll B, and carotenoids in soybean leaves. *Remote sens. environ*, **39**, 239-247.

Chen, F.H. 1988. *Foundations on expansive soils, 2d edition*, Elsevier, New York, 464 pp.

Chen, S. 1986. Tianjin Atlas of Environmental Quality, Sceince Press, Beijing, 270p.

Chen, S., Zeng, S. & Xie, C. 2000. Remote Sensing and GIS for urban growth analysis in China. *Photogrammetric Engineering and Remote sensing,* **66,** 593-598.

Chica-Olma, M. & Arbarca-Hernandez, F. 2000. Computing Geostatistical Image Texture for Remotely Sensed Data Classification. *Computers & Geosciences*, **26,**373-383.

Chidsey T. C. Jr., D.E. Eby, and D.M. Lorenz, 1996b. Geological and reservoir characterization of small shallow-shelf carbonate fields, southern Paradox Basin, Utah. *In:* C. Huffman Jr., W. R. Lund and L. H. Godwin (Eds.), *Geology & resources of the Paradox Basin*, Special symposium, Utah Geological Association & Four Corners Geological Society, pp. 39-56.

Chiu, H.Y. & Collins, W. 1978. A spectroradiometer for airborne remote sensing. *Photogramm. Eng. and Remote Sensing* **44**, 507-517.

Christense, P. et. al. 1992. Thermal emission spectral library of rock forming minerals. *J. Geophys. Res.* **105**, 9735-0739.

Chung, C.J.. & Fabbri, A.G. 1993. The representation of geoscience information for data integration. *Nonrenewable Resources*, **2,** 122-139.

Chung, J.J. 1989. SPOT Pixel Analysis for Urban Ecosystem Study in Salt Lake City, Utah." M.S. Thesis, University of Utah, Salt Lake City. 67pp.

Clark, C.D., Mumby, P.J., Chisholm, J.R.M., Jaubert, J. & Andrefouet, S. 2000. Spectral discrimination of coral mortality states following a severe bleaching event. *Int. J. Remote Sensing* **21**, 2321-2327.

Clark, C.D., Ripley, H.T., Green, E.P., Edwards, A.J. & Mumby, P.J., 1997. Mapping and measurement of tropical coastal environments with hyperspectral and high spatial resolution data. *Int. J. Remote Sensing* **18,** 237-242.

Clark, R.N. 1981. Water Frost and Ice: The Near-Infrared Spectral Reflectance 0.65-2.5 $\mu$m. *J. Geophys. Res.,* **86**, 3087-3096.

Clark, R.N., Gallagher, A., Swayze, G.A. 1990. Material absorption band depth mapping of imaging spectrometer data using a complete band shape least-squares fit with library reference spectra. *In:* R.O Green (Ed.), *Proceedings of the Second Airborne Visible/Infrared Imaging Spectrometer (AVIRIS) Workshop*, Pasadena, USA, NASA-JPL Publication 90-54, pp. 176-186.

Clark, R.N., King, T.V.V., Ager C. & Swayze, G.A. 1997. Vegetation Species and Stress Indicator Mapping in the San Luis Valley, Colorado using Imaging Spectrometer data. *Remote Sens. Environ.,* submitted.

Clark, R.N., King, T.V.V., Ager, C. & Swayze, G.A. 1995. Initial vegetation species and senescence/stress indicator mapping in the San Luis Valley, Colorado using imaging spectrometer data. *In:* Green, R.O. (Ed.), *Summaries of the Fifth Annual JPL Airborne Earth Science Workshop*. January 23-26, California. JPL Publication 95-1, pp. 35-38.

Clark, R.N., King, T.V.V., Kleijwa, M., Swayze, G.A. & Vergo, N., 1990. High spectral Resolution Reflectance Spectroscopy of Minerals. *J. Geophys. Res.* **95,** 12653-12680.

Clark, R.N., Swayze, G. & Gallagher, A. 1992. Mapping the mineralogy and lithology of Canyonlands,
Clark, R.N. & Roush, T.L. 1984. Reflectance spectroscopy: Quantitative analysis techniques for remote
sensing applications. *J. Geophys. Res.*, **89**, 6329-6340.
Utah with imaging spectrometer data and multiple spectral feature mapping algorithm, *Summaries of
the Third Annual JPL Airborne Geoscience Workshop*, Jet Propulsion Laboratory, Pasadena, U.S.A, JPL
Publication 92-14, pp. 11-13.
Clark, R.N., Swayze, G.A., Koch, C. & Ager, C. 1992. Mapping vegetation types with the multiple
spectral feature mapping algorithm in both emission and absorption. *In:* Green, R.O. (Ed.), *Summaries of the
Third Annual JPL Airborne Geoscience Workshop*. June 1-5, California. JPL Publication 92-41, pp. 60-62.
Clark, R.N., Swayze, G.A., Rowan, L., Live, E. & Watson, K. 1996. Mapping Surficial Geology,
Vegetation Communities, and Environmental Materials in Our National Parks: The USGS Imaging
Spectroscopy, Integrated Geology, Ecosystems, and Environmental Mapping Project. Presented at the
airborne geoscience workshop 1996, Pasadena, Ca.
Clevers, J.G.P.W. 1988. The derivation of a simplified reflectance model for the estimation of Leaf
Area Index. *Rem. Sens. Envir.* **25**, 53-69.
Clevers, J.G.P.W. 1989. The application of a weighted infrared-red vegetation index for estimating
Leaf Area Index by correcting for soil moisture. *Rem. Sens. Envir.* **29**, 25-37.
Clevers, J.G.P.W. 1994. Imaging spectrometry in agriculture - Plant vitality and yield indicators. *In:* J.
Hill and J. Mégier (Eds.), *Imaging Spectrometry - A Tool for Environmental Observations*, Kluwer Academic
Publishers, Dordrecht, pp. 193-219.
Clevers, J.G.P.W. 1999. The use of imaging spectrometry for agricultural applications. *ISPRS Journal
of Photogrammetry & Remote Sensing* **54**, 299-304.
Clevers, J.G.P.W. & Buiten, H.J. 1990. *Multitemporal analysis of optical satellite data*, BCRS, Delft,
report 91-02, 64 pp.
Clevers, J.G.P.W., Büker, C., Van Leeuwen, H.J.C. & Bouman, B.A.M. 1994. A framework for
monitoring crop growth by combining directional and spectral remote sensing information. *Rem. Sens. Envir.*
**50**, 161-170.
Clevers, J.G.P.W. & Büker, C. 1991. Feasibility of the red-edge index for the detection of nitrogen
deficiency. *Proceedings 5th International Colloquium on Physical Measurements and Signatures in Remote
Sensing*, Courchevel, France, ESA SP-319, pp. 165-168.
Clevers, J.G.P.W. & Van Leeuwen, H.J.C. 1994. Combining directional and high spectral resolution
information from optical remote sensing data for crop growth monitoring, *Proceedings Sixth International
Symposium on Physical Measurements and Signatures in Remote Sensing*, Val d'Isère, France, 17-21 January
1994.
Clevers, J.G.P.W., Scholte, K., Van der Meer, F., Bakker, W.H., Epema, G.F., De Jong, S.M. &
Skidmore, A.K. 2000. The use of the MERIS standard band setting for deriving the red edge index,
*Proceedings of the ISSSR International Symposium on Sensors and Systems for the new Millennium*, Las
Vegas, Oct 31 – Nov 5, 1999 (in press).
Clevers, J.G.P.W. & Van Leeuwen, H.J.C. 1996. Combined use of optical and microwave remote
sensing for crop growth monitoring. *Rem. Sens. Envir.* **56**, 42-51.
Clevers, J.G.P.W. & Verhoef, W. 1993. LAI estimation by means of the WDVI: a sensitivity analysis
with a combined PROSPECT-SAIL model. *Remote Sensing Reviews* **7(1)**, 43-64.
Clevers, J.G.P.W., Verhoef, W. & Van Leeuwen, H.J.C. 1992. Estimating APAR by means of
vegetation indices: a sensitivity analysis, *International Archives of Photogrammetry and Remote Sensing* Vol.
XXIX, Part B7, Comm. VII, XVIIth ISPRS Congress, Washington D.C., 1992, pp. 691-698.
Collins, W. 1978. Remote sensing of crop type and maturity. *Photogramm. Eng. and Remote Sensing*,
**44**, 43-55.
Collins, W., Chang, S.H., Raines, G., Canney, F. & Ashley, R. 1983. Airborne biogeochemical
mapping of hidden mineral deposits. *Econ. Geology,* **78**, 737-743.
Collins, W.S., Chang, S.H., Koo, J.F. & Mancinelli, A. 1980. Biogeochemical laboratory studies to
establish the spectral properties of mineral-induced stress in plants, NTIS Report PB80-196769. University of
Columbia (NY), 69 pp.
Colvin, J.R. (1980) Biosynthesis of cellulose. *In:* Preiss, J. (Ed.), *The biochemistry of plants, A
comprehensive treatise, Vol.3, Carbohydrates: structure and function*, New York, Academic Press, pp. 543-
570.
Condit, H.R. 1970. The spectral reflectance of American soils. *Photogrammetric Engineering and
Remote Sensing* **36**, 955-966.
Conel J. E. & Alley, R.E. 1985. Lisbon Valley, Utah uranium test site report. *In:* M. J. Abrams, J. E.
Conel, H. R. Lang, and H. N. Paley (Eds.), *The Joint NASA/Geosat Test Case Project: final report*, AAPG
Special Publication, pt.2, **1**: 8-1--8-158.

Conel, J.E., Green, R.O., Carrere, V., Margolis, J.S., Alley, R.E., Vane, G., Bruegge, C.J. & Gary, B.L. 1988. Atmospheric water mapping with the airborne visible/infrared imaging spectrometer (AVIRIS), Mountain Pass, CA, *In:* Vane, G. (Ed.), *Proceedings AVIRIS Performance Evaluation Workshop.* JPL Publ. 88-38, Jet Propulsion Lab., Pasadena, California, pp. 21-26.

Coppin, N.J. & Richards, I.G. 1990. Use of Vegetation in Civil Engineering. London, Butterworths.

Couloigner, I., Ranchin, T., Valtonen, V.P. & Wald, L. 1998. Benefit of the future SPOT-5 and of data fusion to urban roads Mapping. *Int. J. Remote Sensing, 19,* 1519-1532.

Cracknell, A.P. 1999. Remote sensing techniques in estuaries and coastal zones - an update. *Int. J. Remote Sensing 20,* 485-496.

Crist, E.P. & Cicone, R.C. 1984. A Physically Based Transformation of Thematic Mapper Data: The TM Tasselled Cap. *IEEE Transaction on Geosciences and Remote Sensing, 22,* 256-263.

CropScanTM 1993. *Multi-Spectral Radiometer (MSR): Users manual and technical reference,* CropScanTM, Rochester.

Crosta, A.P., do Prado, I.D.M. & Obara, M. 1996. The use of Geoscan AMSS data for gold exploration in the Rio Itapicuru greenstone belt (BA), Brazil. *Proceedings of the Eleventh Thematic Conference and Workshop on Applied Geologic Remote Sensing,* Las Vegas, Nevada, 27-29 February, Vol. II, pp. 205-214.

Crosta, A.P., Sabine, C. & Taranik, J.V. 1998. Hydrothermal alteration mapping at Bodie, California, using AVIRIS hyperspectral data. *Remote Sens. Environ., 65,* 309-319.

Crowley, J.K. 1993. Mapping playa evaporite minerals with AVIRIS data: a first report from Death Valley, California. *Rem. Sens. Environ., 44,* 337-356.

Crowley, J.K., 1991. Visible and near-infrared (0.4-2.5 $\mu$ m) Reflectance spectra of playa evaporite minerals. *J. Geophys. Res.* **96**, 16231-16240.

Crowley, J.K., Brickey, D.W. & Rowan, L.C. 1989. Airborne imaging spectrometer data of the Ruby mountains, Montana: mineral discrimination using relative absorption band-depth images. *Remote Sens. Environ., 29,* 121-134.

Crowley, J.K. & Swayze, G.A. 1995. Mapping minerals, amorphous materials, environmental materials, vegetation, water, ice, and other materials: The USGS Tricorder Algorithm. *In:* Summaries of the Fifth Annual JPL Airborne Earth Science Workshop, JPL Publication 95-1, pp. 39-40.

Crowley, J.K. & Zimbelman, D.R. 1996. Mapping hydrothermally altered rocks on Mount Rainier, Washington: application of airborne imaging spectrometry to volcanic hazards assessment. *Proceedings of the Eleventh Thematic Conference and Workshop on Applied Geologic Remote Sensing,* Las Vegas, Nevada, 27-29 February, Vol. II, pp. 460-461.

Crowley, J.K. & Zimbelman, D.R. 1997. Mapping hydrothermally altered rocks on Mount Rainier, Washington, with Airborne Visible/Infrared Imaging Spectrometer (AVIRIS) data. *Geology, 25,* 559-562.

Cudahy, T., Huntington, J., Okada, K., Yamato, Y. & Hackwell, J. 2000. Mapping skarn alteration mineralogy at Yerrington, Nevada using airborne hyperspectral TIR SEBASS imaging data. *Proc. ERIM 14$^{th}$ Int'l. Conf. Applied Geologic Remote Sensing,* Las Vegas, Nov., pp. 70-78.

Cudahy, T., Whitbourn, L., Connor, P., Mason, P. & Phillips, R. 1999. Mapping surface mineralogy and scattering behavior using backscattered reflectance from a hyperspectral midinfrared airborne CO2 laser system (MIRACO2LAS). *IEEE Trans. Geosci. Rem. Sens.* **37,** 2109-2134.

Cudlip, W., Lysons, C., Ley, R., Deane, G., Stroink, H. & Roli, F. 1999. A new information system in support of landscape assessment: PLAINS. *Computers Environment and Urban Systems, 23,* 459-467.

Curran, P.J. (Eds.), *Environmental Remote Sensing From Regional To Global Scales.* Wiley & Sons, Chichester, pp. 149-166.

Curran, P.J. 1989. Remote sensing of foliar chemistry. *Remote sens. environ, 30,* 271-278.

Curran, P.J. 1994a. Attempts to derive ecosystem simulation models at local to regional scales.

Curran, P.J. 1994b. Imaging Spectrometry. *Progress in Physical Geography, 18,* 247-266.

Curran, P.J., Dungan, J.L. & Gholz, H. L. 1990. Exploring the relationship between reflectance red edge and chlorophyll content in slash pine. *Tree Physiology, 7,* 33-48.

Curran, P.J., Dungan, J.L., Macler, B.A. & Plummer, S.E. 1991. The effect of a red pigment on the relationship between red edge and chlorophyll concentration. *Remote sens. environ, 35,* 69-76.

Curran, P.J., Dungan, J.L., Macler, B.A., Plummer, S.E. & Peterson, D.L. 1992. Reflectance spectroscopy of fresh whole leaves for the estimation of chemical concentration. *Remote sens. environ, 39,* 153-166.

Curtiss, B. & Maecher, A.G. 1991. Changes in forest canopy reflectance associated with chronic exposure to high concentrations of soil trace metals. *Proceedings of the Eighth Thematic Conference on Geologic Remote Sensing,* Denver, Colorado, USA, April 29 - May 2, pp. 337-347.

Cwick, G.J., Bishop, M.P., Howe, R.C., Mausel, P.W., Everitt, J.H. & Escobar, D.E. 1995. Multispectral video data for detecting biogeochemical conditions at an Alabama oil field site. *Geocarto International, 10,* 59-66.

D

Da Costa, L.M. 1979. Surface Soil Color and Reflectance as Related to Physiochemical and Mineralogical Soil Properties. PhD-dissertation, University of Missouri, Columbia, Mo., 154 pp.

Danson, F.M., Steven, M.D., Malthus, T.J. & Clark, J.A. 1992. High-spectral resolution data for determining leaf water content. *Int. J. Remote Sensing*, **13**, 461-470.

***Dawson, T.P. & Curran, P.J. 1998. A new technique for interpolating the reflectance red edge position, Int. J. Remote Sensing 19, 2133-2139.***

Dawson, T.P., Curran, P.J. & Kupiec, J.A. 1995. Causal correlation of foliar biochemical concentrations with AVIRIS spectra using forced entry linear regression. *In:* Green, R.O. (Ed.), *Summaries of the Fifth Annual JPL Airborne Earth Science Workshop.* January 23-26, California. JPL Publication 95-1, pp. 43-46.

De Carvalho L. 2000. Key note presentation, 19th ISPRS Conference and Exhibition, Amsterdam, July 2000.

De Cola, L. 1989. Fractal Analysis of a Classified Landsat Scene. *Photogrammetric Engineering & Remote Sensing*, **55**, 601-610.

De Jong, S.M. 1994, Derivation of Vegetative Variables from a Landsat TM Image for Erosion Modelling. *Earth Surface Processes and Landforms*, **19**, 165-178.

De Jong, S.M., 1994b. Applications of Reflective Remote Sensing for Land Degradation Studies in a Mediterranean Environment. Ph.D. dissertation, Netherlands Geographical Studies 177. KNAG Utrecht, 253 pp.

De Jong, S.M. 1998. Imaging spectrometry for monitoring tree damage caused by volcanic activity in the Long Valley Caldera, California. *ITC Journal*, **98-1**, 1-10.

De Jong, S.M.& Burrough, P.A. 1995. A Fractal Approach to the Classification of Mediterranean Vegetation Types in Remotely Sensed Images. *Photogrammetric Engineering & Remote Sensing*, **61**, 1041-1053.

De Jong, S.M.. & Chrien, T.G. 1996. Mapping Volcanic gas emissions in the Mammoth Mountain area using AVIRIS. Presented at the sixth Earth Science Workshop, March 4-8, 1996, Jet Propulsion Laboratory, Pasadena 1996.

De Jong, S.M., Clevers, J.G.P.W., Pebesma,E. & Lacaze, B. 2000. Assessing Aboveground Biomass of Mediterranean Forests Using Airborne Hyperspectral Images and Geostatistics. *Proceedings of the 4th International Symposium on Spatial Accuracy Assessment in Natural Resources and Environmental Sciences*, Accuracy-2000, Amsterdam 11-13 July, pp 161-169.

De Jong, S.M., Pebesma, E. & Lacaze, B. 2001. Aboveground Biomass Assessment of Mediterranean Forests using Airborne Imaging Spectrometry: the DAIS Peyne Experiment. *Int. J. Remote Sensing*, submitted.

De Jong, S.M., de Roo, A.P.J., Paracchini, M.L., Bertolo, F., Folving, S. & Megier, J. 1999. Regional Assessment of Soil Erosion using the Distributed Model SEMMED and Remotely Sensed Data. *Catena*, **37**, 291-308.

De Jong, S.M. & Riezebos, H.Th. 1994. Imaging Spectroscopy, Geostatistics and Soil Erosion Modelling. *In:* R.J. Rickson (Ed.), *Conserving Soil Resources: European Perspectives.* CAB Int., Oxon, pp.232-245.

Dekker, A.G. 1993. *Detection of optical water quality parameters for eutrophic waters by high resolution remote sensing.* PhD Thesis, Vrije Universiteit, Amsterdam.

Dekker, A.G., Hoogenboom, H.J., Volten, H., Schreurs, R. & De Haan, J.F., 1997. Angular scattering functions of algae and silt: an analysis of backscattering to scattering fraction.*In:* S.G. Ackleson & R. Frouin (Eds.), *Ocean Optics XIII,* SPIE, Bellingford, USA, pp. 392-400.

Dekker, A.G., Malthus, T.J.M. & Hoogenboom, H.J. 1995. The remote sensing of inland water quality. *In:* F.M. Danson & S.E. Plummer (Eds.), *Advances in Environmental Remote Sensing.* Advances in Environmental Remote Sensing. John Wiley & Sons, UK., Chichester, pp. 123-142.

Dekker, A.G., Peters, S.W.M., Rijkeboer, M. & Berghuis, H. 1999. Analytical processing of multitemporal SPOT and Landsat images for estuarine management in Kalimantan Indonesia. *In:* G.J.A. Nieuwenhuis, R.A. Vaughan and M. Molenaar (Eds*.), 18th EARSeL Symposium on Operational Remote Sensing for Sustainable Development.* Balkema, Rotterdam, The Netherlands, Enschede, The Netherlands, pp. 315-324.

Dekker, A.G., Peters, S.W.M., Vos, R.J. & Rijkeboer, M. 2001. Remote sensing for inland water quality detection and monitoring: State-of-the-art application in Friesland waters. *In:* A. van Dijk and M.G. Bos (Eds.), *GIS and Remote Sensing techniques in Land- and Water-management.* Kluwer Academic Publishers, Netherlands, pp. 17-38.

De Oliveira, W.J. & Crosta, A.P. 1996. Detection of hydrocarbon seepage in the Sao Francisco basin, Brazil, through Landsat TM, soil geochemistry and airborne/field spectrometry data integration. *Proceedings of the Eleventh Thematic Conference and Workshop on Applied Geologic Remote Sensing*, Las Vegas, Nevada, 27-29 February, Vol. I, pp. 155-165.

De Roo, A.P.J. (Ed.), 1999. Special Issue: Soil Erosion Modelling at the Catchment Scale. *Catena, 37*, 275-541.

De Roo, A.P.J. & Jetten, V.G. 1999. Calibrating and Validating the LISEM Model for Two Datasets from the Netherlands and South Africa. *Catena, 37*, 477-494.

De Wit, C.T. 1965. *Photosynthesis of leaf canopies*, Agricultural Research Report 663, PUDOC, Wageningen, The Netherlands.

Delécolle, R., Maas, S.J., Guérif, M. & Baret, F. 1992. Remote sensing and crop production models: present trends, *ISPRS J. Photogramm. Rem. Sensing 47*, 145-161.

Demetriades-Shah T.H. & Steven, M.D. 1988. High spectral resolution indices for monitoring crop growth and chlorosis. *Proceedings 4th Int. Coll. on Spectral Signatures of Objects in Remote Sensing*, Aussois (France), ESA SP-287, pp. 299-302.

Demetriades-Shah T.H., Steven, M.D. & Clark, J.A. 1990. High resolution derivative spectra in remote sensing. *Remote Sensing Environ. 33*, 55-64.

Dockter, K., Schellberg, J., Kühbauch, W., von Rüsten, C., Templemann, U. & Kupfer, G. 1988. Spectral reflectance of sugar beet and winter wheat canopies in the visible and infrared during growth. *Proceedings of the 4th International Colloquium of spectral signatures of Objects in Remote Sensing*, Aussois, France, on 18-22 January 1988. ESA SP-287 (Paris: European Space Agency), pp. 211-216.

Doerffer, R. 1992. Imaging spectroscopy for detection of chlorophyll and suspended matter. *In:* F. Toselli & J. Bodechtel (Eds.), *Imaging spectroscopy: fundamentals and prospective applications.* Euro courses remote sensing. Kluwer, Dordrecht, pp. 215-257.

Doerffer, R. & Fischer, J. 1994. Concentrations of chlorophyll, suspended matter and gelbstoff in case II waters derived from satellite coastal zone color scanner with inverse modeling methods. *J. Geophys. Res.* **99**, 7457-7466.

Doerffer, R. & Schiller, H. 1994. Inverse modelling for retrieval of ocean color parameters in Case II coastal waters: an analysis of the minimum error. *In:* J.S. Jaffe (Ed.), *Ocean Optics XII*. Ocean Optics XII. SPIE, Bellingham, Washington, USA, pp. 887-893.

Donaldson, G.W. 1969. The occurrence of problems of heave and the factors affecting its nature, *Proceedings Second International Research and Engineering Conference on Expansive Clay Soils*, Texas A & M Press.

Donovan, T.J. 1974. Petroleum microseepage at Cement, Oklahoma: evidence and mechanism. *AAPG Bulletin*, **58**, 429-446.

Dreccer, M.F. 1999. *Radiation and nitrogen use in wheat and oilseed rape crops*, PhD Thesis Wageningen Agricultural University, The Netherlands, 135 pp.

Drever, J.I. 1973. The preparation of oriented clay mineral specimens for x-ray diffraction analysis by a filter-membrane peel technique. *American Mineralogist*, **58**, 553-554.

Driessen, P.M. & Dudal, R. (Eds), 1989. Lecture Notes on geography, Formation, Properties and Use of major Soils of the World. Agricultural University Wageningen.

Duchscherer, W. 1986. The principle of delta carbonate geochemical hydrocarbon prospecting demonstrated. In: *Unconventional method IV*. Dallas, TX: Southern Methodist University Press, pp. 14-16.

Duchscherer W. Jr. 1980. Geochemical methods of prospecting for hydrocarbons. *Oil and Gas Journal*, **1**, 194-208.

Duchscherer, W. Jr. 1982. Geochemical exploration for hydrocarbons, no new tricks-- but an old dog. *Oil And Gas Journal*, **5**, 163-176.

Dunk, I. & Lewis, M.R. 1999. Seagrass and shallow water feature discrimination using Hymap Imagery. *Proceedings 10th Australian Remote Sensing and Photogrammetry Conference*. University of Adelaide, Australia, pp. 1092-1108.

Duntley, S.Q. 1963. Light in the sea. *J.Opt.Soc.America* **53**, 214-233.

Durand, D., Bijaoui, J. & Cauneau, F. 2000. Optical remote sensing of shallow-water environmental parameters. *Rem. Sens. Environ.* **73**, 152-161.

Durand, D. et al. 1999. Characterisation of inland and coastal waters with space sensors. Nansen Environmental and Remote Sensing Centre (NERSC) Technical Report No. 164, European Commission-Centre for Earth Observation (CEO), Solheimsviken, Norway.Dwyer, J.L., Kruse, F.A. & Lefkoff, A.B. 1995. Effects of empirical versus model-based reflectance calibration on automated analysis of imaging spectrometer data: a case study from the Drum mountains, Utah. *Photogramm. Eng. and Remote Sensing*, **61**, 1247-1254.

**E**

Eliasson, I. 1996. Intra-urban nocturnal temperature differences: a multivariate approach. *Climate Research*, **7,** 21-30.

Elmore, A.J., Mustard, J.F., Manning, S.J. & Lobell, D.B. 2000. Quantifying Vegetation Change in Semiarid Environments: Precision and Accuracy of Spectral Mixture Analysis and the Normalized Difference Vegetation Index. *Rem. Sens. Environ.,***73,** 87-102.

Eloheimo, K. et al. 1998. Coastal monitorng using satellite, airborne and in situ measurement data in the Baltic Sea. *Proceedings 5th International Conference on Remote Sensing for Marine and Coastal Environments.* ERIM-Veridian, San Diego, California, pp. 306-131.

Elvidge, C.D. 1987. Reflectance characteristics of dry plant materials. *Proceedings of the 21st International Symposium on Remote sens. environ,* Ann Arbor, Michigan, 26-30 October, pp. 721-733.

Elvidge, C.D. 1988. Vegetation reflectance feature in AVIRIS data. *Proceedings of the 6th Thematic Conference: Remote Sensing for Exploration Geology,* Houston, Texas, pp. 169-182.

Elvidge, C.D. 1990. Visible and near infrared reflectance characteristics of dry plant materials. *Int. J. Remote Sensing,* **11,** 1775-1795.

Elvidge, C.D., Chen, Z. & Groeneveld, D.P. 1993. Detection of trace quantities of green vegetation in 1990 AVIRIS data. *Remote sens. environ,* **44**, 271-279.

Elvidge, C.D. & Mouat, D.A. 1989. Analysis of green vegetation detection limits in 1988 AVIRIS data. *Proceedings International Symposium Remote sens. environ, 7th Thematic Conference: Remote Sensing for Exploration Geology,* Calgary, Canada.

Elvidge, C.D. & Portigal, F.P. 1990. Phenologically induced changes in vegetation reflectance derived from 1989 AVIRIS data. *Proceedings of 23rd International Symposium Remote Sensing of Environment,* Bangkok, Thailand.

Elvidge, C.D., Portigal, F.P. & Mouat, D.A. 1990. Detection of trace quantities of green vegetation in 1989 AVIRIS data. *Proceedings of the 2nd Airborne Visible/Infrared Imaging Spectrometer (AVIRIS) Workshop,* NASA JPL Publication 90-54, Pasadena, CA., pp. 35-41.

Epema, G.F. 1992. Spectral Reflectance in the Tunisian Desert. Ph.D. dissertation, Wageningen University, The Netherlands.

Epstein, E. & Grant, W.J. 1973. Soil Crust Formation as Affected by Raindrop Impact. *In:* A.Hadas, D. Swartzendruber, P.E. Rijtema, M. Fuchs & B. Yaron (Eds.), Physical Aspects of Soil Water and Salts in Ecosystems. Springer-Verlag, Berlin, pp. 195-201.

ESA 1995. *MERIS: The Medium Resolution Imaging Spectrometer,* ESA SP-1184, 63 pp.

ESA 1997. *Envisat-1 Mission & System Summary,* ESTEC-ESA, Noordwijk, 84 pp.

Escadafal, R. 1989. Charactérisation de la Surface des Sols Arides par Observatios de Terrain et par Télédetection. Ph.D. dissertation, Université Pierre et Marie Curie, Paris, 317 pp.

Escadafal, R. 1992. Soil Spectra and their Relationship with Pedological Parameters. *In:* J. Hill & J. Mégier (Eds.), Imaging Spectrometry - a Tool for Environmental Observations. Dordrecht, Kluwer Academic, 15 pp.

Escadafel, R., Bacha, S. & Delaitre, E. 1997. Desertification Watch in Tunisia: Land Surface Changes during the Last 20 Years and Onwards. *In:* A. Spitieri, Integrated Applications for Risk Assessment and Disaster Prevention for the Mediterranean, Balkema, Rotterdam, pp.35-42.

Estep, L. & Arnone, R.A. 1994. Effect of whitecaps on determination of chlorophyll concentration from satellite data. *Rem. Sens. Environ.* **50,** 328-334.


**F**

FAO, 1979. A Provisional Methodology for Soil Degradation Assessment. Food and Agricultural Organization of the United Nations. FAO, Rome, 47 pp.

FAO, 1983. Guidelines for the Control of Soil Degradation. Food and Agricultural Organization of the United Nations. FAO, Rome, 38 pp.

FAO, 1988. FAO-UNESCO Soil Map of theWorld. Food and Agricultural Organization, Rome.

FAO, 1990. Guidelines for Soil Profile Description. Food and Agricultural Organisation, Rome.

Farmer, V.C. 1974. The layer silicates. *In:* V.C. Farmer (Ed.), *The infrared spectra of minerals,* Mineralogical Society, London, pp. 331-364.

Farrand, W.H. & Harsanyi J.C. 1997. Mapping the distribution of mine tailings in the Coeur d'Alene River valley, Idaho, through the use of a Constrained Energy Minimization Technique. *Remote Sens. Environ.,* **59,** 64-76.

Farrand, W.H. 1997. Identification and mapping of ferric oxide and oxyhydroxide minerals in imaging spectrometer data of Summitville, Colorado, U.S.A., and the surrounding San Juan Mountains. *Int. J. Remote Sensing,* **18,** 1543-1552.

Farrand, W.H. & Seelos, A. 1996. Using mineral maps generated from imaging spectrometer data to map faults: an example from Summitville, Colorado. *Proceedings of the Eleventh Thematic Conference and Workshop on Applied Geologic Remote Sensing*, Las Vegas, Nevada, 27-29 February, Vol. II, pp. 222-230.

Farrand, W.H., Singer, R.B. & Merenyi, E. 1994. Retrieval of Apparent Surface from AVIRIS Data: A Comparison of Empirical Line, Radiative Transfer and Spectral Mixture Methods. *Rem. Sens. Environ.*, **47**, 311-321.

Feldman, S.C. & Taranik, J.V. 1988. Comparison of techniques for discriminating hydrothermal alteration minerals with airborne imaging spectrometer data. *Remote Sens. Environ.*, **24**, 67-83.

Fell, F. 1997. *Validierung eines Modells zur Simulation des Strahlungstransportes in Atmosphäre und Ozean*. PhD Thesis, Freie Universitat, Berlin.

Fell, F. & Fischer, J. in press. Numerical simulation of the light field in the atmosphere-ocean system using the matrix-operator method. *J. Quant. Spec. Radiat. Transfer*.

Fenstermaker, L.K. & Miller, J.R. 1994. Identification of fluvially redistributed mill tailings using high spectral resolution aircraft data. *Photogramm. Eng. and Remote Sensing*, **60**, 989-995.

Ferguson, R.L., Wood, L.L. & Graham, D.B. 1993. Monitoring spatial change in seagrass habitat with aerial- photography. *Photogrammetric Engineering and Remote Sensing*, **59**, 1033-1038.

Ferrari, G.M., Hoepffner, N. & Mingazzini, M. 1996. Optical properties of the water in a deltaic environment: prospective tool to analyze satellite data in turbid waters. *Rem. Sens. Environ.* **58**, 69-80.

Ferrier, G. & Anderson, J.M. 1997. The application of remotely sensed data in the study of tidal systems in the Tay Estuary, Scotland, U. K. *Int. J. Remote Sensing* **18**, 2035-2065.

Ferrier, G. & Wadge, G. 1996. The application of imaging spectrometry data to mapping alteration zones associated with gold mineralization in southern Spain. *Int. J. Remote Sensing*, **17**, 331-350.

Fillela, I. & Peñuelas, J. 1994. The red edge position and shape as indicators of plant chlorophyll content, biomass and hydric status. *Int. J. Remote Sensing*, **15**, 1459-1470.

Finn, J.D. 1974. *A general model for multivariate analysis*, New York, Holt, Rinehart and Winston.

Fischer, J. 1983. *Fernerkundung von Schwebstoffen im Ozean*. PhD Thesis, University of Hamburg, Hamburg, Germany.

Fiumie, L. & Marino, C.M. 1997.   Airborne Hyperspectral MIVIS Data for the Characterization of Urban Historical Environments, *Proceedings of the Third International Airborne Remote Sensing Conference and Exhibition* II, Copenhagen, Denmark, pp. 770-771.

Forget, P., Ouillon, S., Lahet, F. & Broche, P., 1999. Inversion of reflectance spectra of nonchlorophyllous turbid coastal waters. *Rem. Sens. Environ.* **68**, 264-272.

Foster, G.R. 1987. Modelling Soil Erosion and Sediment Yield. *In:* R.Lal (Ed.), *Soil Erosion Research Methods*. Soil and Water Conservation Society, Wageningen, pp.97-118.

Fourty Th., Baret F., Jacquemoud S., Schmuck G. & Verdebout J. 1996. Leaf optical properties with explicit description of its biochemical composition: Direct and inverse problems. *Remote sens. environ*, **56**, 104-117.

Frette, O., Stamnes, J.J. & Stamnes, K. 1998. Optical remote sensing of marine constituents in coastal waters: a feasible study. *Appl.Opt.* **37**, 8318-8326.

Frouin, R., Deschamps, P.Y. & Lecomte, P. 1990. Determination from space of atmospheric total water amounts by differential absorption near 940 nm: Theory and airborne verification. *J. Appl. Meteorol.*, **29**, 448-460.

Fung, T. & Siu, W. 2000. Environmental quality and its changes, an analysis using NDVI. *Int. J. Remote Sensing*, **20**, 1011-1024.


G

Gaddis, L.R., Soderblom, L.A., Kieffer, H.H., Becker, K.J., Torson, J. & Mullins, K. 1996. Decomposition of AVIRIS spectra: extraction of surface-reflectance, atmospheric and instrumental components. *IEEE Trans. Geosci. Rem. Sens.*, **34**, 163-178.

Gamon, J.A. & Field, C.B. 1992. Tracking photosynthetic efficiency with narrow-band spectroradiometry. *In:* Green, R.O. (Ed.), *Summaries of the Third Annual JPL Airborne Geoscience Workshop*. June 1-5, California. JPL Publication 92-41, pp. 95-97.

Gao, B. & Goetz, A.F.H. 1992. A linear spectral matching technique for retrieving equivalent water thickness and biochemical constituents of green vegetation. *In:* Green, R.O. (Ed.), *Summaries of the Third Annual JPL Airborne Geoscience Workshop*. June 1-5, California. JPL Publication 92-41, pp. 35-55.

Gao, B.C. & Goetz, A.F.H. 1990. Column atmospheric water vapor and vegetation liquid water retrievals from airborne imaging spectrometer data. *J. Geophys. Res.*, **95**, 3549-3564.

Gao, B.C. & Goetz, A.F.H. 1995. Retrieval of equivalent water thickness and information related to biochemical components of vegetation canopies from AVIRIS data. *Remote Sens. Environ.*, **52**, 155-162.

Gao, B.-C., Goetz, A.F.H. & Wiscombe, W.J. 1993. Cirrus cloud detection from airborne imaging spectrometer data using the 1.38 μm water vapor band. *Geophys. Res. Lett.*, **20**, 301-304.

Gao, B.C., Heidebrecht, K.B. & Goetz, A.F.H. 1993. Derivation of Scaled Surface Reflectance from AVIRIS Data. *Rem. Sens. Environ.*, **44**, 165-178.

Gates, D.M. 1965. Energy, plants, and ecology. *Ecology*, **46**, 1-13.

Gates, D.M. 1968. Transpiration and leaf temperature. *Annual Review of Plant Physiology*, **19**, 211-238.

Gates, D.M., Keegan, H.J., Schleter, V.R. & Weidner, V.R. 1965. Spectral properties of plants. *Applied Optics*, **4**, 11-20.

Gausman, H.W. 1985a. *Plant leaf optical properties*. Lubbock, Texas Tech. Press.

Gausman, H.W. 1985b. Plant leaf optical properties in visible and near infrared light, Graduate Studies N 29, Texas Tech. University, 78 pp.

Gausman, H.W., Allen W.A., Cardenas R. & Richardson A.J. 1970. Relationship of light reflectance to histological and physical evaluation of cotton leaf maturity. *Applied Optics*, **9**, 545-552.

Gill, J.D., West, M.W., Noe, D.C., Olsen, H.W. & McCarty, D.K. 1996. Geologic control of severe expansive clay damage to a subdivision in the Pierre Shale, Southwest Denver metropolitan area, Colorado. *Clays and Clay minerals*, **44(4)**, 530-539.

Gillespie, A.R. 1992. Enhancement of multispectral thermal infrared images: decorrelation contrast stretching. *Remote Sens. Environ*, **42**, 147-155.

Gillespie, A., Kahle, A. & Palluconi, F. 1984. Mapping alluvial fans in Death Valley, California using multi-channel thermal infrared images. *Geophys. Res. Lett.* **11**, 1153-1156.

Gitelson, A. & Merzlyak, M.N. 1994. Spectral reflectance changes associated with autumn senescence of Aesculus hippocastanum L. and Acer platanoides L. leaves: Spectral features and relation to chlorophyll estimation. *Journal of Plant Physiology*, **143**, 286-292.

Gitelson, A.A. & Merzlyak, M.N. 1996. Signature analysis of leaf reflectance spectra: algorithm development for remote sensing of chlorophyll. *Journal of Plant Physiology*, **148**, 494-500.

Givoni, B. 1991. Impact of planted areas on urban environmental quality: a review. *Atmospheric Environment*, **25b**, 289–299.

Goel, N.S.1989. Inversion of Canopy Reflectance Models for Estimation of Biophysical Parameters from Reflectance Data. *In:* G. Asrar (Ed.), *Theory and Applications of Optical Remote Sensing*. Wiley, New York, pp. 205-251.

Goel, N.S. & Deering, D.W. 1985. Evaluation of a canopy reflectance model for LAI estimation through its inversion. *IEEE Trans. Geosci. Rem. Sens* **23**, 674-684.

Goel, N.S. & Grier, T. 1986. Estimation of canopy parameters for inhomogeneous vegetation canopies from reflectance data. *Int. J. Remote Sensing*, **7**, 665-681.

Goel, N.S. & Grier, T. 1987. Estimation of canopy parameters of row planted vegetation  canopies using reflectance data for only four view directions. *Remote sens. environ*, **21**, 37-51.

Goetz, A.F.H. 1991. Imaging Spectrometry for studying Earth, Air, Fire and Water. *EARSeL Advances in Remote Sensing* **1**, 3-15.

Goetz, A.F.H. 1992. Principles of narrow band spectrometry in the visible and IR: instruments and data analysis. *In:* F. Toselli & J. Bodechtel (Eds.), *Imaging Spectroscopy: Fundamentals and Prospective Applications*. Dordrecht, Kluwer Academic Publishers, pp. 21-32.

Goetz, A.F.H., 1989. Spectral Remote sensing in Geology. *In:* G. Asrar (Ed.), *Theory and Applications of Optical Remote Sensing*, Wiley, New York, pp. 491-526.

Goetz, A.F.H., Gao, B,C., Wessman, C. A. & Bowman, W.D. 1990. Estimation of Biochemical Constituents From Fresh Green Leaves By Spectrum Matching Techniques. *Proceedings 10th International Geoscience & Remote Sensing Symposium, IGARSS'90)*, University of Maryland , 20 - 24 May 1990, pp. 971 - 974.

Goetz, A.H. & Herring, M. 1989. The High Resolution Imaging Spectrometer for EOS. *IEEE Trans. Geosci. Rem. Sens.* **27**, 136-144.

Goetz A.F.H. & Rowan L.C. 1981. Geologic remote sensing. *Science*, **211**, 781-791.

Goetz, A.F.H., Rowan, L.C. & Kingston, M.J. 1982. Mineral identification from orbit: initial results from the Shuttle Multispectral Infrared Radiometer. *Science*, **218**, 1020-1031.

Goetz, A.F.H., Vane G., Solomon, J.E. & Rock, B.N. 1985. Imaging spectrometry for earth remote sensing. *Science*, **228**, 1147-1153.

Gong, P., Pu, R. & Miller, J.R. 1995. Coniferous Forest Leaf Area Index Estimation along the Oregon Transect Using Compact Airborne Spectrographic Imager Data. *Photogramm. Eng. and Remote Sensing*, **61**, 1107-1117.

Gordon, H.R. 1989. Can the Lambert-Beer law be applied to the diffuse attenuation coefficient of ocean water? *Limnol.Oceanogr.* **34**, 1389-1409.

Gordon, H.R. et al. 1988. A semianalytical model of ocean colour. *J.Geophys.Res.* **93,** 10909.

Gordon, H.R., Brown, O.B. & Jacobs, M.M. 1975. Computed relationships between the inherent and apparent optical properties of a flat homogeneous ocean. *Appl.Opt.* **14,** 417-427.

Gordon, H.R. & Wang, M.H.1994. Retrieval of water-leaving radiance and aerosol optical-thickness over the oceans with Seawifs - a preliminary algorithm. *Appl. Opt.* **33,** 443-452.

Gould, R.W. & Arnone, R.A. 1997. Remote sensing estimates of inherent optical properties in a coastal environment. *Remote Sens.Environ.* **61,** 209-301.

Govaerts, Y.M., Verstraete, M.M., Pinty, B. & Gobron, N. 1999. Designing optimal spectral indices: a feasibility and proof of concept study. *Int. J. Remote Sensing* **20**, 1853-1873.

Gower, J.F.R., Doerffer, R. & Borstad, G.A., 1999. Interpretation of the 685 nm peak in water-leaving radiance spectra in terms of fluorescence, absorption and scattering, and its observation by MERIS. *Int. J. Remote Sensing,* **20,** 1771-1786.

Green, A.A., Berman, M., Switzer, P. & Graig, M.D. 1988. A transformation for ordering multispectral data in terms of image quality with implications for noise removal. *IEEE Trans. Geosci. Rem. Sens.,* **26,** 65-74.

Green, A.A. & Graig, M.D. 1985. Analysis of aircraft spectrometer data with logarithmic residuals. *In:* G. Vane & A.F.H. Goetz (Eds.), *Proceedings of the Airborne Imaging Spectrometer Data Analysis Workshop,* Pasadena, USA, NASA-JPL Publication 85-41, pp. 111-119.

Green, E.P., Mumby, P.J., Edwards, P.J. & Clark, C.D. 1996. A review of remote-sensing for the assessment and management of tropical coastal resources. *Coastal Management* **24,** 1-40.

Green, R.O., Chrien, T.G., Sarture, C.M., Eastwood, M.L., Chippendale, B.J., Kurzweil, C., Chovit, C.J. & Faust, J.A. 1999b. Operation, calibration, georectification, and reflectance inversion of NASA's Airborne Visible/Infrared Imaging Spectrometer (AVIRIS) four- and two- meter data acquired from a low-altitude platform, in *Summaries of the Eighth JPL Airborne Earth Science Workshop,* Jet Propulsion Laboratory, Pasadena, U.S.A., 9-11 February 1999, JPL Publication 99-17, pp. 177-188.

Green, R.O., Conel, J.E., Margolis, J.S., Bruegge, C.J. & Hoover, G.L. 1991. An inversion algorithm for retrieval of atmospheric and leaf water absorption from AVIRIS radiance with compensation for atmospheric scattering, *In:* Green, R.O. (Ed.), *Proceedings Third AVIRIS Workshop,* JPL Publ. 91-28, Jet Propulsion Lab., Pasadena, California, pp. 51-57.

Green, R.O., Eastwood, M.L., Sarture, C.M., Chrien, T.G., Aronsson, M., Chippendale, B.J., Faust, J.A., Pavri, B.E., Chovit, C.J., Solis, M., Olah, M. & Williams, O. 1998. Imaging spectroscopy and the Airborne Visible/Infrared Imaging Spectrometer (AVIRIS). *Rem. Sens. Environ.,* **65,** 227-248.

Green, R.O. & Roberts, D.A. 1995. Vegetation species composition and canopy architecture information expressed in leaf water absorption measured in the 1000nm and 2200nm spectral region by an imaging spectrometer. *In:* Green, R.O. (Ed.), *Summaries of the Fifth Annual JPL Airborne Earth Science Workshop.* January 23-26, California. JPL Publication 95-1, pp. 95-98.

Green, R.O., Pavri, B.E., Faust, J.A. & Williams, O. 1999a. AVIRIS radiometric laboratory calibration, inflight validation and a focused sensitivity analysis in 1998, *Summaries of the Eighth JPL Airborne Earth Science Workshop,* Jet Propulsion Laboratory, Pasadena, U.S.A., 9-11 February 1999, JPL Publication 99-17, pp. 161-176.

Greenwood, D.J., Lemaire, G., Gosse, P., Cruz, P., Draycott, A. & Neeteson, J.J. 1990. Decline in percentage N of C3 and C4 corps with increasing plant mass, *Annals of Botany* **66**, 425-436.

Grenon, M. & Batisse, M. 1989. Futures for the Mediterranean Basin: The Blue Plan. Oxford University Press.

Grimmond, C. S. B., Souch, C. & Hubble, M. D. 1996. Influence of tree cover on summertime surface energy balance fluxes, San Gabriel Valley, Los Angeles. *Climate Research,* **6,** 45-57.

Grove, C.I., Hook, S.J., & Paylor II, E.D. 1992. *Laboratory Reflectance Spectra of 160 minerals, 0.4 to 2.5 Micrometers.* NASA-JPL Publication 92-2, Pasadena. 300 pp.

Guyot, G. & Baret, F. 1988. Utilisation de la haute resolution spectrale pour suivre l'etat des couverts vegetaux, *Proceedings 4th International Colloquium on Spectral Signatures of Objects in Remote Sensing,* Aussois, France, 18-22 January 1988 (ESA SP-287, April 1988), pp. 279-286.

Guyot, G, Baret, F. & Major, D.J. 1988. High spectral resolution: determination of spectral shifts between the red and near infrared. *Int. Arch. of Photogr. and Rem. Sensing* **27(11)**, 750-760.

Gwinner, K. & Schaale, M., 1997. A Case Study on the Influence of Shade and Shading on Multispectral Airborne Imaging Data. *Proceedings of the Third International Airborne Remote Sensing Conference and Exhibition,* Copenhagen, Denmark, pp. 409-416.


H

Hackwell, J., Warren, D., Bongiovi, R. & Hansel, S. 1996. LWIR/MWIR imaging hyperspectral sensor for airborne and ground based remote sensing. *SPIE,* V. 2819, pp. 102-107.

Halbouty, M.T. 1980. Geologic significance of Landsat data for 15 giant oil and gas fields. *AAPG Bulletin*, **64**, 8-36.

Hall, P.L. 1987. Clays: their significance, properties, origins and uses. *In:* M.J. Wilson (Ed.), *A handbook of determinative methods in clay mineralogy*, Chapman and Hall, New York, pp. 1-25.

Hall, T. 1988. Urban Geography. Routledge Press, London New-York

Hamilton, M.K., Davis, C.O., Pilorz, S.H., Rhea, W.J. & Carder, K.L., 1993. Estimating chlorophyll content and bathymetry of Lake Tahoe using AVIRIS data. *Rem. Sens. Environ.* **44**, 217-230.

Hapke, B. 1981. Bidirectional reflectance spectroscopy, 1, theory. *J. Geophys. Res.*, **86**, 3039-3054.

Hare, E.W., Miller, J.R., Hollinger, A.B., Sturgeon, D.R., O'Neill, N.T. & Ward, T.V. 1986. Measurements of the vegetation reflectance edge with an airborne programmable imaging spectrometer. *Fifth Thematic Conference on Remote Sensing for Exploration Geology*, Reno, Nevada.

Harman, H.H. 1968. *Modern factor analysis*, 2nd edition, University of Chicago Press.

Hart, S. S. 1974. Potentially swelling soil and rock in the Front Range Urban Corridor, Colorado, *Environmental Geology 7, 4 maps, scale 1:1,000,000, Colorado Geological Survey, Denver, CO*, 23 pp.

Hauff, P.L., Lindsay, N., Peters, D., Borstad, G., Peppin, W., Costik, L. & Glanzman, R. 1999. Hyperspectral evaluation of mine waste and abandoned mine lands, NASA and EPA sponsored projects in Idaho, *Summaries of the Eighth JPL Airborne Earth Science Workshop*, Jet Propulsion Laboratory, Pasadena, U.S.A., 9-11 February 1999, JPL Publication 99-17, pp. 229-238.

Hepner, G.F., Houshmand, B., Kulikov, I. & Brayant, N. 1998. Investigation of the Integration of AVIRIS and IFSAR for urban analysis. *Photogrammetric Engineering and Remote Sensing*, **64**, 813-820.

Hergbert, H.L. 1973. Infrared spectra. *In:* K.V. Sarkanen and C.H. Ludwig (Eds.), *Lignins: Occurrences, Formation, Structure and Reactions*. Wiley Interscience, New York.

Heuvelink, G.B.M. 1993. Error Propagation in Quantitative Spatial Modelling. Applications in Geographical Information Systems. Ph. D. dissertation, Netherlands Geographical Studies163. KNAG, Utrecht, 151pp.

Hick, P. 1997. *Determination of water column chaeracteristics in coastal environments using remote sensing*. PhD Thesis, Curtin University of Technology, Perth.

Hill, J., Lacaze, B., Caselles, V., Coll, C., Hoff, C., de Jong, S.M., Mehl, W., Negendank, J., Riezebos, H., Rubio, E., Sommer, S., Teixeira Filho, J. & Valor, E. 1996. Integrated Approaches to Desertification Mapping and Monitoring in the Mediterranean Basin. Final DeMon-report, EUR 16448, Brussels, 165 pp.

Hill, J., Mégier, J. & Mehl, W. 1995. Land Degradation, Soil Erosion and Desertification Monitoring in the Mediterranean Ecosystems. *Remote Sensing Reviews*, **12**, 107-130.

Hillel, D. 1980, Fundamentals of Soil Physics. Academic Press, New York.

Hillel, D. 1993. Out of the Earth: Civilization and the Life of the Soil. University of California Press, Los Angeles, 321 pp.

Hilton, J. 1984. Airborne remote sensing for freshwater and estuarine monitoring. *Water Res.* **18**, 1195-1223.

Himmelsbach, D.S., Boer, J.D., Akin, D.E. & Barton, E.E. 1988. Solid-state carbon-13 NMR, FTIR and NIR spectroscopic studies of ruminant silage digestion. In: C.S. Creaser and A.M.C. Davies (Eds.) *Analytical Applications of Spectroscopy*, Royal Society of Chemistry, London, pp. 410-413.

Hochberg, E.J. & Atkinson, M.J. 2000. Spectral discrimination of coral reef benthic communities. *Coral Reefs* **19**, 164-171.

Hoerig, B. & Kuehn, F. 2000. HyMAP hyperspectral remote sensing to detect hydrocarbons, *Proceedings of the 14th International Conference on Applied Geologic Remote Sensing*, Las Vegas, Nevada, pp. 206-211.

Hoffer, R.M. 1978. Biological and physical considerations in applying computer aided analysis techniques to remote sensor data. *In:* Swain, P.H. & Davis, S.M. (Eds.), *Remote sensing: the quantitative approach*. McGraw Hill, New York, pp. 227-289.

Holasek, R. et al. 1998. Coral and substrate mapping in Kaneohe Bay, Oahu, Hawaii using the advanced airborne hyperspectral imaging system (AAHIS). *Proceedings Fifth International Conference on Remote Sensing for Marine and Coastal Environment*, San Diego, California, pp. 72-77.

Holben, B.N., Schutt, J.B. & McMurtrey, J. 1983. Leaf water stress detection utilising thematic mapper bands 3, 4 and 5 in soybean plants. *Int. J. Remote Sensing*, **4**, 289-297.

Holden, H. & LeDrew, E. 1997. Spectral discrimination of bleached and healthy submerged corals based on principal component analysis. *Proceedings 4th International Conference on Remote Sensing for Marine Coastal Environments*. Waterloo laboratory for Earth Observations, University of Waterloo, Orlando, Florida, pp. 177-185.

Holden, H. & LeDrew, E. 1998a. Hyperspectral identification of coral reef features in Fiji and Indonesia, *Proceedings 5th international conference on remote sensing for marine and coastal environments*, San Diego, California, pp. II-78-II-84.

Holden, H. & LeDrew, E. 1998b. Spectral Discrimination of Healthy and Non-Healthy Corals Based on Cluster Analysis, Principal Components Analysis, and Derivative Spectroscopy. *Remote Sens.Environ.* **65,** 217-224.

Holden, H. & LeDrew, E. 1999. Hyperspectral identification of coral reef features. *Int. J. Remote Sensing* **20,** 2545-2563.

Holden, H. & LeDrew, E. 2000. Hyperspectral versus multispectral imaging for submerged coral detection, *Proceedings 6th International Conference on Remote Sensing for Marine and Coastal Environments,* Charleston, NC, pp. 186-193.

Hoogenboom, H.J., Dekker, A.G. & De Haan, J.F. 1998. Retrieval of chlorophyll and suspended matter in inland waters from CASI data by matrix inversion. *Canadian Journal of Remote Sensing* **24,** 144-152.

Hoogenboom, H.J. & Roberti, J.R. 2000. An integral approach to retrieve total suspended matter concentrations in the Marsdiep (Netherlands), Ocean Optics XV, Monaco, pp. 10.

Hook, S., Abbott, E., Grove, C., Kahle, A. & Palluconi, F. 1999. Multispectral thermal infrared data in geologic studies. *In:* A. Rencz (Ed.), *Manual of Remote Sensing.* John Wiley & Sons, New York, pp. 59-110.

Hook, S,J., Elvidge, C.D., Rast, M. & Watanabe, H. 1991. An evaluation of short-wave-infrared (SWIR) data from the AVIRIS and GEOSCAN instruments for mineralogical mapping at Cuprite, Nevada. *Geophysics,* **56,** 1432-1440.

Hook, S., Gabell, A., Green, A. & Kealy, P. 1992. Comparison of techniques for extracting emissivity information from thermal infrared data for geologic studies. *Rem. Sens. Environ.* **42,** 123-135.

Hook, S., Karlstrom, K., Miller, C. & McCaffrey, J. 1994. Mapping the Piute Mountains, California with thermal infrared multispectral scanner (TIMS) images. *J. Geophys. Res.* **199,** 15605-15622.

Hoque, E. & Hutzler, P.J.S. 1992. Spectral blue shift of red edge monitors damage class of beech trees. *Rem. Sens. Environ.,* **39,** 81-84.

Horler, D.N.H., Barber, J. & Barringer, A.R. 1980. Effects of heavy metals on the absorbance and reflectance spectra of plants. *Int. J. Remote Sensing,* **1,** 121-136.

Horler, D.N.H., Dockray, M. & Barber J. 1983. The red edge of plant leaf reflectance. *Int. J. Remote Sensing,* **4,** 273-288.

Hornstra, T.J., Lemmens, M.J.P.M. & Wright, G.L. 1999. Incorporating Intra-Pixel Variability in the Multi-Spectral Classification Process of High-Resolution Satellite Imagery of Urbanised Areas. *Carthography,* **28,** 1-9.

Hornstra, T.J., Maas, G.G. & de Jong, S.M. 2000. Classification of Spectroscopical Imagery by Combining Spectral and Spatial Information: the SSC Method. *Proceedings of the 19th ISPRS Congres,* Amsterdam 16-23 July, pp. 550-558.

Howard, J.A., Watson, R.D. & Hessin, T.D. 1971. Spectral reflectance properties of Pinus ponderosa in relation to copper content of the soil - Malachite Mine, Jefferson Country, Colorado, *Proceedings of 7th International Symposium on Remote Sensing of Environment,* Ann Arbor (MI), 17-21 May 1971, Vol. 1, pp. 285-297.

Huete, A.R. 1988. A soil adjusted vegetation index (SAVI). *Rem. Sens. Environ,* **23,** 213-232.

Hunt, G.R. 1977. Spectral signatures of particulate minerals in the visible and near-infrared. *Geophysics* **42,** 501-513.

Hunt, G.R. 1980. Electromagnetic radiation: the communication link in remote sensing. *In:* B. Siegal & A. Gillespie (Eds.), Remote Sensing in Geology, New York, Wiley, 702pp.

Hunt, E.R., Rock, B.N. & Nobel, P.S. 1987. Measurement of leaf relative water content by infrared reflectance. *Remote sens. environ,* **22,** 429-435.

Hunt, G. & Salisbury, J. 1974. Mid-infrared spectral behavior of Igneous rocks. *Environ. Res. Paper 496-AFCRL-TR-74-0625,* Air Force Cambridge Research Lab, Hanson Air Force Base, Mass, 142 pp.

Hunt, G. & Salisbury, J. 1975. Mid-infrared spectral behavior of sedimentary rocks. *Environ. Res. Paper 520-AFCRL-TR-75-0256,* Air Force Cambridge Research Lab, Hanson Air Force Base, Mass, 49 pp.

Hunt, G. & Salisbury, J. 1976. Mid-infrared spectral behavior of metamorphic rocks. *Environ. Res. Paper 543-AFCRL-TR-76-0003,* Air Force Cambridge Research Lab, Hanson Air Force Base, Mass, 67 pp.

Hunt, G. 1980. Electromagnetic radiation: the communications link in remote sensing, *In:* B. Siegal & A. Gillespie (Eds.), *Remote Sensing in Geology,*Wiley, New York, pp. 5-45.

Hunt, G.R. & Evarts, R.C. 1981.The use of near-infrared spectroscopy to determine the degree of serpentinization of ultramafic rocks. *Geophysics,* **46,** 316-321.

Hunt, G.R. & Salisbury, J.W. 1976a. Visible and near-infrared spectra of minerals and rocks: XI. Sedimentary rocks. *Modern Geology* **5,** 211-217.

Hunt, G.R. & Salisbury, J.W., 1976b. Visible and near-infrared spectra of minerals and rocks: XII. Metamorphic rocks. *Modern Geology* **5,** 219-228.

Hunt, G.R., & Salisbury, J.W., 1970. Visible and near-infrared spectra of minerals and rocks, I. Silicate Minerals. *Modern Geology* **1,** 283-300.

Hunt, G.R., & Salisbury, J.W., 1971. Visible and near-infrared spectra of minerals and rocks, II. Carbonates. *Modern Geology* **2**, 23-30.

Hunt, G.R., Salisbury, J.W. & Lenhoff, C.J. 1971a. Visible and near-infrared spectra of minerals and rocks, III. Oxides and Hydroxides. *Modern Geology* **2**, 195-205.

Hunt, G.R., Salisbury, J.W. & Lenhoff, C.J. 1971b. Visible and near-infrared spectra of minerals and rocks, IV. Sulphides and sulphates. *Modern Geology* **3**, 1-14.

Hunt, G.R., Salisbury, J.W. & Lenhoff, C.J. 1972. Visible and near-infrared spectra of minerals and rocks, V. Halides, arsenates, vanadates, and borates. *Modern Geology* **3**, 121-132.

Hunt, G.R., Salisbury, J.W. & Lenhoff, C.J. 1973a. Visible and near-infrared spectra of minerals and rocks, VI. Additional Silicates. *Modern Geology* **4**, 85-106.

Hunt, G.R., Salisbury, J.W. & Lenhoff, C.J. 1973b. Visible and near-infrared spectra of minerals and rocks, VII. Acidic Igneous Rocks. *Modern Geology* **4**, 217-224.

Hunt, G.R., Salisbury, J.W. & Lenhoff, C.J. 1973c. Visible and near-infrared spectra of minerals and rocks, VIII. Intermediate Igneous Rocks. *Modern Geology* **4**, 237-244.

Hunt, G.R., Salisbury, J.W. & Lenhoff, C.J. 1974. Visible and near-infrared spectra of minerals and rocks, IX. Basic and Ultrabasic igneous rocks. *Modern Geology* **4**, 15-22.

Hunt, G.R., Salisbury, J.W. & Lenhoff, C.J. 1975. Visible and near-infrared spectra of minerals and rocks, X. Stoney Meteorites. *Modern Geology* **5**, 115-126.

Huntington J.F., Green, A.A. & Graig, M.D. 1989. Identification, the Goal bbehind Discimination: the Status of Mineral and Lithological Identification from High Resolution Spectrometer Data: Examples and Challenges. *Proceedings of IGARSS'89*, July 1-14, Vancouver, Canada, pp.6-11.

Hurcom, S.J. & Harrison, A.R. 1998. The NDVI and Spectral Decomposition for Semi-Arid Vegetation Abundance Estimation. *Int. J. Remote Sensing.*, **19**, 3109-3125.

Hutsinpiller, A. 1988. Discrimination of Hydrothermal Alteration Mineral Assemblages at Virginia City, Nevada, Using the Airborne Imaging Spectrometer. *Remote Sens. Environ.*, **24**, 53-66.


I

Ichoku, C. & Karnieli, A. 1996. A review of mixture modelling techniques for sub-pixel land cover estimation. *Remote Sensing Reviews*, **13**, 161-186.

IOCCG, 2000. *Remote sensing of ocean colour in coastal, and other optically-complex waters*. 3, International Ocean-Colour Coordinating Group, Dartmouth

Iqbal, M. 1983. *An introduction to solar radiation*, Academic Press, Ontario, Canada, 390 pp.

Irons, J.R., Weismiller, R.A. & Petersen, G.W. 1989. Soil reflectance. *In:* G. Asrar (Editor), *Theory and Applications of Optical Remote Sensing*. Wiley, New York, pp 66-106.

Isaaks, H.S. & Srivastava, R.M. 1989. Introduction to Applied Geostatistics, 1st edn (New York: Oxford University Press).

ISRIC, 1995. World Soils and Terrain Digita Database (SOTER). Technical SOTER Reports 1 to 7, ISRIC, Wageningen.


J

Jackson, R.D. & Ezra, C.E. 1985. Spectral response of cotton to suddenly induced water stress. *Int. J. Remote Sensing*, **6**, 177-185.

Jackson, R.D. 1986. Remote sensing of biotic and abiotic plant stress. *Annual Reviews in Phytopathology*, **24**, 265-287.

Jacquemoud, S. & Baret, F. 1990. PROSPECT: a model of leaf optical properties spectra. *Remote sens. environ*, **34**, 75-91.

Jacquemoud, S., Ustin, S.L., Verdebout, J., Schmuck, G., Andreoli, G. & Hosgood, B. 1996. Estimating leaf biochemistry using the PROSPECT leaf optical properties model, *Remote sens. environ*, **56**, 194-202.

Jain, S.C. & Miller, J.R. 1976. Subsurface parameters: optimization approach to their determination from remotely sensed water color data. *Appl.Opt.* **15**, 886-890.

Jakubauskas, M., Kindscher, K., Fraser, A., Debinski, D. & Price, K.P. 2000. Close-range remote sensing of aquatic macrophyte vegetation cover. *Int. J. Remote Sensing* **21**, 3533-3538.

Jansen, W.T. 1993. General Imaging Spectrometry Interpretation Systems. WTJ-Software San Mateo, CA, 45pp.

Jansinski, M.F. 1990. Sensitivity of the Normalized Difference Vegetation Index to Sub-pixel Canopy Cover, Soil Albedo and Pixel Scale. *Rem. Sens. Environ.*, **32**, 169-187.

Jauregui, E. 1990. Influence of a large park on temperature and convective precipitation in a tropical city. *Energy and Building,* **15/16**, 457-463.

Jensen, J.R. et al. 1986. Remote sensing inland wetlands: a multispectral approach. *Photogrammetric Engineering and Remote Sensing* **52**, 87-100.

Jensen, L.M. 1997. Classifcation of urban land cover based on expert systems, object models and texture. *Computer Environmental and Urban Systems,* **21**, 291-302.

Jetten, V.G. 1994. Modelling the effects of logging on the water balance of a tropical rain forest. A study in Guyana. Ph.D. Dissertation, Utrecht University. Tropenbos Series 6. The Tropenbos Foundation, Wageningen, the Netherlands.

Jetten, V.G., de Roo, A.P.J. & Favis-Mortlock, D. 1999. Evaluation of Field-scale and Catchment-scale Soil Erosion Models. *Catena,* **37,** 521-542.

Jernakoff, P. & Hick, P. 1994. Spectral measurement of marine habitat: simultaneous field measurements and CASI data. *Proceedings 7th Australasian Remote Sensing Conference,* Melbourne, Australia, pp. 706-713.

Jia, X. & Richards, J.A 1993. Binary Coding of Imaging Spectrometer Data for Fast Spectral Matching and Classification. *Remote Sens. Environ.,* **43,** 47-53.

Johnson, P.E., Smith, M.O., Taylor-George, S. & Adams J.B. 1983. A semiempirical method for analysis of the reflectance spectra of binary mineral mixtures. *J. Geophys. Res.,* **88,** 3557-3561.

Johnson, R.D. 1997. Clay and engineering properties of the steeply dipping Pierre Shale and adjacent formations along the Front Range Piedmont, Jefferson and Douglas Counties, Colorado, M.E. thesis, Colorado School of Mines, Golden, CO, 73 pp.

Jones, D.E. & Holtz, W.G. 1973. Expansive soils - The hidden disaster. *Civil Engineering,* **43(8),** 49-51.

Jongschaap, R.E.E. 2000. *Towards a direct nitrogen fertilization decision support system in potato using multi-spectral reflection patterns,* Plant Research International, Wageningen, The Netherlands (in press).

Jongschaap, R.E.E. & Booij, R. 2000. *Application of a direct nitrogen fertilization decision support system in potato using multi-spectral reflection patterns,* Plant Research International, Wageningen, The Netherlands (in press).

Jørgensen, P.V. & Edelvang, K. 2000. CASI data utilized for mapping suspended matter concentrations in sediment plumes and verification of 2-D hydrodynamic modelling. *Int. J. Remote Sensing* **21,** 2247-2258.

Jupp, D.L.B. et al., 1996. *Port Philip Bay Benthic Habitat mapping project Task G2.2.* Task G2.2, CSIRO Land&Water, Melbourne.

Jupp, D.L.B., Kirk, J.T.O. & Harris, G.P. 1994. Detection, identification and mapping of cyanobacteria-using remote sensing to measure the optical quality of turbid inland waters. *Australian Journal of Marine and Freshwater Research* **45,** 801-828.

Jupp, D.L.B., Mayo, K.K. & Al., E. 1985. Remote sensing for planning and managing the Great Barrier Reef of Australia. *Photogrammetria* **40,** 21-42.

Justice, C.O., Vermote, E., Townshend, J.R.G., Defries, R., Roy, D.P., Hall, D.K., Salomonson, V.V., Privette, J.L., Riggs, G., Strahler, A., Lucht, W., Myneni, R.B., Knyazikhin, Y., Running, S.W., Nemani, R.R., Wan, Z., Huete, A.R., van Leeuwen, W., Wolfe, R.E., Giglio, L., Muller, J.P., Lewis, P. & Barnsley, M.J. 1998. The Moderate Resolution Imaging Spectroradiometer (MODIS): Land remote sensing for global change research. *IEEE Trans. Geosci. Rem. Sens.* **36,** 1228-1249.


K

Kahle, A.B. 1987. Surface emittance, temperature, and thermal inertia derived from Thermal Infrared Multispectral Scanner (TIMS) data for Death Valley, California. *Geophysics,* **52,** 858-874.

Kahle, A.B. & Goetz, A.F.H. 1983. Mineralogic information from a new airborne Thermal Infrared Multispectral Scanner. *Science,* **222,** 24-27.

Kahle, A., Palluconi, F. & Christensen, P. 1993. Thermal emission spectroscopy. *In:* M. Pieters & J. Engels (Eds.), *Remote Geochemical Analysis: Elemental and Mineralogical Composition,* s, Cambridge University Press, pp. 99-120.

Kahle, A.B., Palluconi, F.D., Hook, S.J., Realmuto, V.J. & Bothwell, G. 1991. The Advanced Spaceborne Thermal Emission and Reflectance Radiometer (ASTER). *Int. J. Imag. Syst. Techn.* **3,** 144-156.

Kahle, A. & Rowan, L. 1980. Evaluation of multispectral thermal infrared aircraft images for lithologic mapping in the East Tintic Mountains, Utah. *Geology* **8,** 234-239.

Kahle, A., Shumate, M. & Nash, D. 1984. Active airborne infrared laser system for identification of surface rock and minerals. *Geophys. Res. Lett.* **11,** 1149-1152.

Kanari, Y., Mills, F. & Watanabe, H. 1992. Comparison of preliminary results from the airborne ASTER simulator (AAS) with TIMS data, Proc. 3$^{rd}$ Airborne Geoscience Workshop, *JPL Pub. 92-14,* Pasadena CA, pp. 13-15.

Kanemasu, E.T., Asrar, G. & Fuchs, M. 1984. Application of remotely sensed data in wheat growth modelling, in W. Day and R.K. Atkin (eds.), *Wheat growth modelling*, Plenum Press, New York and London, pp. 357-369.

Kariuki, P.C. 1999. Analysis of the effectiveness of spectrometry in detecting the swelling clay minerals in soils, M.Sc. thesis, International Institute for Aerospace Survey and Earth Sciences, Enschede, The Netherlands, 96 pp.

Karlstrom, K., Miller, J., Kingsbury, J. & Wooden, J. 1993. Pluton emplacement along an active ductile thrust zone, Piute Mountains California: interaction between deformational and solidification processes, *Geol. Soc. Amer. Bull.* **105**, 213-230.

Karmanova, L.A. 1981. Effect of Various Iron Compounds on the Spectral Reflectance and Color of Soils. *Sov. Soil Sci.,* **13,** 63-70.

Karpouzli, E. et al. 2001. Underwater light characterisation for correction of remotely sensed images. *Int. J. Remote Sensing,* In press.

Kaufmann, H., Weisbrich, W., Beyth, M., Baerov, Y., Itamar, A., Ronen, S. & Kafri, U. 1991. Mineral identification using GER-II data acquired from Makhtesh Ramon, Negev, Israel. *EARSel Advances in Remote Sensing,* **1,** 82-92.

Kauth, R.J. & Thomas, G.S. 1976. The tasseled cap - A graphic description of the spectral-temporal development of agricultural crops as seen by Landsat, *Proceedings Symposium on Machine Processing of Remotely Sensed Data*, Purdue Univ., W-Lafayette, Ind., 4B, pp. 41-51.

Kealey, P. & Hook, S. 1993. Separating temperature and emissivity in thermal infrared multispectral scanner data: Implications for recovering land surface temperatures, *Geosci. Remote Sens.* **31**, 1155-1164.

Keirein-Young, K.S. 1997. The integration of optical and radar data to characterize mineralogy and morphology of surfaces in Death Valley, California, U.S.A. *Int. J. Remote Sensing,* **18**, 1517-1541.

Keller, J. & Lamprecht, R. 1995. Road dust as an indicator for air pollution transport and deposition: An application of SPOT Imagery. *Rem. Sens. Environ.,* **54,** 1-12.

Kenneth, J., Zhang, H. & Chapin, A. 2000. Bi-directional reflectance distribution functions (BRDF) of benthic surfaces in the littoral zone. *Proceedings Ocean Optics XV.* SPIE, Monaco, France.

Kerekes, J.P. & Landgrebe, D.A. 1991. Parameter Trade-offs for Imaging Spectroscopy Systems. *IEEE Trans. Geosci. Rem. Sens.* **29**, 57-65.

Kim, J.O. & Mueller, C.W. 1978. *Introduction to factor analysis*, Beverly Hills, Saga Press.

Kimes, D. S. & Kirchner, J.A. 1982. Irradiance measurement errors due to the assumption of a lambertian reference panel *Rem. Sens. Environ.*,**12**, 141-149.

King, T.V.V. & Clark, R.N. 1989. Reflectance Spectroscopy (0.2 to 20$\mu$m) As An Analytical Method For The Detection of Organics *In:* Soils: *Proceedings First International Symposium: Field Screening Methods for Hazardous Waste Site Investigations*, EPA, pp. 485-488.

Kirk, J.T.O. 1983. *Light and photosynthesis in aquatic ecosystems.* Cambridge University Press.

Kirk, J.T.O. 1984. Dependance of relationship between inherent and apparent optical properties of water on solar altitude. *Limnol.Oceanogr.* **29,** 350-356.

Kirk, J.T.O. 1989. The upwelling light stream in natural waters. *Limnol.Oceanogr.* **34,** 1410-1425.

Kirk, J.T.O. 1991. Volume scattering function, average cosines, and the underwater light field. *Limnol.Oceanogr.* **36,** 455-467.

Kirk, J.T.O. 1994a. Characteristics of the light field in highly turbid waters: a Monte Carlo study. *Limnol.Oceanogr.* **39,** 702-706.

Kirk, J.T.O.1994b. *Light & photosynthesis in aquatic ecosystems.* University Press, Cambridge, UK, 1-509 pp.

Klepper, O. 1989. *A model of carbon flows in relation to macrobenthic food supply in the Oosterschelde estuary (S.W. Netherlands)*, PhD Thesis Wageningen Agricultural University, Wageningen, The Netherlands.

Klusman, R.W., Saeed, M.A. & Abu-Ali, M. 1992. The potential use of biogeochemistry in the detection of petroleum microseepage. *AAPG Bulletin,* **76**, 851-863.

Knipling E.B. 1970. Physical and physiological basis for the reflectance of visible and near-infrared radiation from vegetation. *Remote sens. environ,* **1,** 155-159.

Koch, B., Ammer, U., Schneider, T. & Wittweier, H. 1990. Spectroradiometer measurements in the laboratory and in the field to analyse the influence of different damage symptoms on the reflection spectra of forest trees. *Int. J. Remote Sensing,* **11,** 1145-1163.

Kohler, D.D.R. & Philpot, W.D. 2000. Comparing in situ and remotely sensed measurements in optically shallow waters. *Proceedings Ocean Optics XV*, Monaco, pp. 8.

Kondratyev, K.Y., Pozdnyakov, D.V. & Pettersson, L.H. 1998. Water quality remote sensing in the visible spectrum. *Int. J. Remote Sensing* **19**, 957-979.

Konovalova, T.I. 1999. Remote sensing analysis of environmental conditions in Siberian cities. *Mapping Sciences and Remote Sensing,* **36,** 92-104.

Kraft, M., Weigel, H., Mejer, G. & Brandes, F. 1996. Reflectance measurements of leaves for detecting visible and non-visible ozone damage to crops. *Journal of Plant Physiology*, **148**, 148-154.

Kratzer, S., Bowers, D. & Tett, P.B. 2000. Seasonal changes in colour ratios and optically active constituents in the optical Case-2 waters of the Menai Strait, North Wales. *Int. J. Remote Sensing* **21**, 2225-2246.

Kruse, F.A. 1996. Mineral mapping for environmental hazards assessment using AVIRIS data, Leadville, Colorado, USA. *Proceedings of the Eleventh Thematic Conference and Workshop on Applied Geologic Remote Sensing*, Las Vegas, Nevada, 27-29 February, Vol. II, pp. 526-533.

Kruse, F.A. 1997. Regional geological mapping along the Colorado Front Range from Ft Collins to Denver using the Airborne Visible/Infrared Imaging Spectrometer (AVIRIS), *Proceedings of the 12th International Conference on Applied Geologic Remote Sensing*, Denver, U.S.A., 17-19 November 1997, pp. II:91-98

Kruse, F.A. 1988. Use of Airborne Imaging Spectrometer data to map minerals associated with hydrothermally altered rocks in the northern Grapevine Mountains, Nevada and California. *Rem. Sens. Environ.*, **24**, 31-51.

Kruse, F.A., Keirein-Young, K.S. & Boardman, J.W. 1990. Mineral Mapping at Cuprite, Nevada with a 63-channel Imaging Spectrometer. *Photogramm. Eng. and Remote Sensing*, **56**, 83-92.

Kruse, F.A., Lefkoff, A.B., Boardman, J.W., Heidebrecht, K.B., Shapiro, A.T., Barloon, P.J. & Goetz, A.F.H. 1993. The Spectral Image Processing System (SIPS) - Interactive Visualization and Analysis of Imaging Spectrometer Data. *Remote Sens. Environ.*, **44**, 145-163.

Kruse, F.A., Lefkoff, A.B. & Dietz J.B. 1993. Expert system-based mineral mapping in Northern Death Valley, California/Nevada, using the Airborne Visible/Infrared Imaging Spectrometer (AVIRIS). *Remote Sens. Environ.*, **44**, 309-336.

Krijgsman, J. 1994. *Optical remote sensing of water quality parameters; interpretation of reflectance spectra*. PhD Thesis, Delft University of Technology, Delft, The Netherlands.

Kumar, R. & Silva, L. 1973. Light ray tracing through a leaf cross section. *Applied Optics*, **12**, 2950-2954.

Kuo, J. & Mc Comb, A.J. 1989. A treatise on the biology of seagrass with special reference to the Australian region. *In:* A.W.D. Larkum, McComb, A.J. & Shepherd, S.A. (Ed.), *Biology of Seagrass.*, Elsevier, Amsterdam, pp. 6-73.

Kutser, T., Herlevi, A., Kallio, K. & Arst, H. 2001. A hyperspectral model for interpretation of passive optical remote sensing data from turbid lakes. *The Science of the Total Environment* **268**, 47-58.

Kutser, T., Parslow, J., Clementson, L., Skirving, W.J. & Done, T. 2000a. Hyperspectral detection of coral reef bottom types. *Proceedings Ocean Optics XV*, Monaco, pp. 12.

Kutser, T. et al. 2000b. Hyperspectral detection of coral reef health. *Proceedings 10th Australasian Remote Sensing Conference*, Adelaide, Australia.

Kupiec, J., Smith, G.M. & Curran, P.J. 1993. AVIRIS spectra correlated with the chlorophyll concentration of a forest canopy. *In:* Green, R.O. (Ed.), *Summaries of the Fourth Annual JPL Airborne Geoscience Workshop*. October 25-29, California. JPL Publication 93-26, pp. 105-108.


L

LaBaw, C. 1987. Airborne Imaging Spectrometer 2: The optical design. *In:* G. Vane, A.F.H. Goetz & D.D. Norris (Eds.), *Proceedings of the Society of Photo-optical Instrumentation Engineers*, SPIE, Vol. 834.

Laflen, J.M. & Colvin, T.S. 1981. Effect of Crop residue on Soil Loss from Continuous Row Cropping. *Transactions of the American Society of Agricultural Engineers*, **24**, 605-609.

Lahet, F., Ouillon, S. & Forget, P. 2000. A three-component model of ocean color and its application in the Ebro River mouth area. *Rem. Sens. Environ.* **72**, 181-190.

Landsberg, H. E. 1981. *The urban climate. In:* J.V. Mieghem, A.L. Hales & W.L. Donn (Eds.) *International Geophysics Series.* Vol. 28. Academic Press, New York.

Lang, H.R., Aldeman, W.H. & Sabins, F.F. Jr. 1985a. Patrick Draw, Wyoming - petroleum test case report. *In:* M.J. Abrams, J.E., Conel, H.R. Lang, and H.N. Paley (Eds.), *The Joint NASA/Geosat Test Case Project: Final Report, AAPG Special Publication,* Lang, H.R. & Cabral-Canoz, E. 1998. Preliminary analysis of AVIRIS data for tectonostratigraphic assessment of Northern Guerrero state, southern Mexico", Presented at the Geoscience workshop, 1996, Pasadena, Ca. (1998).

Lang, H.R., Curtis, J.B. & Kovacs, J.C. 1985b. Lost river, West Virginia - petroleum test site report. . *In:* M.J. Abrams, J.E., Conel, H.R. Lang, and H.N. Paley (Eds.), *The Joint NASA/Geosat Test Case Project: Final Report, AAPG Special Publication,* pt.2, vol.2, pp. 12-1--12-96.

Lauten, G.N. & Rock, B.N. 1992. Physiological and spectral analysis of the effects of sodium chloride on *Syringa vulgaris. Proceedings of the IGARSS'92 Symposium*, May 26-29, Houston, Texas, **1**, pp. 236-238.

Lee, Z., Carder, K.L. & Chen, R.F. 2000. Environmental properties observed from AVIRIS data. *In:* R.O. Green (Ed.), *Proceedings of the 9th JPL Airborne Earth Science Workshop.* NASA/JPL, California, pp. 279-282.

Lee, Z. et al. 1994. Model for the interpretation of hyperspectral remote-sensing reflectance. *Appl. Opt.* **33**, 5721-5732.

Lee, Z., Carder, K.L., Mobley, C.D., Steward, R.G. & Patch, J.F. 1999. Hyperspectral remote sensing for shallow waters: 2. deriving bottom depths and water properties by optimization. *Appl. Opt.* **38**, 3831-3843.

Lee, C. & Landgrebe D.A. 1993. Analyzing High-Dimensional Multispectral Data. *IEEE Trans. Geosci. Rem. Sens.,* **31**, 792-800.

Lee, J.B., Woodhyatt, S. & Berman, M. 1990. Enhancement of high spectral resolution remote-sensing data by a noise-adjusted principal components transform. . *IEEE Trans. Geosci. Rem. Sens.,* **28**, 295-304.

Lehmann, F., Buchert, T., Hese, S., Hofmann, A., Mayer, A., Oschuts, F. & Zhang, Y. 1998. Data fusion of HyMAP heperspectral with HRSC-A multispectral stereo and DTM data: Remote sensing data validation and application in different disciplines. *First EARSeL Workshop on Imaging Spectroscopy,* Zurich, Switzerland, pp. 105-117.

Lehmann, F., Rothfuß, H. & Richter, R. 1990. Evaluation of imaging spectrometer data (GER) for the spectral analysis of an old vegetation covered waste deposit. *Proceedings of the 10th International Geoscience & Remote Sensing Symposium,* IGARSS '90, May 20-24, Washington D.C., USA, pp. 1613-1616.

Le Houérou, H.N. 1981. Impact of Man and his Animals on Mediterranean Vegetation. *In:* F. di Castri, D.W. Goodall & R.L. Specht (Eds.), Ecosystems of the World, 11: Mediterranean-type Shrublands, Elsevier, Amsterdam, pp. 479-517.

Le Houérou, H.N. 1992. Vegetation and Land Use in the Mediterranean Basin by the Year 2050: a Prospective Study. *In:* L. Jeftic, J.D. Milliman & G. Sestini (Eds.), Climatic Change and the Mediterranean, E. Arnold, Londen, pp.175-232.

Lelong, C.C.D., Pinet, P.C. & Poilve, H. 1998. Hyperspectral imaging and stress mapping in agriculture: a case study on wheat in Beauce (France). *Remote Sens. Environ.,* **66**, 179-191.

Lennon, P. & Luck, P. 1990. Seagrass mapping using Landsat TM data: a case study in southern Queensland. *Asian-Pacific Remote Sensing Journal* **2**, 6-9.

Lichtenthaler, H.K., Gitelson, A. & Lang, M. 1996. Non-destructive determination of chlorophyll content of leaves of a green and an aurea mutant of tobacco by reflectance measurements. *Journal of Plant Physiology,* **148**, 483-493.

Lillesand, T.M. & Kiefer, R.W. 1994. *Remote Sensing and Image Interpretation,* New York: John Wiley & Sons., 750 pp.

Lindell, L.T., Pierson, D., Premazzi, G. & Zillioli, E. 1999. *Manual for monitoring European lakes using remote sensing techniques.* European Communities, Luxembourg, 164 pp.

Loercher, G., Endres, S. & Sommer, S. 1994. Mapping hydrothermal alterations in the landmannalaugar area (Iceland) using MAC-Europe 1991 AVIRIS data. *Proceedings Tenth Thematic Conference and Workshop on Applied Geologic Remote Sensing,* San Antonio, Texas, 9-12 May, Vol. I, pp. 333-340.

Looyen, W.J., Verhoef, W. Clevers, J.G.P.W. Lamers, J.T. & Boerma, J. 1991. *CAESAR: evaluation of the dual-look concept,* BCRS, Delft, report 91-10, 144 pp.

Lorenzen, B. & Jensen, A. 1989. Changes in leaf spectral properties induced in Barley by cereal powdery Mildew. *Remote sens. environ,* **27**, 201-209.

Louchard, E.M., Reid, R.P. & Stephens, C.F. 2000. Classification of sediment types and estimation of water depth using spectral libraries. *Proceedings Ocean Optics XV,* Monaco, pp. 13.

Low, P.F. 1973. Fundamental mechanisms involved in expansion of clays as particularly related to clay mineralogy, *Proceedings of Workshop on Expansive Clays and Shales in Highway Design and Construction,* Vol. 1.

Lyon, R. 1965. Analysis of rocks and minerals by reflected infrared radiation. *Econ. Geol.* **60,** 715-736.

Lyzenga, D.R. 1981. Remote sensing of bottom reflectance and water attenuation parameters in shallow water using aircraft and Landsat data. *Int. J. Remote Sensing* **2,** 71-82.


**M**

Maas, S.J. 1988. Use of remotely sensed information in agricultural crop growth models. *Ecological modelling* **41**, 247-268.

Maas, S.J. & Dunlap, J.R. 1989. Reflectance, transmittance, and absorptance of light by normal, etiolated, and albino corn leaves. *Agronomy Journal,* **81**, 105-110.

Maldegem, D.C. 1992. *De slibbalans van het Schelde-estuarium.* NOTA GWAO-91.081, SAWES - NOTA 91.08, Rijkswaterstaat, DGW, Middelburg, The Netherlands.

Malthus, T.J. et al. 1997. Can biophysical properties of submersed macrophytes be determined by remote sensing?, *Proceedings Fourth International Conference on Remote Sensing for Marine and Coastal Environments: Technology and Applications.* ERIM-Veridian, Michigan, Orlando, Florida., pp. 562-571.

Malthus, T.J.M., Best, E.P.H. & Dekker, A.G. 1987. An assessment of the importance of emergent and floating-leaved macrophytes to trophic status in the Loosdrecht lakes (The Netherlands). *Hydrobiologia* **191,** 257-263.

Malthus, T.J. & George, D.G. 1997. Airborne remote sensing of macrophytes in Cefni Reservoir, Anglesey, UK. *Aquatic Botany* **58,** 317-332.

Malthus, T.J. & Madeira, A.C. 1993. High resolution spectroradiometry: spectral reflectance of field bean leaves infected by Botrytis fabae. *Remote sens. environ,* **45,** 107-116.

Marino, C.M., Panigada, C., Galli, A., Boschetti, L. & Buseto, L. 2000. Environmental applications of airborne hyperspectral remote sensing: asbestos concrete sheeting identification and mapping. *Proceedings of the 14$^{th}$ International Conference on Applied Geologic Remote Sensing,* Las Vegas, Nevada, pp. 607-610.

Maritorena, S. 1996. Remote sensing of water attenuation in coral reefs: case study in French Polynesia. *Int. J. Remote Sensing* **17,** 155-166.

Maritorena, S. & Guillocheau, N. 1996. Optical properties of water and spectral light absorption by living and non-living particles and by yellow substances in coral reef waters of French Polynesia. *Marine Ecol.Progr.Ser.* **131,** 245-255.

Maritorena, S., Morel, A. & Gentili, B. 1994. Diffuse reflectance of oceanic shallow waters: influence of water depth and bottom albedo. *Limnol.Oceanogr.* **39,** 1689-1703.

Marshall, T.J. & Holmes, J.W. 1981. Soil Physics. Cambridge University Press, Cambridge.

Marshall, T.R. & Lee, P.F. 1994. Mapping aquatic macrophytes through digital image-analysis of aerial photographs - an assessment. *Journal of Aquatic Plant Management* **32,** 61-66.

Martin, M. & Aber, J.D. 1990. The effect of water content and leaf structure on the estimation of nitrogen, lignin and cellulose content of forest foliage samples. *Proceedings of the International Geoscience and Remote Sensing Symposium (IGARSS 90),* May 20-24, College Park, MD, vol. 1:563.

Mather, P.M. 1976. *Computational methods of multivariate analysis in physical geography,* Wiley, Chichester.

Matson, P., Johnson, L., Billow, C., Miller, J. & Pu, R. 1994. Seasonal patterns and remote spectral estimation of canopy chemistry across the Oregon transect. *Ecological Applications,* **4,** 280-298.

McKeen, R.G. 1992. A model for predicting expansive soil behavior. *Proceedings of the 7$^{th}$ International Conference on Expansive Soils,* Dallas, U.S.A., 3-5 August 1992, pp. 1-6.

McMurtrey III, J.E., Chappelle, E.W., Kim, M.S., Corp, L.A. & Daughtry, C.S.T. 1996. Blue-green fluorescence and visible-infrared reflectance of corn (*Zea mays L.*) grain *in situ* field detection of nitrogen supply. *Journal of Plant Physiology,* **148,** 509-514.

Meinesz, A. 1999. *Killer algae: the true tale of a biological invasion.* University of Chicago Press, Chicago, 360 pp.

Melville, M.D. & Atkinson, G. 1985. Soil Color: its Measurement and its Destination in Models of Uniform Color Space. *Soil Science,* **36,** 495-512.

Metternicht, G.I. & Fermont, A. 1998. Estimating Erosion Surface Features by Linear Mixture Modeling. *Rem. Sens. Environ.,* **64,** 254-265.

Meulstee, C., Nienhuis, P.H. & Van Stokkom, H.T.C. 1986. Biomass assessment of estuarine macrophytobenthos using aerial photography. *Marine Biology* **91,** 331-5.

Miller, J.R., Hare, E.W. & Wu, J. 1990. Quantitative characterisation of the vegetation red edge reflectance 1. An inverted-Gaussian reflectance model. *Int. J. Remote Sensing,* **11,** 1755-1773.

Miller, J.R., Wu, J., Boyer, M.G., Belanger, M. & Hare, E.W. 1991. Seasonal patterns in leaf reflectance red-edge characteristics. *Int. J. Remote Sensing,* **12,** 1509-1523.

Mitchell, J.K. 1993. *Fundamentals of soil behavior,* John Wiley and Sons, New York, 437 pp.

Mobley, C.D. 1994. *Light and water; Radiative transfer in natural waters.* Academic Press, London, 592 pp.

Mobley, C.D. 1995. The optical properties of water. *In:* I. Bass (Ed.), *Handbook of Optics.* McGraw-Hill and Optical Society of America, New York.

Mobley, C.D. & Sundman, L.K. 2000. *Hydrolight 4.1 technical documentation.* Sequoia Scientific, Inc, WA.

Moore, D.M. & Reynolds, R.C.Jr. 1997. *X-ray diffraction and the identification and analysis of clay minerals, 2d edition,* Oxford University Press, New York, 378 pp.

Moore, G.F., Aiken, J. & Lavender, S.J. 1999. The atmospheric correction of water colour and the quantitative retrieval of suspended particulate matter in Case II waters: application to MERIS. *Int. J. Remote Sensing* **20,** 1713-1733.

Morel, A. & Gentili, B. 1993. Diffuse reflectance of oceanic waters. II. Bidirectional aspects. *Appl.Opt.* **32,** 6864-6879.

Morel, A. & Prieur, L. 1977. Analysis of variations in ocean colour. *Limnol.Oceanogr.* **22,** 709-722.

Morgan, R.P.C. 1995. Soil Erosion.& Conservation. Addison Wesley Longman Ltd., Essex, 198 pp.

Morris, H. 1964. Geology of the Eureka quadrangle, Utah and Juab Counties, Utah. *US Geol. Surv. Bull.* **1142-K**, 29 pp.

Moss, D.M. & Rock, B.N. 1991. Analysis of red edge spectral characteristics and total chlorophyll values for red spruce (Picea rubens) branch segments from Mt. Moosilauke, NH, USA, *Proceedings of the 11th International Geoscience and Remote Sensing Symposium (IGARSS'91),* Helsinki (Finland), 3-6 June 1991, pp. 1529-1532.

Moussa, H.B., Viollier, M. & Belsher, T. 1989. Remote-sensing of macrophytic algae in the Molene Archipelago: ground-radiometry and application to SPOT satellite-data. *Int. J. Remote Sensing* **10,** 53-69.

Mueller, A., Lehmann, F. & Rothfuß, H. 1996. The potential of imaging spectrometry (DAIS 7915) for the monitoring of recultivation activities in mining areas. *Proceedings of the Eleventh Thematic Conference and Workshop on Applied Geologic Remote Sensing,* Las Vegas, Nevada, 27-29 February, Vol. II, pp. 350-357.

Mulders, M.A. 1987. Remote Sensing in Soil Science. Developments in Soil Science 15. Elsevier, Amsterdam, 379 pp.

Muller, E., Decamps, H. & Dobson, M.K. 1993. Contribution of Space Remote Sensing to River Studies. *Freshwater Biology,* **29,** 301-312.

Mumby, P.J., Clark, C.D., Green, E.P. & Edwards, A.J., 1998a. Benefits of water column correction and contextual editing for mapping coral reefs. *Int. J. Remote Sensing* **19,** 203-210.

Mumby, P.J., Green, E.P., Clark, C.D. & Edwards, A.J. 1998b. Digital analysis of multispectral airborne imagery of coral reefs. *Coral Reefs* **17,** 59-69.

Mumby, P.J., Green, E.P., Edwards, A.J. & Clark, C.D. 1997. Measurement of seagrass standing crop using satellite and digital airborne remote sensing. *Marine Ecology-Progress Series* **159,** 51-60.

Mumby, P.J. & Harborne, A.R.1999. Development of a systematic classification scheme of marine habitats to facilitate regional management and mapping of Caribbean coral reefs. *Biological Conservation* **88,** 155-163.

Munsell Color, 1975. Standard Soil Colour Charts. Ministry of Agriculture and Forestry, Japan, 35 pp.

Murphy, R.J. 1995. Mapping of jasperoid in the Cedar Mountains, Utah, U.S.A., using imaging spectrometer data. *Int. J. Remote Sensing,* **16,** 1021-1042.

Mustard, J.F. 1993. Relationship of soil, grass, and bedrock over the Kaweah serpentinite melange through spectral mixture analysis of AVIRIS data. *Remote Sens. Environ.,* **44,** 293-308.

Mustard, J.F. & Pieters, C.M. 1986. Quantitative abundance estimates from bidirectional reflectance measurements. *J. Geophys. Res.,* **92,** 617-626.

Mustard, J.F. & Pieters, C.M. 1987. Abundance and distribution of ultramafic microbreccia in Moses Rock dike: Quantitative application of mapping spectroscopy. *J. Geophys. Res.,* **92,** E617-E626.

Myers, M.R., Hardy, J.T., Mazel, C.H. & Dustan, P. 1999. Optical spectra and pigmentation of Caribbean reef corals and macroalgae. *Coral Reefs* **18,** 179-186.

Myers, V.I. 1970. Soil, water and plant relations. *In:* Anonymous, A.A. (Ed.), *Remote sensing with special reference to agriculture and forestry,* National Academy of Sciences, Washington D.C., pp. 253-297.

Myers, V.I. 1983. Remote Sensing Applications in Agriculture. *In:* R.N. Colwell (Ed.), *Manual of Remote Sensing.* American Society of Photogrammetry, Falls Church, Virginia, pp. 2111-2228.


N

Nedeljkovic, V. & Pendock, N. 1996. Finding spectral anomalies in hyperspectral imagery using nonparametric probability density estimation techniques. *Proceedings of the Eleventh Thematic Conference and Workshop on Applied Geologic Remote Sensing,* Las Vegas, Nevada, 27-29 February, Vol. I, 197-204.

Neeteson, J.J. 1989. *Assessment of fertilizer nitrogen requirements of potatoes and sugar beet,* PhD Thesis Wageningen Agricultural University, The Netherlands, 141 pp.

Nichol, J.E. 1998. Visualisation of urban surface temperatures derived from satellite images. *Int. J. Remote Sensing,* **9,** 1639-1649.

Nielsen, A.A. & Larsen, R. 1994. Restoration of GERIS data using the Maximum Noise Fractions Transform. *Proceedings of the First International Airborne Remote Sensing Conference and Exhibition (ERIM),* Strassbourg, France (Ann Arbor: ERIM International), pp. 557-568.

Ninomiya, Y., Matsunaga, T., Yamaguchi, Y., Ogawa, K., Rokugawa, S., Uchida, K., Muraoka, H. & Kaku, M. 1997. A comparison of thermal infrared emissivity spectra measured *in situ,* in the laboratory and similar spectral derived from thermal infrared multispectral scanner (TIMS) data in Cuprite, Nevada, U.S.A. *Int. J. Remote Sensing,* **18,** 1571-1581.

Noe, D.C. 1997. Heaving-bedrock hazards, mitigation, and land-use policy: Front Range Piedmont, Colorado. *Environmental Geosciences,* **4** (2), 48-57.

Noe, D.C. & Dodson, M.D. 1995. The Dipping Bedrock Overlay District (DBOD): An area of potential heaving bedrock hazards associated with expansive, steeply dipping bedrock in Douglas County, Colorado, *Colorado Geological Survey Open-File Report 95-5*, Colorado Geological Survey, Denver, CO, 1 plate, scale 1:500000, 32 pp.

Noe, D.C. & Dodson, M.D. 1999. Heaving bedrock hazards associated with expansive, steeply dipping bedrock in Douglas County, Colorado. *Colorado Geological Survey Special Publication 42*, Colorado Geological Survey, Denver, CO, 2 plates, scale 1:24000, 80 pp.

Novo, E. Gastil, M. & Melack, J. 1995. An algorithm for chlorophyll using first difference transformations of AVIRIS reflectance spectra. In: Green, R.O. (Ed.) *Summaries of the Fifth Annual JPL Airborne Earth Science Workshop*. January 23-26, California. JPL Publication 95-1, pp. 121-124.


O

Oehler, D.Z. & Sternberg, B.K. 1984. Seepage-induced anomalies, false anomalies and implications for electrical prospecting. *AAPG Bulletin*, **68,** 1121-1145.

Okada, K. & Iwashita, A. 1992. Hyper-multispectral image analysis based on waveform characteristics of spectral curve. *Adv. Space Res.*, **12,** 433-442.

Oke, T. R. 1987. *Boundary layer climates*. Methuen Press, London, pp. 252–302.

Oke, T. R. 1995. The heat island of the urban boundary layer: characteristics, causes and effects. *In:* J.E. Cermak, A.G. Davenport, E.J. Plate and D.X. Viegas (Eds.), *Wind climate in cities*, Kluwer Academic Publishers, Dordrecht, pp. 81–107.

Olbert, C. 1998. Atmospheric correction for CASI data using an atmospheric radiative transfer model. *Can.J.Remote Sensing* **24,** 114-127.

Olbert, C. 2000. *Bestimmung raumlicher Verteilungsmuster von Wasserinhaltstoffen in ausgewahlten Berliner und Brandenburger Gewassern mit Methoden der Fernerkundung*. PhD Thesis, Freie Universitat Berlin, Berlin, 172 pp.

Olson, C.E. 1967. Optical remote sensing of the moisture content in fine forest fuels, 1st Report N 8036-1-F, University of Michigan, Ann Arbor (MI), 21 pp.

Olson Jr, C.E. & Zhu, Z. 1985. Forest species identification with high spectral resolution data. *In:* Vane, G. & Goetz, A.F.H. (Eds.), *Proceedings of the Airborne Imaging Spectrometer Data Analysis Workshop*. April 8-10, California. JPL Publication 85-41, pp. 152-157.

Olsen, H.W., Krosley, L., Nelson, K., Chabrillat, S., Goetz, A.F.H. & Noe, D.C. 2000. Mineralogy-swelling potential relationships for expansive shales, *Geotechnical Special Publication No. 99, Advances in Unsaturated Geotechnics. In:* C.D. Shackelford, S.L. Houston, and N-Y Chang (Eds), Sponsored by the GeoInstitute of ASCE, Denver, CO, 3-8 August 2000, pp. 361-378.


P

Palluconi, F. & Meeks, J. 1985. Thermal infrared multispectral scanner (TIMS): an investigator's guide to TIMS data. *JPL Publ. 85-32*, Pasadena CA.

Pasqualini, V., Clabaut, P., Pergent, G., Benyoussef, L. & Pergent-Martini, C. 2000. Contribution of side scan sonar to the management of Mediterranean littoral ecosystems. *Int. J. Remote Sensing* **21,** 367-378.

Pasqualini, V., Pergent-Martini, C. & Fernandez, C. 1997. The use of airborne remote sensing for benthic cartography: advantages and reliability. *Int. J. Remote Sensing* **18,** 1167-1177.

Pasterkamp, R., Peters, S.W.M. & Dekker, A.G. 2000. Environmental baseline mapping of total suspended matter concentrations in the Western Scheldt Estuary supported by the use of SPOT images. *Proceedings Sixth International Conference on Remote Sensing for Marine and Coastal Environments*. Veridian ERIM International, Charleston, South Carolina, US, pp. I-260 to I-267.

Pate, J.S. 1983. Patterns of nitrogen metabolism in higher plants. *In:* J.A. Lee, S. McNeill & I.H. Rorison (Eds.), *Nitrogen as an ecological factor*. Blackwell, Oxford, England.

Penning de Vries, F.W.T. & Van Laar, H.H. 1982. *Simulation of plant growth and crop production*, Simulation Monographs, Pudoc, Wageningen, 308 pp.

Peñuelas, J., Filella, I., Biel, C., Serrano, L. & Save, R. 1993. The reflectance at the 950-970nm region as an indicator of plant water status. *Int. J. Remote Sensing*, **14,** 1887-1905.

Penuelas, J., Gamon, J.A., Griffin, K.L. && Field, C.B. 1993. Assessing Community Type, Plant Biomass, Pigment Composition, and Photosynthetic Efficiency of Aquatic Vegetation From Spectral Reflectance. *Rem. Sens. Environ.* **46,**110-118.

Pesaresi, M. 2000. Texture analysis for urban pattern recognition using fine-resolution panchromatic satellite imagery. *Geographical and Environmental Modeling*, **4,** 43-63.

Peterson, D.L. & Running, S.W. 1989. Applications in Forest Science and Management, *In:* Asrar, G. (Ed.), *Theory and Applications of Optical Remote Sensing*. Wiley Publishers, New York, pp. 429 - 473.

Peterson, D.L., Aber, J.D., Matson, P.A., Card, D.H., Swanberg, N., Wessman, C. & Spanner, M. 1988. Remote sensing of forest canopy and leaf biochemical contents. *Remote sens. environ*, **24**, 85-108.

Peterson, D.L. & Waring, R.H. 1993. Overview of the Oregon transect ecosystem research project. *Ecological Applications*, **4**, 211-225.

Phillips, R.C. & Menez, A. 1988. *Seagrasses*. Smiths. Contr. Mar. Sci., Washington, D.C.

Philpot, W.D. & Vodacek, A. 1989. Laser-induced fluorescence: limits to the remote detection of hydrogen ion, aluminium, and dissolved organic matter. *Remote Sens.Environ.* **29,** 51-65.

Pieters, C.M. & Englert, P.A.J. 1993. Remote Geochemical Analysis: Elemental and Mineralogical Composition, Cambridge University Press, New York.

Pinel, V., Zagolski, F., Gastellu-Etchegorry, J.P., Giordano, G., Romier, J., Marty, G., Mougin, E. & Joffre, R. 1994. An assessment of forest chemistry with ISM. *In:* Chavez, P.S., Marino, C.M. & Schowengerdt, R.A. (Eds.), *Recent Advances in Remote Sensing and Hyperspectral Remote Sensing*. SPIE Vol. 2318, pp. 40-51.

Pinzon, J.E., Ustin, S.L., Hart, Q.L., Jacquemoud, S. & Smith, M.O. 1994. Comparison of multivariate statistical techniques for estimating vegetation parameter. *In: Spectral Analysis Workshop: The Use of Vegetation as an Indicator of Environmental Contamination*, Reno, Nevada, Nov 9-10, 1994.

Plass, G.N. & Kattawar, G.W. 1972. Monte Carlo calculations of radiative transfer in the Earth's atmosphere-ocean system. I. Flux in the atmosphere and ocean. *J. Phys. Oceanogr* **2,** 139-145.

Posselt, W., Kunkel, P., Schmidt, E., Del Bello, U. & Meynart, R. 1996. Process Research by an Imaging Space Mission. *In:* Fuijisada, H. , Calamai, G. & Sweeting, M. (Eds.), *Proceedings of the Advanced and Next-Generation Satellites II*, SPIE, Toarmina, September 1996, Vol. 2957.

Post, J.L. & Noble, P.N. 1993. The near-infrared combination band frequencies of dioctahedral smectites, micas, and illites. *Clays and Clay minerals*, **41**, 639-644.

Pozdnyakov, D.V. & Kondratev, K.Y. 1999. Remote sensing of natural waters in the visible band of the spectrum. 3. Some special questions. *Earth Observation & Remote Sensing* **15**, 365-385.

Pratt, P., Carder, K.L. & Costello, D.K. 1997. Remote sensing reflectance algorithms developed to correct underwater coral imagery for the effect of optical thickness to assist in benthic classification. *Proceedings 4th Remote Sensing Conference for Marine and Coastal Environments*. ERIM, Orlando, Florida.

Press, N.P. 1974. Remote sensing to detect the toxic effects of metals on vegetation for mineral exploration. *Proceedings of the 9th International Symposium on Remote sens. environ.* 15-19 April, Ann Arbor, Michigan. Vol. 3, pp. 2027-2038.

Press, W.H., Flannery, P., Teukolsky, S.A. & Vetterking, A. 1988. *Numerical Recipes in C-The Art of Scientific Computing*. New York, Melbourne, Sidney: Cambridge University Press, 735pp.

Price, J.C. 1995. Examples of High Resolution Visible to Near-Infrared Reflectance Sand a Standardized Collection for Remote Sensing Studies. *Int. J. Remote Sensing,* **16,** 993-1000.

Price, J.C., Steven, M., Andrieu, B. & Jaggard, K. 1996. Visible near-infrared radiation parameters for sugar-beets. *Int. J. Remote Sensing*, **17**, 3411-3418.

Price, W.L. 1979. A controlled random search procedure for global optimisation. *The Computer Journal* **20**, 367-370.


**R**

Ramsey, M.S., Stefanov, W.L. & Christensen, P.R. 1999. Monitoring world-wide land cover changes with ASTER: Preliminary results from the Phoenix AZ Lter site. *Proceedings of the International Conference on Applied Geologic Remote Sensing*, Vancouver BC., pp. 237-244.

Ranchim, T. & Wald, L. 2000. Fusion of  high spatial and spectral resolution images: the ARSIS concept and its implementation. *Photogrammetric Engineering and Remote Sensing,* **66,** 49-61.

Rast, M. 1991. Imaging Spectroscopy and Its Applications in Spaceborne Systems. *European Space Agency Publications* ESA SP-1144, Noordwijk, The Netherlands, 144p.

Rast, M. 1992. ESA's Activities in the field of imaging spectroscopy. *In:* F. Toselli & J. Bodechtel (Eds.), *Imaging Spectroscopy: Fundamentals and Prospective Applications*. Dordrecht, Kluwer, pp. 167-191.

Rast, M. & Bézy, J.L. 1990. ESA's Medium Resolution Imaging Spectrometer (MERIS): mission, system and applications, *SPIE* **1298**, 114-126.

Rast, M. & Bézy, J.L. 1995. The ESA Medium Resolution Imaging Spectrometer (MERIS): requirements to its mission and performance of its system. *In:* Curran, P.J. & Robertson, C. (Eds.), *Proceedings of the RSS'95 Remote Sensing in Action*, 11-14 September 1995, Southampton, U.K. pp. 125-132.

Rast, M., Bézy, J.L. & Bruzzi, S. 1999. The ESA Medium Resolution Imaging Spectrometer MERIS – a review of the instrument and its mission. *Int. J. Remote Sensing* **20**, 1681-1702.

Rast, M., Hook, S.J., Elvidge, C.D. & Alley R.E. 1991. An evaluation of techniques for the extraction of mineral absorption features from high spectral resolution remote sensing data. *Photogramm. Eng. and Remote Sensing*, **57**, 1303-1309.

Ray, T.W., Murray, B.C., Chehbouni, A. & Njoku, E. 1993. The red edge in arid region vegetation: 340-1060nm spectra. *In:* Green, R.O. (Ed.), *Summaries of the Fourth Annual JPL Airborne Geoscience Workshop*. October 25-29, California. JPL Publication 93-26, pp. 149-152.

Rees, W.G. 1996. *Physical principles of remote sensing*. Cambridge University Press, Cambridge, 247 pp.

Resmini, R.G., Kappus, M.E., Aldrich,W.S., Harsanyi, J.C. & Anderson, M. 1997. Mineral mapping with Hyperspectral Digital Imagery collection Experiment (HYDICE) sensor data at Cuprite, Nevada, U.S.A. *Int. J. Remote Sensing*, **18**, 1553-1570.

Richards, J.A. 1986. *Remote Sensing Digital Image Analysis: An introduction*, Springer-Verlag, Berlin.

Richardson, A.J. & Wiegand, C.L. 1977. Distinguishing vegetation from soil background information. *Photogram. Eng. Rem. Sensing* **43**, 1541-1552.

Richers, D.M., Jones, V.T., Matthews, M.D., Maciolek, J., Pirkle, R.J., and Sides, W.C. 1986. The 1983 Landsat soil-gas geochemical survey of Patrick Draw area, Sweetwater county, Wyoming. *AAPG Bulletin*, **70**, 869-887.

Richers, D.M., Reed, R.J., Horstman, K.C., Micheals, G.D., Baker, R.N., Lundell, L. & Marrs, R.W. 1982. Landsat and soil-gas geochemical study of Patric Draw oil field, Sweetwater County, Wyoming. *AAPG Bulletin*, **66**, 903-922.

Richter, R. 1996, Atmopsheric correction of DAIS hyperspectral image data. *Computers and Geosciences*, **22**, 785-793.

Richter, R. 1998. Correction of satellite imagery over mountainous terrain. *Applied Optics*, **37**, 4004-4015.

Ridd, M.K. 1995. Exploring a V-I-S (Vegetation-Impervious Surface-Soil) Model for Urban Ecosystem Analysis through Remote Sensing: Comparative Anatomy for Cities. *Int. J. Remote Sensing*, **16**, 2165-2185.

Ridd, M.K., Ritter, N.D. & Green, R.O. 1997. Neural Network Analysis of Urban Environments with Airborne AVIRIS Data. *Proceedings of the Third International Airborne Remote Sensing Conference and Exhibition*, Copenhagen, Denmark, pp. 197-203

Ripple, W.J. 1986. Spectral reflectance relationships to leaf water stress. *Photogramm. Eng. and Remote Sensing*, **52**, 1669-1675.

Roberts, D.A., Smith, M.O. & Adams, J.B. 1993. Green vegetation, non-photosynthetic vegetation and soils in AVIRIS data. *Remote sens. environ*, **44**, 255-269.

Roberts, D.A., Smith, M.O., Adams, J.B., Sabol, D.E., Gillespie, A.R. & Willis, S.C. 1990. Isolating woody plant material and senescent vegetation from green vegetation in AVIRIS data. *Proceedings of the Second Airborne Visible/Infrared Imaging Spectrometer (AVIRIS) Workshop*, June 4-5. NASA-JPL Publication 90-54, pp. 42-57.

Roberts, D.A., Yamaguchi, Y. & Lyon, R.J.P. 1985. Calibration of Airborne Imaging Spectrometer Data to percent reflectance using field spectral measurements. *Proceedings of the 19th International Symposium on Remote Sensing of Environment*, ERIM, 21-25 October, Ann Arbor, USA, pp. 679-688.

Rock, B.N., Hoshizaki, J. & Miller, J.R. 1988. Comparison of *In-situ* and airborne spectral measurements of the blue shift associated with forest decline. *Remote sens. environ*, **24**, 109-127.

Rock, B.N., Lauten, G.N. & Moss, D.N. 1993. High-spectral resolution reflectance measurements of red spruce and eastern hemlock foliage over a growing season. *Ground Sensing*, SPIE vol 1941, pp. 1-12.

Rock, B.N., Miller, J.R., Moss, D.M., Freemantle, J.R. & Boyer, M.G. 1990. Spectral characterisation of forest damage occurring on Whiteface Mountain (NY) - Studies with the Fluorescence Line Imager (FLI) and ground-based spectrometers. *In:* Vane, G. (Ed.), *Imaging Spectroscopy of the Terrestrial Environment*. SPIE 1298, Orlando (Florida), 16-20 April 1990, pp. 190-201.

Rockwell, B.W., Clark, R.N., Livo, K.E., McDougal, R.R., Kokaly, R.F. & Vance, J.S. 1999. Preliminary materials mapping in the Park City region for the Utah USGS-EPA imaging spectroscopy project using both high and low altitude AVIRIS data, *Summaries of the Eighth JPL Airborne Earth Science Workshop*, Jet Propulsion Laboratory, Pasadena, U.S.A., 9-11 February 1999, JPL Publication 99-17, pp. 365-376.

Roessner, S., Segl, K., Heiden, U., Munier, K. & Kaufmann, H. 1998. Application of hyperspectral DAIS data for differentiation of urban surfaces in the city of Dresden, Germany. 1st EARSeL Symposium on Imaging Spectroscopy, Zurich, pp. 463- 472.

Rohde, W.G. & Olson, C.E. 1971. Estimating foliar moisture content from infrared reflectance data, *Proceedings 3rd Biennial Workshop: Color Aerial Photography in the Plant Science and Related Fields*, American Society of Photogrammetry, Falls Church (VA), pp. 144-164.

Rosaire, E.E. (1940) Geochemical prospecting for petroleum. *AAPG Bulletin*, **24**, 1400-1433.

Ross, J. 1981. *The Radiation Regime and Architecture of Plant Stands*. Dr W. Junk Publishers, London.

Roth, M., Oke, T. & Emery, W. J. 1989. Satellite-derived urban heat islands from three coastal cities and the utilization of such data in urban climatology. *Int. J. Remote Sensing,* **10,** 1699–1720.

Rouse, J.W. Jr., Haas, R.H., Deering, D.W., Schell J.A. & Harlan, J.C. 1974. *Monitoring the vernal advancement and retrogradation (green wave effect) of natural vegetation,* NASA/GSFC Type III Final Report, Greenbelt, Md., 371 pp.

Rouse, J.W. Jr., Haas, R.H., Schell, J.A. & Deering, D.W. 1973. Monitoring vegetation systems in the Great Plains with ERTS, *Proceedings Earth Res. Techn. Satellite-1 Symposium,* Goddard Space Flight Center, Washington D.C., pp. 309-317.

Rowan, L.C., Anton-Pancheco, C., Brickey, D.W., Kingston, M.J., Payas, A., Vergo, N. & Crowley, J.K. 1987. Digital classification of contact metamorphic rocks in Extremadura, Spain using Landsat Thematic Mapper data. *Geophysics,* **52,** 885-897.

Rubio, J.L. 1995. Desertification: Evolution of a Concept. *In:* R. Fantechi, D. Peter, P. Balabanis & J.L. Rubio (Eds.), Desertification in a European Context: Physical and Socio-economic Aspects. EUR Report 15415, Brussels pp.5-14.

Ruff, S., Christensen, P., Barbera, P. & Anderson, D. 1997. A laboratory technique for measurement and calibration. *J. Geophys. Res.* **102,** 14899-14913.

Rundquist, D.C. 1985. Preliminary evaluation of AIS spectra along a topographic/moisture gradient in the Nebraska sandhills. *In:* Vane, G. & Goetz, A.F.H. (Eds.), *Proceedings of the Airborne Imaging Spectrometer Data Analysis Workshop.* April 8-10, California. JPL Publication 85-41, pp. 74-78.

Rutchey, K. & Vilchek, L. 1999. Air photointerpretation and satellite imagery analysis techniques for mapping cattail coverage in a northern Everglades impoundment. *Photogrammetric Engineering and Remote Sensing* **65,** 185-191.

Ryskin, Y.I 1974. The vibrations of protons in minerals: Hydroxyl, water and ammonium. *In:* V.C. Farmer (Ed.), *The infrared spectra of minerals,* Mineralogical Society, London, pp. 137-183.


S

Sabol Jr., D.E., Roberts, D., Smith, M. & Adams, J. 1992. Temporal variation in spectral detection thresholds of substrate and vegetation in AVIRIS images. *In:* Green, R.O. (Ed.), *Summaries of the Third Annual JPL Airborne Geoscience Workshop.* June 1-5, California. JPL Publication 92-41, pp. 132-134.

Sala, M. & Calvo, A. 1990. Response of Four Different Mediterranean Vegetation Types to Runoff and Erosion. *In:* J.B. Thornes (Ed.), *Vegetation and Erosion: Processes and Environments.* Wiley, Chichester, pp.347-362.

Salisbury, J., Walter, L., Vergo, N. & D'Aria, D. 1991. *Infrared (2.1-25 um) spectra of minerals,* Johns Hopkins University Press, Baltimore, 267 pp.

Sandidge, J.C. & Holyer, R.J. 1998. Coastal bathymetry from hyperspectral observations of water radiance. *Rem. Sens. Environ.* **65,** 341-352.

Sanger, J.E. 1971. Quantitative investigation of leaf pigments from their inception in buds through autumn coloration to decomposition in falling leaves. *Ecology,* **52,** 1075-1089.

Saunders, D.F., Burson, K.R., & Thompson, C.K. 1991. Observed relation of soil magnetic susceptibility and soil gas hydrocarbon analyses to subsurface petroleum accumulations. *AAPG Bulletin,* **75,** 389-404.

Saunders, D.F., Burson, K.R., & Thompson, C.K. 1999. Model for hydrocarbon microseepage and related near-surface alterations. *AAPG Bulletin,* **83,** 170-185.

Schaepman, M., Itten, K.I., Schläpfer, D., Kurer, U., Veraguth, S. & Keller, J. 1995. Extraction of ozone and chlorophyll-a distribution from AVIRIS data. *In:* Green, R.O. (Ed.), *Summaries of the Fifth Annual JPL Airborne Earth Science Workshop.* January 23-26, California. JPL Publication 95-1, pp. 149-152.

Schalles, J.F., Gitelson, A., Yacobi, Y. & Kroenke, A.E. 1998. Estimation of Chlorophyll a from time series measurements of high spectral resolution reflectance in an eutrophic lake. *Journal of Phycology* **34,** 383-390.

Schalles, J.F., Rundquist, D.C., Gitelson, A. & Keck, J. 2000. Close range, hyperspectral reflectance measurements of healthy Indo-Pacific and Carribean Corals. *Proceedings sixth International Conference on Remote Sensing for Marine and Coastal Environments.* ERIM-Veridian, Charleston, pp. 431-440.

Schanda, E. 1986. *Physical fundamentals of remote sensing.* Springer-Verlag, Berlin, 187 pp.

Schepers, J.S., Blackmer, T.M., Wilhelm, W.W. & Resende, M. 1996. Transmittance and reflectance measurements of corn leaves from plants with different nitrogen and water supply. *Journal of Plant Physiology,* **148,** 523-529.

Schiller, H. & Doerffer, R. 1999. Neural network for emulation of an inverse model - operational derivation of Case II water properties from MERIS data. *Int. J. Remote Sensing* **20,** 1735-1746.

Schowengerdt, R.A. 1997. Remote Sensing: Models and Methods for Image Processing. Academic Press, San Diego, 522pp.

Schultz, L.G. 1964. Quantitative interpretation of mineralogical composition from x-ray and chemical data for the Pierre Shale, *U. S. Geological Survey professional paper 391-C*, 31 pp.

Schultz, L.G. 1978. Mixed-layer clay in the Pierre Shale abd equivalent rocks, Northern Great Plains region, *U. S. Geological Survey professional paper 1064-A*, 27 pp.

Schumacher, D. 1996. Hydrocarbon-induced alteration of soils and sediments. *In:* Hydrocarbon migration and its near-surface expression, Schumacher D. and Abrams A.A. (Eds.), *AAPG Memoir*, **66,** pp. 71-89.

Schumacher, D. & Abrams, A. A. 1996. Hydrocarbon microseepage and its near-surface expression. *AAPG Memoir 66.*

Seed, H.B., Woodard, R.J. & Lundgren, R. 1962. Prediction of swelling potential for compacted clays. *Journal of American Society of Civil Engineers, Soil Mechanics and Foundations Division*, **88(SM3)**, 53-87.

Segel, D.B., & Merin, I.S. 1989. Successful use of Landsat Thematic Mapper data for mapping hydrocarbon microseepage-induced mineralogic alteration, Lisbon Valley, Utah. *Photogrammetric Engineering and remote sensing*, **55,** 1137-1145.

Segel, D. B., Ruth, M.D. & Merin, I.S. 1986. Remote detection of anomalous mineralogy associated with hydrocarbon production, Lisbon valley, Utah. *Mountain Geologist*, **23**, 51-62.

Segel, D. B., Ruth, M.D., Merin, I.S., Watanabe, H., Soda, K., Takano, O. & Sano, M. 1984. Correlation of remotely detected mineralogy with hydrocarbon production, Lisbon valley, Utah. *Proceedings of International Symposium On Remote Sensing For Exploration Geology, ERIM,* pp. 273-292.

Segl, K., Roessner, S. & Heiden, U. 2000. Differentiation of urban surfaces based on hyperspectral image data and a multi-technique approach. *Proceedings of the IEEE IGARSS 2000,* Honolulu, Hawaii, pp. 1600-1602.

Seinfeld, J.H. 1989. Urban air pollution: states of the science. *Science,* **243,** 745-752.

Sekine, T. 1999. Relationship between living environment and land use for a residential area in Morioka City: RS/GIS analysis of SPOT XS data. *Geographical Review of Japan,* Series A, 2, 75-92

Settle, J.J. & Drake N.A. 1993. Linear mixing and the estimation of ground cover proportions. *Int. J. Remote Sensing*, **14,** 1159-1177.

Shibayama, M., Morinaga, S. & Inoue, Y. 1993. Detection of water status in rice paddies using high resolution field spectra of visible to mid-infrared ranges. *Proceedings of the International Geoscience and Remote Sensing Symposium (IGARSS'93)*, pp. 519-521.

Short, F.T. & Wyllie-Echeverria, S. 1996. Natural and human-induced disturbance of seagrasses. *Environmental Conservation* **23,** 17-27.

Shumate, M., Menzies, R., Grant, W. & McDougal. K. 1981. Laser absorption spectrometer: remote measurement of tropospheric ozone. *Applied Optics* **20,** 545-553.

Siegel, F.R. & Goetz A. 1977. Effect of vegetation on rock and soil type discrimination. *Photogrammetric Engineering and remote sensing,* **43,** 191-196.

Siegel, H. & Gerth, M. 2000. Satellite-Based Studies of the 1997 Oder Flood Event in the Southern Baltic Sea. *Rem. Sens. Environ.* **73,** 207-217.

Sinclair, T.R., Hoffer, R.M. & Schreiber, M.M. 1971. Reflectance and internal structure of leaves from several crops during a growing season. *Agronomy Journal*, **63**, 864-868.

Sinclair, T.R., Schreiber, M.M. & Hoffer, R.M. 1973. Diffuse reflectance hypothesis for the pathway of solar radiation through leaves. *Agronomy Journal*, **65**, 276-283.

Singer, R.B. & McCord, T.B. 1979. Mars: Large scale mixing of bright and dark surface materials and implications for analysis of spectral reflectance. *Proceedings of the 10th Lunar and Planetary Science Conference*, Houston, U.S.A., 19-23 March 1979, pp. 1835-1848.

Singh, R.P. & Sirohim A. 1994. Spectral reflectance properties of different types of soil surfaces. *ISPRS Journal of Photogrammetry and Remote Sensing* **49,** 34-30.

Slater, P.N. 1980. *Remote sensing: optics and optical systems.* Addison-Wesley, London.

Slatyer, R.O. 1967. *Plant-water relationships.* Academic Press, London.

Smailbegovich, A., Taranik, J. & Vaughan, R. 2000a. Geologic utility of ASTER, MASTER, TIMS and SEBASS multispectral thermal data. *Proc. ERIM 14$^{th}$ Int'l. Conf. Applied Geologic Remote Sensing*, Las Vegas, Nov., pp. 141-150.

Smailbegovich, A., Vaughan, R., Calvin, R. & Taranik, J. 2000b. Analysis of hyperspectral thermal data from SEBASS for mapping hydrothermal alteration and hot springs deposits on Geiger Grade and Steamboat Springs, Nevada. *Proc. ERIM 14$^{th}$ Int'l. Conf. Applied Geologic Remote Sensing*, Las Vegas, Nov., pp. 431-437.

Smith, G.M. & Curran, P.J. 1993. The effect of signal noise on the remote sensing of foliar biochemical concentration. *In:* Green, R.O. (Ed.) *Summaries of the Fourth Annual JPL Airborne Geoscience Workshop.* October 25-29, California. JPL Publication 93-26, pp. 161-164.

Smith, M.O., Johnston, P.E. & Adams, J.B. 1985. Quantitative determination of mineral types and abundances from reflectance spectra using principal component analysis. *J. Geophys. Res.,* **90,** 797-804.

Smith, M.O., Roberts, D.A., Hill, J., Mehl, W., Hosgood, B., Venderbout, J., Schmuck, G., Koechler, C. & Adams, J. 1994. A new approach to quantifying abundancies of materials in multispectral images. *In: IGARSS 94: Proceedings International Geosciences Remote Sensing Symposium*, vol. 4, pp. 2372-2374.

Soha, J. & Schwartz, A. 1978. A decorrelation stretch algorithm for multivariate analysis, *Proc. Fifth Can. Symp. On Rem. Sens.*, Victoria, BC, pp. 86-93.

Soil Survey Staff 1988. *Keys to Soil Taxonomy* (fourth printing). SMSS technical monograph no. 6 Cornell University, Ithaca, New York,  280 pp.

**Soils Task Force 1996. *Soils Task Force Report – Draft for Review,* Home Builders Association of Metropolitan Denver, November 1.**

Sombroek, W.W.G. 1985. Introduction to Soil Surface Sealing and Crusting. *In:* F.Callebout, D.Gabriels & M.de Boodt (Eds.), Assessment of Soil Surface Sealing and Crusting. Flanders Research Centre for Soil Erosion and Soil Conservation, Ghent, pp. 1-7.

Spitters, C.J.T., Van Keulen, H. & Van Kraailingen, D.W.G. 1989. A simple and universal crop growth simulator: SUCROS87. *In:* R. Rabbinge, S.A. Ward and H.H. van Laar (Eds.), *Simulation and systems management in crop protection*, Simulation Monographs 32, Pudoc, Wageningen, 420 pp.

Stavn, R.H. & Weidemann, A.D. 1989. Shape factors, two-flow models, and the problem of irradiance inversion in estimating optical parameters. *Limnol.Oceanogr.* **34,** 1426-1441.

Stephens, A., Louchard, E.M., Brand, L.E. & Reid, P. 2000. Effects of micro-algal communities on radiance reflectance of carbonate sediments in optically shallow marine environments. *Proceedings Ocean Optics XV.* SPIE, Monaco.

Steven, M.D., Biscoe, P.V. & Jaggard, K.W. 1983. Estimation of sugar beet productivity from reflection in the red and infrared spectral bands. *Int. J. Rem. Sensing* **2,** 117-125.

Stocking, M.A. 1981. A Working Model for the Estimation of Soil Loss Suitable for Underdeveloped Areas. Development Studies Occasional Papers, No.15, Harare.

Stoner, E.R. & Baumgardner, M.F. 1981. Characteristic variations in reflectance of surface soils. *Soil Science Society of America Journal* **45,** 1161-1165.

Suits, G.H. 1972. The calculation of the directional reflectance of a vegetative canopy. *Rem. Sens. Envir.* **2,** 117-125.

Suits, G. H. 1983. The nature of electromagnetic radiation. *In:* R. N. Colwell (Ed.), *Manual of Remote Sensing, Volume 1.* ASPRS, Falls Church, Virginia.

Sultan, M., Fiske, M., Stein, T., Gamal, M., Abdel-Hady, Y., El-Araby, H., Madani, A., Mehanee, S. & Becker, R. 1999. Monitoring the urbanization of the Nile Delta, Egypt. *Ambio, 28,* 628-631

Swain, P. H. & Davis, S.M. 1978. *Remote sensing: the quantitative approach,* McGraw-Hill International Book Company.


T

Tassan, S. 1994. Local algorithms using SeaWIFS data for the retrieval of phytoplankton, pigments, suspended sediment, and yellow substance in coastal waters. *Appl.Opt.* **23,** 2369-2378.

Tassan, S. 1996. Modified Lyzenga's method for macroalgae detection in water with non- uniform composition. *Int. J. Remote Sensing* **17,** 1601-1607.

Tedesco, S.A. 1995. *Surface geochemistry in petroleum exploration.* New York. Chapman &

Thomas, J.R. & Gausman, H.W. 1977. Leaf reflectance vs. leaf chlorophyll and carotenoid concentrations for eight crops. *Agronomy Journal,* **69,** 799-802.

Thomas, J.R., Myers, V.I., Heilman, M.D. & Wiegand, C.L. 1966. Factors affecting light reflectance of cotton. *Proceedings of the 4th Symposium on Remote Sensing of Environment,* Ann Arbor (MI), 12-14 April 1966, pp. 305-312.

Thomas, J.R., Namken, L.N., Oerther, G.F. & Brown, R.G. 1971. Estimating leaf water content by reflectance measurements. *Agronomy Journal,* **63,** 845-847.

Thomas, S.G. & Middleton, N.J. 1994. Desertification: Exploding the Myth. John Wiley & Sons, New York, 194 pp.

Thompson, C.K., Saunders, D.F. & Burson, K.R. 1994. Model advanced for hydrocarbon microseepage, related alterations. *Oil & Gas Journal,* **14,** 95-99.

Thompson, R.W. 1992. Swell testing as a predictor of structural performance, *Proceedings of the 7ᵗʰ International Conference on Expansive Soils,* Dallas, U.S.A., 3-5 August 1992, pp. 84-88.

Thompson, R.W. 1997. Evaluation protocol for repair of residences damaged by expansive soils, *Proceedings on unsaturated soils at Geo-Logan '97,* Logan, U.S.A., July 1997, pp. 255-276.

Thornes, J.B. (Ed.) 1990a. *Vegetation and Erosion: Processes and Environments.* Wiley, Chichester.

Thornes, J.B. 1990b. The Interaction of Erosional and Vegetational Dynamics in Land Degradation: Spatial Outcomes. *In:* J.B. Thornes (Ed.), *Vegetation and Erosion: Processes and Environments.* Wiley, Chichester, pp. 41-53.

Tolk, B.L., Han, L. & Rundquist, D.C. 2000. The impact of bottom brightness on spectral reflectance of suspended sediments. *Int. J. Remote Sensing* **21**, 2259-2268.

Torrent, J., Schwertmann, U., Fetcher, H. & Alferez, F. 1983. Quantitative Relationships between Soil Color and Hematite Content. *Soil science,* **13**, 354-358.

Tourtelot, H.A. 1973. Geologic origin and distribution of swelling clays, *Proceedings of Workshop on Expansive Clays and Shales in Highway Design and Construction,* **1**.

Tucker, C.J. 1979. Red and Photographic Infrared Linear Combinations for Monitoring Vegetation. *Rem. Sens. Environ.,* **8,** 127-150.

Tucker, C.J. & Garratt, M.W. 1977. Leaf optical system modelled as a stochastic process. *Applied Optics,* **16**, 635-642.


U

Uenk, D. & Booij, R. 2000. *Stikstofbijmestsysteem in aardappelen: effecten op productie en milieu van aardappelen in Noordwest Overijssel,* Plant Research International Series, Nota (1), Wageningen, The Netherlands, 21 pp.

USDA, 1999. Soil Survey Staff, Natural Resources Conservation Service, National Soil Survey Handbook, title 430-VI, Washington, D.C., U.S. Government Printing Office.

Ustin, S.L., Sanderson, E.W., Grossman, Y., Hart, Q.J. & Haxo, R.S. 1993. Relationships between pigment composition variation and reflectance for plant species from a coastal savannah in California. *In:* Vane, G. (Ed.), *Proceedings of the 4th Annual JPL Airborne Geoscience Workshop, Volume 1. AVIRIS Workshop,* 25-29 October 1993, Washington (DC), NASA-JPL Publication 93-26, pp. 181-184.


V

Valentin, C. & Bresson, L.M. 1992. Morphology, Genisis and Classification of Surface Crusts in Loamy and Sandy Soils. *Geoderma,* **55,** 225-245.

Van de Vlag, D. Jetten, V., Nachtergaele, J. & Poesen, J. 2000. Event-based Modelling of Gully Incision and Development in the Belgian Loess Belt. *Proceedings Int. Symposium on Gully Erosion under Global Change,* April 16-19 Leuven.

Van den Bosch, J. & Alley, R. 1990. Application of LOWTRAN 7 as an atmospheric correction to Airborne Visible/InfraRed Imaging Spectrometer (AVIRIS) data. *In:* Green, R.O. (Ed.), *Proceedings Second AVIRIS Workshop,* JPL Publ. 90-54, Jet Propulsion Lab., Pasadena, California, pp. 78-81.

Van der Meer, F. 1994. Calibration of Airborne Visible/Infrared Imaging Spectrometer data (AVIRIS) to reflectance and mineral mapping in hydrothermal alteration zones: an example from the Cuprite Mining District. *Geocarto Int.,* **9,** 23-37.

Van der Meer, F. 1994. Extraction of mineral absorption features from high-spectral resolution data using non-parametric geostatistical techniques. *Int. J. Remote Sensing,* **15,** 2193-2214.

Van der Meer, F. 1994. Sequential Indicator Conditional Simulation and Indicator Kriging Applied to Discrimination of Dolomitization in GER 63-channel Imaging spectrometer data. *Nonrenewable Resources,* **3,** 146-164.

Van der Meer, F. 1995. Estimating and simulating the degree of serpentization of peridotites using hyperspectral remote sensing imagery. *Nonrenewable Resources,* **4,** 84-98.

Van der Meer, F. 1996. Metamorphic facies zonation in the Ronda peridotites mapped from GER imaging spectrometer data. *Int. J. Remote Sensing,* **17,** 1633-1657.

Van der Meer, F. 1996. Performance of the indicator classifier on simulated image data. *Int. J. Remote Sensing,* **17,** 621-627.

Van der Meer, F. 1998. Mapping dolomitization through a co-regionalization of simulated field and image-derived reflectance spectra; a proof-of-concept study. *Int. J. Remote Sensing,* **19,** 1615-1620.

Van der Meer, F. 1999. Imaging spectrometry for geological remote sensing. *Geologie en Mijnbouw,* **77,** 137-151.

Van der Meer, F. 1999. Iterative spectral unmixing. *Int. J. Remote Sensing,* **20,** 3241-3247.

Van der Meer, F., 2000. Geophysical inversion of hyperspectral data. *Int. J. Remote Sensing,* **21,** 387-393.

Van der Meer, F. & Bakker, W. 1997. Cross Correlogram Spectral Matching (CCSM): application to surface mineralogical mapping using AVIRIS data from Cuprite, Nevada. *Remote Sens. Environ,* **61,** 371-382.

Van der Meer, F. & Bakker, W. 1998. Validated surface mineralogy from high-spectral resolution remote sensing: a review and a novel approach applied to gold exploration using AVIRIS data. *Terra Nova,* **10,** 112-118.

Van der Meer, F. & De Jong, S.M. 2000. Improving the results of spectral unmixing of LANDSAT TM imagery by enhancing the orthogonality of end-members. *Int. J. Remote Sensing,* **21,** 2781-2797.

Van der Meer, F., Fan, L. & Bodechtel, J. 1997. MAIS Imaging spectrometer data analysis for Ni-Cu prospecting in ultramafic rocks of the Jinchuan group, China. *Int. J. Remote Sensing.*, **18,** 2743-2761.

Van Keulen, H. & Seligman, N.G. 1987. *Simulation of water use, nitrogen nutrition and growth of a spring wheat crop*, Simulation monograph, PUDOC, Wageningen, The Netherlands.

Van Leeuwen, H.J.C. 1996. *Synergy of optical and microwave remote sensing in agricultural crop growth monitoring*, PhD Thesis, Wageningen Agricultural University, Wageningen, The Netherlands, 233 pp.

Van Leeuwen, H.J.C. & Clevers, J.G.P.W. 1994. Synergy between optical and microwave remote sensing for crop growth monitoring, *Proceedings Sixth International Symposium on Physical Measurements and Signatures in Remote Sensing*, Val d'Isère, France, 17-21 January 1994.

Vanderbilt, V.C. 1985. Urban, forest, and agricultural AIS data: fine spectral feature. *In:* Vane, G. and Goetz, A.F.H. (Eds.), *Proceedings of the Airborne Imaging Spectrometer Data Analysis Workshop*. April 8-10, California. JPL Publication 85-41, pp. 158-165.

Vane, G., Chrisp, M., Enmark, H., Macenka, S. & Solomon, J. 1984. Airborne Visible/Infrared Imaging Spectrometer: An advanced tool for Earth remote sensing, *Proceedings IGARSS '84*, SP215, 751.

Vane, G., Duval, J.E. & Wellman, J.B. 1993. Imaging spectroscopy of the Earth and other solar system bodies. *In:* C.M. Pieters and P.A.J. Englert (Eds.), *Remote Geochemical Analyses: Elemental and Mineralogical Composition*, Cambridge University Press, Cambridge, pp. 121-144.

Vane, G. & Goetz, A.F.H. 1988. Terrestrial Imaging Spectroscopy. *Rem. Sens. Environ.*, **24,** 1-29.

Vane, G., Green, R.O., Chrien, T.G., Enmark, H.T., Hansen, E.G. & Porter, W.M. 1993. The Airborne Visible/Infrared Imaging Spectrometer (AVIRIS). *Remote Sens. Environ.* **44,** 127-143.

Vasilkov, A.P., Burenkov, V.I. & Ruddick, K.G. 1999. The spectral reflectance and transparency of river plume waters. *Int. J. Remote Sensing* **20,** 2497-2508.

Vaughan, R., Calvin, W., Smailbegovich, A. & Taranik, J. 2000. Analysis of mineral alteration in the Virginia  Range, Nevada with SEBASS hyperspectral thermal infrared data. *Proc. ERIM 14$^{th}$ Int'l. Conf. Applied Geologic Remote Sensing*, Las Vegas, Nov., pp. 177-182.

Verdebout, J., Jacquemoud, S. & Schmuck, G. 1994. Optical properties of leaves: modelling and experimental studies, *In:* Hill, J. & Mégier, J. (Ed.), *Imaging Spectrometry - a Tool for Environmental Observations*, ECSC, EEC, EAEC, Brussels and Luxembourg, pp. 169-191.

Verhoef, W. & Bunnik, N.J.J. 1975. A model study on the relations between crop characteristics and canopy spectral reflectance. NIWARS Publication 33, Delft, The Netherlands.

Verhoef, W. & Bunnik, N.J.J. 1981. Influence of crop geometry on multispectral reflectance determined by the use of canopy reflectance models, *Proceedings International Colloquium on Signatures of Remotely Sensed Objects*, Avignon, France, pp. 273-290.

Verhoef, W. 1984. Light scattering by leaf layers with application to canopy reflectance modeling: The SAIL model. *Remote sens. environ*, **16,** 125-141.

Verhoef, W. 1985a. Earth observation modeling based on layer scattering matrices. *Rem. Sens. Envir.* **17,** 165-178.

Verhoef, W. 1985b. A scene radiation model based on four stream radiative transfer theory, *Proceedings 3rd International Colloquium on Spectral Signatures of Objects in Remote Sensing*, Les Arcs, France, ESA SP-247, pp. 143-150.

Verhoef, W. 1998. *Theory of radiative transfer models applied in optical remote sensing of vegetation canopies*, PhD Thesis, Wageningen Agricultural University, The Netherlands, 310 pp.

Verstraete, M.M., Pinty, B. & Curran, P.J. 1999. MERIS potential for land applications. *Int. J. Remote Sensing* **20,** 1747-1756.

Vickers, R. & Lyon, R. 1967. Infrared sensing from spacecraft- a geologic interpretation. *Proc. Thermophysics Special Conf. Amer. Inst. Astron.*, paper 67-284.

Vogelmann, J.E., Rock B.N. & Moss D.M. 1993. Red edge spectral measurements from sugar maple leaves. *Int. J. Remote Sensing*, **14,** 1563-1575.

Vos, J. & Bom, M. 1993. Hand-held chlorophyll meter: a promising tool to assess the nitrogen status of potato foliage, *Potato Research* **36,** 301-308.


W

Wald, L. & Baleynaud, J.M. 1999. Observation air quality over the city on Nantes by means of Landsat thermal infrared data, *Int. J. Remote Sensing,* **20,** 947-956.

Walker, R.E. 1994. *Marine light field statistics*. Wiley series in pure and applied optics. Wiley, New York, 675 pp.

Wallace, C.S.A., Watts, J.M. & Yool, S.R. 2000. Characterizing the Spatial Structure of Vegetation Communities in the Mojave Desert using Geostatistical Techniques. *Computers & Geosciences,* **26,** 397-410.

Warner, T.A. 1999. Analysis of Spatial Patterns in Remotely Sensed Data Using Multi-variate Spatial Correlation. *Geocarto International,* **14,** 59-65.

Welles, J.M. & Norman, J.M. 1991. Photon transport in discontinuous canopies: a weighted random approach. *In:* R.B. Myneni & J. Ross (Eds.), *Photon-vegetation interactions: Applications in optical remote sensing and plant ecology.* Springer-Verlag, Berlin.

Wernand, M.R., Shimwell, S.J. & Munck, J.C.D. 1997. A simple method of full spectrum reconstruction by a five-band approach for ocean colour applications. *Int. J. Remote Sensing* **18,** 1977-1986.

Wessman, C.A. 1989. Evaluation of Canopy Biochemistry. In: R.J. Hobbs and H.A. Mooney, Remote Sensing of Biosphere Functioning. Ecological Studies, 79. Springer-Verlag, New York, pp. 135-156.

Wessman, C.A., Aber, J.D. & Peterson, D.L. 1987. Estimation of forest canopy characteristics and nitrogen cycling using imaging spectrometry. *In:* Vane, G. (Ed.), *Imaging Spectrometry II,* SPIE Vol. 834.

Wessman, C.A., Aber, J.D., Peterson, D.L. & Melillo, J.M. 1988. Foliar analysis using near infrared reflectance spectroscopy. *Canadian Journal of Forest Research* 18, 6-11.

Wessmann, C.A., Aber, J.D., Peterson, D.L. & Melillo, J. 1988. Remote sensing of canopy chemistry and nitrogen cycling in temperate forest ecosystems. *Nature,* **335,** 154-156.

Whitfield, D.M. & Conner, D.J. 1980. Penetration of photosynthetically active radiation into tobacco crops. *Australian Journal of Plant Physiology,* 7, 449-461.

Whitfield, D.M. 1986. A simple model of light penetration into row crops. *Agricultural and Forest Meteorology,* **36,** 297-315.

Whitlock, C.H. et al. 1981. Comparison of reflectance with backscatter and absorption parameters for turbid waters. *Appl.Opt.* **20,** 517-522.

Wiggins, J.H. 1978. *Building losses from natural hazards: Yesterday, today and tomorrow,* Report for National Science Foundation, Grant No. ENV.-77-08435.

Wilcox, D.A. 1995. Wetland and aquatic macrophytes as indicators of anthropogenic hydrologic disturbance. *Natural Areas Journal* **15,** 240-248.

Williams, J.H. & Ashenden, T.W. 1992. Differences in the spectral characteristics of white clover exposed to gaseous pollutants and acid mist. *The New Phytologist,* **120,** 69-75.

Williams, P. & Norris, K. (Eds.) 1987. *Near-Infrared Technology in the Agricultural and Food Industries.* American Association of Cereal Chemists, St. Paul, MN.

Willstätter, A. & Stoll, K. 1918. Untersuchungen uber die assimilation der kohlensaure. Springer-Verlag, Berlin.

Windeler, D.S. & Lyon, R.J.P. 1991. Discriminating Dolomitization of Marble in the Ludwig Skarn near Yerington, Nevada Using High-Resolution Airborne Infrared Imagery", *Photogramm. Eng. and Remote Sensing,* **57,** 1171-1177.

Wilson, W.H. & Kiefer, D.A. 1979. Reflectance spectroscopy of marine phytoplankton. Part 2 A simple model of ocean color. *Limnol.Oceanogr.* **24,** 673-682.

Wischmeier, W.H. & Smith, D.D. 1978. Predicting Rainfall Erosion Losses, a Guide to Conservation Planning. Agriculture Handbook No.282, United States Department of Agriculture, Washington.

Wittlinger, S. & Zimmermann, R. 2000. Hyperspectral remote sensing of subtidal macroalgal assemblages in optically shallow waters. *Proceedings Ocean Optics XV,* Monaco, pp. 6.

Woodcock, C.E., and Strahler, A.H. 1987. The factor of scale in remote sensing *Remote Sens. Environ,* **21,** 311-332.

Woodruff, D.L., Stumpf, R.P., Scope, J.A. & Paerl, H.W. 1999. Remote Estimation of Water Clarity in Optically Complex Estuarine Waters. *Rem. Sens. Environ.* **68,** 41-52.

Woolley, J.T. 1971. Reflectance and transmittance of light by leaves. *Plant Physiology,* **47,** 656-662.


X

Xia, Z. 1993. The Uses and Limitations of Fractal Geometry in Digital Terrain Modelling. PhD Dissertation, City University of New York, Ann Arbor, 252 pp.


Y

Yamada, N. & Fujimura, S. 1988. A mathematical model of reflectance and transmittance of plant leaves as a function of chlorophyll pigment content, *Proceedings of the 8th International Geoscience and Remote Sensing Symposium (IGARSS'88),* Edinburgh (Scotland), 13-16 Sept 1988, pp. 833-834.

Yamaguchi, Y., Kahle, A., Tsu, H., Kawakami, T. & Pniel, M. 1998. Overview of the Advanced Spaceborne Thermal Emission and Reflection Radiometer (ASTER). *IEEE Trans. Geosci. Remote Sens.* **36,** 1062-1071.

Yang, H., Zhang, J., van der Meer, F. & Kroonenberg, S.B. 1999. Spectral characteristics of wheat associated with hydrocarbon microseepages. *Int. J. Remote Sensing,* **20,** 807-813.

Yang, H., Zhang, J., van der Meer, F. & Kroonenberg, S.B. 1999. Geochemistry and field spectrometry for detecting hydrocarbon microseepage. *Terra Nova,* **10,** 231-235.

Yang, H., Zhang, J., van der Meer, F. & Kroonenberg, S.B. 2000. Imaging spectrometry data correlated to hydrocarbon microseepage. *Int. J. Remote Sensing*, **21,** 197-20.

Yang, J. & Prince, S.D. 1997. A theoretical assessment of the relation between woody canopy cover and red reflectance. *Remote sens. environ*, **59**, 428-439.

Yoder, B.J. & Pettigrew-Crosby, R.E. 1995. Predicting nitrogen and chlorophyll content and concentrations from reflectance spectra (400-2500 nm) at leaf and canopy scales. *Remote sens. environ*, **53**, 199-211.


Z

Zhang, X. 1998. On the estimation of biomass of submerged vegetation using satellite thematic mapper (TM) imagery: a case study of the Honghu , PR China. *Int. J. Remote Sensing* **19,** 11-20.

Zhang, B., Wang, X., Liu, J., Zheng, L. & Tong, Q. 2000. Hyperspectral image processing and analysis system (HIPAS) and its applications. *Photogrammetric Engineering and Remote Sensing,* **66,** 605-609.

Zimmermann, R. & Wittlinger, S. 2000. Hyperspectral remote sensing of submerged aquatic vegetation in optically shallow waters. *Proceedings Ocean Optics XV*, Monaco.